ALLE ZEIT WACH 1842

Eugene J. Kucharz

# The Collagens: Biochemistry and Pathophysiology

With 50 Figures and 9 Tables

Springer-Verlag
Berlin Heidelberg New York
London Paris Tokyo
Hong Kong Barcelona
Budapest

EUGENE J. KUCHARZ, M.D., Ph.D.
Fourth Department of Internal Medicine
Silesian School of Medicine
Al. Korfantego 28/86
PL-40-004 Katowice
Poland

ISBN-13:978-3-642-76199-7 e-ISBN-13:978-3-642-76197-3
DOI: 10.1007/978-3-642-76197-3

Library of Congress Cataloging-in-Publication Data. Kucharz, Eugene J. (Eugene Joseph), 1951–. The collagens: biochemistry and pathophysiology / Eugene J. Kucharz. p. cm. Includes bibliographical references and index.
ISBN-13:978-3-642-76199-7

1. Collagen – Metabolism. 2. Collagen – Pathophysiology. I. Title. [DNLM: 1. Collagen – chemistry. 2. Collagen – physiology. 3. Collagen diseases – physiopathology. WD 375 K95c] QP552.C6K68 1992 591.1'85–dc20


Softcover reprint of the hardcover 1st edition 1992



Typesetting: K+V Fotosatz GmbH, Beerfelden
27/3145-5 4 3 2 1 0 – Printed on acid-free paper

*To my daughter Patricia,*
*the light of my life*

# Preface

*Quam magnificata sunt opera tua, Domine!*
*Omnia in sapientia fecīti,*
*plena est terra creaturis tuis.*
*Psalmus CIV.XXIV.*

The days when the answer to the questions "What is collagen and what is it for?" was simple and short are gone. Nowadays, collagen cannot be considered as a single and metabolic inert fibrous protein responsible only for some mechanical properties of tissues. The family of collagen proteins increases rapidly, and new collagens are still being discovered. They differ each from the other, and only the basic amino acid sequence containing proline and hydroxyproline is the one constant and unique feature of proteins which are very different in all other aspects. This heterogeneity is related to the variety of their biological functions. The list of physiological aspects involving collagens as well as the number of disorders associated with collagen abnormalities are expanding rapidly. The structure and function of almost every organ are in some way related to connective tissue and to collagens in particular. Our knowledge in this field is significantly growing but is still far from complete. The efforts to modify collagen gene expression, structure, and metabolism in order to manage some disorders are a new goal of pharmacology and are a natural sequel of the progressing understanding of the pathophysiological role of collagens.

This book provides a comprehensive overview of the pathophysiological aspects of collagens with special emphasis on medical problems, i.e. their involvement in the development, normal and disease-altered states of various organs and systems. The book is designed for everyone interested in matrix pathophysiology. The matrix of connective tissue is one of the basic structures responsible for the integration of organs and tissues although dispersed throughout the whole body. This explains the wide interest in the collagens among all branches of medicine.

Studies on collagens have revealed more than the complexity of their molecular and pathophysiological problems but also the fascinating beauty of their nature. My fascination with these problems has led me to write this book and explains the use of the psalm as the motto.

November 1991 EUGENE J. KUCHARZ

# Acknowledgements

It is my pleasant duty to thank all those who, in one way or another, have helped me in this endeavor.

A great part of the materials for this book was collected during my visit to the Medical College of Wisconsin where I was working in the laboratory of James S. Goodwin, M.D. He surely deserves all my thanks for granting me the opportunity to visit the USA. I would like to express my sincere thanks to Mrs. Ruth M. Sitz, Wauwatosa, for her continuous friendly help and encouragement in the preparation of this book.

I am very grateful to Mrs. Barbara Karas, Whitefish Bay, Marc A. Shampo, Ph.D., Rochester, and Gerard Jonderko, M.D., D.Sc., Tychy, for their help in providing me with several reprints. This book could not have been written without the help of the librarians in the Todd Wehr and Murphy Libraries of the Medical College of Wisconsin and the Central Library of the Silesian School of Medicine, especially Alfred Puzio, Ph.D., Jadwiga Barczyk, M.A., Anna Majewska-Piątkowska, M.A., Mrs. Laura Skopowska, and Mrs. Eugenia Piotrowska.

I would like to thank Dr. F.M. Pope, Harrow, and Dr. Anne De Paepe, Gent, for their kind permission to use pictures of their patients with rare defects of collagen metabolism. I am also grateful to Marc C. Browning, M.D., Winston-Salem, and Ronald R. Minor, V.M.D., Ph.D., Ithaca, who kindly provided me with illustrations.

The preparation of this book involved a great deal of secretarial work. Barbara Nowińska, Ph.D., accomplished it with kind dedication and efficiency. Skilful drawing of the figures by Mr. Zbigniew Muniowski is gratefully acknowledged.

I wish to pay a tribute to the editorial and publisher's staff of Springer Verlag for their constant, invaluable help in the final preparation of the book.

Finally, my gratitude goes to my parents, Bożena Halina Kucharz and Józef Kucharz, Katowice, for their understanding and patience.

November 1991 EUGENE J. KUCHARZ

# Contents


| | | |
|---|---|---|
| 1 | **Introduction: Historical Outline** | 1 |
| 2 | **Structure, Heterogeneity, and Distribution** | 5 |
| 2.1 | Basic Nomenclature | 5 |
| 2.2 | Triple-Chain Helical Structure | 6 |
| 2.3 | Distribution | 7 |
| 2.4 | Classification and General Structural Features of the Collagens | 9 |
| 2.5 | Type I, Type II, Type III Collagens, and Type I-Trimer Collagen | 10 |
| 2.6 | Type IV Collagen | 12 |
| 2.7 | Type V Collagen | 15 |
| 2.8 | Type VI Collagen | 17 |
| 2.9 | Type VII Collagen | 20 |
| 2.10 | Type VIII Collagen | 21 |
| 2.11 | Type IX Collagen | 21 |
| 2.12 | Type X Collagen | 23 |
| 2.13 | Type XI Collagen | 24 |
| 2.14 | Type XII Collagen | 24 |
| 2.15 | Type XIII Collagen | 25 |
| 2.16 | Proteins Containing a Collagenlike Sequence | 25 |
| | 2.16.1 Acetylcholinesterase | 26 |
| | 2.16.2 Mammalian Lectins | 26 |
| | 2.16.2.1 Complement Component C1q | 27 |
| | 2.16.2.2 Conglutinin | 28 |
| | 2.16.2.3 Mannan-Binding Protein | 29 |
| | 2.16.2.4 Lung Surfactant Protein | 29 |
| 3 | **Biosynthesis of Collagen** | 31 |
| 3.1 | Introduction | 31 |
| 3.2 | Procollagen Genes | 31 |
| 3.3 | Cleavage of the Signal Peptide | 34 |
| 3.4 | Hydroxylation of Proline and Lysine Residues | 34 |
| 3.5 | Glycosylation of Hydroxylysine Residues | 39 |
| 3.6 | Glycosylation of Asparagine Residues | 41 |

3.7 Chain Association, Disulfide Bonding, and Formation of the Helical Structure ............ 42
3.8 Conversion of Procollagen to Collagen ........ 42
3.9 Formation of the Supramolecular Structures ... 44
3.10 Cross-Linking of Collagen .................. 45
3.11 Interaction of Collagen with Components of the Extracellular Matrix .................. 49
3.11.1 General Remarks .................... 49
3.11.2 Collagen-Proteoglycan Interactions .... 49
3.11.3 Structural Glycoproteins ............ 51
3.11.3.1 Fibronectin ......................... 52
3.11.3.2 Laminin ......................... 53
3.11.3.3 Entactin and Nitogen ............... 53
3.11.3.4 Chondrionectin ..................... 53
3.11.3.5 Fibromodulin ....................... 53

**4 Degradation** ............................. 55
4.1 Introduction ............................. 55
4.2 General Pathways of Collagen Degradation .... 55
4.3 Depolymerization ......................... 57
4.4 Tissue Collagenases ....................... 58
4.5 Collagenolytic Cathepsins .................. 63
4.6 Gelatinases and Collagen Peptidase .......... 64
4.7 Collagenolytic Enzymes of Microbial Origin ... 66

**5 Turnover and Regulation of Collagen Metabolism** ............................. 69
5.1 Introduction ............................. 69
5.2 Transcriptional and Translational Control of Collagen Synthesis ..................... 70
5.3 Regulation of Collagen Synthesis at the Intracellular Posttranslational Level ..... 71
5.4 Intracellular Degradation of Procollagen ....... 71
5.5 Feedback Inhibition of Biosynthesis of Collagen by Aminoterminal Propeptides .............. 74
5.6 Role of the Free Proline Pool in the Regulation of Collagen Synthesis ..................... 75
5.7 Turnover of Collagen ....................... 75

**6 Collagen in Development, Aging, and Phylogeny** ........................... 77
6.1 Collagen and Development .................. 77
6.2 Collagen and Aging ......................... 79

6.3 Phylogenetic Aspects of Collagen ............. 81
6.3.1 Collagen and Animal Phylogeny ...... 81
6.3.2 Plants and Unicellular Organisms ..... 82
6.3.3 Porifera ............................ 82
6.3.4 Coelenterata ........................ 82
6.3.5 Platyhelmintha ...................... 83
6.3.6 Aschelminthes ...................... 83
6.3.7 Annelida ........................... 84
6.3.8 Arthropoda ......................... 84
6.3.9 Mollusca ........................... 85
6.3.10 Echinodermata ...................... 85
6.3.11 Chordata ........................... 86

**7 Hormonal Regulation of Collagen Metabolism** . 87
7.1 Introduction ............................... 87
7.2 Thyroxine, Triiodothyronine, and Thyrotropic Hormone .................. 87
7.3 Glucocorticoids ............................ 93
7.4 Growth Hormone ............................ 98
7.5 Insulin and Glucagon ...................... 100
7.6 Prolactin .................................. 104
7.7 Histamine ................................. 105

**8 Immunobiology of Collagen** ................ 107
8.1 Introduction ............................... 107
8.2 Antigenicity of Collagen and Procollagen ...... 108
8.3 Immune Response to Collagens and Procollagens .......................... 110
8.4 Effect of Cytokines on Collagen Metabolism ... 111

**9 Hereditary Disorders of Collagen** ............ 113
9.1 Introduction ............................... 113
9.2 The Ehlers-Danlos Syndrome ................ 113
9.3 The Marfan Syndrome ...................... 119
9.4 Osteogenesis Imperfecta .................... 122
9.4.1 Osteogenesis Imperfecta Type I ....... 122
9.4.2 Osteogenesis Imperfecta Type II ....... 123
9.4.3 Osteogenesis Imperfecta Type III ...... 124
9.4.4 Osteogenesis Imperfecta Type IV ...... 125
9.5 Epidermolysis Bullosa ..................... 125
9.6 Chondrodysplasias ......................... 126
9.7 Congenital Dislocation of the Hip ............ 128
9.8 The Nail-Patella Syndrome .................. 129

9.9 Hereditary Progressive Glomerulopathy (Alport's Syndrome) ... 129
9.10 Homocystinuria ... 131
9.11 Alkaptonuria ... 132
9.12 Aspartylglycosaminuria ... 133
9.13 Familial Cutaneous Collagenoma ... 133
9.14 Focal Dermal Hypoplasia ... 134
9.15 Hyalinosis Cutis et Mucosae ... 134
9.16 Fibrodysplasia Ossificans Progressiva ... 135
9.17 Hereditary Disorders of Collagen Metabolism in Animals ... 135

**10 Collagen Diseases** ... 139
10.1 Introduction ... 139
10.2 Scleroderma and Related Conditions ... 139
10.3 Other Collagen Diseases ... 144
10.4 Drug-Induced Lupus-Like Syndrome ... 145

**11 Musculoskeletal System** ... 149
11.1 Introduction ... 149
11.2 Bone ... 149
11.2.1 Structure and Function ... 149
11.2.2 Collagen in Bone ... 151
11.2.3 Collagen in Pathology of Bone ... 154
11.2.3.1 Disturbed Regulation of Calcium and Phosphorus Metabolism ... 154
11.2.3.2 Osteoporosis ... 155
11.2.3.3 Paget's Disease ... 156
11.3 Cartilage ... 157
11.3.1 Structure and Function ... 157
11.3.2 Collagens of the Cartilage ... 157
11.4 Intervertebral Disc ... 158
11.4.1 Structure and Function ... 158
11.4.2 Collagen in the Intervertebral Disc ... 159
11.4.3 Collagen in Pathology of the Intervertebral Disc ... 160
11.4.3.1 Age-Related Degeneration ... 161
11.4.3.2 Chemonucleolysis ... 161
11.4.3.3 Scoliosis ... 161
11.5 Tendons and Related Structures ... 162
11.5.1 Structure and Function ... 162
11.5.2 Collagen in Tendinous Pathology ... 164
11.5.2.1 Dupuytren's Contracture ... 165
11.5.2.2 Carpal Tunnel Syndrome ... 166

11.6 Joints ..... 166
11.6.1 Role of Collagen in the Structure and Function ..... 166
11.7 Skeletal Muscle ..... 167
11.7.1 Role of Collagen in the Structure and Function ..... 167
11.7.2 Collagen in the Pathology ..... 169
11.8 Collagen in Rheumatic Disorders ..... 170
11.8.1 Rheumatoid Arthritis ..... 170
11.8.2 Osteoarthrosis ..... 172
11.8.3 Animal Models of Rheumatic Disorders 173

**12 Cardiovascular System** ..... 177
12.1 Introduction ..... 177
12.2 The Heart ..... 177
12.2.1 General Structural Features ..... 177
12.2.2 Collagen in the Normal Heart Muscle . 178
12.2.3 Collagen in Myocardial Hypertrophy .. 180
12.2.4 Collagen in Myocardial Infarction ..... 182
12.2.5 Collagen in the Valves in Health and Disease ..... 184
12.3 Vessels ..... 187
12.3.1 General Characteristics ..... 187
12.3.2 Collagen in the Normal Vessel Wall ... 188
12.3.3 Atherosclerosis ..... 189
12.3.4 Hypertension ..... 192

**13 Respiratory System** ..... 195
13.1 General Structural Features ..... 195
13.2 Collagen Content in the Lungs ..... 195
13.3 Cells Involved in Collagen Production ..... 197
13.4 Collagen and Lung Disease ..... 198
13.4.1 Introduction ..... 198
13.4.2 Fibrotic Lung Disorders ..... 198
13.4.3 Destructive Lung Disorders ..... 200

**14 The Liver** ..... 203
14.1 Introduction ..... 203
14.2 Collagen Content in the Liver ..... 203
14.3 Collagen Synthesizing Cells ..... 204
14.4 Fibrosis as a Response of the Liver Bioecosystem ..... 205
14.5 Collagen Content in the Fibrotic Liver ..... 206

14.6 Metabolism of Collagen in the Liver .......... 208
14.7 Regulation of Collagen Metabolism and Development of Fibrosis ................ 210
14.8 Collagen in Liver Regeneration .............. 213
14.9 Collagen Metabolism in Diseases of the Bile Ducts .......................... 213
14.10 Indices of Collagen Metabolism in Hepatic Disorders ....................... 214

**15 The Teeth and Periodontal Tissues ........... 219**
15.1 Structure and Collagen Content in Normal Teeth ........................... 219
15.1.1 Dentin ............................ 219
15.1.2 Cementum ........................ 221
15.1.3 Enamel ........................... 221
15.1.4 Dental Pulp ...................... 221
15.1.5 Periodontium ..................... 222
15.2 The Role of Collagen in Development of Teeth . 223
15.3 Collagen and Dental and Periodontal Pathology 223
15.3.1 Periodontal Disease ................ 224
15.3.2 Hydantoin-Induced Gingival Hyperplasia ...................... 226

**16 The Skin and Wound Healing ............... 227**
16.1 Collagen in the Skin ....................... 227
16.2 Collagen in Cutaneous Diseases ............. 227
16.3 Collagen in Keloids ........................ 229
16.4 Wound Healing ............................ 230

**17 The Eye and The Ear ...................... 233**
17.1 The Eye ................................. 233
17.1.1 General Structural Features .......... 233
17.1.2 The Cornea and the Sclera .......... 233
17.1.3 The Uvea ......................... 236
17.1.4 The Lens .......................... 237
17.1.5 The Vitreous Body ................. 237
17.1.6 The Retina ........................ 238
17.1.7 Collagen in Ocular Pathology ........ 238
17.2 The Ear ................................. 241

**18 Urinary and Reproductive Systems ........... 245**
18.1 Introduction ............................. 245

18.2 The Urinary System .......................... 245
18.3 The Ovary .................................. 248
18.4 The Uterus and Pregnancy ................... 250
18.5 Placenta ................................... 257

**19 Collagen in the Nervous System** .............. 261

**20 Collagen in Hematology** ..................... 265
20.1 Collagen and Hematostasis .................. 265
20.2 Collagen in the Bone Marrow ................ 268
20.3 Collagen in the Lymph Nodes ................ 269

**21 Collagen and Neoplasia** ..................... 271
21.1 Introduction ............................... 271
21.2 Collagen in the Stroma of Neoplasms ........ 272
21.3 Collagenolysis and Tumor Invasion .......... 275

**22 Effect of Nutrition** ........................ 279
22.1 Malnutrition and Overnutrition ............. 279
22.2 Scurvy ..................................... 280

**23 Effect of Ionizing Radiation** ............... 283

**24 Pharmacological Control of Collagen Metabolism** ................. 287
24.1 Introduction ............................... 287
24.2 Direct Inhibitors of Collagen Synthesis .... 288
24.2.1 Pharmacological Control of Hydroxylation of Proline and Lysine Residues ..................... 288
24.2.1.1 Structural Analogues of Proline ....... 288
24.2.1.2 Cosubstrates and Cofactors of Hydroxylation ................... 290
24.2.1.3 Fibrostatins ......................... 292
24.2.2 Pharmacological Control of Collagen Secretion ............... 294
24.2.3 Pharmacological Control of the Conversion of Procollagen to Collagen . 295
24.2.4 Pharmacological Control of Extracellular Processing of Collagen ... 295

24.2.4.1 Lathyrism ........................ 296
24.2.4.2 Penicillamine ........................ 298
24.2.4.3 Inhibitors of Lysyl Oxidase with Respect to Pyridoxal ........................ 300
24.2.5 Antifibroblast Serum ................ 300
24.3 Pharmacological Control of Collagen Degradation .................... 300
24.3.1 Induction of Collagenolysis .......... 300
24.3.2 Inhibitors of Collagenase Secretion .... 301
24.3.3 Inhibitors of Mammalian Collagenases . 302
24.4 Drugs That Indirectly Affect Collagen Metabolism ............................ 303
24.4.1 Nonsteroidal Antiinflammatory Drugs . 303
24.4.2 Tissue Extracts ..................... 304
24.4.3 Heavy Metals ....................... 305
24.4.4 Fluoride ........................... 307
24.4.5 Cytostatic Drugs .................... 308
24.4.6 Immunomodulants .................. 309
24.4.7 Antibiotics ......................... 309

**25 Indices of Collagen Metabolism in Clinical Practice** .......................... 311

**References** ..................................... 315

**Subject Index** .................................. 423

# 1 Introduction: Historical Outline

Collagen is a protein present in the majority of species throughout the animal kingdom, including all vertebrates. It is the most abundant protein of human and other mammalian bodies. It is not a single protein, as thought until the 1970s, and the family of collagen proteins has increased rapidly in recent years due to discoveries of new, genetically distinct types. The biological role of collagens is not limited to supporting the scaffold, rather the function of nearly all systems and organs of the body is related to collagenous structures. On the other hand, the biological functions of collagens should not be considered in isolation since they are constituents of the connective tissue and remain in close mutual relationship with other components of the extracellular matrix.

The term "collagen" was introduced into histological works in the nineteenth century and was derived from the Greek words "kolla" and "genos," which mean "glue" and "formation." The earliest observations were related to the ability of collagen to produce glue (gelatin) when tissues, usually hide, were extracted with boiling water. This first property of collagen was known in antiquity (Bogue 1922). In modern times, the first studies on collagen were carried out by chemists involved in leather tanning and gelatin production.

A few fields of interest can be distinguished in the history of collagen research. Beside the studies related to leather and gelatin technology, biochemical and pathophysiological investigations revealed an unexpected biological importance of collagen. Th wide distribution of collagenous fibers within the body was described in the second half of the nineteenth century following the rapid progress in histology based on optical microscopy. In 1900, Zaccharidès described the swelling of tendons inacid solution. This finding was followed by a report of Nageotte (1927) who was the first to isolate a native soluble collagen by extraction of rat tail tendon with cold dilute acetic acid. He also demonstrated aggregation of soluble collagen in vitro when dialysed against water. This observation was the first attempt to apply biochemical methods to the study of collagen. As mentioned earlier, chemists involved in leather technology contributed to the description of its specific physicochemical features, including amino acid composition, the unusual resistance to proteolysis, and the extraordinary mechanical properties of collagenous structures (Grassmann 1955, 1960). Collagen was one of the first proteins investigated by X-ray diffraction (Wyckoff and Corey 1936; Bear 1942). In the early 1950s the general concept of the triple-helical structure of collagen was elaborated (Ramachandran and Kartha 1954; Rich and Crick 1955). The periodicity and other morphological details visible with electron microscopy were reported (Asbury

1940; Schmitt et al. 1942; Hall et al. 1942; Wolpers 1943; Highberger et al. 1950). In these years the relatively slow metabolism of the protein and the high content of glycine and imino acids were noted (Neuberger et al. 1951). It was discovered that hydroxyproline is formed after the incorporation of its precursor into collagen molecules (Stetten and Schoenheimer 1944). Several workers showed the biological importance of its solubility, indicating that the neutral, salt-soluble collagen represents the newly synthesized form of the protein (Orekhovich 1952; Jackson 1957; Jackson and Bentley 1960). Methods for the determination of collagen extractibility were widely introduced into biochemical and medical investigations. These findings were followed by research on the nature of cross-links in collagen molecules as cross-link formation is associated with loss of collagen extractability (Piez 1968).

Investigations on collagen in health and disease began in the early 1960s. A number of physiological and pathological conditions associated with an altered solubility of collagen were reported. Urinary hydroxyproline output became a widely used index of collagen metabolism applicable in clinical practice (Kivirikko 1970). The important discovery then was the description of precursor forms appearing during collagen synthesis (Bellamy and Bornstein 1971). These findings initiated a series of studies which led to the elucidation of a complex, multistep process of collagen biosynthesis. Discovery of several features of the intracellular synthesis of collagen was followed by research on the pharmacological regulation of hydroxylation of proline and lysine residues (Chvapil and Hurych 1968). Immunological properties of collagen were investigated in the 1970s and have been a basis for immunochemical techniques useful in certain aspects of the research of connective tissue.

The discovery of the genetic heterogeneity of collagen (Miller and Matukas 1969) opened a new field. In 1971, Kefalides reported that basement membranes contain a specific collagen, type IV. Kühn et al. (1981) and Timpl et al. (1981) found that type IV collagen differs significantly in its molecular structure and ability to form supramolecular structures from other collagens. More than 12 types of collagen have been described, and the discovery of others is probable (Bornstein and Sage 1980). Reports of the heterogeneity of the primary structure of collagen preceded studies on collagen genes and regulation of their expression. Clarification of the biosynthesis and molecular genetics aided at least partially the elucidation of molecular mechanisms responsible for several inherited diseases. Progress in the biochemistry of collagen facilitated understanding of many changes in collagen structure and metabolism in acquired diseases of various organs (Weiss and Jayson 1982). Additional biological research on collagen begun in the 1970s has been focused on its distribution and heterogeneity in the animal kingdom and the evolutionary as well as paleobiological aspects of this protein (Mathews 1975). Recent efforts have brought about a better comprehension of the enzymes of collagen biosynthesis and in turn led to some discoveries in the pharmacological regulation of collagen metabolism.

The term "collagenase" was used for the first time by Ssadikow in 1927 to describe enzymes cleaving native collagen and gelatin. Until 1962 when Gross and Lapière originally described vertebrate collagenase, only the enzymes of bacterial origin had been known. In recent years, the possible metabolic pathways of collagen breakdown as well as inhibitors and activators of collagenase have been reported (Weiss 1976; Woolley and Evanson 1980).

The amount of original publications on collagen increased rapidly (for early papers, see bibliography in Boransky 1950). In the past few years just the reports on its medical aspects accounted for about 250 papers annually. In addition to original papers many monographic surveys were published in the 1950s–1970s (Asboe-Hansen 1954; Gustavson 1956; Chvapil 1960, 1967; Gross 1961; Harkness 1961; Policard and Collet 1961; Banga 1966; Ramachandran 1967b; Bailey 1968b; Gould 1968; Balazs 1970; Fernandez-Madrid 1970; Böhmer 1971; Traub and Piez 1971; Bańkowski 1972; Grant and Prockop 1972; Slavkin 1972; Bornstein 1974; Burleigh and Poole 1974; Fricke and Hartmann 1974; Gross 1974; Nimni 1974; Otaka 1974; Martin et al. 1975; Khilkin et al. 1976; Kivirikko and Risteli 1976; Ramachandran and Redi 1976; Fessler and Fessler 1978; Gay and Miller 1978; Lasek 1978; Bornstein and Traub 1979; Prockop et al. 1979; Viidik and Vuust 1980; Minor 1980; Eyre 1980) and in the recent decade (Deyl and Adam 1981, 1982; Małdyk 1981; Cârsteanu and Vlădescu 1982; Piez and Reddi 1984; Veis 1984; Bairati and Garrone 1985; Kühn and Krieg 1986; Ramirez et al. 1986; Mayne and Burgeson 1987; Miller and Gay 1987; Uitto and Perejda 1987; Seubert and Hamerman 1988; Nimni 1988). In 1963–1983 ten volumes of the International Review of Connective Tissue Research were published under the editorship of Hall, and since volume 5, Hall and Jackson. In 1981, the first specialistic journal *Collagen and Related Research* appeared, and since 1989 it has been renamed *Matrix* (edited by Gay and Miller). This journal was preceded by a quarterly dedicated to a slightly wider subject, *Connective Tissue Research,* published since 1972.

Current research concentrates on newly discovered types, their molecular and supramolecular structure, metabolism, and biological functions. Our knowledge of the extracellular matrix as a complex system involving collagenous structures is still insufficient for comprehensive understanding of several pathophysiological phenomena. Elucidation of these aspects will provide valuable tools for the management of certain disorders.

# 2 Structure, Heterogeneity, and Distribution

## 2.1 Basic Nomenclature

Collagens represent a number of closely related but chemically distinct macromolecules. All forms are made up of three left-handed polypeptide chains. These chains are coiled in a right-handed direction about a common axis to form the rope-like molecule. The major part of the collagen molecule has triple-helical structure. The nonhelical domains are present at both ends of the chains, and in some types of collagen the helical part is interrupted by nonhelical domains. The presence of a sizeable part of the molecule with the triple-helical conformation is a basic feature for classification of the proteins as collagens. Those containing only a small portion of the molecule in the form of the triple-helical conformation are termed proteins containing a collagen sequence.

The polypeptide chains are termed $\alpha$-chains. The chains with different primary structure are numbered with arabic numerals, e.g., $\alpha 1$, $\alpha 2$. They may be identical or different within the same molecule. The distinct collagens are called "types" and are numbered with roman numerals. The same numerals are added to the chains in brackets, following the arabic number of the chain. For example, the composition of the most common type I collagen is $[\alpha 1(\mathrm{I})]_2 \alpha 2(\mathrm{I})$, i.e., its molecule consists of two identical $\alpha 1(\mathrm{I})$ chains and one chain $\alpha 2(\mathrm{I})$. It is important to remember that chains with the same arabic number of different types of collagen are not identical, i.e., the chain $\alpha 1(\mathrm{I})$ has different primary structure than the chain $\alpha 1(\mathrm{II})$, etc. The term "collagen types" is sometimes substituted with the term "collagen systems" (Miller 1988). The term system has been suggested as more appropriate because the same type of collagen may describe more than one species of molecule. This may result from the association of three different chains in various manners to form different molecules. The term "collagen type" is used in this book as it represents commonly used terminology.

Collagen, as seen in the extracellular space, is synthesized from a precursor protein form. The polypeptide chains in nascent form are termed preprocollagen chains and are abbreviated as prepro $\alpha$ with the appropriate numerals. After cleavage of the signal peptide, the preprocollagen chains form procollagen chains. They are subjected to a number of posttranslational modifications, and three procollagen chains assemble into the procollagen molecule. Conversion of procollagen to collagen is associated with removing the propeptides from all three chains on both ends. Six telopeptides are cleaved to realize one collagen molecule. The cleavage of amino- and carboxyterminal propeptides is done at one time for all three peptides of the single end of the molecule. Procollagen after removal of the aminoterminal propeptide is called pro-C-collagen (proC) and after separation of the carboxylterminal propeptide, pro-N-collagen (proN). The cleavage of propeptides is not an universal property of all types of collagen. In some of them, the propeptides are cleaved from only one end of the procollagen molecules, and in type IV the lack of cleavage on both ends of the molecule has been reported. The final molecule which is the basic unit for fibrils and other collagenous structures is called collagen. This is synonymous with the old term tropocollagen.

## 2.2 Triple-Chain Helical Structure

The structure of the collagen molecule was discovered on the basis of X-ray diffraction studies followed by examination of synthetic polypeptides. The principal features which effect helix formation is a high content of glycine and imino acid residues and the sequence of repeating residues Gly-X-Y, where X and Y can be any amino acid, often proline and hydroxyproline (Piez 1984). The sequence Gly-Pro-Hyp makes up about 10% of the molecule of type I collagen. An individual chain has a left-handed helical structure which is not stable when the chains are separated. The presence of glycine as every third residue is an absolute requirement for the triple-helix formation because glycine is the only amino acid without a side chain. Three glycine residues alternately from three chains form a shallow helix up the center of the superhelical structure. The side chains of residues of amino acids other than glycine are directed outwards. Three left-handed helical chains form a "supercoil" with a pitch of approximately 8.6 nm. The distance between amino acids within each chain is 0.291 nm, and the relative twist is 10°. Thus, the distance between each third glycine residue is 0.87 nm (Fig. 2.1) (Ramachandran 1963; 1967b, 1988; Yonath and Traub 1969; Piez 1984; Nimni and Harkness 1988).

Several factors contribute to stabilization of the helical structure. The three-strand rope structure is stabilized by the high content of imino acid residues, and proline is the major component stabilizing the conformation. Proline has the rigid ring structure which prevents rotation about the N-$C_\alpha$ bond in a polypeptide chain and probably also rotation about the $C_\alpha-C=O$ bond. The position of the proline residue with respect to the glycine residues is decisive (Carvier and Blout 1967). As it has been shown by conformational studies of synthetic polypeptides, only proline residues which follow glycine ones take part in stabilization. Poly(Gly-Pro-Ala) forms a stable helix whereas poly (Pro-Gly-Ala) is unable to make a stable structure (Traub and Piez 1971; Guantieri et al. 1987). This observation raises the question of the role of hydroxyproline residues in stabilization of the helical structure (Némethy and Scheraga 1986).

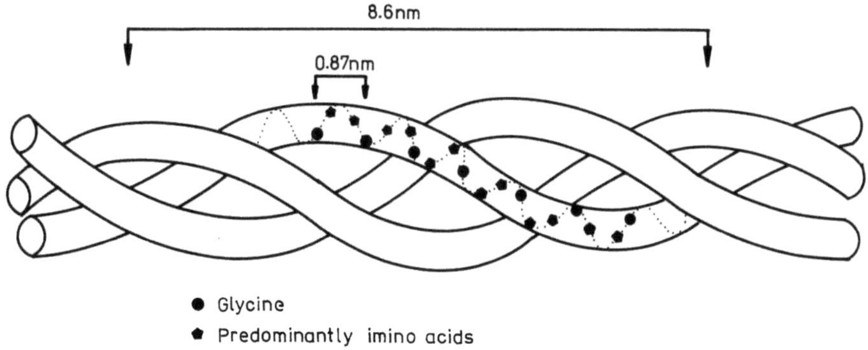


**Fig. 2.1.** The collagen triple helix. (From Nimni and Harkness 1988 with permission)

Experiments with underhydroxylated collagen indicate clearly that the molecules are less stable (Berg and Prockop 1973 a). On the other hand, the determination of the primary structure showed that almost all hydroxyproline residues precede glycine residues. The other mechanism of stabilization of the structure by hydroxyproline suggested involves the hydroxyl groups in water-bridged hydrogen bonds. The ability to bind water by collagen molecules has been also suggested to be a stabilization mechanism of the molecule.

A number of physicochemical studies on a variety of polymers with glycine in every third position have been reported (e.g., Tamburro et al. 1984). These studies are useful in understanding collagen structure. Details are reviewed by Bhatnagar and Rapaka (1976).

## 2.3 Distribution

Distribution of collagen within the body refers to the amounts of collagen which occur in certain organs or parts of the body as well as to the tissue and organ localization. The first aspect was investigated in the early years of biochemical research on collagen. The results obtained are still valuable, although the methods used were based only on simple extraction procedures and hydroxyproline determinations.

The total amount of collagen is estimated at 25% – 30% of total body proteins; thus collagen is the most abundant protein in the mammalian body. In adult mice collagen was found to make up 28.8% of the body weight at death. Measurements carried out in mice showed that about 40% of the total body collagen is in the skin, which forms only 17.6% of the body weight (Harkness et al. 1958). In human skin, collagen constitutes about 75% of total nitrogen content (Eisele and Eichelberger 1945). The musculoskeletal system of mice was found to contain 52%, and viscera accounted for only 3% of the total

**Table 2.1.** Collagen content in tissues of rat or mouse. (From Chvapil 1961 with permission)

| Tissue | Collagen (mg/g dry weight) |
|---|---|
| Liver | 8 – 10 |
| Lungs | 90 – 110 |
| Kidney | 20 – 30 |
| Spleen | 25 – 35 |
| Aorta | 200 – 260 |
| Skin | 650 – 720 |
| Bone | 150 – 250 |
| Cartilage | 310 |
| Tendon | 810 – 850 |

**Table 2.2.** Chain composition and distribution of collagens

| Collagen type | Chain composition | Distribution |
|---|---|---|
| I | $[\alpha 1(I)]_2\ \alpha 2(I)$ | Skin, tendon, bone, cornea, dentin, fibrocartilage, large vessels, intestine, uterus |
| I-trimer | $[\alpha 1(I)]_3$ | Dentin, dermis, tendon |
| II | $[\alpha 1(II)]_3$ | Hyaline cartilage, vitreous, nucleus pulposus, notochord |
| III | $[\alpha 1(III)]_3$ | Large vessels, uterine wall, dermis, intestine, heart valve, gingiva |
| IV | $[\alpha 1(IV)]_2\ \alpha 2(IV)$ | Basement membranes |
| V | $[\alpha 1(V)]_2\ \alpha 2(V)$, $\alpha 1(V)\ \alpha 2(V)\ \alpha 3(V)$, $[\alpha 1(V)]_3$, $[\alpha 3(V)]_3$. | Cornea, placental membranes, bone, large vessels, hyaline cartilage, gingiva |
| VI | $\alpha 1(VI)\ \alpha 2(VI)\ \alpha 3(VI)$, $[\alpha 1(VI)]_3$ | Descemet's membrane, skin, nucleus pulposus, heart muscle |
| VII | $[\alpha 1(VII)]_3$ | Skin, placenta, lung, cartilage, cornea |
| VIII | Unknown | Produced by endothelial cells, Descemet's membrane |
| IX | $\alpha 1(IX)\ \alpha 2(IX)\ \alpha 3(IX)$ | Cartilage |
| X | $[\alpha 1(X)]_3$ | Produced by hypertrophied chondrocytes during the process of endochondral ossification |
| XI | $\alpha 1(XI)\ \alpha 2(XI)\ \alpha 3(XI)$ | Hyaline cartilage, intervertebral disc, vitreous humor |
| XII | $[\alpha 1(XII)]_3$ | Chick embryo tendon, bovine periodontal ligament |
| XIII | Unknown | Fetal skin, bone, intestinal mucosa |

body collagen. The highest relative content of collagen has been found in tendons, which are build up of 85% (Harkness et al. 1958). The collagen content in various tissues is shown in Table 2.1 according to Chvapil (1961). The distribution of collagens of various types has only partially been explored. Details are given in the following sections, while an overview is summarized in Table 2.2.

Microscopic distribution of collagens has been recognized on the basis of the immunofluorescence localization of these proteins. The details can be found in the following reviews: Gay and Miller (1978), Linsenmayer (1982), Von der Mark (1981).

## 2.4 Classification and General Structural Features of the Collagens

Collagen types are commonly classified into the group of fibril-forming and nonfibril-forming collagens. The former includes types I, II, III, V, and XI collagens. Their molecules contain a continuous or almost continuous helical domain, and this feature, probably, affects the way of aggregation into fibrils. The remaining collagens are composed of helical domains separated by variously sized nonhelical domains. These collagens aggregate into different fibres than supramolecular structures.

A more precise classification is, at present, impossible. The newly discovered collagen types are insufficiently characterized, each type of collagen has its own specific properties, and the classification criteria are frequently chosen arbitrarily.

At the time of writing 13 types of collagen have been reported. Their chain composition and overview of distribution are summarized in Table 2.2. The details of chain composition of recently described collagen types are partially obscure, and those available are discussed with the description of the individual types.

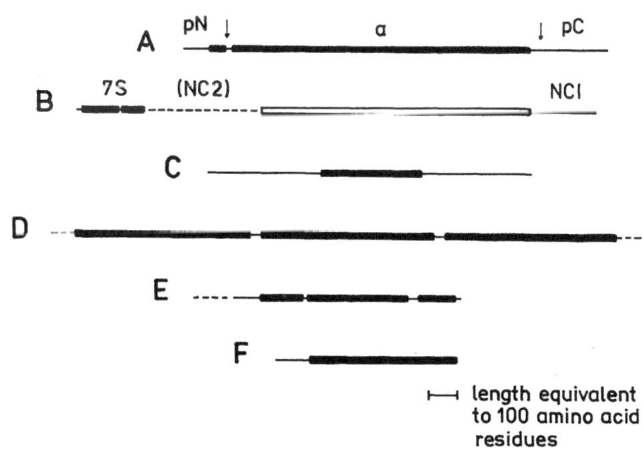


**Fig. 2.2.** Scale drawing indicating the size and topology of structural features for collagen chains. *Solid lines* indicate regions of nontriplet structure; *filled bars* depict regions of repetitive Gly-X-Y triple structure; the *open bar* in *B* represents a region in which the repetitive triplet structure is frequently interrupted; *broken lines* represent regions in which the structure remains to be determined. Details in the text. (From Miller 1988 with permission)

As mentioned above, the major feature which is responsible for supramolecular organization of the collagenous structures and for the physicochemical properties of collagen is the size and localization of the helical domain(s) in the chains of collagen. Figure 2.2 presents the scale drawing indicating the size and topology of collagen chains. Types I, II, and III collagen as well as types V and XI are built up of three chains, all composed mostly of the continuous triple-helical structure composed mostly of the continuous triple-helical structure (Fig. 2.2A). A significantly different structure of the chains is found in type IV collagen (Fig. 2.2B). The regions with the triple-helical conformation are interrupted with large nonhelical domains as well as with short nonhelical peptide interruptions. The chains of type VI collagen have a general design similar to that of type I with respect to a single uninterrupted helical domain, but the nonhelical parts located at the both ends of the molecule are large and account for about two-thirds of the whole molecule (Fig. 2.2C). The chains of type VIII collagen are larger that those of the others and are made up mostly of the interrupted helical structure (Fig. 2.2D). The chains which constitute type IX and X collagens are small and contain some nonhelical domains (Fig. 2.2E, F).

## 2.5 Type I, Type II, Type III Collagens, and Type I–Trimer Collagen

Types I-III collagens are the most investigated proteins of the collagen family. They are known as classic collagens (Miller 1985; Miller and Gay 1987; Kühn 1987). Type I contains two $\alpha 1$ (I) and one $\alpha 2$(I) chains, type II is a homotrimer of $\alpha 1$ (II) chains, and type III is also a homotrimer composed of three identical $\alpha 1$ (III) chains. The molecular structure and amino acid sequence of the four chains of types I–III collagens are well established (Fietzek et al. 1972; Fietzek and Kühn 1976; Allmann et al. 1979; Galloway 1982; Seyer et al. 1989). The chains are different and are products of different genes, but profound similarities in their structure have been shown.

The chains are synthesized in the form of preprocollagen chains (for details on collagen biosynthesis see Chap. 3). The preprocollagen chains consist of the signal peptide, aminoterminal propeptide, proper portion of the chain, carboxyterminal propeptide (Fig. 2.3). The signal peptide is cleaved in an early phase of synthesis. Both propeptides are removed during the conversion of procollagen to collagen. The molecules of types I–III collagens contain a continuous triple-helical domain, forming about 97% of the molecule. At both ends of the collagen molecule, the short, noncollagenous domains are present and are different from the terminal propeptides.

The aminoterminal propeptides of the $\alpha 1$ (I) and $\alpha 1$ (III) procollagen chains are globular domains of 66 and 68 residues, respectively. The propeptide of the

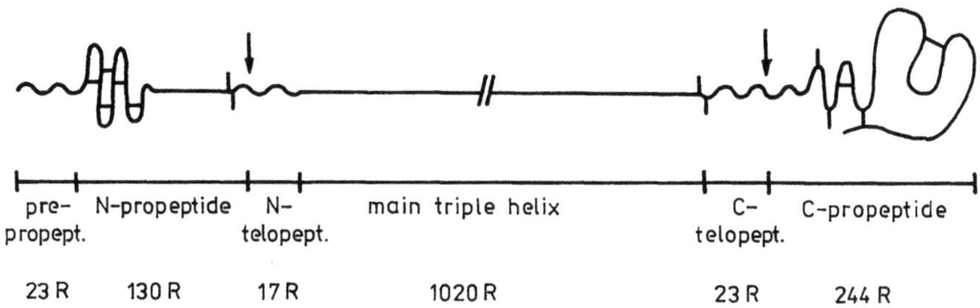


**Fig. 2.3.** The prepro $\alpha 1$ (III) chains of collagen. The length of the individual part of the molecule is indicated as "R" number of amino acid residues. *Arrows* indicate the cleavage sites of procollagen peptides. (From Kühn 1986 with permission)

$\alpha 2$(I) chain has a helical structure. Similarly, the propeptide of the $\alpha 1$(II) chain contains twelve triplets of the helical structure.

The helical domain is slightly longer than 1000 residues. The nonhelical aminoterminal region of the collagen molecule is very short (about 16 residues). The nonhelical carboxyterminal region of the molecule contains about 25 residues. The carboxyterminal propeptides of types I–III procollagens are relatively similar. The propeptides of $\alpha 1$ (I) and $\alpha 1$ (II) chains contain 247 residues, and those of $\alpha 1$ (III) and $\alpha 2$ (I) chains are made up of 246 residues. They contain eight cysteine residues, two involved in interchain bridges and the remaining six forming intrachain disulfide bonds. The amino acid sequence of the propeptides shows a profound homology.

Types I–III collagens are fibril-forming and their supramolecular structure is very similar. They form native fibrils with an identical period of 67 nm and a homologous banding pattern. The assembly into fibrils is regulated by the sequence of the polar charged and hydrophobic amino acid residues. These regions are distributed along the molecule in four homologous portions, labelled D1, D2, D3, and D4, each being 234 residues long. The collagen molecules assemble in a D-staggered parallel array (Fig. 2.4). The molecule is longer by about one-third of the D region than the summed length of four D portions. Thus, the end-to-end regions overlap. The portion of the molecule that overlaps contributes to cross-link formation.

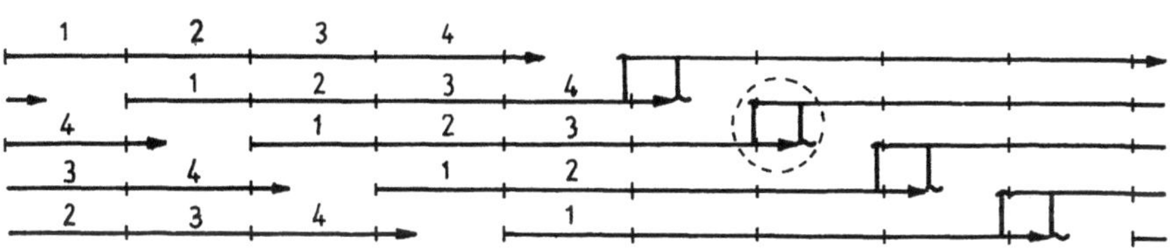

**Fig. 2.4.** Axial arrangement of the interstitial collagen molecules in the 67-nm, cross-striated fibril. The molecules are staggered against each other by the distance of 234 amino acid residues. (From Kühn 1986 with permission)

The molecules built of three $\alpha 1$ (I) chains form so-called type I-trimer collagen. It is stable and occurs in small quantities in various tissues. On the contrary, molecules with two $\alpha 2$(I) chains are not stable and are not present in tissues. Type I-trimer collagen is also synthesized under pathological conditions. Whether the $\alpha 1$ (I) chains in type I-trimer and type I collagen are identical has not been demonstrated. Some investigators have suggested that there are some differences in the primary sequence and that type I-trimer is, in fact, a separate type of collagen (Shupp-Byrne and Church 1982).

The biophysical properties of type I-trimer collagen are practically the same as those of type I collagen, for example, the melting temperature is only 1 °C less (Tkocz and Kühn 1969). Type I-trimer is found to be relatively resistant to human fibroblast and neutrophil collagenases. The enzymes' degradative activity is less than one-fifth of that on type I collagen (Narayanan et al. 1984). The hypothetical type I-trimer-specific collagenase has not been discovered. The biological role of type I-trimer collagen remains unknown.

Type I collagen is the predominant collagen of the body. It is found in skin, ligaments, bone, muscle, blood vessels, and internal organs. Type III collagen forms relatively fine fibrils as compared with type I; they are known as reticulin fibrils (Sandberg et al. 1989a). The distribution of type III collagen is similar to that of type I. Type II collagen is found in hyaline cartilage, the notochord, nucleus pulposus, and vitrous humor. It is a typical cartilage collagen (Von der Mark 1981).

Types I–III collagens are profoundly characterized also in respect of procollagen structure (Bańkowski and Mitchell 1973; Mäkelä et al. 1988), physicochemical properties in vivo and in vitro (Sharimanov et al. 1979; Na 1988; Bender et al. 1983; Nandi et al. 1985; Oxlund 1986; Hulmes et al. 1985), and molecular aggregation and fibril formation (Kühn 1982a; Parry 1988; Glanville 1982; Kadler et al. 1988). The details can be found in the references cited above.

## 2.6 Type IV Collagen

Type IV collagen has several distinct features and is a highly specialized form. It is found only in the basement membranes, where its unique structure plays an important role in the formation of the resilient, three-dimensional network. Therefore, this type is responsible for the functional state of the basement membrane.

The basement membranes are extracellular structures found between the cells and the connective tissue stroma. They are thin (40–60 nm), sheetlike structures which can be seen under the electron microscope. They are found in every organ of the body, and with only a few exceptions all cells are surrounded by them. The majority of the membranes are bilaminar. All contain a lamina densa. The other layer is called the lamina rara or lamina lucida. The trilaminal structure seen in some organs results from fusion of the endothelial basement membranes.

The biological functions of the basement membrane are not limited by its mechanical properties. Owing to the type IV collagen it has significant strength and provides structural support for the tissue. Additionally, it is responsible for cell attachment, probably due to an interaction of the type IV collagen and laminin with specific binding proteins of the cell membrane. The selective filtration through the barrier of the basement membrane is the third function. The biology of basement membranes is a subject of comprehensive reviews, and the reader is referred to the following monographs: Kefalides (1973, 1982); Kefalides and Alper (1988); Martinez-Hernandez and Amenta (1983).

Type IV collagen was the first discovered, non-fiber-forming, collagenous protein. Early X-radiographic studies on intact basement membranes indicated the presence of collagen within them (Pirie 1951). Hydroxyproline was detected in the amino acid composition of whole basement membranes by Kefalides and Winzler (1966), but the new type of collagen was described later (Kefalides 1973). Since this discovery, our knowledge concerning type IV collagen has grown rapidly.

Type IV collagen contains two sorts of chains, $\alpha 1$(IV) and $\alpha 2$(IV). The chain composition, despite several studies, is not fully understood. Several authors suggest the heterotrimeric structure $\alpha 1(IV)_2\alpha 2(IV)$ (Trüeb et al. 1982; Fujiwara and Nagai 1981; Qian and Glanville 1984). Another study has provided strong support for a homotrimeric model $[\alpha 1(IV)]_3$ or $[\alpha 2(IV)]_3$. If true, there are, in fact, two types of collagen known as type IV (Haralson et al. 1985).

Both chains have a similar pattern of helical and nonhelical domains. At the aminoterminal end the helical domain known as 7S is present. This is separated from the main helical domain by a noncollagenous sequence, NC2. The main helical domain is about 326 nm long and is interrupted by short nonhelical domains. It has been reported that 21 interruptions are present in $\alpha 1$(IV) chain (Kühn 1986; Kühn et al. 1985). At the carboxyterminal end of the molecule an 8 nm long, noncollagenous domain, NCl, is present (Fig. 2.5).

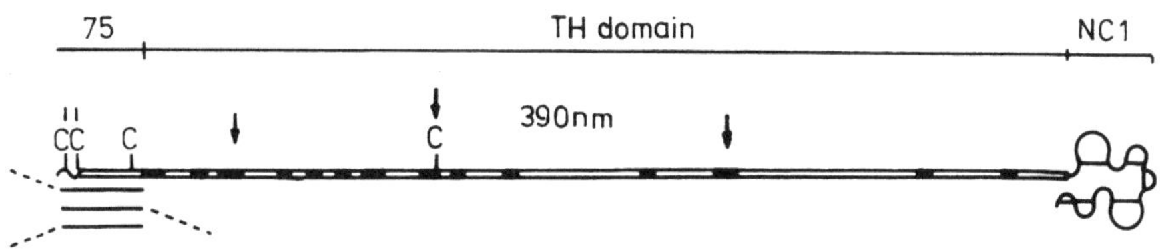


**Fig. 2.5.** The $\alpha 1$(IV) chain consits of the aminoterminal 7S domain, the central triple helical part (*TH*), and the carboxyterminal NC1 domain. Location of the interruptions of the triple helix longer than 5 amino acid residues along the molecule denoted by dark bars. Arrows indicate pepsin cleavage sites. Cysteine residues are indicated as Ç. (From Kühn 1986 with permission)

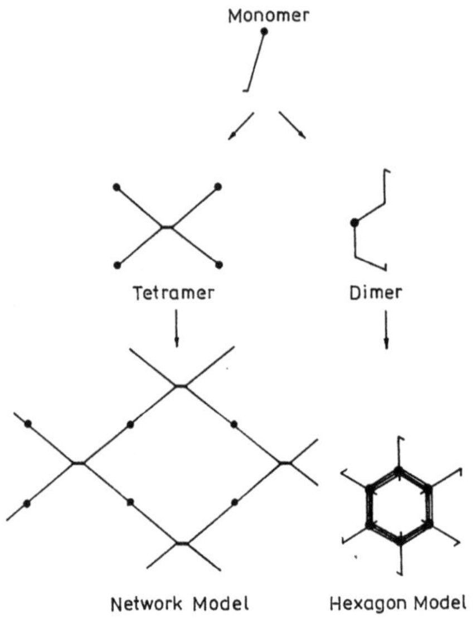


**Fig. 2.6.** Two possible models of type IV collagen association. (From Glanville 1987 with permission)

The amino acid sequence of the polypeptide chains of type IV collagen has been reported (Babel and Glanville 1984; Glanville et al. 1985; Hostikka and Tryggvason 1988). Extensive homology between the noncollagenous domains of $\alpha 1$(IV) and $\alpha 2$(IV) chains has been found (Hostikka et al. 1987).

Interesting features of type IV collagen were found in the protein synthesized by the cultured embryo-derived, parietal yolk sac, carcinoma cells. The collagen produced was a homotrimer with a molecular weight of about 95000 Da and was composed of $\alpha 1$(IV) procollagen chains. This as yet unique property has been shown only in this cellular line (Haralson et al. 1985).

The supramolecular structure of type IV collagen has been extensively investigated. The collagen forms a network, and two models have been proposed to explain the nature of its supramolecular organization. As shown in Fig. 2.6, the pathway of aggregation depends upon whether the tetramers are first formed by self-assembly of the 7S domains or whether dimers are formed by the interaction of two globular domains (Glanville 1987; Dölz et al. 1988).

The thermal stability of type IV collagen and procollagen has been studied with a special interest in the structure of these proteins. The melting temperature of type IV collagen from lamb anterior lens capsule has been found to be 40.0 ° ± 1.0 °C while in situ the melting point varied with tissue and is generally higher than in solution (Gelman et al. 1976; Linsenmayer et al. 1984). Detailed studies of the thermal stability and folding of type IV procollagen have been reported by Davis et al. (1989).

The ability of proteases to degrade type IV collagen is related to the presence of helical and nonhelical domains. The latter are cleaved by less specific proteases, and it is difficult to classify the enzyme as a type IV collagenase only on the basis of its ability to degrade this type of collagen. Type IV-specific collagenase has been isolated from tumor cell cultures (Liotta et al. 1979). The enzyme is a neutral metalloproteinase of 60000–70000 Da and is activated by trypsin. It is specific for type IV collagen and cleaves both $\alpha 1$(IV) and $\alpha 2$(IV) chains at one specific locus localized approximately 90 nm from the 7S domain. The enzyme is inactive against types I, II, III, or V collagens (Glanville 1987). Neutrophil elastase has also been found to degrade type IV collagen (see Chap. 4).

The biological functions of type IV collagen are related to the structural and functional support of the basement membranes. An interesting finding has been reported by Kostrzyńska et al. (1989), who found specific binding of a certain bacterial strain to type IV collagen and suggested that such initial adhesion is a crucial step for the development of an infection.

## 2.7 Type V Collagen

Type V collagen was originally reported as AB-collagen. Most of the early studies on this form referred to the material extracted from tissues after pepsin digestion (Burgeson et al. 1976; Chung et al. 1976). Pepsin removes a significant part of the molecule, i.e., the nonhelical regions, and the preparations obtained after digestion differ in many respects from the native protein.

The chain composition of type V collagen remains obscure. There is strong evidence that it is a group of related proteins, composed of three from four polypeptide chains. The following specific chains have been described: $\alpha 1$(V), $\alpha 1'$(V) $\alpha 2$(V), and $\alpha 3$(V). Different trimeric combinations have been reported. The heterotrimer $[\alpha 1(V)]_2\alpha 2(V)$ has been proposed as the most common form (Burgeson and Hollister 1979; Rhodes and Miller 1978, 1981; Deyl et al. 1979). Another heterotrimer, $\alpha 1$(V), $\alpha 2$(V)$\alpha 3$(V) has been described (Rhodes and Miller 1981; Niyibizi et al. 1984). The homotrimer $[\alpha 1(V)]_3$ was reported in the continuous cell lines derived from hamster lung (Haralson et al. 1980, 1984; Haralson and Mitchell 1981) and human liver (Biempica et al. 1980). The homotrimer $[\alpha 3(V)]_3$ was described by Madri et al. (1982). Investigations in vitro suggested that the $[\alpha 2(V)]_3$ homotrimer is unstable, but they were carried out with reconstituted, pepsin-derived chains. The covalent structure of type V collagen has been partially recognized (Seyer and Kang 1989).

The occurrence of the fourth chain is a unique feature of type V collagen. This chain was described for the first time in 1985 and was named $\alpha 4$(V) (Fessler et al. 1985). Further studies showed that it shares many properties with the $\alpha 1$(V) chain, and the chain was renamed the $\alpha 1'$(V) chain (Fessler and Fessler 1987). Despite the similarities it is believed that the chains are products of separate genes. The other possibility is that the chains are products of the same gene, and the variation is caused by different splicing of the RNA. The main changes between the chains are located in nonhelical regions of the molecule. Thus, after pepsin digestion, it is impossible to distinguish the molecules. The homotrimeric molecule $[\alpha 1'(V)]_3$ and the heterotrimer $\alpha 1(V)\alpha 1'(V)\alpha 2(V)$ have been postulated (Fessler and Fessler 1987).

The physiological role of the various combinations of the chains of type V collagen remains unknown. Combinations other than those discovered up to now are possible. Single reports indicate that the forms of type V collagne differ in their stability and flexibility. Niyibizi et al. (1984) showed that $[\alpha 1(V)]_2\alpha 2(V)$ collagen melts at 39 °C while the heterotrimer $\alpha 1(C)\alpha 2(V)\alpha 3(V)$ dissociated at 36 °C. The latter also has a greater flexibility (Fessler and Fessler 1987).

The helical part of type V procollagen has the same length as that of type I procollagen. Two ends of the molecule have knoblike projections of noncollagenous propeptides. Only a part of the propeptides is removed during processing (Woodbury et al. 1989). In the pro-$\alpha 1(V)$ the aminoterminal end of the molecule has a noncollagenous domain of 85000 Da. Some of which (45000 Da) remains in the final collagen molecule. Its carboxyterminal end contains a 35000 Da noncollagenous propeptide which is totally removed in the conversion to collagen. The helical part has a molecular weight of about 100000 Da. This indicates that a significant part of the nonhelical structure remains in the type V collagen. A similar estimation for the $\alpha 1'(V)$ and $\alpha 2(V)$ chains showed that the aminoterminal end of the former contains a 30000 Da nonhelical domain, and the carboxyterminal end of the latter contains a 20 000 Da nonhelical part (Fig. 2.7).

The physiological function of type V collagen has not been definitely elucidated. Initially, type V collagen was believed to be a part of the basement membranes. It has been found there as an extrinsic component, i.e., it is discov-

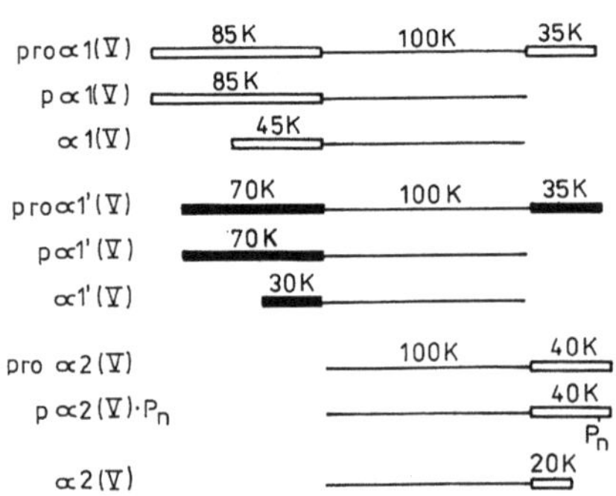


**Fig. 2.7.** Structure of pro$\alpha$(V) chains and their derivatives from chick embryo. The central collagenous sequence is indicated by a thin line and is assigned a nominal molecular size 100 kDa. The nominal molecular size of the flanking, noncoleagenous sequences are indicated. The conversion of pro$\alpha 2$(V) to $\alpha 2$(V) chain is a two step process, and includes formation of the intermediate called p$\alpha 2$(V). Upon conversion of pro$\alpha 2$(V) to p$\alpha 2$(V) two or more peptides called P-peptides ($P_n$), became disulfide linked to the noncollagenous region of p$\alpha 2$(V). (From Fessler and Fessler 1987 with permission)

ered occasionally within the membrane but is not synthesized by the cells resting on the membrane (Martinez-Hernandez and Amenta 1983). This concept is in agreement with controversial results of the immunolocalization of type V collagen in the basement membranes (Gay et al. 1981; Martinez-Hernandez et al. 1982). The cellular origin of type V collagen found in the basement membranes remains unknown, and trapping of the molecule during filtration seems to be a simple explanation.

Type V collagen has been shown in the fibrous stroma of various tissues, including tendon, cornea, and vessel smooth muscle. A comprehensive list of locations in which type V collagen has been found is given by Fessler and Fessler (1987).

Various biological functions have been attributed to type V collagen. It has been suggested to be a connector between basement membranes and the stroma. Its ability to form fibrils suggests other possible biological functions. Pepsin-derived preparations of type V collagen are able to precipitate as thin fibers (Chiang et al. 1980; Broek et al. 1985). It is possible that type V collagen in vivo contributes to fibrillogenesis of type I collagen. This suggestion arose during immunolocalization studies of type V collagen in the cornea (see Chap. 17). Two possible models of fibers containing types I and V collagens have been suggested. The molecules of both types can be intimately mixed throughout the whole cross-section of a fiber, or type V collagen could be restricted to the central part of the fiber (Linsenmayer et al. 1983, 1984, 1985; Fitch et al. 1984). Both models have implications for fibrillogenesis and an influence upon the fiber properties, including its diameter.

The occurrence of type V collagen in the immediate vicinity of the basement membranes suggests an anchoring function. The small type V collagen-containing fibrils are postulated to join the basement membrane and the stroma (Modesti et al. 1984).

## 2.8 Type VI Collagen

Type VI collagen was originally described as an intima or a short chain collagen. This latter term reflects its unique property, the rather short triple-helical domain which accounts for less than half of the total mass of the protein. It is composed of the helical part of the molecule which was solubilized but not degraded after pepsin digestion. The pepsin solubilization causes the loss of about 50% of the protein. The molecular mass of an intact type VI collagen molecule is estimated at 330000–420000 Da. Solubilization with pepsin reduces this to 170000 Da (Aumailley et al. 1985; Gibson and Cleary 1985; Furthmayr et al. 1983).

The chain composition has not been definitely elucidated. It was found that the type VI collagen from the uterus consists of three distinguishable chains. Two of them, $\alpha 1$(VI) and $\alpha 2$(VI), are smaller (140000 Da), and the third

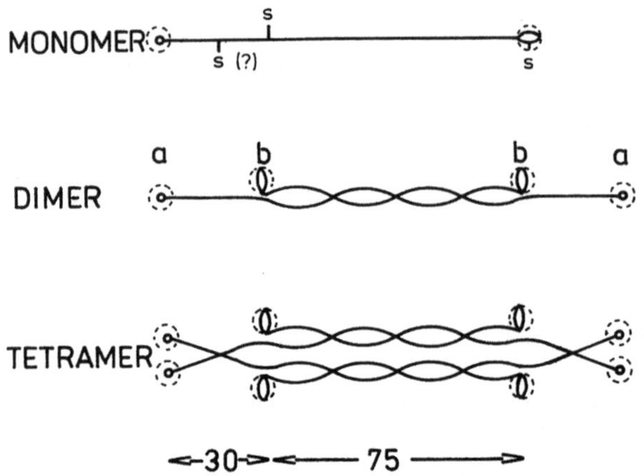


**Fig. 2.8.** Models of type VI collagen monomers, dimers, and tetramers. Dimensions between *arrows* are given in nanometers. Outer globular domains are indicated as *a*, and inner globular domains are shown as *b*. (From Timple and Engel 1987 with permission)

$\alpha$3(VI) is larger with a molecular weight about 200000 Da (Trüeb and Winterhalter 1986). The chains differ in amino acid composition and patterns of collagenase-resistant peptides. The amino acid composition of the intact $\alpha$1(VI) chain and the helical parts of the three chains of type VI collagen have been described (Timpl and Engel 1987; Chu et al. 1988; Trüeb et al. 1989). Chicken $\alpha$2(VI) chain has been found to contain a collagenous segment of 335 residues and two noncollagenous domains, an aminoterminal one and a carboxyterminal one, 228 and 432 residues long, respectively (Koller et al. 1989). The collagenous domain contains seven Arg-Gly-Arg tripeptide units, some of which are likely to be used as cell-binding sites. The large, nonhelical globular domains show homology with von Willebrand's factor and cartilage matrix protein (Koller et al. 1989) and lend type VI collagen the unique ability to aggregate in a different manner to other types of collagen. The single molecule, i.e., the monomer, can form dimers or tetramers. The tetramers are the basic units for further aggregation in the form of fibrils. Formation of oligomers and microfibrils is shown in Fig. 2.8. Formation of tetramers leads to localization of the globular domains in outer or inner positions. The biological role of this two-step fibril formation remains unknown. Oligomer formation increases both thermal stability and protease resistance (Jander et al. 1981; Odermatt et al. 1983).

There are several suggestions that type VI collagen forms microfibrillar aggregates. Furtmayr et al. (1983) isolated aggregates of six tetramers linked to each other in end-to-end fashion. According to the model shown in Fig. 2.9, the globular domains in one tetramer are in close contact with those of the next one. The microfibrils are thought to aggregate into larger structures. Several electron microscopy observations suggesting the occurrence of type VI collagen fibrils in Descemet's membrane, nucleus pulposus, skin, myocardium, and

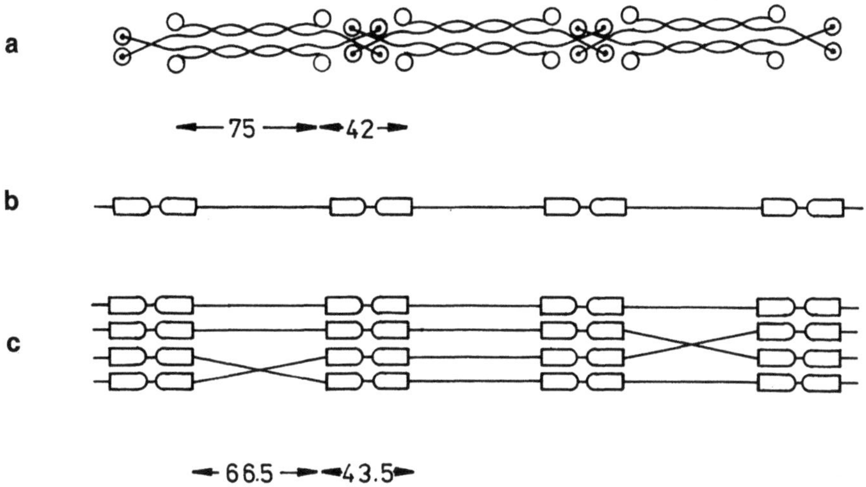


**Fig. 2.9.** Molecular models of linear aggregation of type VI collagen tetramers; *a*, microfibrils; *b*, aggregated microfibrils; *c*, aggregates which are found in cell cultures. (From Timpl and Engel 1987 with permission)

some tumors (schwannoma, lymph node tumors, squamous cell carcinoma) have been reported (Keene et al. 1988). These cross-banded fibers, resembling in morphology lateral aggregates of type VI collagen microfibrils, are reviewed by Timpl and Engel (1987). Their nature has not been fully elucidated.

The biosynthesis of type VI collagen has been described only partially. Posttranslational modifications include in addition to the usual events of collagen synthesis the formation of disulfide bonds. One of the unusual features of type VI collagen is the absence of reducible cross-links and the disulfide bonds form the only known cross-links. This may suggest that aggregates of the collagen are reversible and may be modulated by some, currently unknown factors.

The biological function of type VI collagen is not known. This form is relatively ubiquitous and is hypothesized to form link structures between other collagenous fibers and globular proteins. This concept is supported by a strong interaction with the collagenase-resistant polypeptide termed GP-250 (Carter 1982b), which is probably disulfide-bonded to collagen. Another suggestion for the biological role of type VI collagen has come from the studies of Carter (1982a). He demonstrated spreading and adhesion of fibroblasts to matrices made of native type VI collagen. This interaction is not mediated by fibronectin. Thus, type VI collagen may form a regulatory system for fibroblast cytoarchitectural organization. It has also been put forward that type VI collagen forms a flexible network anchoring some interstitial structures (nerves, vessels) and collagen bundles into surrounding connective tissue (Keene et al. 1988).

## 2.9 Type VII Collagen

Type VII collagen is one of the largest forms. Its triple-helical portion has a length of 450 nm. One interruption has been suggested to be present within the helical domain. The whole molecule is probably slightly longer. The collagen is a homotrimer and aggregates into dimers (Fig. 2.10) (Burgeson 1987b).

Type VII collagen has been identified in a variety of tissues, including skin, chorioamnion, placenta, lung, oral mucosa, tongue, bladder, vaginal mucosa, cartilage, and cornea (Bentz et al. 1983). Immunofluorescence localization indicates that this form occurs in some tissues when ultrastructural investigations show the presence of anchoring fibrils, strongly suggesting that type VII collagen is related to the structure of the anchoring fibrils. The latter are broad, central symmetrically banded structures. Their length is about 800 nm (Palade and Farquhar 1965; Bruns 1969; Keene et al. 1987). Ultrastructural studies have revealed that they form horseshoe-shaped structures with both distal ends inserted into the lamina densa. The remaining portion surrounds

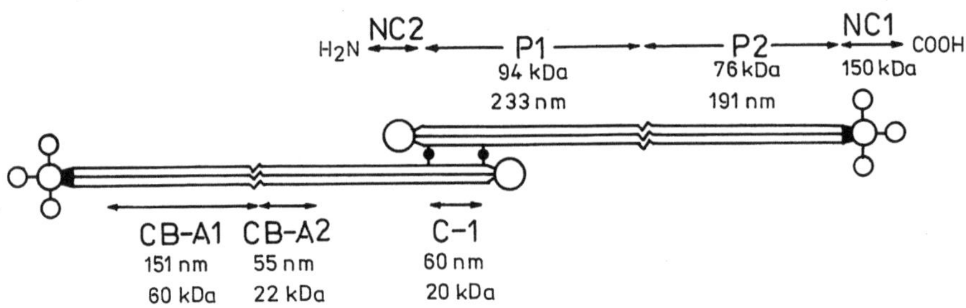


**Fig. 2.10.** The current understanding of type VII procollagen. Disulfide bonds are indicated by *small solid circles.* (From Burgeson 1987b with permission)

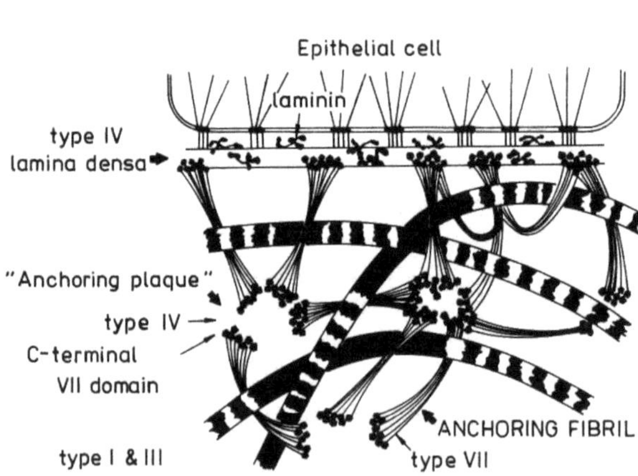


**Fig. 2.11.** The anchoring fibril network. (From Keene et al. 1987 with permission)

large collagen fibrils. It has been suggested that the anchoring fibrils provide mechanical support for the network of fibrils and the basement membranes, as shown in Fig. 2.11. Details of this network are reviewed by Burgeson (1987b).

## 2.10 Type VIII Collagen

Type VIII collagen was primarily discovered as a product of cultured endothelial cells (Sage et al. 1980; Sage et al. 1983; Sage and Bornstein 1987) and was originally termed endothelial collagen. Our knowledge of its structure and distribution is relatively scanty.

The polypeptide chains of type VIII collagen are extraordinarily long, 180000 Da. It has been suggested that the helical portion of the molecule is interrupted twice, at one-third and two-thirds along the length. This feature is associated with sensitivity to pepsin digestion. Pepsin cleaves the chains and yields 50000–60000 Da fragments. Vertebrate collagenase digests type VIII collagen to a few fragments.

Its composition is unknown. The protein is secreted as three molecular forms which are not interchain disulfide bonded. The role of these forms in the formation of the supramolecular structure remains unclear.

The ability to secrete type VIII collagen has been found in a number of endothelial cell cultures (Sage and Bornstein 1987). It is of interest that rapidly proliferating cells plated at low density produce high amounts of the collagen, while confluent cells do not (Alitalo et al. 1983). This type of collagen is also found in the bovine corneal Descemet's membrane (Benya 1980; Kapoor et al. 1986).

The biological function has not been elucidated. It has been suggested that this collagen may be associated with the cell surface, contributing to the regulation of cell differentiation.

## 2.11 Type IX Collagen

Type IX collagen is present in cartilage and was the first link protein discovered between the fibrils and other matrix components. The molecule is composed of three chains: $\alpha 1$ (IX), $\alpha 2$ (IX), and $\alpha 3$ (IX), and consists of three triple-helical domains (COL1, COL2, COL3) and four noncollagenous domains (NC1, NC2, NC3, NC4). The domains are numbered from the carboxyterminal end of the molecule (Van der Rest et al. 1985; van der Rest and Mayne 1987). The size of the domains varies for each chain, although the collagenous parts are composed in such a way that they form helical structures (Duance et al. 1985). The number of amino acid residues in the domains is depicted in Fig. 2.12.

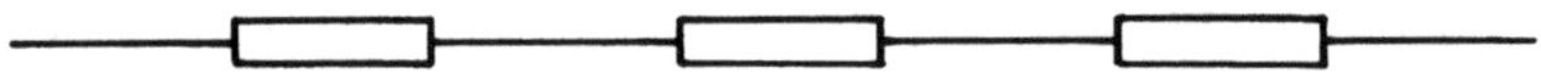

| | NC4 | COL3 | NC3 | COL2 | NC2 | COL1 | NC1 |
|---|---|---|---|---|---|---|---|
| α1 (IX) | unknown | 137 | 12 | 339 | 30 | 115 | 21 |
| α2 (IX) | unknown | unknown | unknown | unknown | 30 | 115 | 15 |
| α3 (IX) | unknown | unknown | unknown | unknown | unknown | ~115 | unknown |

**Fig. 2.12.** Size of collagenous and noncollagenous domains in the chains of type IX collagen

There are several interesting features. The COL1 domains of the α1 (IX) and α2(IX) chains have two imperfections in the collagenous sequence: Gly-Arg-Gly, located 61–63 residues from the aminoterminal end, and Gly-X-Y-X-Y, located at residues 78–82 and 81–85, respectively. Another imperfection has been reported in the COL3 domain of α1 (IX) chain at residues 87–89 in the form of Gly-Lys-Gly.

It is of interest that the NC3 domains of the α1 (IX) and α3 (IX) chains are resistant to proteolysis by pepsin. The NC3 domain of the α2(IX) chain is sensitive to such digestion. The nature of the pepsin-resistant region is not known. The molecule of type IX collagen contains interchain disulfide bridges, found in the COL1, NC1, and NC3 domains.

An interesting feature of type IX collagen is its relationship to proteoglycans. After isolation of a collagenase-sensitive proteoglycan from chick embryo epiphyseal cartilage by Noro et al. (1983), it was found to be identical with type IX collagen. Further studies from various groups revealed that the α2 (IX) chain is bound to one or two molecules of chondroitin sulfate (Vaughan et al. 1985, 1988; Bruckner et al. 1985; Smith et al. 1985). The precise binding site for the glycosaminoglycan side chain is unknown. It is possible that the NC3 and/or NC4 domains contribute in attachment. Type IX collagen is unable to form fibrils. The molecule is highly flexible at the NC3 domain. It has been suggested that type IX collagen is involved in organizing or maintaining the network of collagen fibrils in cartilage (Eyre et al. 1987a; Wotton et al. 1988).

Type IX collagen has been shown to be covalently cross-linked to type II collagen (Van der Rest and Mayne 1988). A cross-link between the aminoterminal end of the COL2 domain of the α2(IX) chain and the noncollagenous aminotelopeptide of type II collagen chains has been reported (Eyre et al. 1987a). It takes the form of a trifunctional pyridinoline link. A similar hydroxypyridinium cross-link has been identified between the α3(IX) and α1(II) chains (Mayne 1989). The role of type IX collagen as the linking protein between the fibrils and/or glycosaminoglycans has been proposed, but the exact nature of this process remains unclear.

## 2.12 Type X Collagen

Type X collagen is one of the so-called minor cartilage collagens and is characterized by several unique properties distingushing it from other collagens. It is a homotrimer of $\alpha 1 (X)$, with a molecular weight of 59000 Da (Schmid and Linsenmayer 1987). The molecule consists of two noncollagenous domains, NC1 and NC2, located at both ends of the collagenous domain of 460 amino acid residues. The NC1 domain is located at the carboxyterminal end of the molecule and is substantially larger than the NC2 domain; it contains 170 amino acid residues. The helical domain of the molecule has two types of interruptions in the Gly-X-Y sequence. Four positions have a Gly-X-Gly sequence, and three others have a Gly-X-Y separated from the next triplet by two additional amino acid residues. These deviations to not cause any significant alteration in the helical structure and do not make it susceptible to pepsin digestion (Schmid and Conrad 1982). The supramolecular structure of type X collagen remains unclarified.

Type X collagen is suspectible to cleavage by vertebrate collagenase. It displays the unique property of being cleaved at two sites within the tripe-helical domain (Schmid et al. 1986). The exact bonds that are cleaved are unknown. Analysis of the genomic sequence indicates that the Gly-Leu bond 92 residues from the aminoterminal end of the triple helix and the Gly-Ile bond 40 residues from the carboxyterminal end of the helical domain are the most likely candidates (Schmid and Linsenmayer 1987).

The molecule has been visualized and is built up of a knoblike globular region located at one end of the rodlike helical domain. The knob, which is observed under electron microscopy, represents the carboxyterminal part of the molecule (Schmid et al. 1984).

Type X collagen belongs to the so-called short collagens. Its molecule is 138 nm in length. Despite the small triple-helical domain, it evidences a greater thermal stability than type I collagen ($T_m$ of 47 °C).

The occurrence of type X collagen is restricted to hypertrophic chondrocytes during the process of endochondral ossification and thus is present in the cartilage that is about to be replaced by bone. In adults, molecules of type X collagen occur in the zone of calcified cartilage which separates the hyaline cartilage from the subchondral bone.

The biological function of type X collagen is not fully understood. It has been suggested that it makes up a temporary matrix in the process of transition from cartilage to bone. Due to its susceptibility to collagenase digestion, it may facilitate the removal of the hypertrophic cartilage by "opening up" the cartilage matrix to degradative enzymes. It has been also hypothesized that type X collagen takes part in formation of the matrix which is amenable to calcification since its deposition slightly precedes calcification. The mechanism of this phenomenon remains unknown.

## 2.13 Type XI Collagen

Type XI collagen, known also as 1α2α3α collagen or K collagen, was discovered as another minor collagen of cartilage (Burgeson and Hollister 1979). It forms small fibrils and thus is similar to type V collagen. There are also other physicochemical similarities between those two types. Studies on the chain composition indicate that type XI collagen is a heterotrimer, α1(XI)α2(XI)α3(XI). It cannot be excluded that preparations of type XI collagen consist of other combinations of chains such as $[\alpha 1(XI)]_2\alpha 2(XI)$, $[\alpha 2(XI)]_2\alpha 3(XI)$, etc. (Eyre and Wu 1987; Morris and Bächinger 1987). On the other hand, there are indirect suggestions that only the composition α1(XI)α2(XI)α3(XI) occurs in tissues: equal amounts of all three chains were isolated from the cartilage of the lathyritic chicken, and only one kind of segment long spacing crystallite was observed in electron microscopic investigations (Eyre and Wu 1987).

The α1(XI) and α2(XI) chains gave unique cyanogen bromide peptide patterns with electrophoresis and chromatography. The α3(XI) chain is very similar to the α1(II) chain. It has been suggested that these chains have the same primary structure and differ in the degree of posttranslational modifications or are products of different but closely related genes (Eyre and Wu 1987).

Type XI collagen was isolated from various hyaline cartilages, including epiphyseal growth cartilage and growth plate as well as articular, costosternal, nasal septum, and laryngeal cartilages (Hartmann 1983; Mayne 1989). Its contribution to the structure of type V collagen fibrils in fetal and adult bone has been recently proposed (Niyibizi and Eyre 1989). It was also isolated from rat chondrosarcoma and human and bovine intervertebral discs (Ayad et al. 1981). Other reported sources of type XI collagen are vitreous humor, elastic cartilage of the ear, and notochordal sheath of lamprey (Eyre and Wu 1987). In all these materials, type XI collagen associated with types II and V collagens. Interestingly, in the knee meniscus which contains predominantly type I collagen, type XI was not detected (Eyre and Wu 1983). In ther intervertebral disc type XI collagen is found in the nucleus pulposus where type II collagen predominantly occurs in contrast to the type I-rich annulus fibrosus (see Chap. 11).

## 2.14 Type XII Collagen

Type XII collagen is a functional analogue of type IX collagen. The discovery of type IX collagen which is able to interact covalently with type II collagen fibrils instigated a search for a collagen with similar properties in type I-containing tissues. Gordon et al. (1987) isolated from a chick embryo tendon cDNA library a cDNA clone that encodes a collagen showing a region of sequence similarity to type IX collagen. The newly discovered chain has been named α1(XII). This finding was followed by isolation of collagen fragments from chick embryo tendon and bovine periodontal ligament (Dublet and van der Rest 1987; Dublet et al. 1988, 1989).

Type XII collagen is a homotrimer, $[\alpha 1\,(XII)]_3$. This differentiates it from type IX collagen. It is not a proteoglycan since its electrophoretic mobility is not affected by chondroitinase ABC digestion. The presence of carbohydrates in type XII collagen has been shown although their nature has not been characterized. The molecule was visualized by rotary shadowing electron microscopy and was found to have a "lollipop" appearance. When analyzed under nondenaturating conditions, the molecules appeared as crosses with one thin, 75-nm tail and a central globular domain from which three 60-nm arms extended. It is possible that the cross structures represent doublets. A large globular domain was shown to be composed of a nontriple-helical structure while the short, approximately 75-nm long tail was collagenase-sensitive and appeared kinked. Type XII collagen contains two triple-helical domains of 16000 and 10000 Da. The molecular weight of the whole molecule is probably 220000–270000 Da. The supramolecular arrangement of the molecule is not yet known.

The biological function is not clear although similarities to type IX collagen provide the hypothesis that type XII collagen is associated with type I collagen-containing fibrils.

## 2.15 Type XIII Collagen

Type XIII collagen is a newly described, short-chain collagen. The chains are found in several different forms which are generated through alternative splicing of the primary transcript (Pihlajaniemi et al. 1987; Tikka et al. 1988). This phenomenon is not seen with other types of collagen. The length of the chains varies between 654 and 566 amino acid residues, consisting of three collagenous domains joined by noncollagenous segments and short terminal domains. The collagenous domains contain 95, 172, and 235 amino acid residues (Sandberg et al. 1989a).

Type XIII collagen has been found in bones, cartilage, intestine, skin, and striated muscles of human fetuses. In the skin, it is seen in the epidermis, hair follicles, and nail root cells (Sandberg et al. 1989b).

Its biological functions are unknown.

## 2.16 Proteins Containing a Collagenlike Sequence

The presence of a triple-helical structure is a specific feature of collagens. This structure must predominate in a molecule in order to classify it into the collagen family. The occurrence of relatively small triple-helical fragments have been shown in proteins which are not categorized as collagens: acetylcholinesterase and certain mammalian lectins.

### 2.16.1 Acetylcholinesterase

Acetylcholinesterase (acetylcholine hydrolase, EC 3.1.1.7.) is an enzyme which catalyzes the hydrolysis of the neurotransmitter acetylcholine. It is involved in cholinergic synaptic transmission and is found in neuronal ganglia and the neuromuscular junction. Acetylcholinesterase also occurs in nonneuronal cells, where it appears to be associated with membranes. Most of the cells contain more than one form of the enzyme. In general, acetylcholinesterase forms are classified as “asymmetric” ($A_n$) and “globular” ($G_n$). The number “n” describes the catalytic subunits. Only A forms contain a collagenlike tail (Rosenberry et al. 1982). For a detailed description of acetylcholinesterase, readers are referred to the review of Rosenberry (1975).

The best characterized acetylcholinesterease is an enzyme isolated from the electric organ of some fishes (electric rays, *Torpedo marmorata, T. californica,* and electric eel, *Electrophorus electricus*). The molecule categorized as an $A_{12}$ form is composed of 12 catalytic subunits and three collagenlike peptides that are interwound as a collagen like “tail”. It is believed that the $A_{12}$ form isolated from rat diaphragm is identical with acetylcholinesterase from the electric organ. The structure of the enzyme is depicted in Fig. 2.13.

The $A_{12}$ form has a molecular weight of about 1100000 Da. The twelve globular catalytic units have a 75000 Da molecular weight each. The catalytic units are arranged in tetramers. The “tail” has a molecular weight of approximately 100000 Da.

Hydroxyproline and hydroxylysine occur only in the “tail.” This part of the enzyme contains 7.2% 4-hydroxyproline, 6.8% hydroxylysine, and 30.6% glycine, indicating that more than 90% of the structure can be considered as a collagenlike, -Gly-X-Y-repeating triplet sequence. There is no direct evidence for such a structure although the “tail” can be digested with bacterial collagenase. The collagenlike part of the enzyme anchors the molecule to the basement membrane. Digestion with bacterial collagenase has been shown to release the enzyme.

### 2.16.2 Mammalian Lectins

The term “lectins” describes proteins with the ability to bind polysaccharides and glycoproteins. In recent years, a few lectins have been described. In this way, the family of proteins with collagenlike sequences increased to five. The best known representative of the group of lectins is the complement component C1q. As shown in the recently published review of Thiel and Reid (1989), the structure of all lectins with collagenlike sequences is similar, and all are

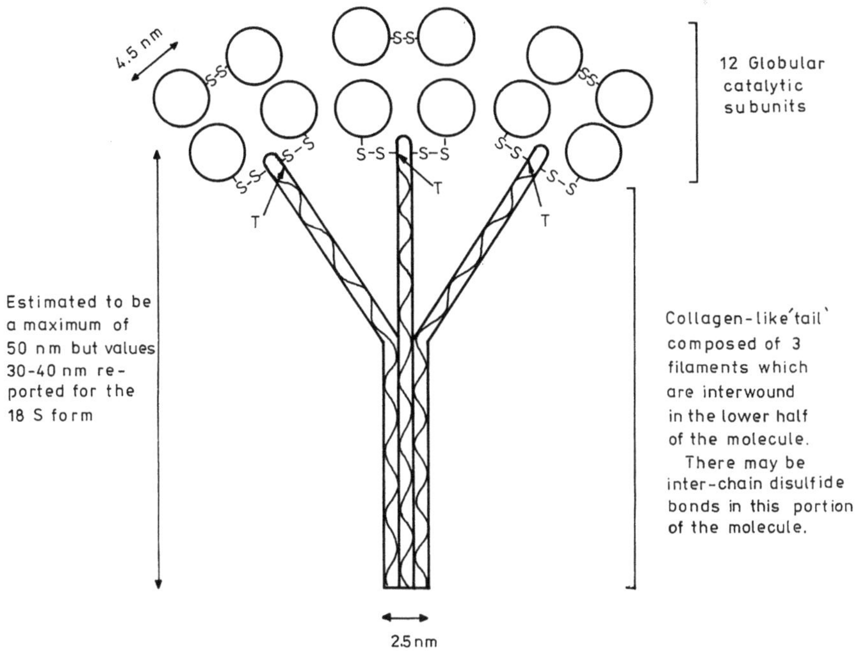


**Fig. 2.13.** Structure of acetylcholinesterase from electric eel. (From Reid 1982 with permission)

composed of chains with a globular domain and collagenlike sequence (Fig. 2.14).

### 2.16.2.1 Complement Component C1q

The complement system is part of the major host defense and inflammatory system of the body. It consists of a complex of plasma proteins that are normally inactive, and which upon activation play an important role in opsonic function, release of inflammatory mediators, and cytotoxic activity. The complement system activation occurs in a cascading manner. C1q is a component of the classic pathway of activation and is a portion of the Cl protein.

C1q is a serum glycoprotein which is present in serum at a concentration of 60–80 ng/l. It has a molecular weight of 410000 Da and is composed of 18 polypeptides. There are three types of polypeptides, classified as A, B, C. The chains have a molecular weight of about 23000 Da and are arranged into dimers. There are six A-B dimers and three C-C dimers. These nine dimers are coupled into three structural units. The structure of the C1q molecule has the form of a bunch of flowers. The "stems" of the bunch consist of the collagenlike fragments, which account for approximately 40% of the molecule. Each chain consists of a short aminoterminal sequence followed by approximately 80 residues of the collagenlike sequence. The remaining portion of the

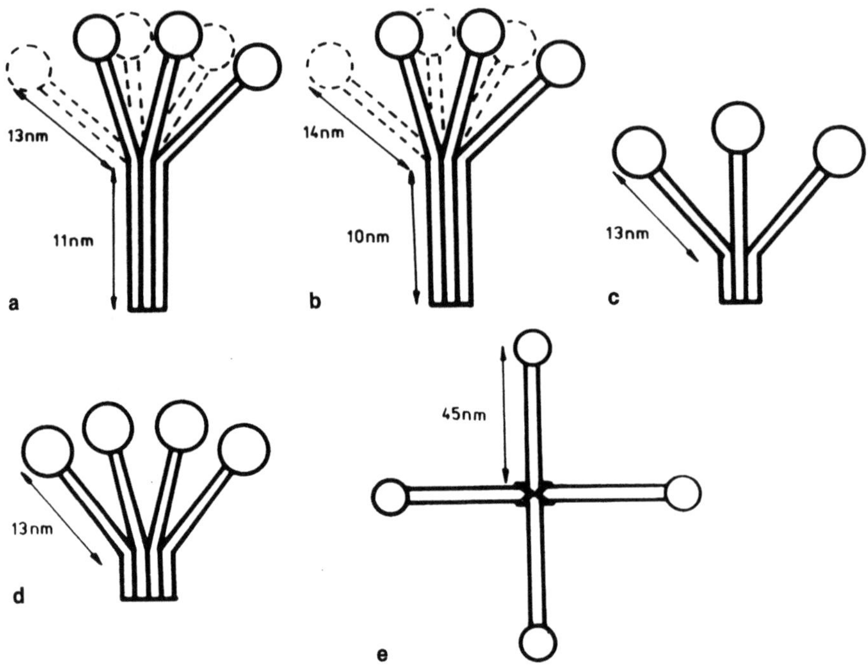


**Fig. 2.14 a – e.** Diagrams based on the electron microscopy pictures of **a** C1q, **b** pulmonary surfactant apoprotein, **c** and **d** mannan binding protein (trimeric and tetrameric forms), and **e** bovine conglutinin. (From Thiel and Reid 1989 with permission)

chains builds the globular sequence (about 110 residues). The collagenous sequences of the A, B, and C chains contain 14%, 23%, and 26% imino acid residues, respectively, and they are broken at one point. The helical region of the C1q molecule is resistant to pepsin digestion, but it can be cleaved with bacterial collagenase (Reid 1989).

The globular regions are responsible for activation of the complement cascade. The role of the collagenlike portion is not fully recognized. The free helical portion may interact with platelets and lymphoid cells. It is also of interest that it shows cross-reactivity in the immune response to collagen. The pathophysiological consequences of these properties of C1q are unknown.

### 2.16.2.2 Conglutinin

Conglutining is a plasma lectin which binds to *N*-acetyl-D-glucosamine and mannose-terminated carbohydrate structures. Its physiological function is only partially elucidated. This lectin is able to bind to a degradation product of the major complement component C3; the ligand is an inactivated form of the C3b component, known as iC3b. The binding is a calcium-dependent process and occurs between conglutinin and polymeric mannose structures (Hirani et al. 1986; Strang et al. 1986). Conglutinin has a structure similar to C1q. Its polypeptide chains are composed of a 20000 Da collagenlike region. This domain has on the aminoterminal end a short noncollagenous sequence and on the carboxyterminal end a large globular part with a molecular weight of 20000 Da.

One of possible functions of conglutinin is contributing to leukocyte bactericidal activity. Its globulin domain is suggested to bind to the iC3b component attached to the bacterial cell surface. The collagenlike domain binds to leukocytes via the C1q or similar receptor (Hawgood et al. 1985; Reid 1989).

#### 2.16.2.3 Mannan-Binding Protein

Mannan-binding protein occurs in the body in soluble and membrane-associated forms. The soluble form has been found in human, cow, rat, and rabbit plasma, while the membrane-associated one has been shown to be present on the surface of hepatocytes of certain species (Drickamer et al. 1986). The lectin specifically binds yeast mannan in a calcium-dependent reaction with the mannose and *N*-acetyl-D-glucosamine domains of the ligand. It occurs as polymeric molecules composed of chains with globular and collagenlike sequences. The biological functions remain unknown although it has been suggested to be involved in complement activation (Reid 1989).

#### 2.16.2.4 Lung Surfactant Protein

Pulmonary surfactant is responsible for the reduction of surface tension in the alveoli. It is composed of surfactant protein component A, a lectin protein with a collagenlike sequence, made up of polypeptide chains with a molecular weight of 28000–38000 Da. Binding of glycolipids via the mannose-frucose domains is probably involved in the formation of the complex structure of physiologically active lung surfactant (Hawgood et al. 1985; Voss et al. 1988).

# 3 Biosynthesis of Collagen

## 3.1 Introduction

The unique structure of collagens, as seen in the extracellular matrix, results from a number of cotranslational and posttranslational modifications of polypeptide chains of preprocollagen. Some of them take place within the cells which synthesize collagen, and some are unique to collagen and a few collagenlike proteins. These modifications are catalyzed by highly specific enzymes. The remaining modifications are similar to those occurring during the synthesis of other secretory proteins and require less specific enzymes. In general, one of features of collagen metabolism is a multistage, complex pathway of biosynthesis, including intracellular and extracellular modifications. The early events are similar to those of other proteins and include transcription of particular genes and translation. Intracellular processing includes the following stages leading from preprocollagen polypeptide chains to the procollagen molecule: cleavage of the signal peptide, hydroxylation of proline residues in position 4 or 3, hydroxylation of lysine residues, incorporation of carbohydrates, chain association, disulfide bond formation, and formation of triple-helical structure of procollagen. After the transportation of procollagen molecules to the extracellular space, procollagen is converted to collagen. Then the collagen molecules aggregate to form supramolecular structures. These structures are stabilized by cross-links and achieve their functional state after interaction with other components of the matrix.

## 3.2 Procollagen Genes

Significant progress has been made in the past few years in understanding the structural organization and chromosomal localization of collagen genes (Cheah and Grant 1982). It has also been found that the heterogenicity and the intron-exon arrangement of the gene family are more complex than expected. Collagen genes are termed COL with two figures separated by a letter "A". The first number codes the type of collagen and the second one the chain number. For example, COL2A1 means a gene for the $\alpha 1$(II) chain of type II collagen.

Human collagen genes were localized using cloned DNA probes. The gene COL1A1 was found at q21–q25 bands of chromosome 17 (Huerre et al. 1982; Solomon et al. 1984; Retief et al. 1985). The gene COL1A2 is located on chromosome 7 approximately within bands q21–q22 (Henderson et al. 1983; Solomon et al. 1983; Junien et al. 1983; Myers and Emanuel 1987). The cartilage collagen, type II, is coded by the gene COL2A1 on chromosome 12, bands q13.1–q13.2 (Strom et al. 1984; Huerre-Jeanpierre et al. 1986; Arheden et al. 1989; Vikkula and Peltonen 1989). The type III collagen gene has been located on chromosome 2, bands q24.3–q31 (Emanuel et al. 1985). The genes coding both chains of the basement membrane collagen, type IV, are found on chromosome 13. The gene COL4A1 is sited on band q32.3–q34 (Boyd et al. 1986; Emanuel et al. 1986; Griffin et al. 1987; Kaytes et al. 1988). Localization of the COL4A3 gene remains unknown. The gene coding the $\alpha 2$(V) chain lies at the same place as the COL3A1 gene, chromosome 2 in bands q24.3–q31 (Emanuel et al. 1985; Tsipouras et al. 1988). The loci for the $\alpha 1$(VI) and $\alpha 2$(VI) chains are located on the same chromosome, chromosome 21, within band q22.3. The $\alpha 3$(VI) chain gene, COL6A3, is found on chromosome 2, in region q37 (Weil et al. 1988). Recently, human collagen $\alpha 1$(XIII) chain gene (COL13A1) has been located on chromosome 10 in the q22 band (Shows et al. 1989). The COL9A1 gene lies on human chromosome 6 (Kimura et al. 1989). The chromosomal siting of other collagen genes remains obscure.

The size of collagen genes varies from 39000 base pairs (bp) (COL1A2 in chicken) to 18000 bp (COL1A1 in humans). Like other eukaryotic genes, their

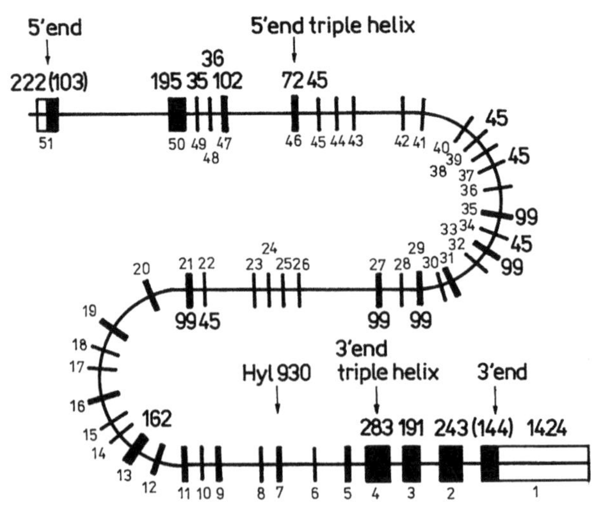


**Fig. 3.1.** Schematic representation of the human $\alpha 1$(I) gene. *Bars* reflect the positions of the exons along the gene. The length of the individual exons of the carboxy and amino propeptide regions are indicated. In the triple-helical domain, *thin bars* reflect exons with 108 nucleotides. Exons with lengths of 45, 99, and 102 bp are indicated. (From Kühn 1986 with permission)

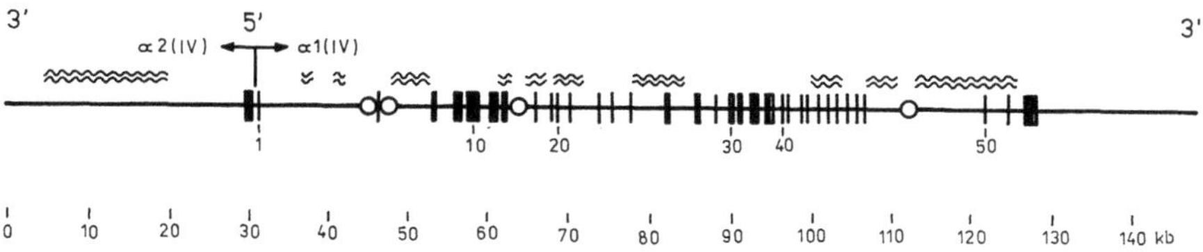


**Fig. 3.2.** Structure of the human $\alpha$ 1 (IV) collagen gene. Location of exons is shown by *vertical bars* and of introns, by *horizontal lines.* Exons are numbered starting from the 5′ end of the gene. Intron sequences of unknown size that are not contained in the genomic clones are indicating by *circles,* and sections containing repetitive intron sequences are depicted by a *wavy pattern.* Position of the adjacent 5′ end of the $\alpha$ 2 (IV) gene is shown. (From Soininen et al. 1989 with permission)

structure consists of several exons separated by noncoding introns. The coding sequence makes up about 10% – 30% of the entire gene. It has been shown that the number and length of the exons are very similar in all genes of the fiber-forming collagens. On the other hand, a substantial variability in the introns has been noted. The schematic structure of the human COL1A1 gene is shown in Fig. 3.1. The exons coding the triple-helical domain are numbered 45 – 5. All of them have a related size, suggesting that they developed from the basic primordial gene unit. A total of 21 exons are 54 bp long and repeat the primordial gene unit. Nine exons are duplicated primordial gene unit (108 bp), and one is a triplicate (162 bp). Five exons are the modified primordial unit with the deletion of 9 bp, and another five constitute duplicates of this modified unit. The deletion of 9 nucleotides corresponds to one triplicate unit (Gly-X-Y).

The aminoterminal section of preprocollagen chains is coded by several exons. They coded the signal peptide, the globular part and the subsequent junction segment, and a triple-helical part of the region. In the COL1A1 gene each part of the aminoterminal region is coded by one exon.

The structural organization of the human COL4A1, the gene of the basement membrane, has been recently reported by Soininen et al. (1989). It is considerably larger than any collagen gene characterized so far. The large size of the gene may be explained, at least partially, by the frequent presence of repetitive sequences. The entire gene is at least 100000 bp long and contains 52 exons whose size and distribution are completely different from that of the genes for fibrillar collagens (Fig. 3.2). The gene is located head-to-head with the COL4A2 on chromosome 13.

Regulation of the expression of the collagen genes is controlled at both a developmental and a tissue-specific level. The regulation mechanisms have only partially been elucidated. Recently, the collagen $\alpha$ 1 (I) promoter was characterized (Brenner et al. 1989b). The coordination of the expression of two type I collagen genes has been reported (Olsen and Prockop 1989). Other mechanisms involved in collagen gene expression have also been investigated (Bornstein et al. 1988a, b).

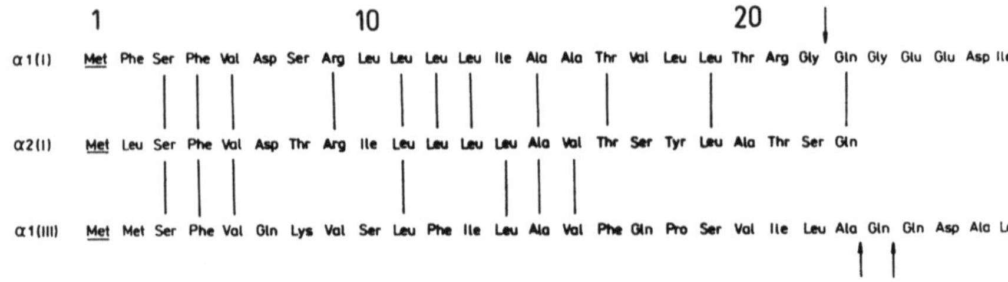


**Fig. 3.3.** Sequence of the signal peptide. (From Yamada et al. 1983)

## 3.3 Cleavage of the Signal Peptide

It is typical for secreted proteins to be synthesized in precursor form, i.e., as a longer polypeptide chain than the proper protein. The extra aminoterminal peptide, called the signal peptide, signal sequence, or leader peptide, usually contains about 15–30 amino acid residues. The sequence of signal peptides for nascent chains of some collagens have been determined (Yamada et al. 1983) (Fig. 3.3). They contain 22–24 amino acid residues, and their structure resembles the signal sequence of other secretory proteins: positively charged amino acid residues at the aminoterminus are followed by hydrophobic amino acid residues.

The enzyme or enzymes responsible for cleavage of the signal peptide remain obscure. It is possible that "signal peptidase" is a single nonspecific enzyme, which hydrolyzes several kinds of proteins. There is no specific sequence between the signal peptide and proper polypeptide chain. The bond usually lies between an amino acid with a small chain and any other amino acid. In preprocollagen $\alpha 1$(I) and preprocollagen $\alpha 2$(I) polypeptides, this bond has the sequence Gly-Gln and Ser-Gln or Ser-Glu, respectively.

The physiological role of the signal peptide is not fully established. It has been suggested that the peptide takes part in the transfer of preprocollagen polypeptide chains across the membrane of the rough endoplasmic reticulum. When the cleavage of the signal peptide is impaired due either to an altered sequence or to incorporation of nonphysiological amino acids, the production of the protein decreases or is inhibited. The inhibition is caused mostly by defective translocation across the membrane or is due to impaired release from the membranes (Kivirikko and Kuivaniemi 1987). The defects of collagen metabolism associated with an abnormal signaling peptide or impaired cleavage have not been yet discovered.

## 3.4 Hydroxylation of Proline and Lysine Residues

Three specific hydroxylases, prolyl 4-hydroxylase, prolyl 3-hydroxylase, and lysyl hydroxylase, are required for the formation in procollagen of 4-hydroxy-

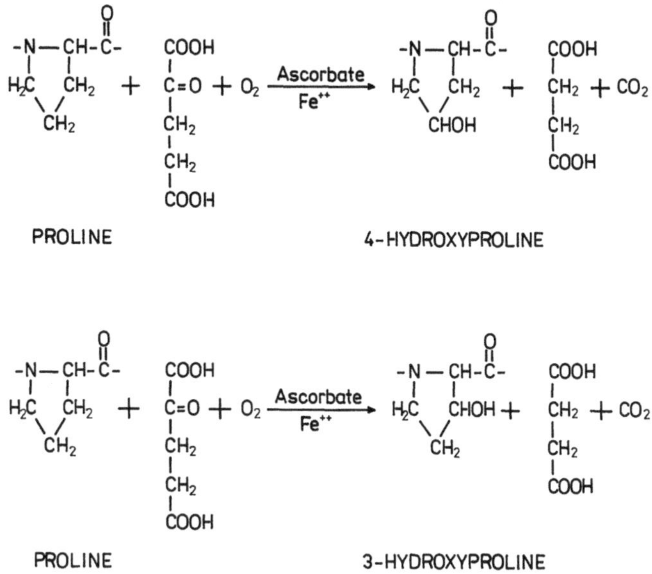


**Fig. 3.4.** Hydroxylation of proline residues in procollagen chains

proline, 3-hydroxyproline, and hydroxylysine, respectively. These compounds are synthesized only from precursor residues incorporated into polypeptide chains. The products of hydroxylation are found almost exclusively in collagen, and this hydroxylation is not only a unique feature of collagen biosynthesis but probably also a key regulatory step in the formation of collagen. Although the catalyzed reactions of hydroxylation show some similarities, they require separate enzymes. The hydroxylases have been purified and characterized over the past decade. The exception is prolyl 3-hydroxylase whose properties still have not been determined (Kivirikko and Myllylä 1980, 1982a, b, 1987).

Prolyl 4-hydroxylase (EC 1.14.11.2, procollagen-proline 2-oxyglutarate 4-dioxygenase) is usually termed prolyl hydroxylase. The enzyme catalyzes hydroxylation of proline residues in polypeptide chains in position 4 (Fig. 3.4). This process is essential for the formation of the triple helix and the stabilization of this structure under physiological conditions. Nonhydroxylated polypeptide chains of procollagen are able to form triple-helical structures only at low temperatures, under 24 °C. Thus, hydroxylation of proline is needed for the formation and stabilization of the helical structure at normal body temperature.

The reaction requires 2-oxyglutarate and $O_2$. The first substrate is decarboxylated and oxidized with one atom of the $O_2$ molecule. The second atom is incorporated into the hydroxyl group in the hydroxyproline residue. Two cofactors are required, $Fe^{2+}$ and ascorbate. Iron is supposed to form a highly reactive iron-oxygen complex, ferryl ion, which hydroxylates the proline residue. It has been hypothesized that $Fe^{2+}$ is not released between most of the catalytic cycles.

The exact role of ascorbate has not been elucidated. This compound is not consumed stoichiometrically. The hydroxylase is able to catalyze some cycles of reaction in the absence of ascorbate. It is thought that ascorbate reduces the enzyme-bound iron and possibly other groups of the enzyme.

The enzymatic protein has been isolated from several tissues or cell cultures (Halme et al. 1970; Pänkäläinen et al. 1970; Rhoads and Udenfriend 1970; Berg and Prockop 1973b; Kuutti et al. 1975; Tuderman et al. 1975; Risteli et al. 1976; Chen-Kiang et al. 1977). The active enzyme has the form of a tetramer, consisting of two types of monomers: two $\alpha$-subunits (molecular weight about 64000) and two $\beta$-subunits (molecular weight about 60000). The separate subunits are inactive. The molecular weight of the tetramer ($\alpha_2\beta_2$) is about 240000 (Pänkäläinen et al. 1970; Halme et al. 1970; Berg and Prockop 1973; Kuutti et al. 1975; Tuderman et al. 1975; Risteli et al. 1976; Majamaa et al. 1979). Electron microscopy investigations showed that the monomers are rod-shaped and that the dimers form V-shaped structures. The tetramer is in the form of two interlocked V-shaped dimers (Olsen et al. 1973b, c). It is believed that prolyl 4-hydroxylase is not specific for any type of collagen. No data about its tissue or organ specificity or its isozymes have been published. This problem, however, needs further exploration. Its only known heterogeneity is the existence of the $\alpha$-subunit in two forms differing in carbohydrate content. The larger ($\alpha'$-subunit) contains two oligosaccharides, linked to asparagine. The oligosaccharides contain eight mannose units each. The smaller $\alpha$-subunit contains a single oligosaccharide with seven mannose units, also linked to asparagine. Kedersha et al. (1985a) found that chick embryo tendon cells synthesize both subunits simultaneously, and no precursor-product relationship exists between them. The same group showed the existence of two separate pathways of synthesis of oligosaccaride units (Kedersha et al. 1985b). It is possible that the variants of the $\alpha$-subunits are products of homologous genes. Some minor differences in the primary structure of the protein part of subunits are also suspected.

No microheterogeneity in the $\beta$-subunits has been detected. This monomer is also a glycoprotein, but carbohydrates are present only in trace amounts. The subunit consists of 491 amino acid residues. Interestingly, it is closely related or possibly identical with protein disulfide isomerase, a nonspecific enzyme involved in the formation of the procollagen structure. The $\beta$-subunit obtained from dissociation of the hydroxylase has the same activity as the protein disulfide isomerase. Both proteins seem to be products of one gene, but very minimal differences may exist as a result of different posttranslational modifications.

The mechanism of hydroxylation has been investigated and is reviewed by Kivirikko and Myllylä (1982a) and Kivirikko and Juvainiemi (1987). The model for the binding of the substrate and cofactors at the active site of prolyl 4-hydroxylase is shown in Fig. 3.5. The sequence of events in hydroxylation is shown in Fig. 3.6.

The inhibitors of prolyl hydroxylase excite interest as potential antifibrotic agents (Chap. 24). Competitive inhibitors may act with respect to iron ions,

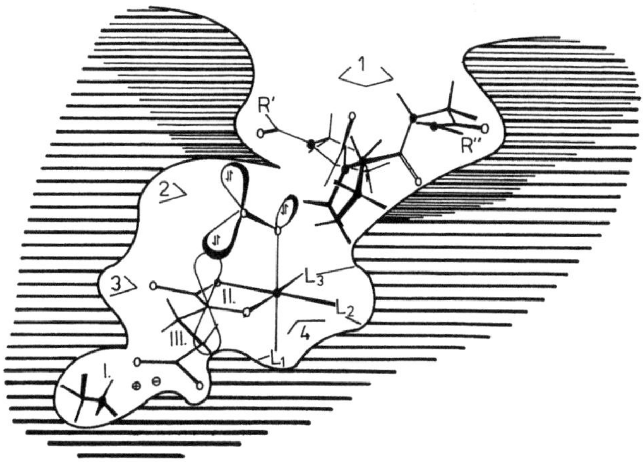


**Fig. 3.5.** Binding of the polypeptide substrate, ferrous ion, and cosubstrates at the active site of prolyl 4-hydroxylase at the beginning of the catalytic cycle. The peptide substrate (*1*) is sterically oriented to participate in the hydroxylation reaction. Molecular oxygen (*2*) is bound end-on in an axial position, yielding a superoxide-like structure. The 2-oxoglutarate (*3*) binds via subsites I, II, and III, and the iron (*4*) is located in a pocket. (From Majamaa et al. 1984 with permission)

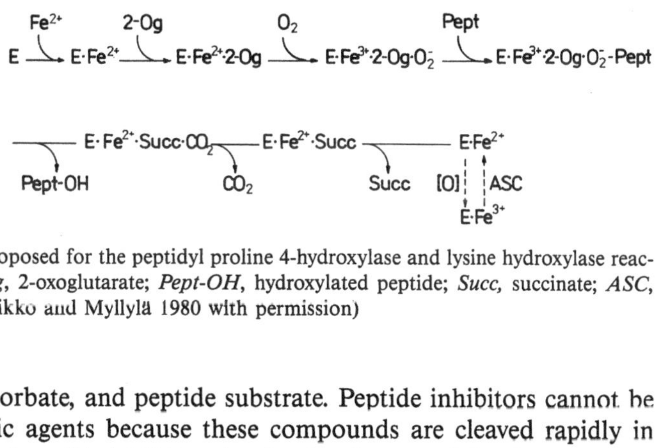


**Fig. 3.6.** Mechanism proposed for the peptidyl proline 4-hydroxylase and lysine hydroxylase reactions. *E*, enzyme; *2-Og*, 2-oxoglutarate; *Pept-OH*, hydroxylated peptide; *Succ*, succinate; *ASC*, ascorbate. (From Kivirikko and Myllylä 1980 with permission)

2-oxyglutarate, ascorbate, and peptide substrate. Peptide inhibitors cannot be used as antifibrotic agents because these compounds are cleaved rapidly in vivo. Application of peptide inhibitors in vitro has been useful to better understanding of the catalytic properties of proline 4-hydroxylase. The following peptides have been found to be competitive inhibitors of the enzyme: poly-(L-proline), (Pro-Ala-Gly)$_n$, (Gly-Pro-Gly)$_n$ (Kivirikko and Prockop 1967b, c; Kivirikko et al. 1967, 1969; Hutton et al. 1968).

There are numerous studies on the sequence requirements of prolyl 4-hydroxylase. It does not hydroxylate free proline (Prockop and Juva 1965; Kivirikko and Prockop 1967a; Risteli et al. 1977); the residue to be hydroxylated must be situated in a peptide, and the position and the minimum sequence required have been investigated using synthetic substrates. The primary structure of collagen was a good source of data on suggested synthetic substrates.

**Table 3.1.** Effect of peptide chain length on the Michaelis constant in the hydroxylation of polypeptides by prolyl 4-hydroxylase. (From Kivirikko and Myllylä 1980)

| Peptide substrate | Michaelis constant (μM) |
|---|---|
| Pro-Pro-Gly | 20000 |
| Acetyl-Pro-Pro-Gly | 23000 |
| Pro-Pro-Gly-$NHCH_3$ | 44000 |
| $(Pro\text{-}Pro\text{-}Gly)_5$ | 1750 |
| $(Pro\text{-}Pro\text{-}Gly)_{10}$ | 280 |
| $(Pro\text{-}Pro\text{-}Gly)_{15}$ | 50 |
| $(Pro\text{-}Pro\text{-}Gly)_{20}$ | 50 |
| Protocollagen | 0.2 |

Hydroxyproline residues in collagen precede the glycine residues, which occur as every third amino acid in the polypeptide chain. The tripeptides Pro-Pro-Gly and Ala-Pro-Gly are the simplest substrates for hydroxylation (Kikuchi et al. 1969; Kivirikko et al. 1969; Suzuki and Koyama 1969). Other sequences such as Gly-Pro-Pro, Gly-Pro-Ala, or Pro-Gly-Pro are not hydroxylated (Kivirikko and Prockop 1967a; Hutton et al. 1968). On the basis of these findings, the minimal sequence required of prolyl 4-hydroxylase has been reported as the -X-Pro-Gly- triplet. The relationship of the peptide length and the Michaelis constant of the hydroxylation is shown in Table 3.1 for peptides with the structure Pro-Pro-Gly. The amino acid residue in the X position of the triplet affects the hydroxylation. The highest velocity of hydroxylation was found in polytripeptides with proline in the X position. An alanine residue also leads to a high velocity (Kivirikko et al. 1968; Rao and Adams 1978). The presence of leucine, arginine, or valine in the X position decreased the velocity of the reaction but did not inhibit hydroxylation (Kivirikko et al. 1972; Rapaka et al. 1978; Prockop et al. 1976; Okada et al. 1972). Synthetic polytripeptides with glycine or sarcosine in the X position are not hydroxylated (Kivirikko et al. 1969; Rao and Adams 1978; Rapaka et al. 1978).

The biological role of prolyl 3-hydroxylase (EC 1.14.11.7, procollagen-proline 2-oxyglutarate 3-dioxygenase) remains unknown. The presence of hydroxyproline with a hydroxyl group at the third carbon atom was reported in several collagens. The reaction catalyzed is similar to that with prolyl 4-hydroxylase (Fig. 3.4). The molecular weight of the enzyme is about 160000, and it requires the same substrates and cofactors as the enzyme responsible for hydroxylation in position 4. Its molecular and subunit structure remain unknown. Hydroxylation in position 3 of proline residues requires the polypeptide sequence -Pro-4Hyp-Gly. This suggests that the enzyme action follows the hydroxylation by prolyl 4-hydroxylase as the triplet -Pro-Pro-Gly- is probably not hydroxylated. The detailed sequences required are not known.

Lysyl hydroxylase (EC 1.14.11.4, procollagen-lysine 2-oxyglutarate 4-dioxygenase) is an enzyme catalyzing the synthesis of hydroxylysine in polypeptide chains of procollagen as depicted in Fig. 3.7. Its general scheme of action is similar to that of prolyl hydroxylases, and the same cofactors are required. The

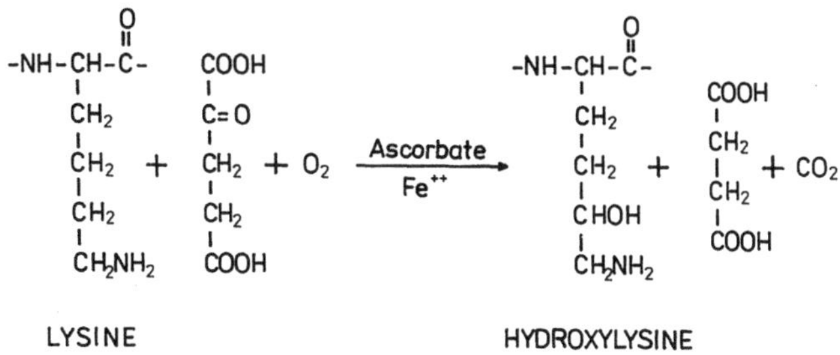


**Fig. 3.7.** Hydroxylation of lysine residues in procollagen chains

enzyme is a glycoprotein, and carbohydrates are required for the hydroxylase activity. Lysyl hydroxylase thus differs from the prolyl hydroxylases, which keep their activity after enzymatic cleavage of the carbohydrates.

## 3.5 Glycosylation of Hydroxylysine Residues

Mammalian collagen contains only two kinds of carbohydrate units linked to hydroxylysine: galactose and glucosylgalactose. The carbohydrate unit is attached to hydroxylysine residues with O-glycoside linkage. Glycosylation of certain hydroxylysine residues is catalyzed by two specific enzymes, hydroxylysyl galactosyltransferase (EC 2.4.1.50, uridine diphosphate (UDP)-galactose-collagen galactosyltransferase, UDP-galactose: 5-hydroxylysine-collagen galactosyltransferase) and galactosylhydroxylysyl glucosyltransferase (EC 2.4.1.66, UDP-glucose-collagen glucosyl-transferase, UDP-glucose: galactosylhydroxylysine-collagen glucosyltransferase) (Kivirikko and Myllylä 1979). The former links galactose, and the latter elongates the carbohydrate domain by adding a glucose unit. The carbohydrates are donated by corresponding UDP-glycosides. The reactions of glycosylation are shown in Fig. 3.8. The enzymes have been characterized partially. Glucosyltransferase isolated from chick embryos is a single polypeptide with a molecular weight of about 70000 Da. The substrate specificity of the transferases reflects the sequence of their action in posttranslational modifications of collagen. Hydroxylysine is formed only from lysine residues in the polypeptide chain, and only when bound in a peptide may it serve as a substrate for glycosylation. Free hydroxylysine is not used as a substrate, and longer peptide chains are better substrates than shorter ones. The amount of -X-Hyl-Gly- triplets is probably an important factor for overall glycosylation, and the amino acid residues in positions close to hydroxylysine are only of relative importance. Glycosylation occurs only in single chains, and the formation of a triple-helical structure comptely prevents the incorporation of carbohydrates to hydroxylysine residues. Galactosyltransferase seems to be more substrate-specific than glucosyltransferase. The last

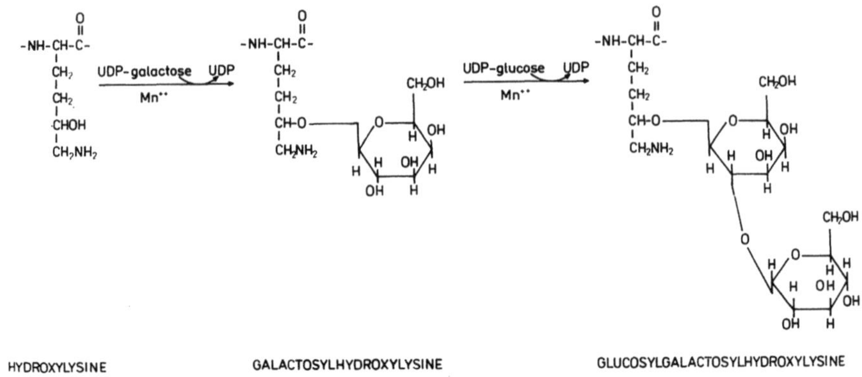


**Fig. 3.8.** Glycosylation of the hydroxylysine residues in procollagen chains

enzyme is able to use free galactosylhydroxylysine as a substrate, but such a situation is only possible in artificial systems in vitro.

The catalytic properties of glucosyltransferase have been partially recognized, mostly by Kivirikko and associates. Optimal conditions for enzyme activity are pH 7.0–7.4 (Kivirikko and Myllylä 1982b). The bivalent cation manganese is required; it can be at least partially substituted by $Fe^{2+}$ and $Co^{2+}$, although the concentration of cobalt required to serve as cofactor for the enzyme is higher than that occurring naturally in tissues (Myllylä et al. 1979). The enzyme can bind two $Mn^{2+}$ ions per molecule. The sites of binding differ in dissociation constants, as the first has a $K_d = 3-5\ \mu M$ while the second site has $K_d = 50-70\ \mu M$. Due to the high dissociation constant the second site probably does not take part in the catalytic process in vivo. The proposed schema of the reaction in the presence of high and low concentrations of manganese is shown in Fig. 3.9. At high concentrations of $Mn^{2+}$, the enzyme binds two cations before binding UDP-glucose. In vivo, the enzyme binds one ion of manganese, then UDP-glucose, and subsequently the peptide substrate. The products are released in inverse order, and manganese can remain in association with the enzyme for another catalytic cycle.

Glycosylation is a stage of biosynthesis subsequent to the hydroxylation of lysine residues, but within the cell glycosylation occurs immediately after

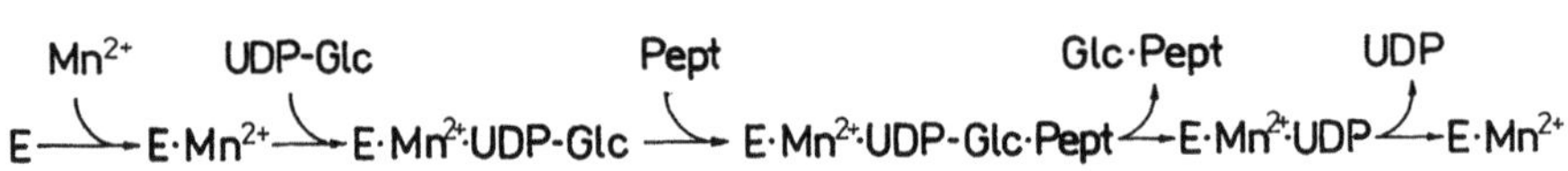


**Fig. 3.9.** Mechanism proposed for the peptidyl galactosylhydroxylysine glucosyltransferase reaction. At high $Mn^{2+}$ concentrations, the enzyme probably binds two $Mn^{2+}$ ions. This situation is not shown. $Mn^{2+}$ ion is bound before the binding of uridine diphosphate (UDP)-glucose. The ion need not leave the enzyme between each catalytic cycle and is shown to remain enzyme-bound. *E*, enzyme; *UDP-Glc*, UDP-glucose; *Glc-Pept*, glucosylated peptide. (From Kivirikko and Myllylä 1984 with permission)

hydroxylation as the polypeptide chains are still being assembled at the ribosomes (Kivirikko and Myllylä 1982a). Both hydroxylation and glycosylation are continued after release of the polypeptide chains into the cisternae of the rough endoplasmic reticulum and end through the formation of the helix.

The biological function of glycosylation remains unknown. It leads to extensive heterogeneity. Not only collagen of different types but the same type of collagen from distinct tissues differ in respect to the extent of hydroxylation, including various ratios of galactosylhydroxylysine to glucosylgalactosylhydroxylysine. It has been proposed that the precise mechanism regulating this process must exist but remains completely obscure. There are no data indicating type-specific or tissue-specific isozymes of the transferases; however, further evaluation of this question is needed. It is possible that small sequence differences among various types of collagen affect glycosylation, but other unknown regulatory mechanisms are assumed to be involved in the variation in glycosylation of the same type of collagen.

It is suggested that glycosylation of collagen affects the diameter of fibril formation. This is consistent with an inverse relationship between carbohydrate content and fibril diameter. There are hints that glycosylation regulates the susceptibility to degradation by collagenases as well as participates in the transportation of procollagen outside the cell. Localization of hydroxylysine glycosides is thought to contribute to corneal transparency (Ibrahim and Harding 1989). This last role is also proposed for asparagine-bound carbohydrates present in the terminal peptides of procollagen. Hydroxylysine residues with bound carbohydrates are able to form cross-links. The influence of carbohydrate units in hydroxylysine residues upon the stability of the cross-links is unknown.

## 3.6 Glycosylation of Asparagine Residues

Collagen molecules contain only carbohydrates bond to hydroxylysine residues. In addition, terminal peptides (propeptides) of procollagen contain different carbohydrate units which are removed when the collagen molecules are formed. Asparagine-linked carbohydrates occur in both amino- and carboxy-terminal peptides. Types I and III procollagens contain oligosaccharides mostly in the carboxy terminal peptide while type II procollagen contains such oligosaccharides in both terminal peptides. There are not data concerning asparagine-bound oligosaccharides in basement membrane collagens, but there is evidence suggesting that such units are present in corresponding procollagens and possibly even in nonhelical parts of the collagen molecule.

The asparagine-linked oligosaccharides of procollagen are synthesized in a similar manner to their analogues in other glycoproteins. The oligosaccharide chains are synthesized on a lipid carrier and then transferred as a whole unit to procollagen. The unit consists of glucose, mannose, and acylglucose according to the formula $Glc_3Man_9GlcNa_2$. Asparagine residues which are subjected to glycosylation are present in the sequence -Asn-Ile-Thr- in human pro $\alpha$1(I) and pro $\alpha$2(I) chains and the chicken pro $\alpha$2(I) and pro $\alpha$1(III) chains. The sequence -Asn-Val-Thr- is found as the binding sequence for oligosaccharide in chick pro $\alpha$1(I) chain.

The biological function of asparagine-bound carbohydrates remains unknown. There are controversial reports that this kind of glycosylation of procollagen is associated with the ability to transport the molecule across the cell membrane.

## 3.7 Chain Association, Disulfide Bonding, and Formation of the Helical Structure

Formation of disulfide bonds between the propeptides of procollagen chains is important in chain association. The disulfide bonds are located at the aminoterminal propeptides of types I and II collagens, and in type III collagen also in the carboxyterminal propeptides. Type IV contains disulfide bonds only between the aminoterminal propeptides. The occurrence and location of the bonds in other types of collagen remain unknown.

Formation of the disulfide bonds provides covalent linkages that stabilize the structure of the protein. There are also data on the role of the bonds in the triple-helix formation. It seems likely that the peptide chains of procollagen associate immediately after hydroxylation and glycosylation, the process arising spontaneously and not requiring the enzyme.

The mechanism of disulfide bond formation is not fully understood. The role of protein disulfide isomerase (EC 5.3.4.1) in this process has been put forth. The enzyme has been purified, cloned, and sequenced from a variety of tissues (Bassuk and Berg 1989). It is very interesting that protein isomerase is identical with the $\beta$-subunit of prolyl 4-hydroxylase (Pihlajaniemi et al. 1987; Koivu and Myllylä 1987; Koivu et al. 1987). The enzyme is also thought to be involved in the nonnuclear binding of thyroid hormone.

## 3.8 Conversion of Procollagen to Collagen

The molecules of procollagen are converted to collagen (old nomenclature: tropocollagen) in the extracellular space. This process includes removal of the propeptides (the telopeptides) from both ends of the molecule. At least two enzymes are required for each type of procollagen molecule, procollagen N-proteinase and procollagen C-proteinase. In recent years, rapid progress has been made in the purification and characterization of these enzymes. It has been found that procollagen N-proteinase is a collagen-type-specific enzyme. The specificity of procollagen C-proteinase has not been definitely established as the partially purified preparation of it acts on procollagen types I, II, III, and V, but it is not known whether there is a mixture of type-specific enzymes or a single type-unspecific proteinase.

Procollagen N- and C-proteinases are endopeptidases requiring $Ca^{2+}$ ions for maximal activity. They operate at a neutral pH and probably have no obligatory sequence for the removal of the propeptides, i.e., amino- or carboxy-terminal propeptides may be cleaved first. On the other hand, there are variations in the activity of procollagen N- and C-proteinases which may result in different kinetics of propeptide removal. It has been found that in many cell and organ cultures in the synthesis of type I collagen, the aminoterminal propeptide is removed before the carboxy terminal propeptide, while in the synthesis of type II collagen the order of cleavage is reversed (Kivirikko and Kuivaniemi 1987; Uitto et al. 1979a). A significant difference was shown in the activity of the proteinases which act on type III procollagen. The cleavage of the aminoterminal propeptide is a much slower process than the similar removal of the carboxyterminal propeptide. This may be a cause of the occurrence of partially processed type III collagen (pN-collagen) which has the propeptide at the amino-end only (Fleischmajer et al. 1981b).

The procollagen N-proteinase which removes the aminoterminal propeptide from type I procollagen has been identified and isolated from various materials, chick embryo tendon, calf tendon, chick embryo calvaria, cultured chick embryo fibroblasts, cultured human fibroblasts, and fetal calf skin (Tuderman and Prockop 1982; Fessler et al. 1975; Kohn et al. 1974). The enzyme isolated from the chick embryo tendons has a molecular weight about 260000 Da, and the $K_m$ for a native type I procollagen is $0.3-0.5$ $\mu M$. It cleaves procollagens types I and II but not types III or IV. The type I procollagen N-proteinase is resistant to inhibitors which affect termolysin (phosphoaramidon) or angiotensin-converting enzyme (3-mercaptopropamoyl-leucine). The important distinction between this and vertebrate collagenases is resistance to inhibitors contained in fetal calf serum (Tuderman and Prockop 1982). Interestingly, the denatured pN-collagen but not denatured collagen inhibited the enzyme activity, and 1 $\mu M$ of denatured substrate caused about 60% inhibition.

There are some proposals that the type I procollagen N-proteinase requires additional (second) metal ions as inhibition cause by preincubation of an almost homogenous preparation of the enzyme with ethylenediaminetetraacetic acid (EDTA) cannot be reversed completely by calcium ions. The nature of the second metal cofactor is unknown.

The biological function of the cleavage of propeptides probably is not limited to allowing the formation of fibers. This activity seems to be basic as procollagen is not able to assemble into fibrils, and the propeptides prevent the intracellular fibrillogenesis. The cleavage of the carboxyterminal propeptide is an absolute requirement for fibrillogenesis (Fleischmajer et al. 1987). The pN-collagen can form fibrils but not thick ones. It is possible that the aminoterminal propeptide takes part in the regulation of fibril size. The free aminoterminal peptide is also a feedback regulatory compound in the control of collagen synthesis (Chap. 5). It is possible that the cleaved propeptides have other, still undiscovered functions.

## 3.9 Formation of the Supramolecular Structures

Collagens are capable of ordered self-assembly and the formation of supramolecular structures. The majority of studies have been focused on type I, II, and III collagens, and the term fibrillogenesis is commonly used to describe formation of filamentous structures. As reviewed in Chap. 2, certain nonfibrillar types of collagens assemble in specific ways to form a three-dimensional network. This process is not fibrillogenesis but represents a close analogue.

Our knowledge of the mechanisms of fibrillogenesis is based mostly on in vitro investigations (Woodhead-Galloway 1982). It is clear that the in vivo system is much more complex than the in vitro one. Investigations in vitro are carried out with purified solutions of collagen molecules, in the absence of cells and other matrix components. Under in vivo conditions, collagen molecules are deposited by the fibroblasts in an ordered manner, and proteoglycans and so-called structural glycoproteins are secreted into the extracellular space. Usually, more than one type of collagen molecule is secreted. Another difference between in vivo and in vitro fibrillogenesis systems is the formation of fibrils in vivo from molecules which are continuously secreted and converted from procollagen to collagen molecules. On the contrary, in vitro collagen molecules are extracted from tissues and derived from the depolymerization of mature and cross-linked proteins (Gelman et al. 1979; Hulmes 1983).

It was found that the in vitro aggregation and fibril formation of type I collagen required a temperature higher than 20 °C and pH about 7.4. Type I molecules are stable in low ionic strength acidic solutions. Fibril formation can be initiated through an increase in pH and ionic strength followed by heating or through heating followed by an increase in pH and ionic strength. The two procedures described, known as the cold neutral route and warm acid route, yield varying kinetics of fibril formation (Fig. 3.10). The final products of fibrillogenesis are different in respect to fibril diameter despite the same conditions at the end-point of the process. The mixed route, i.e., simultaneous increase in temperature and pH, has also been investigated (Veis and Payne 1988).

Kinetic studies on fibrillogenesis provide two hypothetical models of this phenomenon. The first of them is known as the nucleation and growth pathway. According to this, a critical number of molecules leads to the formation of a nucleus and then additional molecules accrue (growth phase), leading to the formation of fibrils. The other model has been described as a multistep assembly; collagen monomers aggregate into intermediates which then form large fibrils. It is possible that these models occur simultaneously in fibrillogenesis as they are not mutually exclusive.

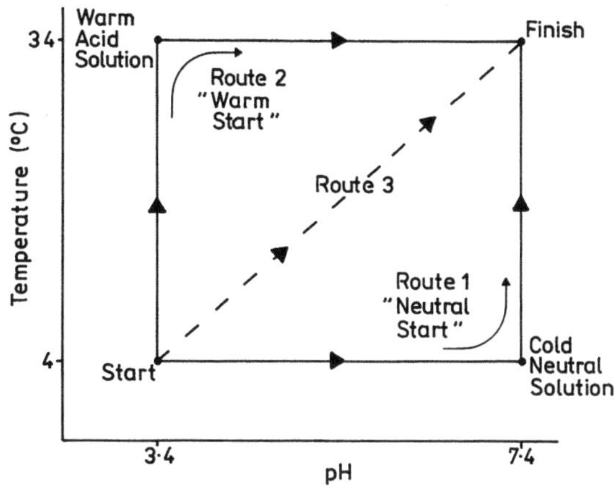


**Fig. 3.10.** Initiation procedures for fibril formation in vitro: three routes are possible by which the pH and temperature of a cold, acidic solution (pH = 3.4; T = 4 °C) of collagen molecules can be adjusted to reach standard fibril-forming conditions (pH = 7.4; T = 34 °C). (From Holmes et al. 1986 with permission)

Fibrillogenesis was found to depend upon added salts. Several cations and anions were reported to lengthen the lag phase of fibrillogenesis. This effect is related to the ionic strength, changes in pH, and the nature of the salt. Phosphate ions were shown to give the highest prolongation of the lag phase. Thiosulfates and sulfates have been found to be less effective than phosphates. Barium and magnesium cations have the greatest effect. Some other small molecular compounds (urea, arginine) were reported to prolong the lag phase of fibrillogenesis. The influence of glycosaminoglycans upon fiber formation is reviewed in Sect. 3.11.2.

## 3.10 Cross-Linking of Collagen

The formation of cross-links in collagen is a basic phenomenon of maturation. As discovered in the early 1950s, newly synthesized collagen can be extracted due to its high solubility in neutral salt solutions or dilute acid solutions. The progressive loss of solubility is acquired during so-called maturation, and this process is associated with the formation of covalent bonds within the collagen structures.

Various amino acids and their derivatives have been implicated. The cross-links can be intramolecular, linking two $\alpha$-chains within the same molecule, and intermolecular, the covalent bonds between chains in different molecules. The cross-links are classified into reducible and nonreducible, the former being reduced by borohydrate under mild conditions. This procedure is a common method applied in investigations of collagen cross-linking.

The chemistry and biology of cross-link formation have been reviewed extensively in recent years (Tanzer 1976; Eyre et al. 1984b; Eyre 1987; Ricard-Blum and Ville 1988; Yamauchi and Mechanic 1988). Despite progress in this field, several aspects of the nonreducible cross-links remain unclear.

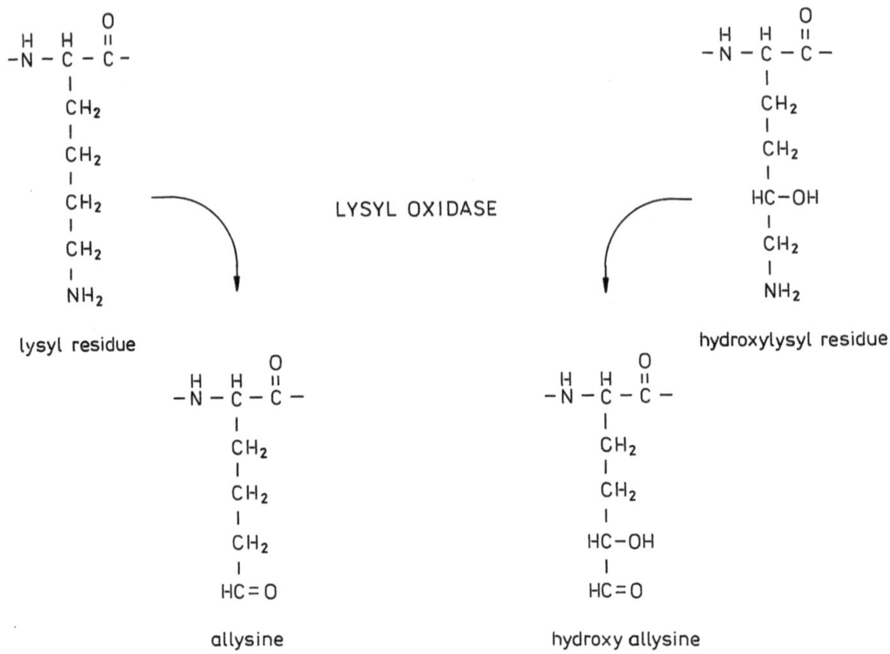


**Fig. 3.11.** Oxidative desamination of lysine and hydroxylysine

The basic mechanism of cross-linking is oxidative desamination of the ε-amino groups in lysine or hydroxylysine residues, catalysed by lysyl oxidase. The details of this enzyme are described later in this chapter. It is believed that oxidative desamination is the only step in the cross-link formation which is catalysed enzymatically. The desamination leads to the formation of aldehydes, called allysine and hydroxyallysine (Fig. 3.11). The aldehydes formed contribute to further chemical reactions, and two pathways of cross-linking can be distinguished, one based on lysine aldehydes and the other on hydroxylysine aldehydes.

Allysine may condensate with lysine, hydroxylysine, or another allysine (Fig. 3.12). Condensation of allysine with the ε-amino group of lysine or hydroxylysine leads through the formation of the reducible iminium cross-links to dehydrolysinonorleucine and dehydrohydroxylysinonorleucine, respectively. Dehydrohydroxylysinonorleucine can then proceed to form via the stable nonreducible cross-link, i.e., condensation with histidine, histidinohydroxylysinonorleucine. Aldol condensation of allysine with allysine forms allysine aldol (aldol condensation product, $\alpha, \beta$-unsaturated aldol). This product reacts with the histidine residue in a process of the Michael addition of imidazole nitrogen-3 atom to the $\beta$-carbon atom of the double bond of allysine aldol. This reaction leads to dehydroaldohistidine, a compound containing the aldehyde group that is able to react with the ε-amino group of hydroxylysine and probably of lysine to form dehydrohistidinohydroxymerodesmosine or de-

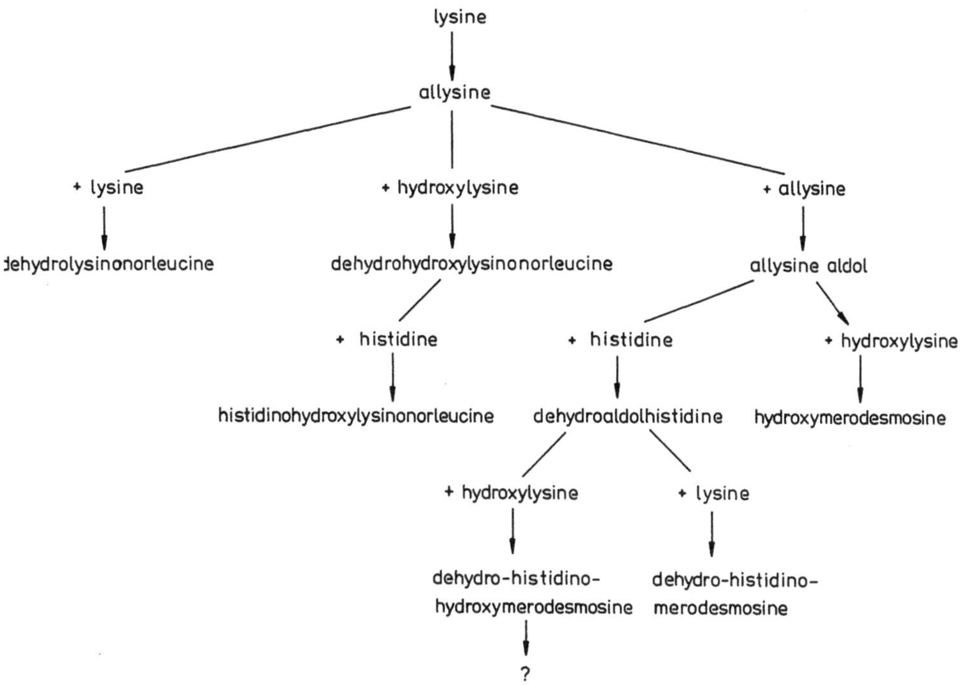


**Fig. 3.12.** Condensation of lysine and allysine

hydrohistidinomerodesmosine. In this way a tetrafunctional cross-link is formed. The last pathway, i.e., condensation of dehydroaldolhistidine with lysine, is not sufficiently documented. Dehydrohistidinohydroxymerodesmosine has been suggested to undergo further reactions, but their chemical nature is unknown. Alternatively, the allysine aldol condensates with hydroxylysine and forms hydroxymerodesmosine.

The second pathway of the cross-linking originates from hydroxylysine. Hydroxyallysine may condensate with lysine or hydroxylysine and forms hydroxylysinonorleucine and dihydroxylysinonorleucine, respectively. These compounds undergo the Amadori rearrangement and form lysino-5-ketonorleucine and hydroxylysino-5-ketonorleucine, respectively (Fig. 3.13). The formed ketoamines are the substrates for the formation of nonreducible pyridinoline cross-links (Fujimoto 1980). The process of formation of pyridinoline cross-links has been only partially elucidated. The most probable mechanism is condensation of two ketoamines (hydroxylysino-5-ketonorleucine). The alternatives that have been suggested are condensation of ketoamine with hydroxyallysine or simple condensation of two hydroxyallysine residues with hydroxylysine.

The above-described pathways of cross-link formation do not explain all the features that have been observed in the collagens. It is very difficult to distinguish the natural products from those formed in vitro during the preparation.

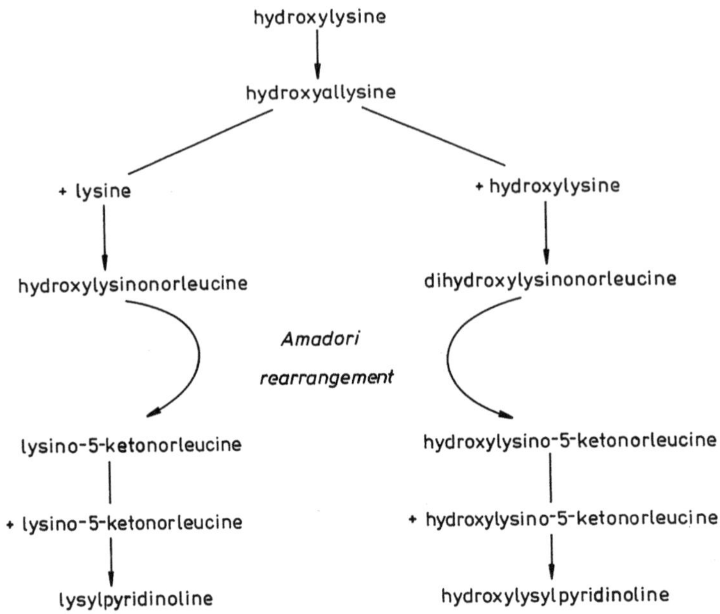


**Fig. 3.13.** Condensation of hydroxylysine and hydroxyallysine

Other mechanisms of cross-linking have been also postulated. In 1987 Barnard and coworkers reported the occurrence of a new compound, designated M, and suggested that it is involved in the formation of stable cross-links.

It is generally accepted that the cross-link formation depends more on tissue localization than on the type of collagen. Allysine aldol condensation is postulated to provide the major intramolecular cross-links in types I–III collagen (Bornstein and Piez 1966; Light 1979). These cross-links have been identified only between aminoterminal propeptides (Eyre et al. 1984a). The intermolecular cross-links based on lysine aldehydes predominate in the skin. The cross-links formed via the hydroxylysine pathway predominate in bone and cartilage. The biological differences between various cross-links are poorly understood. It has been reported that bonds formed from hydroxyallysine are more stable than those of allysine. The role of glycosylation in this process remains unknown.

Lysyl oxidase (Siegel 1979; Prockop and Tuderman 1982) is an enzyme catalyzing formation of allysine and hydroxyallysine from lysine and hydroxylysine, respectively. It has been purified and characterized. In all tissues studied, it occurs as a heterogenous group of isoenzymes, usually in four forms. The exact nature of this heterogeneity remains unknown. It is possible that the forms are products of different genes, but variations in the posttranslational processing of the product of the same gene were suggested as well. All forms of the enzyme catalyze the same reaction and share structural and immunological properties. The molecular weight of all enzymes is about 30000 Da. In some preparations higher weights were reported, probably due to aggregation of the enzyme subunits. The specificity of lysyl oxidase is not restricted to the type of collagen. Moreover, the same forms of enzyme catalyzed cross-link formation in both in collagen and elastin. The same enzyme is able to act on lysine and hydroxylysine residues. The substrate for lysyl oxidase must be aggregated and possess a helical structure. There is no catalytic activity with single collagen molecules, denaturated $\alpha$-chains, or native procollagen.

## 3.11 Interaction of Collagen with Components of the Extracellular Matrix

### 3.11.1 General Remarks

Fibrillar and nonfibrillar structures of collagen reach their full biological activity after interacting with other matrix components. The nature of these interactions is not fully understood. The majority of data have been accumulated from experiments in vitro with purified preparations of collagen and other compounds. It is, however, difficult to conclude from them the complex mechanism of the interaction in vivo. Two groups of components of the extracellular matrix are generally distinguished, proteoglycans and so-called structural glycoproteins.

### 3.11.2 Collagen-Proteoglycan Interactions

The term proteoglycan usually refers to complexes of protein and glycosaminoglycans. Glycosaminoglycans were previously known as mucopolysaccharides, although this term has not been rigorously defined and sometimes was applied to polysaccharide-protein complexes as well. Glycosaminoglycans are linear heteropolysaccharides consisting of hexosamine alternating with another sugar. The disaccharide units of a single glycosaminoglycan always contain only one hexosamine (either glycosamine or galactosamine). The other sugar may be glucuronic acid, iduronic acid, or galactose. It is possible that the same chain contains a mixture of uronic acids, but usually a clear preponderance of one type is shown. The disaccharide basic structural unit is substituted by more complex (four or five sugars) ones in heparan sulfate and heparin, respectively. An almost constant feature of glycosaminoglycans is the substitution of some groups with sulfate. The presence of sulfate causes their highly anionic properties. Their molecular weight varies within the range 10000–100000 Da.

In recent years significant progress in the recognition of proteoglycan structure was made. The majority of studies were carried out on cartilage, but the structure of proteoglycans of other tissues showed considerable similarities.

The proteoglycan monomer consists of the protein core and glycosaminoglycan chains linked to it. Each proteoglycan molecule contains more than 50 glycosaminoglycans. At least two types of glycosaminoglycans are bound to the same core protein. The monomer structure resembles the "bottle brush" configuration. The proteoglycan molecule is attached to hyaluronic acid and forms the proteoglycan aggregate. The aggregate contains at least 100 monomers.

Much research on collagen-glycosaminoglycan interactions was done in vitro. The major point of interest has been fibrillogenesis. It is very difficult to interprete some of the results of these studies since the collagen and proteoglycan or glycosaminoglycan preparations were insufficiently characterized. Isolation and purification of native macromolecules are difficult procedures, and one can assume that the majority of the reported investigations have been carried out on an unspecified mixture of collagens and glycosaminoglycans. Nevertheless, there is evidence that glycosaminoglycans and proteoglycans modify the kinetics of collagen fibrillogenesis in vitro. Wood (1960) was one of the first who reported that chondroitin sulfate A and C accelerated collagen fibril formation. Hyaluronic acid had no effect on this phenomenon. These findings have been confirmed in a number of reports (Németh-Csóka 1974; Németh-Csóka and Kovácsay 1979a, b; Öbrink et al. 1975; Gelman and Blackwell 1973; Herbage et al. 1974). Their summarized conclusions state that all glycosaminoglycans except hyaluronate and keratan sulfate bind to collagen by electrostatic interactions under physiological conditions of pH and ionic strength. The presence of sulfate groups, charge density, and larger molecular size increase the affinity. Similar relations have been found in their effect on fibrillogenesis (Lindahl and Höök 1978).

A delayed effect of proteoglycan monomer and aggregate but not of protein core on the precipitation of collagen fibers was reported by Oegema et al. (1975). Proteoglycan was isolated from bovine nasal cartilage. They also showed that the effect of proteoglycan on fibrillogenesis is related to its binding activity and concluded that it has two distinct effects: it retards collagen assembly if present early enough on in the process, and it affects the fibril diameter. Additionally, Snowden and Swann (1980) reported that proteoglycan from adult bovine articular cartilage reduced the thermal stability of the in vitro assembled collagen fibrils.

Vogel et al. (1984) showed that the low molecular weight proteoglycan isolated from bovine tendon increased fibril diameter when incubated with types I and II collagens. This effect depended on the core protein; glycosaminoglycan alone had no influence, and neither did the high molecular weight proteoglycan from tendon. These findings were confirmed in ultrastructural investigations (Merrilees et al. 1987; Vogel and Trotter 1987). It has been suggested that proteoglycan is similar to the glycoprotein studied by Anderson et al. (1977). This glycoprotein had been found to inhibit fibril formation in vitro. Ultrastructural studies of the fibrils formed in the presence of proteoglycan form another line of evidence of the proteoglycan-collagen interaction. Parry et al. (1982) presented a hypothesis of the regulation of the collagen fibril di-

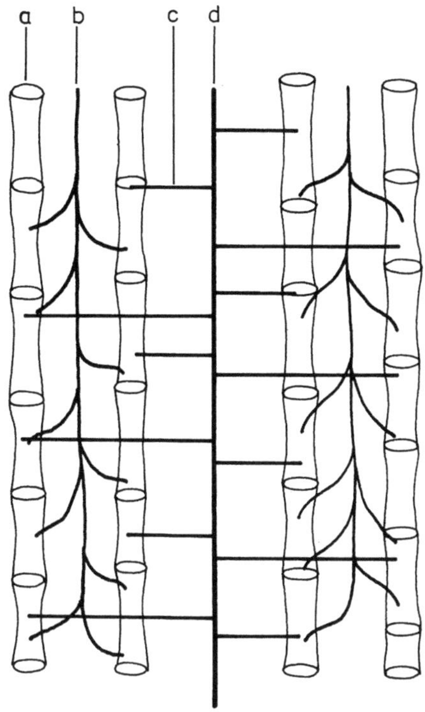


**Fig. 3.14.** Hypothetical model of collagen, proteoglycan, and structural glycoprotein interactions. *a*, collagen; *b*, structural glycoproteins; *c*, glycosaminoglycans; *d*, protein core of proteoglycans. (From Jackson and Bentley 1968 with permission)

ameter by glycosaminoglycans. It was found that tissues with a high relative content of hyaluronic acid contained the smallest mass-average diameter collagen fibrils. A high content of dermatan sulfate was associated with large collagen fibrils. It has been suggested that the lateral growth of fibrils beyond a diameter of about 60 nm is inhibited by the presence of an excess of hyaluronate, but this inhibition may be removed by chondroitin sulfate. A high concentration of chondroitin sulfate also restricts fibril growth to a mass-average diameter of about 150 nm. In turn, this inhibition may be overcome by dermatan sulfate. This concept is consistent with other morphological studies (Steven et al. 1969; Scott et al. 1981; Scott 1980, 1986; Junquiera and Motes 1983; Junquiera et al. 1981).

On the basis of these findings models of the interaction of collagen fiber with proteoglycans have been proposed (Mathews 1965) (Fig. 3.14). It is very difficult to prove the structural nature of proteoglycans in vivo.

### 3.11.3 Structural Glycoproteins

The term structural glycoprotein is used for all glycoproteins (old name: neutral mucopolysaccharides) that are present in the extracellular matrix and are believed to contribute to the structure of the matrix (Labat-Robert and

Robert 1988). The list of structural glycoproteins includes fibronectin, laminin, entactin, nitogen, and a large group of poorly characterized glycoproteins.

### 3.11.3.1 Fibronectin

Fibronectin is a high molecular weight glycoprotein synthesized in the liver as the circulating blood plasma form and also produced by several mesenchymal cells as the extracellular matrix form. It is composed of two protein chains linked together by disulfide bonds at their carboxyterminal ends. The molecular weight of fibronectin is approximately 220000 Da.

Fibronectin was described under different names (e.g., cold insoluble globulin, antigelatin factor, surface fibroblast antigen). This resulted from the variety of its biological functions, the most recognized one being its ability to interact with a number of molecules, including collagen, heparin, fibrinogen gangliosides, and fibronectin itself. The binding is probably associated with such activities as mediation of cell adhesion, regulation of the opsonization activity of macrophages, and inhibition of collagen fiber formation.

Fibronectin binds strongly and specifically to collagen. Native and denatured collagen are both bound, with the highest affinity for denatured collagen. Types I–V collagens are known to bind to fibronectin. It is unclear whether or not the type of collagen affects the binding affinity, although in all investigated forms, denatured collagen binds much better than the native one. The collagenous portion of complement and acetylcholinesterase also bind fibronectin. The rate of binding is higher in the presence of heparin. The binding activity of fibronectin is resistant to pepsin digestion.

The binding site for fibronectin on collagen is probably a relatively small portion of the amino acid chain. According to Kleinman and Wilkes (1982) a synthetic peptide of 20 amino acids (residues 766–785) and even an octapeptide (residues 773–780) of the $\alpha 1$ (I) chain are active in binding. The integrity of residues 775–776, which form the cleavage site for collagenase, is necessary for the binding. It has been suggested that the binding site is less tightly coiled due to a lack of hydroxyproline, proline, and carbohydrates. The hydrophobic properties of the binding region facilitates the interaction, and artificially augmented hydrophobic properties enhanced the efficiency (Kleinman and Wilkes 1982). Fibronectin contains a specific region which is responsible for collagen binding. This region probably has a collagen-like sequence.

The biological role of the collagen-fibronectin interaction has been demonstrated under several pathophysiological conditions. Fibronectin is considered the major molecule for cell-matrix adhesion in fibroblast-type cells. This is of great importance for cell growth and differentiation as well as repair (Cidadão 1989; Cidadão et al. 1988).

Fibronectin has been found to influence collagen fibrillogenesis. Kleinman et al. (1981) shown that it delayed the self-assembly of tail tendon type I collagen. On the contrary, Speranza et al. (1987) reported that the fibrillogenesis of type I collagen was accelerated but that of type III delayed. These effects were concentration-dependent, and the lag phase of fibril formation was more altered than the growth phase in the presence of fibronectin. The explanation of these contradictory results is unknown. It is also unclear why fibronectin causes opposite effects on types I and III collagen fibrillogenesis, possibly due to a much greater affinity for type III (Jilek and Hörmann 1978; Engvall et al. 1978).

#### 3.11.3.2 Laminin

Laminin is a structural glycoprotein with a molecular weight of about 1 000 000 Da. Laminin binds collagen, and this property is limited to native collagen only. The binding site for laminin on type IV collagen is different to that for fibronectin. Binding of collagen type IV to laminin is probably an important mechanism for cell adhesion and organization of basement membranes.

#### 3.11.3.3 Entactin and Nitogen

Entactin and nitogen are believed to be derived from the same precursor protein. Nitogen binds to the globular NCl domain of type IV collagen and to laminin. The physiological function of this interaction remains obscure.

#### 3.11.3.4 Chondrionectin

Chondrionectin is a glycoprotein that mediates the interaction between chondrocytes and type II collagen in cartilage. The binding is significantly enhanced by proteoglycans. Details of the mechanism of the binding are unknown.

#### 3.11.3.5 Fibromodulin

The term fibromodulin was introduced by Oldberg et al. (1989) for a 59 000-Da protein able to bind collagen and which modulates collagen fibrillogenesis (Hedbom and Heinegård 1989). Fibromodulin has been found in cartilage, tendon, skin, sclera, and cornea has a significant sequence homology with proteoglycans PG-S1 and PG-S2, decorin.

# 4 Degradation

## 4.1 Introduction

Collagenous structures whose formation is a complex, multistage process, are a heterogenous part of the connective tissue architecture. Thus, turnover of collagen requires a complex, precisely regulated degrading mechanism, which is highly specific for the variety of collagens occurring in the body and adequate for the physiological conditions of the tissue. In contrast to biosynthesis, which although very complex has the same general pattern in all tissues synthesizing collagen, degradation is assumed to be associated with several alternative pathways. Some of them are more characteristic for physiological conditions or tissues with slow collagen turnover while others are more common in tissues are undergoing rapid physiological changes or occur under pathological conditions. The biological function of connective tissue demands particularly precise mechanisms of regulation of collagen degradation in association with the catabolic processes of other components. Only a small part of these mechanisms are understood, and collagen catabolism despite numerous efforts remains much more obscure than biosynthesis.

## 4.2 General Pathways of Collagen Degradation

Collagenous structures which occur in tissues are more complex than those investigated in vitro. Enzymes which are able to cleave collagen in solution in vitro are unable to break down insoluble polymeric collagen fibers which are associated in tissues with proteoglycans and structural glycoproteins. The first stage of collagen degradation is depolymerization of collagen structures, and its separation from other matrix components. Depolymerized collagen can be cleaved by so-called tissue collagenases, which are able to attack the helical portion of the molecule. These enzymes, labelled as "true collagenases", drew the main attention of investigators. The alternative pathway of degradation by collagenolytic cathepsins secreted into the limited pericellular space of acid pH has been suggested to occur under some conditions of rapid degradation.

Nonhelical parts of collagen molecules can be degraded by certain proteolytic enzymes, and the biological role of this phenomenon requires further exploration. It is unclear which of these enzymes occur in the tissues under physiological conditions.

Breakdown of the helical portion into large fragments is sufficient for their thermal denaturation at normal body temperature. This loss of nativity is required for further degradation by a group of enzymes other than collagenases. This process may occur extracellularly as well as within the cells. In the extracellular space, it involves relatively specific gelatinases and/or collagen peptidase. Intracellular degradation is preceded by phagocytosis of large fragments which are then cleaved by cathepsins within the secondary lysosomes. Despite the route of degradation (extracellular or intracellular) the end products are reutilized in vast amounts in metabolism. A small portion of them are excreted, and it is possible to determine those which contain hydroxyproline due to the specificity of hydroxyproline occurrence in collagen.

Enzymes of bacterial origin have been known for many years to be capable of degrading collagen. With the exception of some forms of infection, there is no involvement in the turnover of collagen in vivo. Their mechanism of action differs from that of vertebrate collagenases, and therefore only a short overview of bacterial collagenases is presented within this chapter.

The general scheme of collagen degradation is shown in Fig. 4.1 and is described in the next section. Physiological and pathological aspects of degradation in specific tissues are described in chapters dealing with organ-specific collagen metabolism.

The nomenclature of the enzymes involved in collagen breakdown remains controversial. Here, the following terms are used: tissue collagenase (matrix

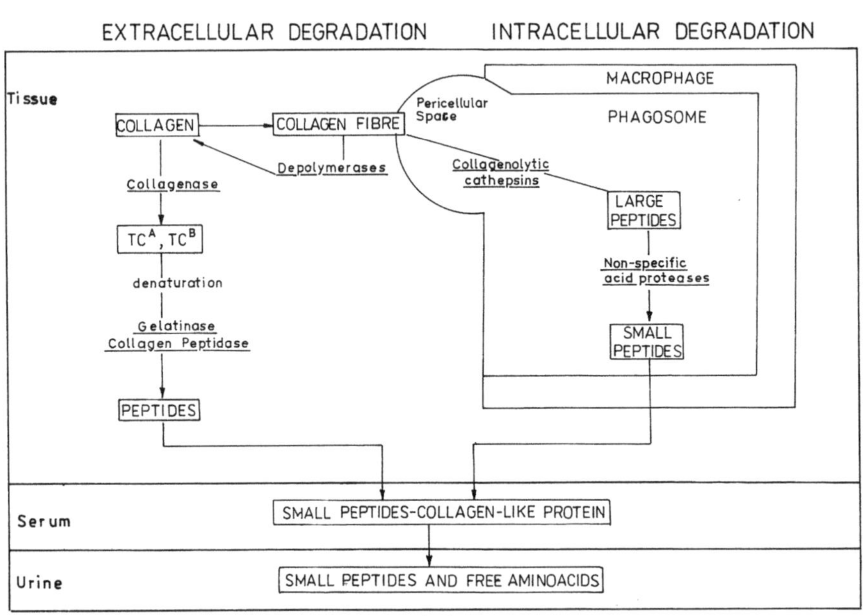


**Fig. 4.1.** General scheme of collagen degradation

metalloendoproteinase 1, vertebrate or mammalian collagenase), gelatinase (matrix metalloendoproteinase 2), collagenolytic cathepsin, collagen peptidase (probably identical with matrix metalloendoproteinase 5), microbial collagenase. The term telopeptidase (matrix metalloendoproteinase 4) refers to poorly characterized enzyme(s) involved in the depolymerization of collagen. Stromelysin (matrix metalloendoproteinase 3) as a nonspecific proteinase is not reviewed here (see Okada et al. 1986).

## 4.3 Depolymerization

Collagen is generally recognized as a protein resistant to the majority of mammalian proteolytic enzymes. Until 1962, when collagenase of tadpole tails was discovered, bacterial collagenase was the only enzyme known to cleave it. On the other hand, many observations of a rapid disappearance of collagen under certain pathophysiological conditions (e.g., uterus involvement after parturition, destruction of skin in trophic ulcers, or joint destruction in rheumatoid arthritis) suggested the existence of highly effective specific mechanism(s) able to breakdown collagen.

A high portion of collagen molecules occurs in helical form. This structure is particularly resistant to most tissue proteases. Factors contributing to the stability of the helix or supramolecular structure including the proteoglycan association increase the resistance to proteolysis.

The first step of degradation takes place extracellularly. Depolymerization of collagen fibrils in vivo remains obscure. In experiments, various enzymes are able to depolymerize collagen, but the occurrence of these enzymes in tissues is doubtful. The role of degradation of nonhelical parts of collagen in the initial steps is unclear. It has been found that pepsin, trypsin, chymotrypsin, thermolysin, and elastase are able to cleave nonhelical terminal regions of the interstitial collagens. These enzymes, however (except elastase), are unlikely to be present in tissues. Depolymerization in vivo is attributed to enzymes active at a neutral pH, including granulocyte elastase and some less well characterized, nonspecific proteases. It is uncertain whether or not the thiol proteases active at an acid pH take part in depolymerization of collagen in the extracellular space. Lysosomal enzymes such as cathepsins G, B, N, H, and L cleave types I and III end regions and can depolymerize collagen fibers. It has been found that under conditions of rapid collagen breakdown some cells adhere to collagenous structures, and the space between the cell membrane and the collagen fiber can be acidified by metabolic products of the cell such as lactic acid. In such pericellular spaces, the enzymes active at a low pH can contribute to depolymerization of collagen.

Some studies indicate that a collagenase preparation produces a slow solubilization of collagen fibers. These reports are difficult to analyze, as many preparations of collagenase may be contaminated with nonspecific proteases serving as depolymerases in the first stages of degradation.

The role of proteoglycans and structural glycoproteins which are associated with collagen in the regulation of collagenolysis remains unclear. It is hypothesized that these compounds protect collagen from degradation or delay breakdown. It is very awkward to verify this hypothesis as it is difficult to isolate intact collagen-proteoglycan complexes and use them to study collagen breakdown in vitro. In vivo, the complexes are attacked by a group of enzymes, and it is possible that some of the enzymes separate or degrade the protective proteoglycans and structural glycoproteins associated with the collagenous supramolecular structures.

## 4.4 Tissue Collagenases

Tissue collagenases (known also as vertebrate collagenases, EC 3.4.24.3) are widely distributed enzymes capable of specifically cleaving a native collagen molecule. The preference for native collagen and the ability to attack the helical portion are their most important common properties. The presence of collagenases has been reported in a variety of tissues or, more frequently, in the culture medium of numerous cells of various vertebrate species (for review, see Eisen et al. 1970; Woessner 1973; Harris and Krane 1974; Woolley and Evanson 1980; Krane 1982; Weiss 1976; Pérez-Tamayo 1982; Mainardi 1987; Stricklin and Hibbs 1988; Harris 1986; Sorsa et al. 1987).

In general, tissue collagenases are classified into two groups. The first includes "classic" collagenases; i.e., enzymes that cleave interstitial collagens. The second group is termed the "type-specific" collagenases; these enzymes are able to degrade one or more collagens other than the interstitial type. It is important to indicate that the term "collagenase" is limited to those enzymes which cleave the helical domain of collagens. Some types of collagen contain several helical domains separated by nonhelical parts which are susceptible to attack by nonspecific proteases. In fact, these proteases act as collagenolytic enzymes but cannot be included into the group of collagenases. It can be assumed that further discoveries will provide more detailed knowledge about their mode of action and type-specificity.

The cellular origin of collagenases is not always known. In many descriptions the tissue is reported as a source of the enzyme. In mammals, including humans, the ability to produce collagenases was reported in granuloma, uterus, skin, several parts of the gastrointestinal system, corneal tissue, synovium, and many kinds of tumors. Collagenases are produced by fibroblasts, chondroblasts, osteoblasts, as well as polymorphonuclear leukocytes, macrophages, and melanoma cells. Many of these enzymes have been partially characterized although they can only be isolated in very small amounts, and their detailed characteristics are difficult to obtain. There are common properties which are found in all or nearly all collagenases.

The molecular weight of some collagenase preparations has been reported. The enzyme isolated from human fibroblasts has 55000–60000 Da as a proen-

**Table 4.1.** Amino acid composition of four vertebrate collagenases. All values are given as residues per 100 residues. (From the review of Stricklin and Hibbs 1988 with permission)

| Amino acid | Human fibroblast | Human neutrophil | Tadpole skin | Rat uterus |
|---|---|---|---|---|
| Asp | 11.7 | 10.0 | 13.9 | 9.5 |
| Glu | 10.5 | 10.4 | 9.8 | 11.5 |
| Thr | 5.5 | 5.1 | 5.7 | 4.6 |
| Ser | 5.4 | 6.7 | 5.7 | 16.5 |
| Pro | 6.1 | 5.1 | 6.5 | 4.5 |
| Gly | 7.8 | 10.2 | 8.3 | 17.7 |
| Ala | 6.6 | 9.6 | 6.7 | 7.2 |
| Val | 5.6 | 5.8 | 5.4 | 2.9 |
| Leu | 6.1 | 8.1 | 7.8 | 5.1 |
| Ile | 3.9 | 3.8 | 4.6 | 1.8 |
| Cys/2 | 1.3 | 3.8 | 0.5 | 1.9 |
| Met | 1.6 | 1.2 | 1.0 | 1.0 |
| Tyr | 3.9 | 2.9 | 4.4 | 3.7 |
| Phe | 7.1 | 3.8 | 7.0 | 3.1 |
| His | 3.6 | 2.1 | 2.6 | 5.2 |
| Lys | 6.4 | 4.6 | 5.2 | 3.8 |
| Arg | 5.4 | 7.0 | 3.4 | 2.8 |

zyme and 45000–50000 Da as the trypsin-activated enzyme (Stricklin et al. 1977). Similar values have been shown for rat uterine collagenase and rabbit synovial collagenase (Roswit et al. 1983; Nagase et al. 1981). The molecular weight of human neutrophil collagenase is estimated to be 70000–91000 Da (MacCartney and Tschesche 1983; Hasty et al. 1986).

The preliminary data about the protein structure of collagenases show them to be cationic. The amino acid composition of a few have been described, and strong similarities are found despite that the enzymes were isolated from such different materials as human fibroblasts and neutrophils and tadpole skin. The amino acid composition of four vertebrate collagenases is shown in Table 4.1.

An interesting finding is the discovery that collagenases occur in two forms, one of which is highly glycosylated. Detailed studies of rabbit synovial enzymes showed that collagenase is present as 61000 and 60000 Da proteins. The former contained carbohydrates linked in a posttranslational glycosylation. Both forms have the same catalytic efficiency. The molecular nature of the enzyme-linked carbohydrates and their functional significance remain unknown (Wilhelm and Eisen 1985; Nagase et al. 1981; Stricklin et al. 1978).

Collagenases are metalloproteinases. Their catalytic activity is associated with the presence of calcium and/or zinc ions. This property explains the inhibitory activity of chelating agents on collagenase activity. Most collagenases are inhibited by EDTA, 1,10-phenantroline, or compounds which react with sulfhydryl groups. Barium and strontium ions have been reported to be able to substitute for calcium ions (Seltzer et al. 1976; Woolley et al. 1978b). Calcium ions are required for activity and structural stability of the enzymes. Zinc ions are probably required to stabilize the active site of the enzyme (Seltzer et al. 1977; Swann et al. 1981).

Tissue collagenases are highly specific. They catalyze cleavage at one specific site located at about one-quarter the length of the molecule from the aminoterminal end. The sequence of amino acids of the site has been established. In $\alpha 1$(I) chain it is between residue 772 of glycine and residue 773 of isoleucine. Tissue collagenases cleave only the bond between glycine and isoleucine. Among the three such bonds present in type I collagen, only those between amino acid residues 772 and 773 in the $\alpha 1$(I) chain and similarly in the $\alpha 2$(I) chain are susceptible to degradation. It is clear that the collagenase recognizes other stereochemical features of the molecule than just the primary sequence of the amino acids which form the bond. Although the native state of the protein, with an unchanged helical structure, is required for cleavage by collagenases, the site of their action has no full helical structure. This part of the molecule is described as semihelical. Every third residue is glycine but imino acids, proline, or hydroxyproline are not present in sufficient amounts to build and stabilize the helical structure. It is suggested that the conformation of this region is the key mechanism for specificity (Burleigh et al. 1977; Gross et al. 1980).

The type-specificity of collagenases is not fully known. It is clear that enzymes obtained from various sources differ in their specificity. The majority of tissue collagenases cleave intersitial collagens, but the rate of degradation varies with the sort of collagen substrate.

In many tissues, separate enzymes specific for basement membrane collagen have been found. The susceptibility of various types of collagen to cleave by collagenases can be regulated by factors other than the primary structure and conformation, e.g., amount of bound carbohydrates.

Several reports have described collagenases that are able to degrade other collagens than interstitial ones (Sorgente et al. 1977; Liotta et al. 1979, 1982a; Mainardi et al. 1980; Puistola et al. 1989; Stettler-Stevenson et al. 1989). The molecular characterization of these enzymes requires further investigation. Some research indicates that leukocyte collagenase differs from the enzyme isolated from mesenchymal tissues. The nature of these variations remains controversial (Sorsa et al. 1987).

It was quite amazing when Gadek et al. (1980) reported that human neutrophil elastase functions as a type III collagen "collagenase." Elastase is capable of degrading types III, IV, V, and X (Mainardi et al. 1980; Sage et al. 1981; Gadher et al. 1988, 1989). The nature of loci in the amino acid sequence vulnerable to neutrophil elastase are different from those known to be attacked by collagenase (Gadher et al. 1989). Birkedal-Hansen et al. (1985) showed that type III collagen in solution was readily cleaved by elastase while the fibrils reconstituted from neutral salt solution, at 35 °C, were highly resistant. Alterations in the helical structure and local folding of the molecule have been suggested as the causes. The physiological role of the susceptibility of certain collagen types to neutrophil elastase remains unknown.

Regulation of collagenase secretion, activation, and inhibition is thought to be one of the most important key mechanisms of the control of collagen degradation. Although a number of papers have been published, our understanding of these phenomena remains poor (Harris et al. 1984).

The ability to stimulate secretion of collagenase is attributed to some mediators of inflammation and the immune response, including interleukin-1 and interferons. These phenomena are reviewed in Chap. 8.

Prostaglandins are believed to be another group of physiological stimulatory compounds of collagenase secretion (Wahl et al. 1977). Carrageenin, an algal polysaccharide, commonly used to elicit an inflammatory granuloma, has been reported to stimulate peritoneal macrophages to release collagenase and gelatinase (López-Escaera and Pardo 1987).

The best known chemical agent that produces enhanced collagenase secretion is phorbol myristate acetate, a tumor-promoting compound. It leads to direct activation of protein kinase C and bypasses the signal transduction of physiological stimuli via diacylglycerol and free intracellular calcium. A significant increase in collagenase production is found in cultured fibroblasts treated with phorbol myristate acetate (Brinckerhoff et al. 1979; Brinckerhoff and Harris 1981; Chua et al. 1987). Recently, Brinckerhoff et al. (1989) reported isolation of two proteins secreted by phorbol myristate acetate-stimulated fibroblasts that are able to induce collagenase synthesis. These autocrine modulators of collagenase production have been shown to be similar to serum amyloid A and $\beta_2$-microglobulin. The role of autocrine induction of collagenase in the regulation of collagen degradation remains unknown.

A proteinase of *Bacteroides gingivalis* has been reported to induce secretion of tissue collagenase. This is suggested to be involved in the etiology of periodontal disease (see Chap. 15; Uitto et al. 1989).

Reduced secretion of collagenase is achieved in cells with an elevated intracellular level of cyclic AMP (Koob and Jeffrey 1974; McCarthy et al. 1980). Glucocorticoids, retinoids, and phenytoin were reported to diminish collagenase secretion (see Chap. 24).

It is well-known that many, if not all, collagenases are secreted in an inactive form. The nature of this inactive form remained a subject of long-lasting controversy. Some investigators claimed that the enzyme was secreted as an enzyme-inhibitor complex, and others suggested that it was produced as a proenzyme. In recent years, several lines of evidence supported the latter view (Nagase et al. 1981). The enzyme is produced as a proenzyme, procollagenase. The enzyme-inhibitor complexes which occur in tissues or culture media have been shown to be secondary products, formed after the interaction of the free active collagenase with specific inhibitors. Several enzymes activate procollagenase: proteolytic enzymes such as trypsin, plasmin, plasminogen activator, cathepsin, and kallikrein are only some of them. It is clear that some proteases which occur in tissues are able to activate collagenase (Capodici et al. 1989). The nature of the enzymes responsible for activation of procollagenase under physiological conditions is not known. A cascade system similar to that of coagulation has been proposed (Vaes 1972). Further investigations are needed to support this hypothesis. Procollagenase activator has been isolated from rabbit synovial fibroblasts (Vater et al. 1983, 1986). The activator, as has recently been shown, is identical with stromelysin, a metalloproteinase that cleaves proteoglycan protein core (Murphy et al. 1988). Activation of procollagenase by collagenase is an interesting example of a positive feedback system accelerating collagenolysis (Jeffrey 1986).

Activation of procollagenase by organic mercurial compounds was demonstrated for the first time by Sellers et al. (1977). The most effective activators are the following: *p*-amino-phenylmercuric acetate, *p*-hydroxymercuribenzoate, phenylmercuric chloride, 3-[2-(carboxymethoxy)-benzamido]-2-methoxypropylhydroxymercury monosodium salt (mersalyl), and 4-chloromercuribenzoate. The mechanism of this activation is unknown. According to Stricklin and Hibbs (1988), the process of activation is most likely not related to the compound's affinity for -SH groups but rather is proportional to its hydrophobicity. Activation requires the presence of high concentrations of the compounds, and progress depends upon the continuous presence of the activator (Stricklin et al. 1983).

Several groups were able to isolate physiological inhibitors of collagenases. The reports varied in the characterization of the molecular nature of the inhibitors, their properties and specificity, and their cellular origin. Therefore, it is difficult to generalize the role of the inhibitors in the regulation of collagen breakdown. On the other hand, it is clear that almost all tissues contain more or less specific inhibitors of collagenolysis. According to Stricklin and Hibbs (1988) inhibitors are classified into three groups, high-molecular-weight forms that occur in plasma and are a part of the plasma antiproteolytic system, small cationic proteins which are present in tissues and are probably specific inhibitors of collagenases, and the tissue inhibitor of metalloproteinases, TIMP.

About 95% of the total collagenase inhibitory activity of plasma results from the presence of $\alpha_2$-macroglobulin. $\alpha_2$-Macroglobulin has a large molecular weight, 725000 Da, and is believed to be unable to penetrate into tissues. Thus, its inhibitory function is limited to the plasma (Sottrup-Jensen and Birkedal-Hansen 1989). The remaining 5% of the total inhibitory activity of plasma arises from a specific $\beta_1$-anticollagenase. This inhibitor was discovered by Woolley et al. (1976). It has a molecular weight of 40000 Da and possibly has access to the extravascular space. There are suggestions that it is a form of TIMP, to be discussed later.

Some reports describe low-molecular protein inhibitors of collagenase. Small cationic inhibitors have been extracted from aorta and cartilage (Kuettner et al. 1976; Murphy et al. 1977; Killackey et al. 1983); their molecular weight was about 11000 Da. These compounds have not been well characterized.

TIMP (tissue inhibitor of metalloproteinases) has drawn the attention of several investigators. Human TIMP is a basic glycoprotein with a molecular weight of 28500 Da (Welgus et al. 1979; Stricklin and Welgus 1983). It is stable and resistant to a temperature of 90 °C and treatment with 6 *M* urea or low pH (Stricklin and Welgus 1983). Its primary sequence has been reported and contains 184 amino acids. Some amino acid residues contain N-linked oligosaccharides (Stricklin and Hibbs 1988).

TIMP inhibits collagenase through the formation of a very tight complex with the enzyme (Weglus et al. 1979, 1985). This complex cannot be disrupted by organomercurials or trypsin. TIMP is able to bind to procollagenase or active enzyme. It is possible that the complex due to its stability represents a terminal event in the life of collagenase.

The presence of TIMP was shown in skin, bone, smooth muscle, tendon, and cartilage. It is also found in body fluids (Aggeler et al. 1981). Some authors suggest that the $\beta_1$-anticollagenase of plasma is identical with TIMP (Stricklin and Hibbs 1988). The molecular weight of $\beta_1$-anticollagenase has been reported to be considerably greater than that of TIMP. Thus, it is possible that $\beta_1$-anticollagenase is a form of TIMP or TIMP-related protein.

Physiological sequestration of active collagenase has been suggested to be an interesting and probably efficient mechanism for controlling collagen degradation. It has been shown that collagenase binds to some components of the extracellular matrix, including sulfated glycosaminoglycans (Halme et al. 1980; Roswit et al. 1983). This finding has been the basis for a hypothesis that some components of the matrix can bind the enzyme and in this way prevent its contact with collagen. Some studies suggest that collagen fibers may bind collagenase (Harper et al. 1971; Stricklin et al. 1978). The ability to bind native collagen fiber is most likely limited to active forms of the enzyme (Welgus et al. 1985). The effect of such binding on the activity of collagenase remains unknown.

## 4.5 Collagenolytic Cathepsins

Etherington (1972) was the first to discover that homogenates of certain tissues incubated under acid pH with insoluble collagen are able to degrade (solubilize) collagen. On the basis of this finding, he introduced the term collagenolytic cathepsin.

A high activity of collagenolytic cathepsin was found in the kidneys, spleen, bone marrow, and liver of rats. Lower activities were reported in other rat tissues, including lungs, skin, heart, and skeletal muscle (Etherington 1972, 1974). An elevated activity was shown in carrageenin-induced granuloma (Nakagawa et al. 1977).

Further studies have revealed that lysosomal cathepsins, proteases operating under acid conditions, are involved in collagen degradation. Their role is not fully understood, and two possible mechanisms have been suggested. The enzymes may contribute to collagenolysis of collagen fragments taken into the cell by phagocytosis. The molecular nature of these fragments and their native/denatured state are unknown. The other mechanism involves, as mentioned earlier (Sect. 4.2), pericellular spaces, i.e., local sites of reduced pH. In such spaces cathepsins may operate outside the cell and may be active against native aggregates of collagen. Pericellular spaces have been shown to occur close to the macrophages attached to the collagen fibers and the osteoclasts involved in bone resorption (Etherington et al. 1981; Baron et al. 1985).

The ability to degrade collagen and/or gelatin has been attributed to several cathepsins. Maciewicz and Etherington (1985) purified lysosomal cathepsins B, L, N, and S. The enzymes were isolated from rabbit spleen and were found to exhibit collagenolytic activity below pH 5.0. Although all enzymes have been found to be cysteine proteases, they have a different specificity toward synthetic and natural substrates and diverse physicochemical properties. The enzymes vary in the kinetics of insoluble collagen degradation. Cathepsins L and N have been reported to be more effective than B and S. The mechanism is unknown. According to Maciewicz et al. (1987) the cathepsins act as depolymerases, i.e., liberate collagen monomers from the fibers. This process is caused by removal of the telopeptides from the collagen molecules. The monomers, i.e., single collagen molecules without telopeptides, denaturate under the acidic conditions that occur at resorption in the pericellular space or in the phagosomes (Bailey and Etherington 1980). Denatured collagen serves as a substrate for the cathepsins responsible for further degradation of this protein (Capodici and Berg 1989). It is unknown whether or not the same cathepsins serve as depolymerases and gelatinases or occupy specific places in the pathway of the lysosomal breakdown of collagen.

In vitro, insoluble collagen is solubilized by cathepsins at a linear rate initially and without a lag phase (Etherington 1972, 1977; Kirschke et al. 1982). The solubilization follows Michaelis-Menten type kinetics. Cathepsin L has the highest preference for insoluble collagen and has been reported to be 5, 30, and 200 times more effective than cathepsin N, B, and S, respectively (Maciewicz et al. 1987).

The cathepsins have also been shown to differ in their susceptibility to chicken egg-white cystatin. It is interesting to note that the enzyme with the greatest potential to degrade collagen has been reported to be most effectively inhibited. It is thought that endogenous cystatins are involved in the regulation of enzyme activity and control of collagen degradation via the lysosomal cathepsin pathway (Maciewicz et al. 1987). Alterations in collagenolytic cathepsin activity were shown under some pathological conditions.

## 4.6 Gelatinases and Collagen Peptidase

Denaturation of collagen is associated with the loss of its helical structure. The product of collagen denaturation is known as gelatin. The susceptibility of gelatin to attack by proteolytic enzymes depends on its structural features. The loss of the helical form makes gelatin susceptible to cleavage in a manner more or less similar to that of the globular proteins. On the other hand, the extremely unusual primary sequence and presence of a high content of imino acids reduce the ability of proteases to degrade gelatin.

Digestion of gelatin with pepsin at a low enzyme-to-substrate ratio yields two high-molecular-weight fragments and a few small fragments. The mode of action of the enzyme is similar to that found in the digestion of globular proteins (Weiss 1976). Chymotrypsin, thermolysin, and papain also cleave gelatin into characteristic fragments (Dixon and Webb 1964; cited in Weiss 1976). Details on the typical cleavage sites of gelatin digested by nonspecific proteases are summarized by Weiss (1976).

Enzymes that are able to degrade gelatin at physiological pH are termed gelatinases. It is believed that gelatinase is an enzyme which catalyzes specifically the cleavage of gelatin formed from $TC^A$ and $TC^B$ fragments of collagen. These fragments result from collagenase digestion and are spontaneously denaturated under physiological conditions.

It is clear that gelatinase activity is almost constantly found in collagenase preparations. This feature of collagenase may be caused either by impurities of the enzyme or by the dual specificity of, at least, some collagenases. The activity of enzymes which are not able to degrade native collagen but can break down gelatin has been shown in tissues, but the molecular nature and properties remain unknown. Difficulties in the preparation of pure, native, and well-characterized substrates as well as isolation of pure enzyme preparations are the major causes of difficulties. Collagenase and gelatinase share several properties: both enzymes are inhibited by metal chelating agents, by certain sulfhydryl reagents, and by similar protein inhibitors. Inhibition with eriochrome black T can be used to differentiate them, because this reagent inhibits collagenase but not gelatinase (Seltzer et al. 1987). Different substrates may be employed for this purpose (Stack and Gray 1989).

Several authors introduced synthetic substrates with a sequence similar to that of collagen. They are resistant to digestion by nonspecific proteases. Synthetic substrates provide evidence of an enzyme that is able to degrade them but is inactive against native collagen. The enzyme is known as collagen peptidase, PZ-peptidase, or collagenase-like peptidase. In studies on proteases of microbial origin, the term pseudocollagenase is used to describe this enzyme (see Sect. 4.7). The relationship between collagen peptidase and gelatinase is unknown. It is possible that the enzyme is able to degrade both gelatin and smaller fragments of digested collagen as well as synthetic peptides that resemble these small fragments.

The best known substrate for collagen peptidase was synthesized by Wünsch and Heidrich (1963). It is known as PZ-peptide and has the following structure, 4-phenylazobenzyl-carbonyl-L-proline-L-leucylglycine-L-proline-D-arginine. The sequence is analogus to that of gelatin and like gelatin has a random-coil configuration. A similar substrate was designed by Nagai et al. (1960) and has the following sequence, carbobenzoxy-glycyl-prolyl-leucyl-glycylproline.

Activity of collagen peptidase has been found in several tissues and body fluids, and determination of the enzyme has been postulated to be useful in the diagnosis of certain disorders (Strauch and Vencelj 1967; Lindner et al. 1974; Kucharz and Nowak 1986).

## 4.7 Collagenolytic Enzymes of Microbial Origin

Bacterial enzymes which are able to degrade collagen have been known since the turn of the century. Their nature and substrate specificity remain controversial. The best known source is cultures of *Clostridium histolyticum.* Other clostridia also excrete collagenolytic enzymes, including *C. perfringens* type A, *C. capitovale, C. integumentum,* and probably *C. tetani.* Collagenolytic enzymes were isolated from culture filtrates of *Pseudomonas aeruginosa, Staphylococcus aureus, Bacteroides melaninogenicus,* and *Achromobacter iophagus.* Some molds also excrete the enzymes, as has been shown in cultures of *Trichophyton schoenleinii* (Mandl 1961; Nordwig 1971). It is of interest that a common member of the normal microbial flora of the human mouth, *Candida albicans,* elaborates an extracellular enzyme capable of attacking undenaturated predentine collagen. The enzyme has been also found to degrade bovine tendon collagen. A possible role of this collagenolytic enzyme in the pathogenesis of yeast infections has been suggested (Kaminishi et al. 1986; Hagihara et al. 1988).

It is clear that culture filtrates contain more than one collagenolytic enzyme. The most thoroughly investigated enzyme system is that excreted by *C. histolyticum.* It consists of at least three enzymes or groups of enzymes. The first of them, known as clostridiopeptidase A or collagenase 1, degrades specifically soluble or insoluble collagen and synthetic polypeptides with a collagen-like structure (substrates for collagen peptidase). It is inactive against such proteins as casein or hemoglobin. The second enzyme or group degrades native and denatured collagen but is unable to cleave the model collagen-like peptides. It is termed collagenase 2. The third enzyme or group is capable of digesting denatured collagen and synthetic peptides but not native collagen. It is classified as pseudocollagenase but in fact is a gelatinase (Nordwig 1971). Studies of Bond and Van Wart (1984a, b, c) have shown that preparations of "collagenase" from *C. histolyticum* are a mixture of six zinc metalloproteases, classified as collagenase I alpha, beta, and gamma, and collagenase II delta, epsilon, zeta. Collagenase I has a high activity against collagen and modest activity against synthetic peptides. Collagenase II has the opposite specificity. The enzymes differ in their susceptibility to certain ketone inhibitors (Grobelny et al. 1989). Other inhibitors have also been described (Ronmestand et al. 1989). Two forms of collagenase have been isolated and characterized in cultures of *Achromobacter iophagus* (*Vibrio alginolyticus*) (Reid et al. 1980; Keil-Dlouha and Keil 1978; Vrany et al. 1988).

The mode of action of collagenolytic enzymes isolated from culture filtrates is not fully understood. Several preparations used in investigations have not been pure, and the presence of contamination has significantly affected the results. It is generally accepted that clostridiopeptidase A is an exopeptidase which cleaves every third bond in the collagenous sequence, as shown by -Gly-X-Y/Gly-X-Y-. Digestion of the native molecules occurs from their ends. This process has been shown at low temperatures while at higher temperatures (about 20 °C) the enzyme also exhibits endopeptidase properties (Nordwig 1971).

The pathophysiological role of collagenolytic enzymes of microbial origin is limited to disorders caused by or associated with bacterial or fungal infections. The details of the contribution of the enzymes to tissue damage and spreading of the microorganisms remain obscure.

# 5 Turnover and Regulation of Collagen Metabolism

## 5.1 Introduction

The biological functions of collagen depend upon precise mechanisms regulating the extracellular matrix and cells of the tissue. Homeostasis of the matrix and its adaptational processes in association with ontogenesis or environmental changes are regulated by a complex local and systemic feedback mechanism. There are many reports indicating the role of various compounds in the regulation of collagen metabolism, but the majority of studies describe a single phenomena only, and there is no systemic understanding.

The regulation of collagen metabolism and structure can be divided into the following groups of mechanisms:

1. Collagen gene expression mechanisms which regulate the ability, rate, and type of synthesized procollagen. These mechanisms are responsible for the expression of collagen production by cells other than the connective tissue cells (e.g., hepatocytes) as well as changes in collagen type composition in the matrix in health and disease.
2. Regulatory mechanisms of intracellular collagen biosynthesis, involving posttranslational modifications. One of the biological functions of these processes is to eliminate defective collagen molecules through intracellular degradation before further processing. Some amounts of procollagen without an adequate level of posttranslational modifications are eliminated in this way during the normal biosynthesis of collagen. Intracellular stages of synthesis are also target mechanisms for various systemic regulatory processes (e.g., hormonal control of collagen synthesis) as well as for pharmacological agents affecting collagen metabolism.
3. Extracellular processing of collagen is probably a subject of a precise control mechanism regulating the size and shape of collagenous structures. It is thought that proteoglycans and structural glycoproteins are at least partially responsible for these phenomena (see Chap. 3). Maturation of collagen is also a target point of a number of endo- and exogenous chemical compounds influencing collagen metabolism (see Chap. 24).
4. Intra- and extracellular stages of collagen biosynthesis remain in functional relationship, and regulatory circuits responsible for the coordination of synthesis within the cell and maturation in the extracellular matrix have been partially elucidated.

5. A number of activators or inhibitors of collagen degradation have been described, and it is clear that regulation of break down is an important part of the regulation of collagen metabolism. These processes include regulation of expression of genes of collagenase and enzymatic activators or inhibitors, liberation of the enzyme from the cells, and the participation of various types of cells in degradation under normal and pathological conditions. There is still only fragmentary knowledge of this regulation, mostly limited to reports of single inhibitors or activators of collagenase, and the systemic regulation of break down remains obscure (see Chap. 4). There also also single data about feedback mechanisms regulating degradation in relationship to the synthesis of collagen.

This chapter provides a survey on regulatory mechanisms thought to be common for all cells able to produce collagen. Hormonal and pharmacological control of collagen metabolism is described in Chaps. 7 and 24.

Activators and inhibitors of collagenolysis are discussed in the description of collagen catabolism. Regulatory processes believed to be limited to certain organs or particular pathological conditions are described with the pathophysiology of collagen in specific organs or systems of the body.

## 5.2 Transcriptional and Translational Control of Collagen Synthesis

Despite rapid progress in molecular biology and structural analysis of collagen genes and the transcribed mRNA thereof, the majority of data about transcriptional regulation of collagen biosynthesis represents phenomenological descriptions without sufficient elucidation of the basic mechanisms. The expression of various collagen type genes is probably regulated at the transcriptional level and is not a result of the differential expression of multiple genes as there is some evidence that the haploid set of chromosomes in mammals contains only single copies for each chain of collagen type. Thus, the amount of mRNA is presumably a rate-limiting factor for the quantity of collagen synthesized. In many experimental systems changes in the synthesis of different collagen types have been found to correlate with changes in the amount of specific mRNA.

There is evidence that certain factors are able to alter the composition of collagen types (as well as other matrix constituents) produced by the cell. Most of the experiments were done with chondrocytes, which produce under normal conditions exclusively type II collagen. Such factors as conditions of culturing (monolayer versus suspension), cell density, as well as transformation with ontogenous virus or treatment with chemical agents affecting nucleic acid synthesis may produce transition of the type II collagen to other types (Nerlich et al. 1986). Some of these changes are associated with cell differentiation, as it has been postulated that the ability to synthesize type II collagen reflects a differentiated state more sensitive to environmental alterations than the relatively stable ability to synthesize types I and III collagens. It is highly probable that disturbed regulation at the transcriptional level is involved in the pathological synthesis of collagen in such disorders as osteoarthrosis or cirrhosis, but the exact mechanism of these processes is still unclear (Nerlich et al. 1986).

Translation is a target for some regulatory mechanisms. The best recognized one involves aminoterminal propeptides. The propeptides affect the rate of translation as described in Sect. 5.5.

## 5.3 Regulation of Collagen Synthesis at the Intracellular Posttranslation Level

The unique posttranslational modifications of collagen (Chap. 3) are a target mechanism for various regulatory factors. They include changes in the activity of specific enzymes caused by an altered access to cofactors of modified substrate. Scurvy, the depletion of ascorbic acid, is the best known impairment of collagen synthesis due to the altered level of an enzymatic cofactor. The biochemical mechanism of scurvy as well as available data on the influence of other nutritional deficiencies on collagen biosynthesis are reviewed in Chap. 22.

Modification of these substrates is achieved mostly by substitution of proline or lysine with analogues which due to their chemical structure cannot be hydroxylated. This efficient method of inhibiting collagen synthesis is described in the chapter devoted to pharmacological control of collagen metabolism (Chap. 24).

## 5.4 Intracellular Degradation of Procollagen

Intracellular degradation of procollagen is probably a common feature of all secretory proteins. This process constitutes degradation of the newly synthesized protein before its secretion from the cell (Berg 1986; Neblock and Berg 1987). It is believed that this process occurs in all cells producing collagen and has been demonstrated in fibroblasts, chondrocytes, muscle cells, hepatocytes, and some tumor cells (Bienkowski et al. 1978; Berg et al. 1980). The biological role of degradation of newly synthesized procollagen is not completely understood. It is generally accepted that the process is a form of "quality control" of the cell and prevents the secretion of defective collagen. There are data supporting this concept, based on the experimental induction of synthesis of collagen with an impaired structure. On the other hand, there is a constant rate of intracellular degradation even when the cells enjoy optimal conditions.

It is hypothesized that this fraction of intracellular degradation is associated with the regulation of collagen synthesis. This amount of collagen may represent a reserve of protein which can be processed into collagen under conditions of rapid activation of synthesis. On the basis of this hypothetical concept, intracellular degradation is classified into two modes: basal and enhanced. In fact, the experimental data fall into these two subclasses, and there is much evidence that they differ qualitatively.

Basal degradation is a process which seems to be independent of the presence abnormalities in the procollagen structure. It is hypothesized that a random selection of normal molecules assigns some amount of newly synthesized protein to the secretory pathway or to intracellular degradation. On the other hand, it is possible that certain molecules are "marked" or altered and are nonrandomly segregated for degradation. There is no experimental system to isolate and characterize the molecules of procollagen destined for basal degradation. It is impossible to compare their structure with those which are processed for secretion. The cellular location of the proteolysis indicates that the Golgi apparatus is the site for sorting of the molecules for intracellular degradation. Basal degradation is independent of lysosomal proteolysis or at least is not directly related to lysosomes since inhibitors of lysosomal proteolysis have no effect on the process of basal procollagen degradation.

In contrast to basal degradation, enhanced intracellular break down of procollagen appears under a variety of conditions. All of them lead to the synthesis of defective procollagen. The list of such factors includes incorporation of nonphysiological amino acids (e.g., canaverine) or their synthetic analogues (e.g., *p*-fluorophenylalanine), premature termination of protein chain synthesis or error-containing peptides due to specific mutations, incorporation of *cis*-4-hydroxylproline which prevents the formation of a helical structure, and inhibition of hydroxylation of proline and lysine residues. In general, factors responsible for the diminished stability of the collagen triple-helix increase the enhanced intracellular degradation of procollagen. Enhanced degradation takes place in the lysosomal system and is inhibited by agents reducing lysosomal proteolysis.

The two systems remain in a functional relationship although details of this mutual regulation are unknown. The hypothetical model is shown in Fig. 5.1. Inhibition of secretion of procollagen at the level of the Golgi apparatus by monensin leads to the accumulation of the protein associated with an increase in basal degradation which is probably due to the increase in the amount of procollagen to be selected for degradation (Berg 1986). This observation supported the view that the branch-point between the secretion and basal degradation is located in the Golgi apparatus as monensin is the ionophore inhibiting secretion at the level of the *cis*-Golgi cisternae. Destabilization of the procollagen structure as described above is a signal for enhanced lysosomal degradation. Involvement of lysosomes in the enhanced intracellular degradation of collagen has been shown in experiments employing the inhibition of lysosomal proteases. Inhibition of cathepsin B or D by leupeptin and pepstatin A led to more than 98% inhibition of proteolysis (Berg et al. 1984). On the other hand,

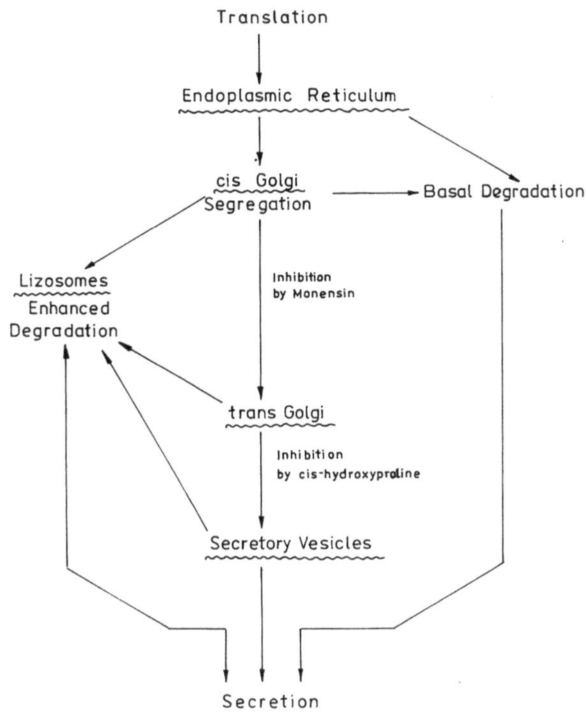


**Fig. 5.1.** Intracellular degradation of newly synthesized collagen. (Modified from Berg 1986 with permission.)

lysosomal enzymes including cathepsin degrade collagen (Chap. 4.5). Thus, together these findings suggest the role of lysosomes in the enhanced degradation of collagen.

Intracellular degradation of collagen is modulated by several factors. It is possible that this is an additional mechanism of regulation of collagen production. Baum et al. (1980) reported that the basal degradation of collagen in cultured human lung fibroblasts increased from 10% to 50% following treatment with prostaglandin $E_1$, prostaglandin $E_2$, or stimulation of $\beta$-adrenergic receptors by isoproterenol. It is possible that all these agents lead to an increase in cyclic AMP, which is thought to be a second messenger for the stimulation of intracellular degradation.

Insulin has been found to reduce intracellular degradation. In diabetic animals a significant increase in degradation of collagen has been reported (Schneir et al. 1982).

Intracellular degradation of collagen may provide an important mechanism for modulation of the amount of collagen available for the extracellular space matrix formation. This mechanism is able to respond to several intracellular and extracellular signals.

## 5.5 Feedback Inhibition of Biosynthesis of Collagen by Aminoterminal Propeptides

Aminoterminal propeptides of the interstitial collagens are separated from the molecule of procollagen when it is converted to collagen (see Chap. 3). The peptides have been found to be relatively stable in biological fluids and culture media, and determination of propeptides is used in the diagnosis of collagen abnormalities (Müller et al. 1986). Two biological functions are attributed to the peptides, regulation of the fiber diameter during fibrillogenesis and regulation of collagen biosynthesis. The role of the propeptides in fibrillogenesis is covered in Chap. 3. Here, the regulatory function of the peptides in the synthesis of collagen is reviewed.

It has been found that the aminoterminal propeptides from types I and III collagens exert an inhibitory activity on collagen synthesis. A dose-dependent inhibition was found in cultured fibroblasts, with 50% inhibition being achieved at a concentration of 6 μ*M*. Carboxyterminal peptides and peptides from other portions of the collagen molecule showed no inhibitory activity. The effect of the aminoterminal propeptide was specific, and the synthesis of other proteins was unaltered. There were also no changes in the hydroxylation of collagen and its intracellular degradation (Wiestner et al. 1979; Paglia et al. 1981; Kühn et al. 1982; Müller et al. 1986).

Paglia et al. (1979) reported that the propeptides inhibited translation of mRNAs for procollagen. The mechanism of this inhibition has only partially been elucidated. The aminoterminal propeptide of type I collagen was found to interfere with elongation and/or termination of translation. It is interesting to note that the tetrapeptide, Pro-Thr-Asp-Glu, obtained by digestion of the aminoterminal propeptide still retained the inhibitory activity but acted as a general inhibitor of translation (Hörlain et al. 1981).

The type-specificity of the aminoterminal propeptide from types I, II, and III collagens is not fully understood. Those from types I and III collagens inhibit collagen synthesis in fibroblasts, i.e., the cells secreting types I and III collagens. There is no inhibition in chondroblasts secreting type II collagen. On the other hand, the type I aminoterminal propeptide has been found to inhibit translation of pro-$\alpha$1 (II) RNA in cell-free lysate. This finding suggests that the resistance of the chondrocytes may be caused by an inability of the propeptide to penetrate the cells. The inhibitory activity of the type II aminoterminal propeptide remains unknown (Müller et al. 1986). The role of propeptides with other types of collagen also is uncharacterized.

It is clear that aminoterminal propeptides form the feedback mechanism that regulates the early stages of biosynthesis in relation to the amount of collagen deposited in the extracellular space. It is also possible that disturbances in this regulatory mechanism may be responsible for the accumulation of an excess of collagen as has been hypothesized in patients with systemic sclerosis (see Chap. 10).

## 5.6 Role of the Free Proline Pool in the Regulation of Collagen Synthesis

It has been found that the intracellular pool of free proline affects the rate of biosynthesis of collagen. Rojkind and Diaz de Leon (1970) have shown that an increased level of proline within the cell enhances collagen synthesis in the liver. This observation is consistent with elevated free proline levels in the liver tissue and serum of patients and animals with cirrhosis (Mata et al. 1975; Kershenobich et al. 1970; Baich and Chen 1979; Tyopponen et al. 1980). Similar results have been obtained in other tissues (Phang et al. 1971).

On the basis of these findings, a hypothetical feedback mechanism has been suggested. Proline and hydroxyproline are relatively specific end products of collagen break down. Imino acids, proline, and hydroxyproline constitute about one-fifth of the amino acids in collagen. The increased collagen degradation leads to an increase in imino acid level. In turn, hydroxyproline has been found to affect amino acid transportation to the cell (Finerman et al. 1967). A linear correlation between transport and the pool of proline has been shown. A direct relationship between the size of the intracellular pool of free proline and collagen synthesis has been noted. Thus, according to the hypothesis, the relationship between the degradation and synthesis of collagen exists in the form of a long axis as compared with the short axis of feedback regulation via propeptides (Kucharz 1978). The mechanism of the effect of free proline on synthesis is unknown. Modification of tRNA has been suggested (Rojkind and Kershenobich 1975).

Most of the data on the hypothetical regulation have been accumulated in studies on the liver, and it is unclear whether the same regulatory phenomena apply to other collagen-synthesizing cells.

## 5.7 Turnover of Collagen

Collagen constitutes a significant portion of the body mass. Traditionally, it has been believed that its turnover is slow and that it is an inert body protein. In contrast, studies on certain tissues revealed rapid turnover of collagen, especially accelerated under such conditions as postpartum uterine involution, inflammation, or wound healing.

One of the major difficulties in estimating collagen turnover are the two pathways of degradation, intracellular and extracellular. Thus, the measurement of hydroxyproline in the urine or tissues following injection of radiolabelled proline reflects intracellular degradation and does not correspond to turnover of collagen present in the extracellular matrix. It has been found that collagen synthesis rates based on uptake of radiolabelled proline into collagen are underestimated due to degradation of newly synthesized collagen during the time of measurement (Laurent 1982; Robins 1982). The intracellular degradation of collagen is a very rapid process, and a significant portion of recently synthesized collagen is degraded (Barnes et al. 1970; Kibrick and Singh 1974; Laurent and McAnulty 1983; Schneir et al. 1984). The rate of this degradation varies widely between organs.

According to McAnulty and Laurent (1987), the rate of collagen synthesis in the lungs, heart, skeletal muscle, and skin of adult rats are 9.0%, 5.2%, 2.2%, and 4.4% per day, respectively. The degradation of newly synthesized collagen in the same experiment has been determined as 37.2%, 53.4%, 37.9%, and 8.8% per day for the lungs, heart, skeletal muscle, and skin, respectively. Organs with a high rate of synthesis (the lungs and heart) have been found to have a high intracellular degradation rate. In the skin, the rate of synthesis is 4.4% per day and the intracellular degradation, 8.8%. It is important to relate these findings to the proportion of total collagen in the body in certain organs. The majority of collagen, as reviewed in Chap. 2, is found in the skin and skeletal musculature, and the internal organs contain only a small amount. Thus, the cumulative turnover of collagen in the body may be slow despite rapid synthesis and degradation in some organs.

Comprehensive data about the effect of physiological factors (age, sex) on collagen turnover are not available. The modern view on the turnover of collagen has to take into account its type-specificity and the occurrence of pools of collagen with rapid or slow turnover. The details of these phenomena are unknown.

# 6 Collagen in Development, Aging, and Phylogeny

## 6.1 Collagen and Development

In recent decades, rather a lot of data has accumulated on the association of the extracellular matrix and development, cell division, and differentiation. It is generally accepted that the extracellular matrix contributes to embryonic and postnatal development, including morphogenesis and the formation of organs. The process of development leads from the spherical fertilized egg to a mature adult human being whose cells all possess the same genome. Thus, it can be considered as a precise regulation of cellular phenotype expression in relation to time (phase of development) and space (location within the body).

The role of the extracellular matrix, and collagen in particular, in development can be considered along the following lines: the occurrence and heterogeneity of collagen in the developing embryo and fetus, influence of the matrix on phenotypic expression and cellular behavior, and the role of collagen in organogenesis. It is also important to point out that it is almost impossible to separate the function of collagen from the role of the whole extracellular matrix.

Collagen can be detected in the very early stages of embryological development. The earliest stage of the mammalian embryo at which collagen has been detected is the 4-cell embryo (Sherman et al. 1980). In the mouse, at this stage of development types IV and V collagens have been found. Type III collagen also appears very early; it can be found in the blastocyst. The early embryo contains several basement membranes, including Reichert's membrane, and all of them are rich in type IV collagen (Minor et al. 1986). Interstitial collagens appear in association with mesodermal differentiation, except for the earlier appearance of type III collagen as mentioned above. They can be found in the mesenchymal structures of the head, heart, and somites and are present within the dermis. In the mouse embryo, types I and III collagens are detected as early as the 8th day of gestation. The appearance of type II collagen is associated with differentiation of the mesenchymal cells to chondroblasts.

Several aspects of the role of collagen in organogenesis have been reported. Mesenchymal cells proliferate and migrate to specific regions of the fetal body and form a miniature model of the future skeletal system. The process of chondrogenesis is related to the transformation of the cells to chondroblasts. As the cartilage matrix increases in amount, the chondroblasts are found to be surrounded by it, containing type II collagen. The vertebral cartilage is formed from the somite sclerotome. Chondrogenesis in the sclerotomal cells is induced by the notochord and neural tube. This process leads to formation of the cartilage model of the vertebral column. It has been found that chondrogenesis in vitro is augmented by collagens (especially type II), and removal of the matrix from the notochord results in a decrease in the chondrogenic stimulation (Kosher and Lash 1975; Kosher and Church 1975). Collagen has been found to stimulate synthesis of proteoglycans during chondrogenesis (Lash and Vasen 1978). In the limb skeleton, the mesenchymal cells transform into cartilage and finally into bone (endochondral osteogenesis). The early stages of this process are associated with type I collagen production by prechondrogenic mesenchyme. The appearance of cartilage is marked by type II collagen secretion, and at the onset of bone formation type I collagen is once more detected (Von der Mark et al. 1976).

Collagen contributes to muscle formation, and collagen-dependent myotube differentiation has been reviewed by Reddi (1984). Although the nervous system is low in collagen, the protein contributes to its development; myelination of the nerves involves contact with the extracellular matrix (Bunge and Bunge 1978). Basement membrane formation is involved in the development of the synaptic region, and migration of the neural crest cells is also regulated by collagens.

The morphogenesis of other organs, including the heart, lungs, mammary gland, kidney, and eye, is regulated by the matrix. The nature of these phenomena is only partially understood (for review see: Adamson 1982; Reddi 1984).

The majority of studies on the effect of collagen on cell behavior have been done in vitro. It is well known that many kinds of cells proliferate on collagen gel significantly better than on other substrates, although the nature of the interaction between them has not been elucidated. It is possible that attachment to collagen provides a signal for proliferation as well as facilitating cell-to-cell contact. The responsiveness of the cells to several growth factors is modified by the conditions of the culture, including the presence of the collagen substratum.

Collagenous structures have been suggested to contribute to cell motility. A characteristic feature of the embryological development is the translocation of cells to distant places. This process is very precise and very important for organogenesis. One of the postulated mechanisms regulating the cell migration is so-called contact guidance. This phenomenon has been described in cells migrating along oriented fibers (Wessells 1977; Elsdale and Bard 1972). This mechanism is proposed for migration of the neural crest cells (Adamson 1982). The additional (or alternative) mechanism includes periodical adhesions and withdrawals along the pathway of migration. Fibronectin is thought to contribute to the process.

The extracellular matrix is involved in the tissue interactions during morphogenesis. One of the best explored models of such phenomena are epithelial-mesenchymal interactions. It has been found that the embryonic epithelial and mesenchymal cells do not grow in separate cultures but are able to grow when some contact (e.g., through micropores of the filter) is allowed. Under natural conditions, the matrix including the basement membranes may perform the function of the filter and regulate cell-to-cell contact and transportation of humoral factors involved in the control of cell proliferation (Reddi 1984). There are numerous data on the amount and type of collagens produced during organogenesis. It is clear that the quality and quantity of collagens in many aspects is decisive in organ formation.

## 6.2 Collagen and Aging

Collagen appears a traditional subject in studies on aging. A number of reports have been published on changes in its content, structure, and metabolism in relation to age. Another substantial group of studies deals with the relation between age and the mechanical properties of tissues (Vogel 1973, 1974), which are related, at least partially, to the collagen content and structure. Most of these studies were done before collagen heterogeneity had been discovered. Since the primary structure of collagen has been elucidated and several types distinguished, more detailed investigations have been undertaken. Although new fields in collagen research are opening, aging is not so attractive a subject as it was earlier. Therefore, the relationship between age and some newly discovered features of collagen remains unknown.

An increase in the total collagen content in the skin, lungs, and liver of aging rats has been reported by Drożdż et al. (1979c). Its level rises from fetal to 12 months of age. In very old rats, a decrease compared with 1-year-old animals was found. A similar trend of changes has been reported in other studies (Svojtková et al. 1972; Mays et al. 1988).

Changes in collagen type pattern associated with age are related to the type I/III ratio. In the skin, the relative increase of this ratio has been noted (Deyl and Adam 1977). According to Mays et al. (1988), at 2 weeks of age type III collagen represents about one-third of types I and III collagens in the skin, lungs, or heart. After this age, the proportion of type III increases in the heart and the lungs and decreases in the skin. The biological role of these changes remains unclear.

In pioneer biochemical studies on collagen it has been found that the amount of collagen extractable from tissues with nondenaturating solvents decreases with age. On the basis of reduced solubility, the concept of increased cross-linking of collagen with age has been suggested (Verzár 1956). The force required to inhibit the thermal contraction of collagen in rat tail tendon was found to increase with age. Verzár (1964) interpreted this as indicative of an increase in collagen cross-linking. Several physical properties of collagenous structures have been shown to be related to aging, for example, thermal denaturation and fibril formation (Nagy et al. 1974; Brooks and Simons 1978; Harrison and Archer 1978; Niedermüller et al. 1977).

Studies on the solubility of collagen showed a lowering when examined at ages ranging from fetal to very old. The determination of cross-links has sometimes provided controversial results despite numerous studies (e.g., Deyl et al. 1971; Light and Bailey 1979; Fujimoto 1984). An age-related decrease in the allysine content has been found (Davison and Patel 1975). Changes in reducible and nonreducible cross-linking of skin and lung collagen have been reported by Reiser et al. (1987b). Biological aging has been shown to be related to several changes. In the first part of the life span, the virtual disappearance postnatally of dihydroxylysinonorleucine in skin, the rapid decrease in the difunctional cross-links of collagen during early growth, and the gradual rise of hydroxypyridinium and lysylpyridinium content have been shown. Changes in cross-linking in the second half of the life span have not been found to be so consistent. The nonreducible cross-links continue to increase in lung tissue. In contrast, these cross-links decrease in the skin during the second part of the life span. These findings are consistent with the results of Moriguchi and Fujimoto (1978) who investigated the age-related changes in the content of pyridoline in human and rat costal cartilage and the Achilles tendon. The content of pyridoline increased during growth and in rats continued to increase after the animal had reached maturity. In human tissue, it begins to decrease after about 30 years of age.

Change in another nonreducible cross-link, histidinohydroxylysinonorleucine, in human and bovine skin have been reported by Yamauchi et al. (1988). The amount of the cross-link increases rapidly from birth through maturation. Subsequently, a steady increase occurs with aging, approaching 1 mole/mole of collagen. The compound is related to collagen solubility, and higher amounts are found in the insoluble fraction.

The cross-linking theory is one of the popular current theories of aging (Yamauchi et al. 1988a). It seems to be clear that solubility and cross-linking alter with age. A progressive reduction of collagen solubility due to an increased amount of cross-links is accepted as the general trend of changes. On the other hand, several reservations must be made in the interpretation of biochemical measurements. Cross-linking of collagen is a complex process of formation of various covalent bonds. Some of them are probably insufficiently characterized. The distribution of the cross-links in different tissues represents a variety of phenomena. Therefore, it is difficult to conclude about cross-linking on the basis of the determination of a few compounds. This may explain controversies in the results obtained. Additional factors may also affect the solubility, for example, alterations in the matrix proteoglycans and, related to them, changes in hydration of the matrix. The role of cross-linking in collagen metabolism is also insufficiently explored. It has been shown that cross-linked collagen is significantly resistant to collagenase digestion. Thus, an increased formation of cross-links probably slows collagen turnover. The relation of this phenomenon to the general mechanism of aging, including the reduction of protein turnover, remains unknown. It is possible that collagen changes either contribute to the process of aging or are a result of the other processes related to aging.

## 6.3 Phylogenetic Aspects of Collagen

### 6.3.1 Collagen and Animal Phylogeny

Collagen is widely distributed in animals. A progressive amount of data has accumulated on the occurrence, amino acid composition, and some structural features of collagen in various species. Despite numerous studies, our understanding of the phylogenetic aspects is very poor. Investigations have been carried out on randomly selected species. Difficulty of access to biological material and application of different methods preclude the systemic analysis of collagens of the animal kingdom. It is noteworthy that the number of species is so great that it is difficult to plan systemic investigations which could be carried out in a short time. Thus, we will have to wait for further accumulation of enough details to form the general theory of collagen evolution.

Despite these limitations, the data available provide interesting conclusions. It seems that collagen is a constant component of multicellular animals. It is found in such primitive animals as sponges and in so highly developed ones as mammals. Collagen occurs in the intercellular space, and it can be hypothesized that the existence of collagen (or of a mixture of matrix components) between cells is a basic requirement for the coordination of cells within the multicellular body. If this is true, the existence of all multicellular animals depends upon the occurrence of collagen. It is difficult to estimate its biological functions. Beside the mentioned coordinatory function and structural function (the latter is per se a coordinatory function for the multicellular organism), other functions are possible. The role of collagen cannot be considered without the analysis of other components of the intercellular substance. Our knowledge of the amorphous matrix in various animals is still very scanty.

Several aspects are open for further studies, including the relationship between collagen genes, its molecular and supramolecular structure, and the phylogenetic relationship of species; that between the structure of collagens and the environmental conditions under which the animals are living; collagen changes in the development of animals; and collagen metabolism and phylogeny.

Biochemical methods applied to the animal world offer other opportunities than the methods of traditional biology. Investigations of the structure-function relationship of collagen in lower animals may be a key for understanding some mechanisms of the complex system of the matrix of the human body under normal and pathological conditions. Therefore, studies on such distant species as sponges may be useful for human pathology.

The phylogenetic aspects of collagens have been comprehensively reviewed (Mathews 1965; Adams 1978; Garrone 1978; Bairati and Garrone 1985; Tanzer and Kimura 1988; Van Ness et al. 1988). The following sections provide a short outline of collagens which are found in the major animal taxonomic units.

### 6.3.2 Plants and Unicellular Organisms

Hydroxyproline has been found in plants, but there is no evidence for a helical structure. Hydroxyproline-containing proteins have been isolated from different plants (Van Holst et al. 1980, 1986; Van Holst and Klis 1981; Kubota et al. 1983; Sauer and Robinson 1985; Wienecke et al. 1982).

The occurrence of collagen in protozoa has not been shown, although some investigators suggested that collagen-like proteins may be synthesized by the mastigophoran *Hymenomones* (Baccetti 1985).

### 6.3.3 Porifera

Porifera or sponges are aquatic animals characterized by a unique organization. They are relatively primitive, multicellular animals inhabiting all seas; only one family is found in fresh water (Garrone 1985).

Sponges are covered by an outer dermal membrane perforated with pores that allow water to enter. This structure forms a filter-feeding system. The water moves through channels in the body, and the sponges feed upon the minute organisms and particles carried in with the current. The water movement is provided by the coordinated whipping action of flagella on the choanocytes. Between the dermal membrane and the choanocytes, several cells are present which are responsible for the production of collagen-like materials.

The intercellular material of sponges is similar to that of the vertebrate extracellular matrix as seen under the light microscope (Garrone 1978). Collagen fibrils are made up of the substance named "spongin A," and they are a constant feature of all sponges. The fibrils are long, slender cylinders of about 20 nm diameter. The banding pattern is slightly visible. The diameter of the fibrils is constant within the given species. The fibrils are either grouped into bundles or randomly dispersed (Garrone 1985). Sponge collagen is highly insoluble. It contains 3-hydroxyproline, 4-hydroxyproline, and hydroxylysine in amounts of 27, 71, and 16 residues per 1000, respectively (values for *Chondrosia*) (Garrone et al. 1975). The collagen is highly hydroxylated and glycosylated. It is interesting that the fibrils also contain polysaccharides.

Spongin B is the another collagenous material of sponges. It forms fine fibrils (Vacelet and Garrone 1985; Garrone 1985). Its chemical composition indicates its collagenous structure. The details of spongins concerning chain composition and supramolecular structure as well as their biological functions are unknown.

### 6.3.4 Coelenterata

Coelenterata or Cnidaria is a phylum of primitive animals including such forms as medusae (jellyfish), sea anemones, corals, marine hydroids, freshwater hydras, and sea fans. The phylum may be defined as the group of radiate animals that usually bear tentacles and possess intrinsic nematocysts. The basic body forms among coelenterates are the polyp and the medusa. The polyp is an upright sessile form, usually of columnar shape. It is attached at its base to the substratum, and the oral end is free and surrounded by tentacles. The medusa is the free-swimming coelenterate popularly known as the jellyfish, with an umbrella-shaped body, called a bell.

The body is composed of two layers, ectoderm and endoderm, separated by the mesogloea. The mesogloea is probably the earliest phylogenetic structure that is an analogue of connective tissue. The presence of collagenous structures within the mesogloea has been shown (Franc 1985).

There are three classes of coelenterates: Hydrozoa (e.g., marine hydroids, freshwater hydras), Anthozoa (e.g., sea anemones, true corals, sea fans, sea feathers, soft corals), and Scyphozoa (e.g., jellyfish).

A large amount of hydroxyproline and a high content of glycosylgalactosylhydroxylysine have been found in the mesogloea of the hydra, *Pelmatodydra pseudoligactis* (Barzansky et al. 1975). Collagenlike protein, composed of three identical $\alpha$-chains, has been demonstrated in sea anemones, *Actinia equina* and *Metridium dianthus* (Nordwig et al. 1973). This protein is very similar to vertebrate collagens as seen under the electron microscope (Tanzer and Kimura 1988).

In Scyphozoa, it has been shown that the mesogloea is rich in hydroxylysine and its glycosides (Kimura et al. 1983). Fibrils without cross-striations have been observed. The protein is composed of three different $\alpha$-chains forming a helical structure.

### 6.3.5 Platyhelmintha

Platyhelmintha is a phylum of soft-bodied animals known as flatworms. The phylum includes both free-living and parasite members. They are bilaterally symmetrical and usually flattened animals with unsegmented bodies. Parenchyma, the tissue that occurs among their internal organs, is a form of connective tissue. The presence of typical collagen fibrils has been reported in the parenchyma (Nordwig and Hayduk 1969). The molecular weight of this collagen is great, about 500000 Da. The supramolecular structure is not known, and the presence of dimers has been suggested. A high content of carbohydrates has been reported (Baccetti 1985).

### 6.3.6 Aschelminthes

Aschelminthes, known as roundworms, include free-living animals and plant and animal parasites. They have an unsegmented, often cylindrical body with a pseudocoele, i.e., a body cavity without an epithelial wall. The body is covered by resilient cuticula. Among aschelminthes, the class Nematoda is characterized in respect to the occurrence of collagen (Ouazana 1985).

The cuticle collagen is the most abundant form in nematodes. The cuticle consists of layers, cortical and basal, separated by a fluid-filled median space with columnal structures (Ouzana 1985). The cuticle collagen is synthesized by the subcuticular tissue. Cuticle collagen contains glycine (27% – 30%) and proline and hydroxyproline (22% – 34%). Half-cystine accounts for 3%. It is interesting that cuticle collagen almost lacks hydroxylysine. Its molecular weight is about 900000 Da. The collagen of *Ascaris lumbricoides* and *A. suum* is made up of three units (Adams 1978). *Ascaris* is the earliest phylogenetic form containing basement membranes (Hung et al. 1981). On the other hand, in more primitive animals, some similarities in collagen structure to proteins of the basement membrane have been suggested. The collagen of *Caenorhabditis elegans* is more complex than that of *Ascaris* (Ouzana 1985).

### 6.3.7 Annelida

Annelides are the segmented wormlike animals and are named for the transverse rings or annulations of the skin. Their body has a cavity (coelom) and movable bristles (sectae) and is segmented. The body structure can be described as a tube within a tube.

Collagen in annelides is found in the cuticle and around the internal organs. Annelidal cuticle collagen is characterized by very long molecules. Segment long spacing crystallites of *Nereis virens* cuticle collagen is 2400 nm long (Murray and Tanzer 1983). This collagen contains a high ratio of hydroxyproline to proline. Aldehyde-derived cross-links are not detected in this collagen (Murray and Tanzer 1985). The interstitial collagen of *Lumbrians* has been described by Vitellaro-Zuccarello et al. (1985).

### 6.3.8 Arthropoda

Arthropoda is the largest phylum of the animal kingdom. The arthropods form about three-quarters of the known species of animals. All of them possess a bilaterally symmetrical segmented body. Each segment may bear a pair of limbs. The most characteristic feature of arthropods is a covering that, when thickened, forms an articulated armor (exoskeleton) over the body and limbs. The phylum includes insects, spiders, mites, scorpions, crustaceans, centipedes, millipedes, and other less familar groups.

The chitinous exoskeleton is made up of collagen. Collagen is found around the internal organs and under the cuticles. In arthropods, specialization of the connective tissue is found. For example, cartilage-like structures and tendons have been reported (Bairati 1972; Brown 1975; Person and Philpott 1969). The collagen contains more than one distinct chain; the heterotrimer $(\alpha 1)_2\alpha 2$ has been reported (Kimura and Matsuura 1974). A high content of hydroxylysine that is glycosylated has been found. A clear heterogeneity of the body collagens is found in arthropods. Types I and IV (the basement membrane form) collagens have been reported (Tanzer and Kimura 1988).

### 6.3.9 Mollusca

Mollusca is one of the largest phylum of animals, and species of molluscs vary in body shape and size. Their body is usually soft and enclosed in a protective shell. All molluscs have a mantle, a thin capa or covering that hangs over the dorsal surface of the body. The mantle secretes the shell. In most molluscs the mantle cavity, the space between the mantle and the visceral mass and foot, contains the gills. The foot is typically a broad, flat, muscular organ, developed for creeping or plowing.

The shell is of epithelial derivation and does not contain collagen. Connective tissue is found in several forms. It occurs around the internal organs, in the muscle of the foot, and in cartilaginous structures. The cartilaginous structures enclose the brain. The presence of type I collagen in the mantle skin of the octopus, *Octopus vulgaris*, has been reported (Kimura et al. 1981). Similar collagen has been found in the arm muscle and the sucker. Polymeric collagen isolated from the squid, *Loligo peallii* has been characterized (Hunt et al. 1970; Gosline and Shadwick 1983a, 1983b). The fibrillar structure of the collagens has been documented in numerous studies. The presence of basement membranes containing collagenous structures has also been reported (Baccetti 1985).

An interesting form of so-called secreted collagen is a protein found in the byssus. The byssus is an extracorporeal device by which bivalves attach themselves to the littoral substratum. The presence of collagen-like structures in the byssus have been suggested (for review see Vitellaro-Zuccorello et al. 1985).

### 6.3.10 Echinodermata

Echinodermata are marine invertebrates such as sea stars (starfish), sea urchins, and many other prickly animals. The fundamental characteristics of echinoderms include five-rayed symmetry and an internal calcareous skeleton.

Collagen fibers from echinoderms exhibit typical 64–68 nm periodicity (Matsumura 1974). According to Matsumura et al. (1979) the amino acid composition of collagen isolated from 15 species of echinoderms is similar to that of mammalian type I collagen. This observation has been confirmed by Bailey et al. (1982). Studies on chain composition are few, and a homotrimer form has been suggested (Bailey 1985).

A unique feature of echinodermal collagen is its rapid changability, known as a "catch mechanism". This mechanism is responsible for the quick stiffening of the sea-cucumber in response to light mechanical stimuli. This feature is related to collagen structure and is not fully understood. The hypothetical mechanism of the phenomenon is reviewed by Bailey (1985).

### 6.3.11 Chordata

Chordata are animals with an elongate rod, or notochord, which stiffens the body. They have a single, hollow nerve cord located on the dorsal side, blood contained within vessels and propelled by a heart located on the ventral side, and a tail extending beyond the anus. Chordates are divided into cephalochordata, tunicata, and vertebrata.

Only single reports deal with collagen in cephalochordates and tunicates. The presence of collagen has been shown in *Amphioxus* and the tunicate of *Ascidiella* (Hall and Saxl 1961; Pikkarainen et al. 1968; Kimura et al. 1972, cited in Baccetti 1985).

Vertebrates are the animals having vertebral columns or backbones. They are unique in possessing an internal skeleton. The vertebral column and limb skeleton provide support for the body as a whole. Connective tissue contributes to the formation of the body skeleton and local support of several organs. Members include the fishes, amphibians, reptiles, birds, and mammals.

Vertebrate collagen has a periodic banding of 60–70 nm. The chain composition remains a subject of controversy (Kühn 1982b). It has been suggested that collagen from the lower vertebrates is composed of a single chain, but this has not been confirmed in a subsequent study (Pikkarainen 1968; Sato et al. 1989). The relationship between the shrinkage temperature and environmental conditions is an interesting evolutionary aspect of the adaptation of collagen in various species (Kimura 1985; Tanzer and Kimura 1988). Despite numerous studies, our knowledge of vertebrate collagens is limited to a few species only, and further characterization of the collagens of lower vertebrates is needed.

# 7 Hormonal Regulation of Collagen Metabolism

## 7.1 Introduction

Despite numerous research efforts, the relationship between the endocrine system and connective tissue remains unclear (Asboe-Hansen 1966; Borel et al. 1984; Kucharz 1988b). Most of the studies have focused on the influence of a single hormone on selected indices of connective tissue metabolism.

Endocrine gland secretion together with many humoral factors secreted by single cells make up one of the main regulatory systems which enable the mutual cooperation of various organs during the continuous adaptation of the body to environmental changes. The relationship between the nervous and endocrine systems is well recognized: The former supervises and cooperates with endocrine secretion in many ways. The connective tissue, analysed as an assemblage of various morphological structures, can be understood as another regulatory and coordinatory system of the body. The reactivity of connective tissue is relatively slower than that of the nervous or endocrine systems, but the tissue takes part in inflammatory and reparatory processes, and the maintenance of integrity of nearly all organs. Moreover, it assists in the transportation of nutrients and other substances between blood vessels and parenchymal cells and constitutes the majority of skeletal components. In the light of this hypothesis, the functional relationships between connective tissue and the endocrine system form a section of the physiological mechanism which assures homeostasis of the complex structure of the human body. The effects of hormones on collagen metabolism are only some of the interactions.

The present chapter reviews the effect of certain hormones on collagen metabolism. The influence of parathormone and calcitonin are described in Chap. 11, and the action of sex hormones is reviewed in Chap. 18.

## 7.2 Thyroxine, Triiodothyronine, and Thyrotropic Hormone

Thyroxine and triiodothyronine, the hormones secreted by the thyroid gland, are dipeptides containing 4 and 3 atoms of iodine, respectively, in each molecule. Both are normally stored in the colloid vesicles of the gland as thyroglobulin, and their release is controlled by thyrotropic hormone (thyrotropin). The latter is secreted by the pituitary gland, which in turn is controlled by the hypothalamus-derived thyrotropin-releasing hormone. Thyroid hormones are involved in a negative feedback mechanism which suppresses the pituitary secretion of thyrotropin.

Thyroid hormones, despite their relatively small size, act directly on most tissues of the body to increase metabolism, and triiodothyronine acts more rapidly and strongly than thyroxine. Thyroid hormone has been shown both in vivo and in vitro to be an important regulator of connective tissue metabolism. It is unclear to what extent thyroxine and/or triiodothyronine directly affect collagen structure and metabolism and which alterations are caused by unspecific changes in the cellular metabolism of the connective tissue.

From early clinical and experimental studies it was known that thyroid dysfunctions altered urinary hydroxyproline output. Numerous reports have shown that hydroxyproline excretion increased in hyperthyroid patients (Araszkiewicz and Styszewska 1972; Benoit et al. 1963; Counte et al. 1970; Eisner and Kondraciuk 1967; Iwańska et al. 1975; Keiser and Sjoerdsma 1961; Kivirikko et al. 1964, 1965; Kocher et al. 1965; Kruze et al. 1973; Laitinen et al. 1966a; Radom et al. 1971; Rubegni et al. 1964; Uitto et al. 1968; Zatońska and Widomska-Czekajska 1970). Similar results were found in animals treated with thyroxine (Drożdż et al. 1979a; Kivirikko et al. 1963, 1964, 1967; Matsumura et al. 1967). Increased urinary output of total hydroxylysyl glycosides, glucosylgalactosyl-hydroxylysine, and galactosyl-hydroxylysine was demonstrated in patients with thyreotoxicosis, but the mutual ratio of various hydroxylysyl glycosides was not significantly different from the values found in normal urine (Askenasi and Demeester-Mirkine 1975; Segrest and Cunningham 1970). An increase in the level of free and peptide-bound hydroxyproline in urine and free hydroxyproline in serum were described (Eisner and Kondraciuk 1967; Kivirikko et al. 1964, 1965; Laitinen et al. 1966a). On the contrary, hypothyroidism leads to a decreased urinary output and serum level of hydroxyproline. The above changes are normalized after treatment of the hyper- or hypothyroidism, and some investigators have found a correlation between hydroxyproline level and other indices of thyroid function (Iwańska et al. 1975; Uitto et al. 1968). The determination of urinary hydroxyproline excretion has been suggested as a useful supplementary laboratory test for the diagnosis of disturbances of thyroid function and control of the effectiveness of the treatment. The reliability of this test is high for hyperthyroidism in adult patients and hypothyroidism in children (Uitto et al. 1968).

The effect of thyroid hormone on the collagen content in tissues is a subject of controversy. The contradictory results are, at least partially, caused by the application of various models of connective tissue activation to study the influence of thyroid hormone on collagen. Thyroxine and triiodothyronine seem to affect both collagen biosynthesis and degradation. Described changes in the collagen content in tissues reflect a cumulative effect of thyroid hormone and other connective-tissue-activating factors on collagen turnover.

Studies with radiolabeled proline indicate that collagen degradation is increased in hyperthyroidism and decreased in hypothyroidism (Kivirikko et al.

1967; Laitinen et al. 1966a). On the other hand, it has been reported that thyroxine increased the incorporation of proline into collagen (Mikkonen et al. 1966) and the maturation of collagen fibers (Drożdż et al. 1979a). Hyperthyroidism produced by thyroxine administration to rats was described as a cause of an increase in soluble collagen content in the skin and liver and a decrease of insoluble collagen in the skin and bone samples. In methylthiouracil-treated animals a decrease in collagen content in the skin and liver and an increased collagen content in bone were observed (Drożdż et al. 1979a; Freihoffer and Wellband 1963).

The direct effect on collagen maturation and catabolism may relate to the effect of thyroid hormone on wound healing. It has been recognized that the addition of exogenous hormone to animals resulted in a decrease in wound's tensile strength (Freihoffer and Wellband 1963; Kivirikko et al. 1967). Collagen deposition due to activation of fibroblast proliferation is the key to wound repair. These are direct effects of thyroid hormone because thyrotropin has no effect on wound repair of thyroidectomized animals (Kohn 1987). Contrary to the physiological process of wound healing, Kucharz et al. (1986b) described a decrease in collagen content in the liver of rats with carbon tetrachloride-induced hepatic fibrosis treated with thyroid extract and an aggravation of the fibrosis in hypothyroid rats. The reported changes in collagen content are presumably related to altered collagen degradation in the fibrotic liver (Kucharz 1987b).

Thyroid hormone is required for the development of epinephrine-induced sclerosis of the arterial system (Kohn 1987). The changes in catecholamine-induced sclerosis do not occur in hypothyroid animals. Epinephrine is presumed to act via activation of adenylate cyclase and the accumulation of cyclic adenosine monophosphate. Thyroid hormone inhibits phosphodiesterase activity of the smooth muscle cells and potentiates the catecholamine effects. Hypothyroidism prevents the accumulation of cyclic adenosine monophosphate and thereby inhibits the process of sclerosis.

Clinical observations indicate that connective tissue changes in hypo- and hyperthyroidism have various organ localizations. Hypertrophic osteoarthropathy occurs sometimes in hyperthyroidism, and its most common form is the clubbing of the phalanges termed thyroid acropachy. The histologic picture includes increased fibroblast proliferation and increased production and deposition of collagen. The mechanism of the disturbances is not understood, and direct hormonal or immunological factors accompanying some thyroid diseases are suggested as a cause. Skeletal changes in patients with thyroid dysfunction are associated primarily with disturbed calcium metabolism, but secondarily the connective tissue is involved in their pathomechanism (Kohn 1987).

In humans, localized pretibial myxedema of hyperthyroidism is related to collagen changes. Collagen fibrils swell and undergo degenerative changes. The role of excess thyroid hormones in the development of these phenomena is unknown, and it has been hypothesized that specific antibodies are a causative factor for the myxedema (Kohn 1987).

Cellular components of the connective tissue are thought to be the target cells of thyroid hormones, and the existence of nuclear receptors for 5,5,3′-triiodothyronine has been demonstrated in human fibroblasts (DeGroot et al. 1978; Refetoff et al. 1972; Shambaugh 1986). De Rycker et al. (1984) reported a decrease of newly synthesized collagen in cultured skin fibroblasts after the addition of $10^{-7}-10^{-9}$ $M$ of triiodothyronine. The depression was dose-related, but no changes in type of collagen secreted were found.

Classic studies on the effect of thyroid hormone on the degradation of collagenous structures in tadpoles (*Rana catesbiana, R. pipiens*) indicate hormonal regulation of collagenase activity (Etkin 1935, cited in Etkin 1968). Spontaneous amphibian metamorphosis is accompanied by extensive alterations, including significant collagen degradation. This process has been proposed to be regulated by thyroxine and prolactin. Thyroxine induces tailfin resorption and stimulates increases in degrading enzyme activity in the resorbing tissue. The amphibian larval tadpole has been adapted as a model for the assessment of thyroid hormone regulation of collagenase, a specific neutral proteinase which cleaves collagen. Thyroid hormones do not initiate this event but, rather, regulate the rate of degradation. Thus, tadpole tails held at 37 °C undergo resorption in the absence of thyroxine, while at 25 °C thyroxine is necessary for resorption. In resorption of the tadpole tail the collagenase level is increased by thyroid hormones. The activity of hyaluronidase, which may be involved in the degradation of the interfibrillar ground, and of other lysosomal enzymes is also increased, and it has been suggested that thyroid hormones might initiate a series of events called a cascade of proteolysis which lead to degradation of collagen and noncollagenous proteins (Davis et al. 1975). Thyroxine added to a culture of tailfin tadpole tissue stimulates progressive degradation, and collagenase activity appears in the medium. Thyroid hormone control of the synthesis of collagen during metamorphosis is organ-specific. It has been reported that thyroxine increased the biosynthesis of collagen in the thigh bone of the frog (*R. catesbiana*), and thyroid hormone seems to reduce the prolactin sensitivity of connective tissue during metamorphosis. These effects are opposite to those observed in tailfin tissue (Yamagushi and Yasumasu 1977a, b). The role of thyroid hormone in the regulation of collagenolytic enzymes in mammals has not been investigated, and only indirect hints have been cited.

Human thyrotropin (thyroid-stimulating hormone) is a glycoprotein that contains 211 amino acid residues, plus hexoses, hexosamines, and sialic acid. It is made up of two subunits, designated alpha and beta. The hormone is produced by the anterior part of the pituitary gland. Its basic function is the stimulation of endocrine activity of the thyroid gland.

The direct effect of thyrotropin upon connective tissue metabolism has been suggested on the basis of investigations of experimental exophthalmos. Thyrotropin injected into fish or guinea pigs induced exophthalmos and injected into mice or rats induced analogous biochemical and histological changes in the retrobulbar space but without proptosis (Smelser 1937; Winard and Kohn 1970). This is a direct effect of thyrotropin because thyrectomized animals respond similarly. The histologic picture is characterized by an increase of connective tissue, with proliferation of all its elements and cellular infiltrations. The connective tissue changes in exophthalmos include accumulation of proteoglycans and collagen deposition with subsequent fatty degeneration and atrophy of cellular elements. Fibrosis due to overproduction of collagen has been observed also in the extraocular muscles.

It was found then that thyroid-stimulating activity of a thyrotropin preparation could be separated from the exophthalmogenic activity, and molecule derivatives containing an effectively intact $\beta$-chain and a carboxyl terminated fragment of the $\alpha$-chain of thyrotropin, which appear to be an artifact of the hormone preparation, are responsible for the exophthalmogenic activity (Bolonkin et al. 1975). Nevertheless, this observation emphasizes the fact that connective tissue changes are involved in thyroid ophthalmopathy, and thyrotropin receptors may be related to the above-described phenomena. The exophthalmogenic factor associated with some thyroid diseases has been documented in further studies (Kohn and Winand 1975). It has been reported that the sera of patients with Graves' disease contain gamma-globulin, different from the long-acting thyroid stimulator, which can induce exophthalmos in some experimental animals. This gamma-globulin was found to increase the binding of thyrotropin to retro-orbital tissue but not to thyroid membranes, and an additive effect of thyrotropin and specific gamma-globulin on the induction of adenylate cyclase in retro-orbital tissue was reported. Immunoglobulin from the sera of patients with Graves' ophthalmopathy stimulated collagen biosynthesis in cultured human fibroblasts. This activity correlated with the severity of exophthalmos and disappeared in remission. The activity is presumably directed against the thyrotropin receptor, and some monoclonal antibodies against certain domains of the thyrotropin receptor (different from the receptor in the thyroid membrane) also induced collagen synthesis in human fibroblasts. The fibroblast collagen production assay seems to be a valid means of measuring exophthalmogenic autoantibodies. The pathophysiological role of the increased collagen biosynthesis in the retro-orbital tissue of patients with exophthalmos is unclear as collagen is a minor component of its connective tissue.

Experimental studies indicate that thyroid-induced accumulation of the connective tissue components is not limited to the retro-orbital tissue (Kohn 1987). Analogous changes have been demonstrated in the perirenal adipose tissue, small intestine, ovary, inner ear, lacrimal gland, and harderian gland (accessory lacrimal gland at the inner corner of the eye) of guinea pigs (Dandona and El Kabir 1970).

The mechanisms of the direct activity of thyrotropin on connective tissue remains unclear. Two groups of observations have been described: immunological, i.e., the binding of an antibody to a specific receptor thus mimicking the action of the hormone, and metabolic, the binding of partially degraded hormone to the retro-orbital tissue. It is possible that both of the phenomena occur in patients with exophthalmos and are additive.

The collagen content in the thyroid gland has not been investigated in detail; however, histological studies indicate that fibrosis of the gland occurs under certain pathological conditions. The role of collagen in the embryological development of the thyroid gland has also been postulated. The histogenesis of thyroid follicles in the chick embryo begins with penetration by cells of the mesenchymal capsule into the solid epithelial primordium. Before the mesenchymal cells invade, channels which contain collagen appear between the epithelial cells towards the center of the primordium (Bradway 1929; Shain et al. 1972). This resembles basal laminae collagen and is produced by the epithelial cell layers (Hilfer 1972). Inhibitors of collagen biosynthesis (L-azetidine-2-carboxylic acid or $\alpha,\alpha'$-dipyridyl) prevent the formation of fibrils in cultured chick primordia or whole thyroids. The drugs inhibit penetration of the mesenchymal cells. It is suggested that collagen, in some way, plays a role in mesenchymal penetration of the epithelial primordium, and the epithelium (producer of collagen) is responsible for the pattern of lobulation within the development of the thyroid gland (Hilfer and Pakstis 1977).

The regulatory function of thyroid hormones on connective tissue structures has been evidenced in many investigations, both in vivo and in vitro (Fig. 7.1). The resulting controversies suggest that the mechanisms are not unique, and reported alterations are produced as the accumulation of effects upon the biosynthesis and degradation of collagen as well as on other components of the tissue. The reactivity of the connective tissue of various organs suggests different susceptibilities of the collagenous structures to thyroid hormone and different results of imbalance between thyroxin-induced synthesis and degrada-

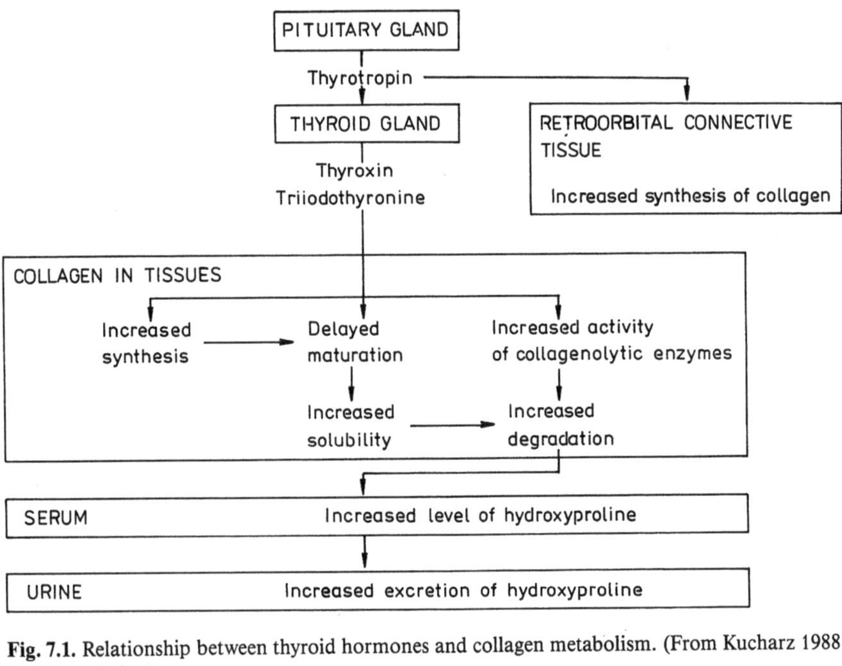


**Fig. 7.1.** Relationship between thyroid hormones and collagen metabolism. (From Kucharz 1988b with permission)

tion of collagen. The sum of the specific and direct influences of thyroid hormones with indirect metabolic alterations is also likely to occur under conditions of hypo- and hyperthyroidism. The indirect effects include such mechanisms associated with thyroid dysfunctions as autoimmune phenomena.

The present knowledge of thyroid hormone control of collagen metabolism is insufficient for evaluation of the molecular bases of these processes, but it is assumed that changes in the connective tissue may be responsible for some clinical symptoms occurring in patients with disorders of the thyroid gland.

## 7.3 Glucocorticoids

Glucocorticoids are a group of steroid hormones produced by the adrenal cortex. Natural forms (cortisol, corticosterol) and their potent synthetic analogues regulate certain biological processes, including protein, carbohydrate, and lipid metabolism, immune and inflammatory responses, hematopoiesis, formation and degradation of bone, and state of the mucous membrane of the digestive tract.

Glucocorticoids have a widespread clinical use and are highly effective in the therapy of various disorders or symptoms. Contrary to other hormones, they produce effects in almost all tissues and cells of the body. The long-term use of glucocorticoids is associated with a number of side-effects, and many of them are thought to be caused by alterations of the connective tissue (Baxter and Forsham 1972; Fukuhara and Tsurufuji 1969; Leung and Munck 1975). These untoward effects include skin and subcutaneous atrophy resulting in thin skin, increased bruising, striae and telangiectases, and poor healing (Lenco et al. 1975; MacDonald 1971; Reiser and Last 1983; Gallant and Kenny 1986). Numerous alterations in skeletal structures have been reported; osteoporosis, avascular necrosis, inhibition of growth, and fragility of tendons are the most frequent (Loeb 1976; Wells and Kendell 1940). Commonly described changes in the location of fat deposits can also be included.

Clinical reports about alterations in connective tissue structures in patients receiving glucocorticoids had suggested an effect of these hormones on collagen metabolism. In 1950 Taubenhaus and Amromin were the first to report an inhibitory effect of cortisone upon granulation tissue formation. In early investigations in this field, systemic indices reflecting collagen metabolism were applied. It was found that cortisone decreased urinary excretion of hydroxyproline, an imino acid almost specific for collagen (Kibrick and Singh 1974; Liakakos et al. 1975; Smith and Allison 1965). Animal studies showed that injection of cortisone decreased incorporation of proline into total skin proteins and decreased hydroxyproline formation from proline (Kivirikko and Laitinen 1965), as well as decreased total collagen content in normal and inflamed tissues (Bavetta et al. 1962; Cutroneo et al. 1986; Drożdż et al. 1984b; Hollenberg 1977; Saarni 1977; Siuko et al. 1959; Sultan et al. 1986). On the basis of these findings, it was concluded that cortisone decreased collagen synthesis. These initial results were confirmed, and depression of hydroxyproline formation by various natural and synthetic glucocorticoids was reported to be dose-dependent in human skin explants (Uitto et al. 1972). This decrease was observed with various experimental designs (e.g., chick embryos, granulation tissue) but was usually associated with diminished biosynthesis of non-collagenous proteins and thus frequently was interpreted as a result of a non-specific antianabolic activity of the hormones (Kowalewski 1969; Uitto and Mustakallio 1971). In 1971 Nakagawa et al. were the first to note the selective decrease in collagen synthesis in granuloma after a single injection of beta-methasone. This selectivity was confirmed by the others using different experimental models (Cutroneo et al. 1971; Newman and Cutroneo 1978; Weiss et al. 1988).

Several studies have shown that glucocorticoids decrease incorporation of proline into proteins (Kivirikko and Laitinen 1965; Uitto et al. 1972). Although proline is not only incorporated into collagen, these observations suggest that glucocorticoids affect the early stages of collagen synthesis. Further studies on their effect on the transcription of collagen genes confirmed these hypotheses (Cutroneo and Counts 1975).

The general mechanism postulated for the activity of glucocorticoids assumed that transcription of some parts of the genome is a target site of their action. Glucocorticoids are believed to enter the cell and associate intracellularly with the receptor. The hormone-receptor complex is transported into the nucleus, and there the glucocorticoids exhibit their activity (Oikarinen and Uitto 1987).

Most studies have concentrated on the mRNA of type I procollagen chains. A decrease in amount of these mRNA induced by glucocorticoids was observed with various experimental models (Oikarinen et al. 1983a; Oikarinen and Ryhanen 1981; Rokowski et al. 1981; Sterling et al. 1983a, b; Weiner et al. 1987) and was similar for mRNA coding pro$\alpha$1(I) and pro$\alpha$2(I) chains of procollagen. According to Oikarinen et al. (1986) there are several possible explanations for a decreased procollagen mRNA level: (a) inhibition of transcription; (b) inhibition of the processing of mRNA to functional mRNA; and (c) increased degradation of mRNA. Hämäläinen et al. (1985) suggested that the last possibility arose in human fibroblasts cultured in the presence of glucocorticoids. They found an increase in the latent RNAse, the enzyme capable of degrading mRNA. There is no explanation as to why degradation is specific for procollagen mRNA. Single reports were published on the regulation of the synthesis of types II and III procollagens. The decrease induced by glucocorticoids was also seen (Shull and Cutroneo 1983). There are few data about glucocorticoid effects on the basement membrane collagen. No changes in the ultrastructure of the basement membrane was found in skin after topical treatment with glucocorticoids (Oikarinen et al. 1983b), but in some cell cultures an increase in type IV collagen production was observed (Salamon et al. 1981). This elevation may be caused by a decrease in activity of the specific collagenase for type IV (Oikarinen et al. 1986).

Hydroxylation of proline residues is one mechanism responsible for the regulation of collagen biosynthesis. Most investigations concerning the role of glucocorticoids in the regulation of hydroxylase activity were carried out before the discovery of prolyl 4-hydroxylase, and altogether the enzyme activity was called "prolyl hydroxylase." In 1971, Cutroneo et al. described a decrease in prolyl hydroxylase in rats receiving triamcinolone or hydrocortisone. In their next paper, they reported a decrease in prolyl hydroxylase activity in various organs (liver, lung, heart, aorta, skin) of newborn and adult rats receiving multiple injections of triamcinolone (Cutraneo et al. 1975). It was found that 12 h after a single injection, the enzyme activity was only slightly decreased, but 24 h after three injections of the same dose of the hormone, the enzyme activity was decreased by 35% (Cutraneo et al. 1971). The decrease was dose- and time-dependent, and it was proposed that the activity of glucocorticoids is parallel to their antiinflammatory potency (Cutraneo et al. 1971). The effect was reversible. The decrease in the enzyme activity is probably caused by decreased synthesis of the enzymatic protein as glucocorticoids diminished the amount of enzyme proteins as measured by immunoassay (Cutraneo et al. 1975). The decrease in prolyl hydroxylase was confirmed in other reports, including studies on human skin (Oikarinen 1977; Blumenkrantz and Asboe-Hansen 1976; Oikarinen and Hannuksela 1980; Risteli 1977b; Trupin et al. 1983).

A decrease in other enzymes responsible for intracellular posttranslational modifications of collagen has also been found. Diminished activity of lysyl hydroxylase and glucotransferases was observed in the skin and liver of rats and chick tendon cells (Newman and Cutraneo 1978; Oikarinen 1977; Risteli 1977b). This decrease is relatively specific as it was not accompanied by a decrease in other enzymes unrelated to collagen synthesis. Decreased hydroxylation of proline and lysine is not a cause of the intracellular accumulation of underhydroxylated collagen, and collagen synthesized in diminished amounts by glucocorticoid-treated cells has the same levels of hydroxylation of prolyl residues as in controls which do not receive this treatment (Counts et al. 1979).

The effect of glucocorticoids on the transport of procollagen into the extracellular space and on the conversion of procollagen to collagen remains unknown. The subsequent extracellular processing of collagen has been reported to be glucocorticoid-sensitive. Its solubility, which indirectly reflects so-called maturation due to cross-link formation, is altered by glucocorticoids. The obtained results are contradictory. It has been reported that glucocorticoids increase the tensile strength of skin (Vogel 1970), a feature associated with an increased content of insoluble collagen (Kowalewski 1969; Kühn et al. 1964). This phenomenon was not observed in cutaneous wound healing, and cortisone treatment reduced wound tensile strength (Sandberg 1964b). On the other hand, an increase in soluble collagen and a decrease in the insoluble form were found in the liver of rats receiving glucocorticoids (Kucharz 1981, 1985). This discrepancy can be explained by differences in collagen turnover in various organs. In the skin and tendons, turnover is slow, and inhibition of synthesis by glucocorticoids is compensated for by a slower degradation of insoluble collagen. This leads to a relative increase in the insoluble collagen content. In the liver, collagen turnover is more rapid; thus, the inhibitory effect on extracellular collagen maturation overwhelms the inhibition of collagen synthesis. This leads to the opposite ratio of soluble/insoluble forms. The duration of treatment with glucocorticoids may also affect the final ratio of collagen fractions with varying solubility (Günther and Carsten 1964; Hirayama et al. 1971). Direct evidence which suggests the inhibition of cross-link formation in collagen maturation was provided in a report showing that glucocorticoids decrease the activity of lysyl oxidase, an enzyme participating in cross-link formation (Benson and LuVulle 1981). Additionally, Dombi (1986) found that dexamethasone inhibited the in vitro fibrillogenesis of collagen. This finding suggests that glucocorticoids act upon the nonenzymatic stage of aggregation of tropocollagen. Regulation of this process in vivo remains unknown.

Formation of collagenous structures is a complex process of interaction of collagen with proteoglycans and structural glycoproteins. The effect of glucocorticosteroids on these phenomena was not investigated. It has been reported that they alter glycosaminoglycan and proteoglycan biosynthesis, and therefore one can assume that they may influence the mutual interaction of amorphous ground substances with collagen structures (Drożdż et al. 1984b; Lorenzen 1969; Oikarinen et al. 1986; Oikarinen and Uitto 1987).

Degradation of collagen under the influence of glucocorticoids is the subject of a few studies only. The results are contradictory and difficult to compare as various experimental designs have been used. Some studies proposed an increase in collagen degradation in granuloma tissue (Ohno and Tsurufuji 1970) while others reported decreased catabolism after glucocorticoid treatment (Nakagawa and Tsurufuji 1972). A decrease in collagen degradation associated with reduced collagenase activity in skin explants was described (Koob et al. 1974). On the other hand, no changes in skin collagenase activity after triamcinolone treatment were reported (Jeffrey et al. 1971, 1985), and Houck et al. (1967) described the appearance of extracellular free collagenolytic activity in rats treated with cortisol. A reduction of collagenase in other tissues obtained from animals receiving glucocorticoids was noted as well (Jeffrey et al. 1971; Werb 1978). A decrease of collagenolytic cathepsin, a lysosomal enzyme capable of cleaving collagen, was reported in the liver of rats treated with glucocorticoids (Kucharz 1984).

The intracellular pool of proline is one of the regulatory factors postulated to be responsible for collagen synthesis, and it has been found that cells containing a greater free proline pool synthesize more collagen (see Chap. 5). There is a single report suggesting that glucocorticoids diminished amino acid transport to the cells, including that of proline (Ryan 1964). This may be a way to depress collagen biosynthesis. On the other hand, one study with fibroblasts cultured in vitro showed that glucocorticoids increased the transportation of some amino acid precursors (Hollenberg 1977), and Cutroneo et al. (1975) were not able to find alterations in the proline pool in rat tissues after treatment with triamcinolone. Thus, the role of this mechanism in collagen regulation by glucocorticoids is unknown.

There are numerous studies on the regulation of collagen metabolism in cultured cell lines, mostly fibroblasts or hepatocytes (Cutroneo et al. 1981;

Oikarinen et al. 1988). The results varied in the extent of the response of cells to glucocorticoids, but generally the hormones decrease collagen metabolism in the same ways as shown in animal or tissue explant studies.

Many aspects of connective tissue metabolism are regulated by products of the cells involved in inflammation. The anti-inflammatory activity of glucocorticoids may be associated with indirect effects on collagen metabolism (Ehrlich et al. 1973). Glucocorticoids decrease the number of cells in the inflamed area, formation of granuloma, and amount and activity of cells (e.g., polymorphonuclear cell, neutrophils) which are able to secrete collagenolytic enzymes.

Trials using glucocorticoids to prevent or reverse fibrosis have been investigated in many experimental and clinical studies, even before the mechanisms of their action on collagen metabolism had been recognized. Animal models of fibrosis have been commonly employed. Liver fibrosis was produced by long-term treatment with carbon tetrachloride. Application of glucocorticoids decreased the total collagen content and increased collagen solubility (Kucharz 1981, 1984). The relative decrease was found to be greater in animals with fibrosis than in those with a normal liver. It is possible that the impaired liver function led to higher levels of glucocorticoids as a result of slower hormone degradation. The experimentally obtained results do not support the clinical usage of glucocorticoids as antifibrotic agents in patients with hepatic fibrosis.

Lung fibrosis produced with bleomycin treatment formed another animal model of fibrosis. Application of glucocorticoids produced partial diminution or could prevent the development of fibrosis (Phan et al. 1981; Sterling et al. 1983b). A decrease in collagen content after glucocorticoid administration was also observed in animals with lung fibrosis due to ozone exposure (Hesterberg and Last 1981).

Clinical application of glucocorticoids as antifibrotic agents is mostly limited to the local treatment of keloids. Studies on keloid fibroblast cultures confirmed their inhibitory effect, and this fibrotic skin disease constitutes a useful model for investigations on regulatory phenomena in fibrosis as reviewed by Abergel and Uitto (1987).

Well-known side-effects of long-term treatment with glucocorticoids are related to bone changes. Osteoporosis is believed to be caused by a decreased content of collagen in bone. Most of the studies using embryonal bones confirmed a depression of collagen synthesis by glucocorticoids; however, molecular mechanisms occurring in adult bone remain obscure and probably are a complex process (for review, see Cutraneo et al. 1986). These skeletal effects are responsible for the delayed body growth in children treated with the hormones.

It is difficult to believe that inhibition of collagen metabolism is the only mechanism responsible for the antifibrotic activity of these hormones. Liver or lung fibrosis is a complex phenomenon associated with many immune and inflammatory mechanisms. Glucocorticoid effects on these processes are probably supplemented only by direct inhibition of collagen synthesis. The complexity of glucocorticoid action upon collagen in pathologically altered tissues can be illustrated by studies on pulmonary fibrosis induced in mice with butylated hydroxytoluene. This agent significantly enhances collagen synthesis in the lungs (Hakkinen et al. 1983; Kehrer et al. 1984; Kehrer and Witschi 1981). Administration of prednisolone after the induction of lung damage results in potentiation of fibrotic lessions (Kehrer 1982). When prednisolone was given during butylated hydroxytoluene treatment, the rate of collagen synthesis decreased, but 2 days after cessation of prednisolone and toxic agent administration, a significant increase was observed and was higher than in mice receiving butylated hydroxytoluene alone (Kehrer et al. 1984). According to Kehrer et al. (1984) impaired degradation of collagen due to prednisolone may be responsible for this increase. On the other hand, in the same study increased levels of free proline were reported in the lungs, and this may influence the regulatory feedback on collagen synthesis as described in Chap. 5.

All these findings indicate a need to be very cautious when interpreting the mechanisms responsible for the antifibrotic activity of glucocorticoids administered in vivo. Application of glucocorticoids for any purpose in clinical practice has to be preceded by an analysis of the benefits and risks, and altered collagen metabolism may be either an adverse or beneficial reaction.

## 7.4 Growth Hormone

The pituitary gland or the hypophysis is a small, complex, endocrine gland located on the underside of the hypothalamus of the brain. It consists of three parts: posterior, median, and anterior lobes. It produces and secretes several hormones and stores and secretes hormones produced in the hypothalamus. Some of them are tropic hormones, i.e., they affect the secretion of the other endocrine glands. Thus, the influence of these hormones on connective tissue is mediated by the effector hormone. These hormones are described in other parts of this chapter (e.g., thyrotrophin, corticotrophin). In the present description the influences of the growth hormones produced by the acidophilic cells of the anterior lobe of the pituitary gland are reviewed.

Human growth hormone is a protein with a molecular weight of 21500 Da. It expresses various activities including promotion of growth. This complex process is associated with acceleration of chondrogenesis in the epiphyseal plates, which can be seen only in individuals without closed epiphyses. These actions are accompanied by various metabolic effects including an increase in protein synthesis (anabolic action) as well as diabetogenic activity due to an anti-insulin effect on the muscle. Its effect on cartilage is mediated by somatomedins. There is a group of peptide somatomedins usually called the growth factors.

Growth hormone is generally accepted to be an anabolic agent and to increase collagen synthesis. Scow (1951) was the first to report the specific enhancement of collagen synthesis under the influence of growth hormones. This intitial finding was confirmed in various experimental models. Growth hormones increased the incorporation of proline into slices of granulation tissue and other tissues in vitro (Vaes and Nichols 1962; Daughaday and Moriz 1962; Mikkonen et al. 1966). An enhanced level of soluble collagen in the tissues of animals treated with growth hormone was reported (Banfield 1958; Kowalewski and Young 1968). An increased rate of collagen synthesis was also observed in the metamorphosis of the tadpole (Yamaguchi and Yasumasu 1977a, b).

Systemic indices of collagen metabolism have been found to be altered: Serum hydroxyproline level and urinary output of hydroxyproline increase in animals receiving growth hormone (Kivirikko et al. 1958; Smiley and Ziff 1964; Aer et al. 1968). Similar results were seen in rats inoculated with MtT-W15 tumor that produces high amounts of growth hormone (Åkerblom et al. 1972). In humans, patients with active acromegaly excrete increased amounts of hydroxyproline. The urinary output of collagen metabolites was normal in patients with inactive acromegaly (Jasin et al. 1962; Benoit et al. 1963; Dull and Henneman 1963; Lee and Lloyd 1964; Kocher et al. 1965; Hioco et al. 1967; Recchia et al. 1967). Halse and Gordeladze (1981) showed that patients with acromegaly excreted enhanced amounts of nondialyzable hydroxyproline in urine, and this index correlated with the mean fasting growth hormone level. A reduced excretion of hydroxyproline was found in children with hypopituitary dwarfism (Jasin et al. 1962; Chiumello and del Guerico 1965; Kivirikko and Laitinen 1965; Faglia et al. 1966; Job et al. 1966).

Hypophysectomy in animals was reported to lead to profound changes in collagen content and solubility in tissues. According to Deyl et al. (1972b) some time after hypophysectomy the absence of growth hormone is of prime importance while directly after hypophysectomy other factors may prevail. They found an increase in the content of total collagen in the heart, kidneys, spleen, aorta, and testes. In the skin, the prevalence of insoluble collagen was shown. These observations are concomitant with changes in the shrinkage properties of tendon collagen in hypophysectomized rats which resembled aging (Steinetz et al. 1966).

The effect of growth hormone on collagen maturation remains unexplored. There is only a single report by Shoshan et al. (1972) that indicates inhibition of the formation of aldehydes from lysyl residues in collagen which may impair collagen cross-link formation.

One of the hypothetical mechanisms of the effect of growth hormone is stimulation of fibroblast proliferation in vitro (Olsen et al. 1980; Weidman and Balo 1980; Clemmonds and Van Wyk 1981). The role of this phenomenon in the hormonal control of collagen metabolism in vivo remains unknown.

In conclusion, the present knowledge of the effects of growth hormone on collagen is based mostly on studies done a few years ago. Modern aspects (e.g., type pattern of collagen in tissues) were not investigated. It is, therefore, unclear to what extent the increase in collagen synthesis is a specific and direct influence of the hormone or results from its unspecific and/or indirect anabolic activity.

## 7.5 Insulin and Glucagon

The pancreas is a duct gland that serves the digestive system. It consists of exocrine and endocrine divisions. The latter is built up of the islands of Langerhans, which produce insulin, glucagon, and at least two other hormones, somatostatin and pancreatic polypeptide.

Insulin and glucagon are the main hormones regulating carbohydrate metabolism, although profound regulatory effects are produced by other hormones as well. Insulin is a polypeptide containing two chains of amino acids linked by disulfide bridges. It is synthesized in the endoplasmic reticulum of the B cells of the islands of Langerhans in the form of preproinsulin. This compound is converted to proinsulin and subsequently to insulin and then is released into the blood stream. Insulin acts on the cell membrane of certain cells by binding to specific receptor proteins. It facilitates glucose uptake in muscles (skeletal, smooth, and cardiac), adipose tissue, fibroblasts, and some other cells. It produces a variety of metabolic effects. The nature and the extent of its actions depend on the kind of effector cell and are modulated by other factors.

Glucagon is a linear polypeptide which in humans contains 29 amino acid residues. It is produced by the A cells of the island of Langerhans. The actions of glucagon are generally opposite to those of insulin.

The most common disorder of the carbohydrate metabolism is diabetes mellitus. In general, the disease is characterized by hyperglycemia and an absolute or relative lack of insulin. When these conditions are present for a number of years, degenerative vascular and nervous changes appear. It is estimated that about 2.5% of the population in the USA is diabetic. Diabetes mellitus is not a single disorder and is usually classified into type 1 or insulin-dependent (juvenile-onset diabetes) and type 2 or noninsulin-dependent (maturity-onset diabetes). The other categories of hyperglycemia include diabetes associated with certain conditions or syndromes (secondary diabetes) and gestational diabetes.

Investigations of the effect of insulin on collagen metabolism have been mostly focused on clinical and experimental diabetes mellitus. Experimental diabetes mellitus in animals is produced by injections of alloxan or streptozocin.

Urinary excretion of hydroxyproline is normal in most patients with diabetes mellitus (Benoit et al. 1963; Bonadonna et al. 1965; Laitinen et al. 1966a; Recchia et al. 1967, cited in Kivirikko 1970; Kivirikko 1970). Increased values were found in insulin-dependent diabetes (Recchia et al. 1967; Mani and Mani 1986). On the other hand, the urinary excretion of hydroxylysylglycosides was shown to be decreased in patients with diabetes (Sato et al. 1980). An elevated serum level of an aminopropeptide of type III procollagen was found in diabetic patients and was shown to correlate with the development of microangiopathy. This index was suggested to be a good noninvasive marker for measuring vascular changes in patients with diabetes mellitus (Okazaki et al. 1988). An increase in the serum level of so-called 7*S* collagen, the cross-

cross-linking domain of type IV collagen, was reported in patients with diabetes. The values found in patients without microangiopathy were lower than those in patients with microangiopathy (36.0±4.4 μg/l and 48.7±5.0 μg/l, respectively; controls 7.8±0.6 μg/l). Similar results were reported in diabetic rats (Brock et al. 1985; Risteli et al. 1982; Hasslacher et al. 1987). The possible role of circulating collagen fragments in the development of microvascular damage via the activation of platelet aggregation has been proposed (Högemann et al. 1986).

Cutaneous manifestations of diabetes mellitus result from vascular complications, neuropathy, and a direct effect of the disease on the cells and matrix of the skin. Diabetic scleroderma, nonpitting swelling and induration of the skin, is the most common cutaneous complication of type 2 diabetes mellitus. Digital sclerosis associated with joint contractures is characteristic of type 1 diabetes (Fleischmajer et al. 1970; Perejda 1987).

Early studies on rats with alloxan-induced diabetes showed a decrease in the collagen content in the skin (Berenson et al. 1972). However, Behera and Patnaik (1979) reported an increase in the collagen content in the skin of diabetic mice and an absence of changes in total collagen content in bone and tendon.

In humans, accumulation of collagen in the skin has been found in diabetic sclerotic complications (Fleischmajer and Lara 1965; Rosenbloom et al. 1981; Konohana et al. 1985). Accumulation is associated with thickening of the dermis and collagen deposition in the subcutaneous tissue. Thus, the amount of collagen per square unit of the skin is significantly enhanced while the concentration of collagen, expressed as amount per unit of weight of dry skin, is unaltered (Perejda 1987). There was no morphological change in the diameter of the collagen fibers in the skin. The amount of hexose bound nonenzymatically to collagen in diabetic cutaneous sclerosis was shown to be dramatically increased as compared with healthy controls (Buckingham et al. 1984). The role of nonenzymatic glycosylation of collagen in the pathogenesis of collagen disturbances in diabetes mellitus is discussed later in this chapter.

It is generally accepted that experimental diabetes is accompanied by a decreased solubility of collagen (Hamlin et al. 1975; Galeski et al. 1977; Behera and Patnaik 1979, 1981 ab, 1982; Mishra and Behera 1986). The increased resistance of diabetic collagen to collagenase indicates an increase in stability of the supramolecular structures (Hamlin et al. 1975; Behera and Patnaik 1982). The effect of diabetes on the cross-linking of collagen remains unclear. Andreassen and Oxlund (1987) were not able to detect any changes in the amount of reducible cross-links in collagen from the aorta of diabetic rats. The indirect suggestion of an increase in stable cross-links was obtained from the biomechanical studies of Andreassen et al. (1981), who reported increased stiffness and strength of the skin of diabetic rats.

The rate of collagen synthesis in the gingival tissue of diabetic rats was found to be decreased (Schneir et al. 1986). The turnover in other tissues in diabetes has not been reported.

Changes in the collagen composition in tissues of patients or animals with diabetes mellitus have been investigated only partially. Kern et al. (1979)

showed a rise in the type III/I collagen ratio in the skin of hereditary diabetic mice. The increase of type III was found in conjunctival biopsies obtained from diabetic humans (Kern et al. 1986). The increase was reported in patients with type 1 and type 2 diabetes. This finding indicates that diabetes mellitus mimics physiological aging in respect of interstitial collagens in the conjunctiva. A positive correlation between the percentage of type III collagen in the conjunctiva and age was found in normal human beings (Kern et al. 1986). In the same study, an increased type III collagen content was reported in prediabetic patients which has been interpreted as a genetic predisposition of the connective tissue to the acquired metabolic derangements observed in diabetes. No changes in the content of interstitial collagen were found in the fetal membranes of patients with overt or gestational diabetes as compared with healthy controls (Leushner and Clarson 1986). In this study, however, the amount of individual types of collagen (I–III) was not determined. An increase in type IV collagen content in the placental villi from patients with overt and gestational diabetes was reported. These changes were associated with a decreased level of type V collagen and an elevated level of type VI collagen. The carbohydrate levels of types IV, V, and VI collagens were shown to be higher than in controls (Leushner et al. 1986).

Thickening of the basement membrane is one of the major matrix alterations in diabetes mellitus. This process is associated with cardiovascular, renal, ocular, skeletal, and neuropathic complications. Thickening of the basement membrane is known to progress with the duration of diabetes and correlates with proteinuria and a reduced glomerular filtration rate (Mogensen et al. 1979; Bloodwork 1980). Thickening results from an increased accumulation of type IV collagen. The mechanism of this phenomenon is not fully understood. The amino acid composition of type IV collagen isolated from normal and diabetic glomerular membrane does not demonstrate any significant difference. Early investigations suggested an altered lysine : hydroxylysine ratio, but these changes were probably caused by insufficient isolation procedures and have not been confirmed (Beisswenger and Spiro 1973; Westberg and Michael 1973; Kefalides 1974). Recently, it was found that experimental hyperglycemia in rats caused an increase in the content of type IV collagen in the glomerular mesangium. Insulin treatment of rats with streptozocin-induced diabetes was associated with a new expression of type III collagen (Abrass et al. 1988).

Diabetic neuropathy affects a significant portion of diabetic patients. Its pathogenesis remains unknown, and the connective tissue of peripheral nerves in diabetes has received little attention. Thickening of the basement membrane of the perineural, endothelial, and Schwann cells have been reported on the basis of ultrastructural investigations (Johnson et al. 1981, 1982). Muona et al. (1989) found that besides type IV collagen the fibrillar collagens are also affected within the nerves. They have shown that collagen fibrils in the sciatic nerves of diabetic rats are significantly thicker than in controls. The mechanism of this abnormality is unknown; changes in the type I/III collagen ratio, glycosylation of collagen, and proteoglycan alterations have been suggested.

Through diabetes mellitus, attention has been focused on a phenomenon known as nonenzymatic glycosylation of proteins (Monnier and Cerami 1982; Kennedy and Baynes 1984). Collagen is one of the proteins which can be nonenzymatically glycosylated under the conditions presented by diabetes. Two metabolic pathways are thought to underlie the pathogenesis. The first of them is known as the Maillard reaction. The sugars react with protein amino groups and form Schiff base-mediated adducts that rearrange into stable glycoconjugates. The formation of a Schiff base intermediate of glucose occurs via condensation of the aldehyde group of glucose with free ε-amino groups of lysine or hydroxylysine residues of collagen. The subsequent transformation is known as the Amadori rearrangement and leads to stable keto-amine-linked 1-deoxyhexose. Formation of the Schiff base intermediate is a rapid and reversible process, while the Amadori rearrangement is an irreversible and relatively slow conversion (Perejda 1987). The bound hexose probably undergoes further transformation and the poorly characterized final products are fluorescent structures. It has been suggested that these fluorescent moieties contribute to cross-link formation in collagen (Suárez et al. 1988). The second hypothetical metabolic pathway of nonenzymatic glycosylation is based on the sorbitol pathway. This pathway results in the conversion of glucose to fructose with the formation of sorbitol as an intermediate metabolite. Accumulation of sorbitol is thought to cause several cellular complications, and a substantial increase in the fructose level in tissues leads to the formation of protein-bound fluorescent products. These products are formed at a rate ten times greater than when glucose is used as the glycosylating agent. Although the mechanism of fructosylation remains unclear, it has been shown that administration of aldolase reductase inhibitor (sorbinil), known to lower the tissue fructose concentration, promotes a decrease in the collagen fluorescence related to nonenzymatic fructosylation in diabetic rats. Application of the aldolase reductase inhibitor does not affect the glucose level and therefore glycosylation of collagen (Suárez et al. 1988). Salicylates have also been found to inhibit glycosylation (Yue et al. 1983, 1984).

Nonenzymatic glycosylation is probably a cause of several changes in the functional properties of collagen. An increase in the thermal stability of rat tail tendon was shown after incubation in vitro in glucose solutions (Andreassen and Oxlund 1985). Similar results were found in vivo (Yue et al. 1983). Lengthening of the breaking time of tendon fibers incubated in urea after nonenzymatic glycosylation was also reported (Kohn et al. 1984). It is possible that other features of collagen isolated from diabetic animals and humans (e.g., resistance to enzymatic digestion, lack of transparency of the ocular tissues) result from nonenzymatic glycosylation (Rosenberg et al. 1979; Monnier et al. 1984, 1988).

Human fibroblasts in culture have been used as a model system to investigate glucose and insulin effects on collagen production. Increased collagen synthesis was demonstrated in cultured fibroblasts from healthy individuals when the cells were grown in a high glucose medium. Greater amounts of collagen were produced in fibroblasts cultured in the glucose concentration range

1–5 mg/ml, but above this range no further increase was noted (Ville and Powers 1977). This finding is in agrement with the results of Kjellström (1986) who reported that fibroblasts from patients with type 1 diabetes produced collagen in a dose-dependent manner. Fibroblasts from patients with type 1 diabetes synthesized collagen at all glucose levels in amounts higher than in the controls. There was no difference in collagen production by fibroblasts obtained from patients with type 2 diabetes and from nondiabetic controls. Similar results were reported by Rowe et al. (1977). They also found a varying response to hydrocortisone in regard to collagen synthesis by fibroblasts from diabetic patients and controls. Enhanced collagen production observed in cultured fibroblasts is consistent with findings of an increase in the secretion of other matrix components (Silbert and Kleinman 1979). The process of matrix secretion is affected by insulin added to the culture medium, and fibroblasts from normal and diabetic subjects have been shown to respond in different manners to insulin (Kjellström and Malmquist 1984).

Cartilage obtained from hypophysectomized rats, i.e., that which is in a resting state due to a lack of growth-hormone-stimulated growth, has also been used as an in vitro model to study collagen metabolism regulation. In this model, it was found that serum from diabetic rats produced a profound decrease in collagen production while serum from normal animals increased synthesis. Addition of insulin did not reverse defective collagen production, and addition of glucose or ketones to normal serum did not induce the changes in collagen production seen after the addition of diabetic serum. Chromatographic separation of serum revealed that the inhibitor or inhibitors are high-molecular-weight factors (Spanheimer 1988). This finding is consistent with the description of the somatomedin inhibitor in diabetic rat serum (Phillips et al. 1979).

## 7.6 Prolactin

Prolactin is a hormone produced by the anterior pituitary. Its major action is to prepare the breast for and sustain lactation. Human prolactin is a protein containing 198 amino acid residues. It is secreted during pregnancy and during lactation.

There are few data about the effect of prolactin on collagen metabolism. It was found that prolactin increased collagen synthesis in the tadpole (Yamagushi and Yasunasu 1977a). Intraperitoneal injection of the hormone caused an increase in collagen synthesis during the premetamorphic and early metamorphic stages, but there was no effect in the metamorphic stage. These effects are different from those of the growth hormone.

The effect of prolactin on collagen metabolism in mammals has not been reported.

## 7.7 Histamine

Histamine is classified as autacoid, i.e., a tissue hormone. It is a derivate of histidine. It is produced by the mast cells and contributes to inflammatory and allergic responses.

The effect of histamine on collagen metabolism has been investigated in relation to wound healing. It has been found to increase the tensile strength of surgical wounds (Kahlson and Zederfeldt 1960; Maśliński et al. 1965) and to quicken the healing process (Dąbrowski et al. 1975; Sandberg 1964a, b). An increase in collagen formation was shown after histamine treatment (Sandberg 1962; Dąbrowski and Maśliński (1970). Similar results were reported after injection of histamine-liberating substances directly into the wounded area (Dąbrowski et al. 1975). It was also found that an increase in collagen formation occurs after administration of low doses of histamine. High doses decrease collagen synthesis (Dąbrowski 1987). Dąbrowski and Szczepanowska (1984) showed that aminoguanidine, a histamine inhibitor, decreased the total collagen content in the skin and bones of chick embryos. The decrease was associated with an increase in the soluble collagen fraction. Opposite changes were found in embryos treated with histamine liberator, a condensation product of *p*-methoxy-*N*-methylophenetylamine with formaldehyde (compound 48/80) (Dąbrowski 1978). The mechanism of action on collagen has not been elucidated. Dąbrowski and Maśliński (1970) showed that histamine stimulated the maturation of collagen. Details of this phenomenon remain unknown. It was also suggested that histamine stabilizes collagen-glycosaminoglycan complexes. Dąbrowski (1987) found that it increased the thermal stability of collagen and elevated the stability of complexes of chondroitin-6-sulfate and collagen in vitro. This ability to affect the stability of complexes has not been investigated in vivo.

Histamine was shown to activate leukocyte collagenase in vivo. A similar effect was obtained with the histamine-releasing compound 48/80. Mepyraminc, cimetidine, and timegadine inhibited activation of latent collagenase by compound 48/80 (Wojtecka-Łukasik and Maśliński 1984).

# 8 Immunobiology of Collagen

## 8.1 Introduction

In comparison with other high-molecular-weight proteins, collagen had been considered for a long time as a weak antigen. Data of recent decades indicate that native as well as denatured collagens are capable of inducing both humoral and cellular immune responses, and the antigenic properties of collagens and procollagens are relatively well characterized (for review see Kirrane and Glynn 1968; Beard et al. 1977; Linsenmayer 1982; Furtmayr and Timpl 1976; Gay and Kresina 1982; Michaeli 1977; Timpl et al. 1973).

The first evidence for the immunogenicity of collagen was provided by Watson et al. in 1975. Since then, its antigenic properties have been intensively investigated with particular emphasis on the relation of structure to antigenicity and, more recently, on the immunological characterization of the different types of collagen or procollagen. It is difficult to interpret early studies due to the fact that different preparations of collagen could be contaminated with small amounts of associated, highly immunogenic proteins (fibronectin, laminin, and other structural glycoproteins) or impurities derived from plasma (e.g., albumin). These proteins were able to stimulate antibody production at higher rates than collagen itself, and studies which applied antibody-containing sera led to false results. On the other hand, the discovery of a number of genetic types of collagen with varying antigenic properties and the production of monoclonal antibodies against them enabled significant progress. Specific antibodies to collagens and procollagens are a valuable tool for cellular and developmental biology. Immunohistochemical localization of collagenous structures is widely used in research on the extracellular matrix (Timpl et al. 1977; Linsenmayer 1982).

The role of the immune response against collagen is considered as one of the pathogenic mechanisms in certain disorders. The occurrence of antibodies against collagen in rheumatic disorders and the pathogenesis of so-called collagen-induced arthritis are covered in Chap. 11. The hypothetical contribution of collagen-specific antibodies to diseases of the ear are reviewed in Chap. 17. It is also possible that various cytokines are involved in the development of fibroplasia in some organs. Preliminary data on the role of cytokines in collagen metabolism are presented in the last part of this chapter.

## 8.2 Antigenicity of Collagen and Procollagen

The antigenic determinants of proteins are divided into two general classes, sequential and conformational (Sela et al. 1967). Sequential determinants depend exclusively upon the arrangement of the amino acid residues which form the polypeptide chain of a part of or of a whole protein. The antigenic properties of sequential determinants are not susceptible to denaturation; moreover, in the denaturated stage antibodies have unlimited access to the determinant while in the native proteins, the secondary, tertiary, and quarternary structure may limit their access and prevent their binding. Antibodies against sequential determinants are useful in the differentiation of changes in the primary structure of proteins, e.g., between species or even between organs. Conformational determinants require a certain three-dimensional protein structure interaction with antibodies. Only proteins with intact tertiary or quarternary structures express their conformational determinants, and denaturation leads to the loss of their antigenic properties. Conformational determinants may be formed by single polypeptide chains which are folded, leading to the juxtaposition of different fragments. The determinants are also formed by parts of several polypeptide chains if they make up a united protein structure. The complex structure of collagen or procollagen molecules contains both sequential and conformational determinants, which are divided into the following groups: (1) helical sites, (2) terminal sites, (3) central determinants, and (4) determinants of terminal peptides of procollagen (Fig. 8.1). Along the native triple-helical domain of the molecule are numerous antigenic determinants of group 1. These are of the conformational type and are destroyed by denaturation. They are resistant to limited digestion with nonspecific proteases (pepsin or pronase) which remove the nonhelical terminal parts of polypeptide chains and have no influence on the spatial structure of the helical domain. In denaturated molecules this type of antigenic property remains unchanged only if the renatured

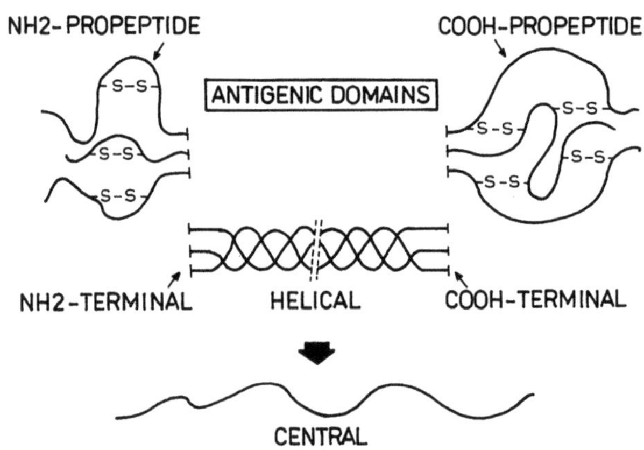


**Fig. 8.1.** Antigenic domains of procollagens and collagens. (From Beil et al. 1973 with permission)

molecule is composed of the original chains. Antibodies against helical sites are able to differentiate common type I collagen composed of $[\alpha 1(I)]_2\alpha 2(I)$ chains from renaturated trimer $[\alpha 1(I)]_3$ (Hahn et al. 1974). Digestion with bacterial collagenase has revealed that the determinants of helical sites consist of segments of three chains which form triple helices since the position of the antibody can be retained in different polypeptide chains. Although helical sites are conformational antigenic determinants and most of the antibodies lose their reactivity against denaturated collagen, some of antibodies produced by immunization of rabbits with the helical regions of native type II collagen react with denatured protein. This finding suggests that sequential determinants are also present in the helical part of the native collagen molecule.

A small section of the native collagen molecule contains a nonhelical structure, and these fragments are called "terminal nonhelical extensions." They are present in the amino- and carboxyterminal ends and are antigenic. It is believed that antigens of terminal sites are sequential. A similar immunogenicity was found in the terminal parts of native and denatured collagen molecules as well as in isolated purified $\alpha$-chains. Determinants of the terminal sites can be removed by limited digestion with nonspecific proteases. Antibodies against the terminal nonhelical parts of collagen are able to differentiate the amino- and carboxy terminal ends and are different for various $\alpha$-chains. These antibodies react with collagen but not with procollagen, suggesting that the conversion of procollagen to collagen creates new antigenic determinants. It is possible that the determinants require pyrrolidone carboxylic acid at the aminoterminal end, and this single amino acid is formed in the separation of terminal peptides of procollagen. It is also possible that the short-lasting, limited proteolysis which removes larger parts of nonhelical segments of procollagen than during conversion to collagen under natural conditions creates new determinants which are not present under physiological conditions. This phenomenon may be responsible for the unsuccessful production of antibodies against "normal" collagen which have been induced by collagen "damaged" during isolation.

Central antigenic determinants are of the sequential class and are located along single chains. In the native state most of these determinants are masked by the triple-helical structure, and denaturation associated with destruction of the helical form of the molecule and separation of the polypeptide chains makes possible the expression of their antigenicity. Antibodies against central determinants are relatively low in specificity probably due to similarities in the amino acid sequence of collagen of different types and from various species.

The procollagen molecule is larger than the collagen one and contains a peptide extension which is removed during conversion to collagen (Chap. 3). These terminal peptides occurred at both ends of the molecule and are highly antigenic. There is no cross-reactivity between amino- and carboxyterminal propeptides.

Most of the reported data refer to the interstitial collagens, types I, II, and III. The basement membrane collagen, type IV, has a different structure and varies in certain antigenic properties from the interstitial collagens. The major antigenic determinants are located in the 7*S* region and NC1 domain. Minor

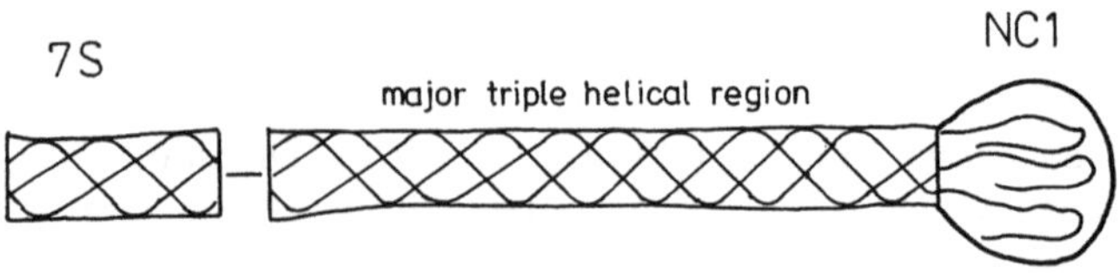


**Fig. 8.2.** Antigenic domains of type IV collagen

antigenic determinants have been found in the major triple-helical portion of the molecule. The antigenicity of the NC2 domain remains unknown (Risteli et al. 1980, 1981). Distribution of the antigenic determinants in the type IV collagen molecule is shown in Fig. 8.2.

The distribution and characteristics of antigenic determinants in other types of collagen are unknown. Due to the diversity of their molecular structure it can be assumed that they possess a number of distinct pattern of determinants.

## 8.3 Immune Response to Collagens and Procollagens

Immunization with a solubilized collagen together with complete Freund's adjuvant leads to the production of autoantibodies. Soluble interstitial collagens stimulate an antibody response in the range 100–500 µg/ml of antiserum. Types I, II, and III procollagens and type IV collagen have been found to be better immunogens, and the immunization results in antibody levels about 1 mg/ml of antiserum. On the contrary, denaturation decreases immunogenicity. Insoluble collagen usually is insufficient to elicit antibodies. There are some differences in the ability of various species to produce antibodies against collagens. According to Timpl (1984), guinea pigs, rats, mice, sheep, and chickens recognize conformational antigenic determinants located within the triple helix. Rabbits produce antibodies mainly to the terminal regions. Chickens also show a strong antibody response against denaturated collagen (Timpl 1982, 1984). Monoclonal antibodies against certain collagens or fragments of collagen molecules have been described.

The nature of the generally low immunogenicity of collagens remains obscure. It has been suggested that their molecular and supramolecular structure interfer with proper handling of the antigen by immune cells (Timpl 1984).

The collagens and procollagens belong to the T-cell-dependent antigens. This means that a response can be achieved only when functionally active T and B cells are present (Nowack et al. 1976).

Cell-mediated immunity to collagen has been examined in various models, including delayed type skin hypersensitivity (Adelmann et al. 1972). Types I, II, and III collagens, individual chains, and cyanogen-bromide-derived peptide fragments of collagen have been shown to induce cell-mediated immunity (Beard et al. 1978; Stuart et al. 1980).

## 8.4 Effect of Cytokines on Collagen Metabolism

The immune response is regulated by a complex network of humoral factors produced by various cells (Old 1987). Some of these factors have been found to affect the metabolism of collagen. Most studies have focused on interleukin 1 and the interferons. Other cytokines have been investigated only in single reports.

Interleukin 1 is a protein produced by stimulated macrophages and monocytes as well as several other cells, including keratocytes, astrocytes, placental cells, mesangial cells, corneal cells, endothelial cells, gingival cells, fibroblasts, and Epstein-Barr virus-infected B cells. It exerts a number of biological actions, including regulation of immunological functions, acting as an endogenous pyrogen, and stimulation of hepatocytes to produce acute-phase proteins (Dinarello 1984). One of its nonimmunological functions is its influence on fibroblast growth and collagen and collagenase production.

Laato and Heino (1988) showed that interleukin 1 decreased the weight and hydroxyproline content of sponge-induced granulomas in rats. A decrease in collagen production was also found in cultured rat granulation tissue exposed to interleukin 1 and in fibroblasts cultured in vitro in a three-dimensional collagen lattice (Gillery et al. 1989). On the other hand, Duncan and Berman (1989a) reported that interleukin 1 and tumor necrosis factor increased collagen production in cultures of fibroblasts. This finding is consistent with the report of Matsushima et al. (1985) who showed that interleukin 1 increased type IV collagen production by mammary epithelial cells. There are a number of reports providing evidence for interleukin-1-induced stimulation of collagenase production. An enhanced collagenase production is found in synovial cells, dermal fibroblasts, chondrocytes, and bone-derived cells (Mizel et al. 1981; Gowen et al. 1983; Postlethwaite et al. 1983; Dayer et al. 1986; Pasternak et al. 1986, 1987; Katsura et al. 1989). The interleukin-1-induced procollagenase production and secretion in human fibroblasts is regulated by calmodulin. Ojima et al. (1989) showed that the calmodulin inhibitor, *N*-(6-aminohexyl)-5-chloro-L-naphthalenesulfonamide, significantly increased collagenase production. Calmodulin was suggested to act as a suppressor of the interleukin-1-induced collagenase production. These results were obtained in cultures of human uterine cervical fibroblasts, and it is unknown whether or not these findings apply to other tissues. It has also been reported that type II collagen is a potent stimulator of interleukin 1 release from human monocytes. The stimulatory determinant was found to be cyanogen bromide 11 peptide. The same peptide is a major arthritigenic epitope for collagen-induced arthritis (Goto et al. 1988). Interleukin 1 is believed to play an important role in the pathogenic progression of rheumatoid arthritis (Dayer and Demczuk 1984) (see Chap. 11).

Tumor necrosis factor, a cytokine secreted by macrophages in response to inflammation, infection, or cancer, is found to stimulate collagenase gene transcription (Brenner et al. 1989a). It is possible that the enhanced secretion of collagenase contributes to the cachexia and tissue remodeling observed after injection of the tumor necrosis factor.

Interferons are cytokines produced by stimulated T cells involved in inflammatory reactions. Investigation of the effect of interferons on fibroblast growth and collagen production have provided controversial results. $\gamma$-Interferon was found to cause a dose-dependent proliferation of fibroblasts but an inhibition of soluble types I and III collagens. There were no changes in the level of insoluble collagen in the cell layer (Rosenbloom et al. 1984; Melin et al. 1987, 1989; Clark et al. 1989). Heckmann et al. (1989) reported that treatment of human skin fibroblasts with increasing doses of $\gamma$-interferon produced a distinct reduction of steady-state levels of the $\alpha$3(VI) chain mRNA but not of the $\alpha$1(VI) and $\alpha$2(VI) chain mRNAs. A similar reduction was also observed for types I and III collagen mRNAs. Reduced deposition of type VI collagen was noted.

In another study, $\alpha_2$-interferon and $\beta$-interferon were seen to cause a significant increase in collagenase production. In contrast, $\gamma$-interferon had no effect on collagenase production by cultured fibroblasts (Duncan and Berman 1989b).

Other lymphokines, interleukin 2, and granulocyte-macrophage colony-stimulating factor were reported to have not effect on collagen and collagenase production in cultured fibroblasts (Duncan and Berman 1989a).

It is very difficult to interpret the results of the described experiments. Under pathophysiological conditions, several cytokines and other factors associated with inflammation modulate the activity of the matrix-secreting cells. The discrepancies in the result obtained are probably caused by imbalances in the stimulation of growth and collagen production and collagenase secretion. Better understanding of the regulatory processes of the immune and inflammatory response will provide the explanation of the role of cytokines in connective tissue metabolism.

# 9 Hereditary Disorders of Collagen

## 9.1 Introduction

Defects in the connective tissue had been suspected as a causative mechanism of certain disorders before the first molecular abnormalities of collagen were discovered, and these diseases have been classified as hereditary disorders of collagen. They form a heterogenous group, still insufficiently explored and, in part, still recognized on the basis of clinical symptomatology. The clinical manifestations, with a few exceptions, are not a reliable guide to the mechanism of the disease. Impairments in connective tissue can be caused by a great range of disturbances involving numerous stages of the complex metabolism and structure of collagen. The final result, as seen by clinical signs and symptoms, is more homogenous than the actual or possible molecular defects because various abnormalities produce the same clinical picture. Inborn defects of collagen are "the experiment of the nature", a tool to understand the pathophysiological role of the mechanisms involved in the metabolism of the most abundant protein of the mammalian body.

In recent years, the clinical, genetic, and molecular aspects of hereditary disorders of collagen were reviewed in many papers (McKusick 1972, 1983; Drożdż et al. 1975; Kivirikko and Risteli 1976; Bornstein and Byers 1980; Hollister et al. 1982; Uitto and Bauer 1982; Byers 1983; Pinnell 1983; Prockop and Kivirikko 1984; Cheah 1985; Prockop and Kuivaniemi 1986). The progress in this field is rapid and substantial.

## 9.2 The Ehlers-Danlos Syndrome

The Ehlers-Danlos syndrome, also called cutis hyperelastica or Meekerem-Ehlers-Danlos syndrome, is a heterogenous group of disorders characterized by abnormalities of the connective tissue. The most prominent changes are found in the skin, joints, and bones, but internal organs and the circulatory system are involved as well. The disorder was known to ancient physicians, and the first detailed descriptions appeared during the Renaissance. Joint laxity and skin hyperextensibility were emphasized in an initial description by Edvard Ehlers (1901) and Henri Alexander Danlos (1908) as well as others (see

**Table 9.1.** The Ehlers-Danlos syndrome: clinical manifestations and inheritance (modified from Byers et al. 1983 with permission)

| Type | General description | Clinical features | Inheritance |
|---|---|---|---|
| I | Gravis | Marked skin hyperextensibility, soft velvety skin; easy bruisability, fragility; cigarette paper-like scars; molluscoid pseudotumors; hypermobility and dislocations of large and small joints; aneurysms and rupture of the large arteries; frequent venous varicosities; mitral valve prolapse; hernias; prematurity due to ruptured fetal membranes is common | Autosomal dominant |
| II | Mitis | Clinical picture resembling type I but significantly less severe | Autosomal dominant |
| III | Benign familiar hypermobility | Skin soft, otherwise minimally affected; marked hypermobility of large and small joints; dislocations are common | Autosomal dominant |
| IV | Ecchymatic or arterial | Skin is thin, translucent; veins are readily visible; severe bruisability and skin fragility; repeated ecchymosis with minimal trauma; minimal hyperextensibility of skin and hypermobility of joints limited to the digits; bowel rupture (usually affecting the colon) and rupture of great vessels are common cause of death | Autosomal dominant or autosomal recessive |
| V | X-linked | Changes mostly limited to skin hyperextensibility and minimal hypermobility of joints | X-linked recessive |
| VI | Ocular | Severe skin changes; hyperextensibility, fragility, and bruisability; kyphoscoliosis; hypermobile joints; ocular fragility and keratoconus | Autosomal recessive |
| VII | Arthrochalasis | Minimal skin changes: soft skin, scars near normal; marked joint hypermobility; congenital hip dislocation | Autosomal dominant or autosomal recessive |
| VIII | Periodontal form | Marked skin fragility with abnormal, atrophic, pigmented scars; slow wound healing; minimal skin extensibility; moderate joint laxity; aesthenic habitus; generalized severe periodontitis | Autosomal dominant |
| IX | Cutis laxa-like form | Skin extensibility; minimal joint hypermobility; skeletal abnormalities; hernias; bladder diverticulae | X-linked recessive (?) |
| X | Unclassified | General symptomatology of the syndrome, cases that cannot be classified into types I–IX | Not known |

Kucharz 1976b). In the past 2 decades, the clinical and genetic heterogeneity of the syndrome has been recognized, and it has been classified into subsets or types. Molecular defects in certain types have been discovered.

Clinical manifestations are different in the various forms of the syndrome and summarized together with the mode of inheritance in Table 9.1. Commonly, hyperelasticity of the skin and hypermobility of joints are involved. The skin is soft and can be extended a few inches away from attachment sites (Fig. 9.1). The skin returns to its original shape and does not flag. Fragility of

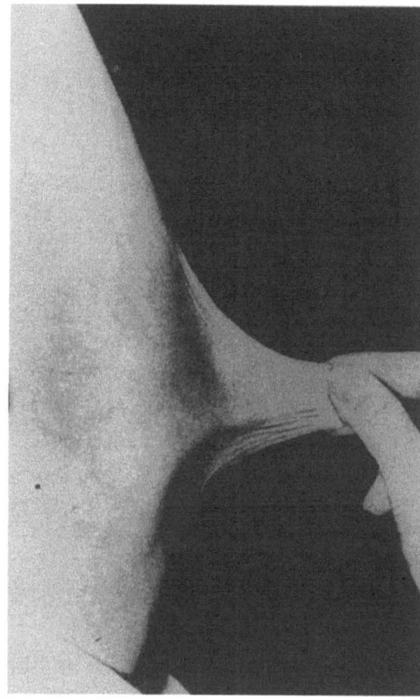

**Fig. 9.1.** Skin hyperextensibility in patient with the Ehlers-Danlos syndrome. (Kindly provided by Dr. Anne De Paepe, Gent, Belgium)

the skin is common in the syndrome, and microtraumas result in wounds and atrophic scars. Joint hyperextensibility is associated with skeletal abnormalities (scoliosis, pes planus, hip displasia) although the bones are not significantly changed (Fig. 9.2). Internal organ involvement includes the cardiovascular system, mitral valve regurgitation, vessel ruptures, easy bleedings, gastrointestinal abnormalities, hemorrhages, and ocular changes (Byers et al. 1983; Child et al. 1986; Krieg et al. 1988).

The molecular defect of the Ehlers-Danlos syndrome type I is unknown. Histologically, the dermal structure is altered, and large collagen fibrils in irregular shapes are observed. The average fibril diameter is 110–140 nm as compared with 90–100 nm in controls (Cupo et al. 1981 (Fig. 9.3). The mechanism which leads to these changes is unknown. It is possible that a defective interaction of collagen with other components of the extracellular matrix produces abnormal fibrils. Irregularities in proteoglycan biosynthesis and fibronectin structure have been described (Arneson et al. 1980). An increased synthesis of glycoproteins was found in fibroblasts cultured in vitro (Shinkai et al. 1976). On the other hand, changes in the region of the collagen molecule responsible for intermolecular interaction are suggested. It is possible that the type I syndrome is not a homogenous disease and is a phenotypic manifestation of a number of biochemical defects which alter collagen fibril formation and stability. Recently, De Paepe et al. (1987) reported a significant reduction

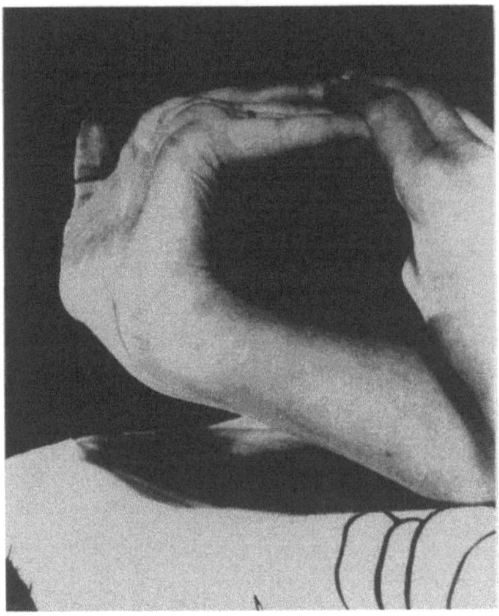

**Fig. 9.2.** Joint hypermobility in patient with the Ehlers-Danlos syndrome. (Kindly provided by Dr. Anne De Paepe, Gent, Belgium)

of type III collagen synthesis in fibroblasts cultured in vitro. The molecular weight of the $\alpha$ 1 (III) chains was unaltered.

The clinical features of the type II syndrome are similar to those of type I but are milder. The molecular defect remains obscure. It is possible that it involves similar mechanisms as postulated in type I, and both forms are subsets of a heterogenous group of disorders of impaired collagen fibril formation (Wordsworth et al. 1985; Rizzo et al. 1987).

The type III syndrome is similar to types I and II but is relatively mild. Joint hypermobility is a predominant symptom. The ultrastructural changes also resemble those of types I and II. The nature of the molecular defect has not been discovered.

The molecular defect in the type IV syndrome is associated with structural errors of the type III collagen. These lead to reduced amounts or the absence of this protein in affected tissues (Pope et al. 1975, 1980; Byers et al. 1979; Stolle et al. 1985; Superti-Furga et al. 1989). Histologically, collagen bundles in the skin are smaller than normal, and ultrastructural studies show dilatation of the rough endoplasmic reticulum of dermal fibroblasts in situ. Fibroblasts obtained from patients and cultured in vitro did not secrete type III procollagen (Pope et al. 1975; Gay et al. 1976). The impaired secretion is associated with intracellular storage of the protein (Byers et al. 1981 a, c; Holbrook and Byers 1981). Aumailley et al. (1988) reported that the decrease in synthesis of type III procollagen was not associated with a reduction in the type III procollagen mRNA. On the basis of their observations two distinctive subsets of the type IV syndrome have been identified: One with an unaltered ratio of types I/III procollagen mRNA, and the other which has a higher proportion of type III procollagen mRNA. Reduced amounts of type III procollagen were

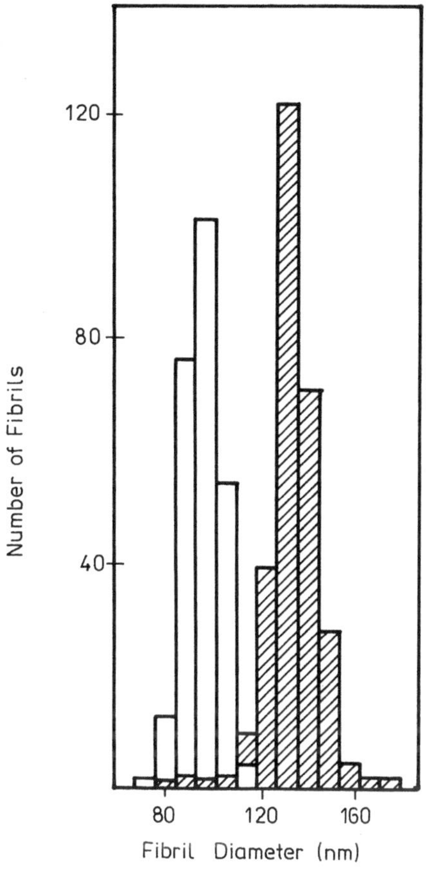


**Fig. 9.3.** Collagen fibril diameters of normal individuals (*open bars*) and patients with Ehlers-Danlos syndrome type I (*striped bars*). (From Byers et al. 1983 with permission)

described in patients of both groups (Aumailley et al. 1988; Superti-Furga and Steinmann 1988).

The type V syndrome is inherited as an X-linked trait. The biochemical basis of this form remains controversial. Di Ferrante et al. (1975) found a decreased activity of lysyl oxidase in the medium of cultured dermal fibroblasts. This was confirmed by Siegel et al. (1979), who were not able to find alterations in the cross-linking of collagen. The reduced activity of lysyl oxidase due to copper deficiency is a causative defect of the type IX syndrome, and it is possible that some cases classified as type V were forms of type IX.

The defect in the type VI syndrome was recognized as early as 1972 when Pinnell et al. described the deficiency of lysyl hydroxylase. Subsequently, more cases of this form of the syndrome have been reported, and some heterogeneity in clinical symptomatology has been shown (Ihme et al. 1984; May and Beauchamp 1987). It is possible that the clinical variability of symptoms is associated with different levels of lysyl hydroxylase. In some patients a total lack of the enzyme has been found, while others note only diminished activity (Pinnell et al. 1972; Ihme et al. 1983; Krieg et al. 1979; Steinmann et al. 1975, 1979). In the majority of patients virtually no hydroxylysine was found in affected tissues (Krane et al. 1972). The presence of a mutant enzyme has been shown; it has been characterized to have an increased Michaelis constant for ascorbic acid (vitamin C) (Miller et al. 1979). This was the reason for using high doses of vitamin C in the treatment of patients, in whom an early diagnosis was reached (Elsas et al. 1978). The efficiency of this method of treatment has not been fully determined. Deficiency in the hydroxylation of lysine impairs cross-link formation (Eyre and Glimcher 1972). Defective cross-linking of collagen has been demonstrated in affected tissues and is responsible for the majority of clinical features of the syndrome. Cultures of fibroblasts obtained from patients with the type VI syndrome displayed other abnormalities in collagen synthesis, probably secondary to impaired hydroxylation of lysine residues. Both types I and III collagen synthesis were increased, and the percentage of type III collagen was found to be lower than in control fibroblasts (Champson et al. 1987).

The type VII syndrome is characterized by extreme joint laxity. This form is heterogenous and was classified into types VIIA and VIIB (Krieg et al. 1988). The defect in subgroup A is associated with impaired conversion of procollagen to collagen. Reduced activity of the procollagen N-proteinase was found in cultured fibroblasts, and extracts from affected tissues are not able to cleave pN-collagen (Lichtenstein et al. 1973). A similar defect, impaired cleavage of the aminoterminal propeptide of type I procollagen, has been found in patients with the subgroup B form of the syndrome. This is not associated with reduced enzyme activity but is caused by structural abnormalities in the $\alpha2$(I) chain and possibly in the $\alpha1$(I) chain (Weil et al. 1989). An interstitial deletion of 18 amino acids in the pro-$\alpha2$(I) chain of procollagen was found; this removes the normal sites of cleavage of the procollagen N-proteinase and also a critical cross-linking lysine residue (Sasaki et al. 1987a). In patients investigated by Wirtz et al. (1987), the ratio of normal $\alpha1$(I):mutant $\alpha2$(I):normal $\alpha2$(I) was 4:1:1. The defect is the result of an exon missing in the gene coding the aminoterminal region of the $\alpha2$(I) chain (Steinmann et al. 1980).

The biochemical defect of the type VIII syndrome is unknown. The most prominent symptoms are associated with periodontal structures, but joints and the skin are also involved (Stewart et al. 1977; Linch and Aston 1979; Nelson and King 1981).

The type IX syndrome has been described in recent years. It includes a group of syndromes which had previously been classified as Menkes' syndrome and the X-linked form of cutis laxa. The defect is associated with copper deficiency, and the connective tissue is involved secondarily.

All forms of cutis laxa except the X-linked one are considered as inherited disorders of elastin on the basis of ultrastructural investigations. The X-linked form, now classified as the Ehlers-Danlos syndrome type IX, is associated with collagen abnormalities, although the elastic fibers are probably altered, too (Taïeb et al. 1987; Byers et al. 1976, 1980). Menkes' kinky hair disease or

Menkes' syndrome is a rare X-linked genetic disorder characterized by mental retardation, imperfect keratinization of the skin, and depigmentation of the hairs. Skeletal and vascular abnormalities have been reported as well. The mechanism of the copper deficiency remains unknown. The deficiency of the two proteins is responsible for the clinical variations, which share features characteristic of the Ehlers-Danlos syndrome and cutis laxa (the loose, easily stretchable skin which does not recoil) (Sartoris et al. 1984; Kuivaniemi et al. 1982, 1985). These are preliminary reports suggesting the occurrence of the type X syndrome although more detailed characteristics are not available (Holbrook and Byers 1982). Type X is employed for those patients who cannot be classified into the other groups. A defect in fibronectin has been postulated in some cases classified as the type X syndrome (Arneson et al. 1980; Viljoen et al. 1987; Krieg et al. 1988).

## 9.3 The Marfan Syndrome

The Marfan syndrome refers to the generalized defect of connective tissue characterized by skeletal, cardiovascular, and ocular abnormalities (McKusick 1972; Pyeritz and McKusick 1979). The first description of a sufferer, a 6-year-old girl with skeletal manifestations, was published in 1896 by the French pediatrician Antoine Bernard Jean Marfan (1896). He called the syndrome dolichostenomelia (long thin extremities), and later the condition was renamed arachnodactyly by Achard (1902). The first reports were followed by more detailed studies which listed ocular and cardiovascular manifestations (Salle 1912; Boerger 1914; Baer et al. 1943) as well as the discovery of earlier, unrecognized reports of similar changes (Williams 1873–1879). Finally, the condition was renamed the Marfan syndrome and included in the inherited connective tissue disorders (Weve 1931).

The most visible abnormalities are the skeletal features. Patients are taller than unaffected sibs or a comparable population. Their limbs are disproportionately long (dolichostenomelia); they have long fingers (arachnodactyly) and anterior chest deformities, produced by longitudinal overgrowth of the ribs as well as vertebral column deformities. The ocular manifestations, subluxation of the lens (ectopia lentis), flat corneas, and myopia occur in 80%–100% of patients (Pyeritz 1983) (Figs. 9.4, 9.5).

Life-threatening manifestations are associated with the cardiovascular system. The two most common features are mitral valve prolapse and dilatation of the ascending aorta. Cardiovascular complications, including dissection or rupture of the aorta and chronic aortic regurgitation with congestive heart failure, accounted for over 90% of the deaths in patients with Marfan syndrome (Berenson and Geer 1963; Pyeritz 1986). The mean age of death is about 32 years (Murdoch et al. 1972). Other abnormalities commonly seen include skin changes (striae distensae) and lung alterations leading to pneumothorax (Pyeritz 1986).

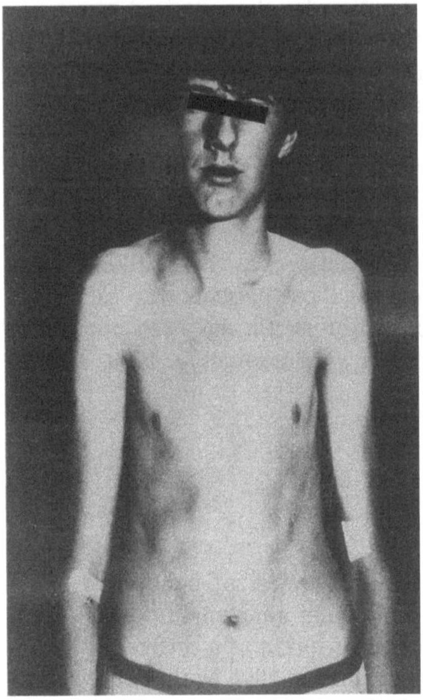

**Fig. 9.4.** Typical body habitus in Marfan syndrome; tall stature, relatively long arms (dolichostenomelia), anterior chest deformity, and scoliosis are characteristic. (Kindly provided by Dr. F.M. Pope, Harrow, UK)

The diagnosis of Marfan syndrome is based on clinical features. The syndrome is probably heterogenous. Several variants thought to be associated with distinct molecular defects have been described (Pope et al. 1987), including the aesthenic and nonaesthenic forms, contractural arachnodactyly, and the Marfanoid hypermobility syndrome. Other classifications have been proposed although all of them are based solely on clinical findings. It may be possible that atypical forms of other inherited disorders of the connective tissue were diagnosed as variants of the Marfan syndrome.

The Marfan syndrome is relatively common among the inherited defects of collagen. The prevalence is estimated as 1–6 cases per 100000 (Lynas and Merritt 1958; Pyeritz and McKusick 1979). The syndrome occurs in all races and is found equally in males and females. The vast majority inherit it as an autosomal dominant trait.

Despite numerous investigations the underlying biochemical defect remains unknown. Alterations in collagen, elastin, and proteoglycans in tissues from these patients have been described, but no one of them is recognized as a common and basic molecular element (Matalon and Dorfman 1968; Lamberg and Dorfman 1973; King and Starcher 1979; Halwe et al. 1985; Yoon et al. 1985; Pope et al. 1987).

The first biochemical abnormality associated with collagen metabolism was reported by Sjoerdsma et al. (1958) who found that patients with Marfan syndrome excreted excess amounts of hydroxyproline in the urine. This observa-

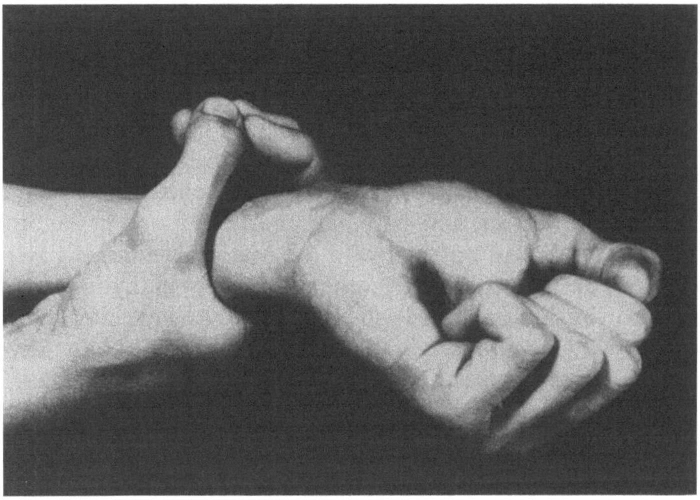

**Fig. 9.5.** Positive wrist sign in patient with Marfan syndrome. (Kindly provided by Dr. F.M. Pope, Harrow, UK)

tion was confirmed by others (Laitinen et al. 1968; Heilmann et al. 1972). Early studies on tissues and cultured fibroblasts from patients with the syndrome demonstrated an increased collagen solubility (Macek et al. 1966; Laitinen et al. 1968; Priest et al. 1973; Francis et al. 1976b; Müller et al. 1987), which suggests impaired cross-linking in collagen. This concept has been supported by the similarity of some of these pathological changes to those found in animals treated with lathyrogens, drugs which inhibit collagen cross-link formation. There is only one report indicating a deficiency of a chemically stable cross-link in collagen obtained from the tissue of patients with the syndrome. Both dihydroxylysinonorleucine and hydroxypyridinium were diminished in the skin and aorta of the patients (Boucek et al. 1981).

The mechanism of impaired maturation of collagen remains unknown. It is unlikely that the Marfan syndrome is an enzymopathy. The vast majority of enzymopathies with gross phenotype are mendelian recessive traits. An enzymatic defect associated with collagen synthesis, especially extracellular processing, has not been detected (Hollister et al. 1982). The activity of lysyl oxidase, an enzyme responsible for cross-link formation, has been found to be normal (Layman et al. 1972; Royce and Danks 1982). Francis et al. (1976a) investigated the stability of polymeric skin collagen. The mean stability of polymeric skin collagen. The mean stability in a group of 14 patients with the Marfan syndrome was not significantly altered, although there were individuals with a low stability. This may suggest that the disturbances in maturation of collagen occur only in some forms of the syndrome.

The profound abnormalities in the vascular tissue suggest changes in the collagen metabolism or composition in this system. Krieg and Müller (1979) found that the adventitia of the aorta lacked type I collagen. This may decrease the mechanical properties of the tissue and make it prone to aneurysm formation (Scheck et al. 1979). On the other hand, Perejda et al. (1985) were not able to find any changes in the ultrastructure of collagen fibrils, content, and solubility in the aortic media of six patients with the Marfan syndrome. Their observations indicated primary or secondary involvement of elastin in the reduced mechanical resistance of the aortic wall.

The disturbances in type I collagen mentioned above were consistent with the structural defect, a 38-base-pair insertion, found in the pro $\alpha 2$(I) collagen gene (COL 1A2) of patients with the Marfan syndrome (Byers et al. 1981 b; Henke et al. 1985). Further investigations showed that this insertion is a common polymorphism (Dalgleish et al. 1986). Recent investigations on collagen genes excluded mutations in the COL 1A2 and COL 3A1 genes, responsible for coding chains in types I and III collagens (Dalgleish et al. 1987; Francomano et al. 1988).

## 9.4 Osteogenesis Imperfecta

Osteogenesis imperfecta, also known as fragilitas ossium, osteopsathyrosis or Ekman-Lobstein-Vrolik disease, is a heterogenous group of disorders characterized by brittle bones. The term "osteogenesis imperfecta" was introduced by Willem Vrolik (cited in Sillence 1983) in 1840. The variability of clinical severity of the disease has been known for many years; however, only recently were basic molecular defects disclosed at least partially, and clinical variants formerly attributed to different degrees of gene expression have been found to be separate disorders. At least four distinct types have been recognized within the osteogenesis imperfecta syndrome (Sillence 1981; Smith 1980, 1986; Smith and Sykes 1985; Smith et al. 1983; Sykes and Smith 1985).

### 9.4.1 Osteogenesis Imperfecta Type I

Osteogenesis imperfecta type I is characterized by osteoporosis leading to bone fragility, presenile hearing impairment, and distinctly deep blue-black sclerae. Patients have deformities of the limbs as a result of many fractures as well as deformities of other origins (genu valgum, flat feet, metatarsus varus). About one-fifth of the patients have a severe progressive kyphoscoliosis. Some extraskeletal abnormalities are also observed, including hyperlaxity of ligaments and premature arcus senilis. The occurrence of hereditary opalescent dentin (dentinogenesis imperfecta) is a basis for the distinction of form A (without opalescent teeth) and form B (with opalescent dentin) (Levin et al. 1980). In patients with form B, a distinctive yellowing of the teeth and apparent transparency are seen. Teeth have short roots with a constricted coronoradicular junction; some of them have a particularly grey-blue hue (Lukinmaa 1988).

Type I is the most common form of the syndrome, and its frequency is estimated at 1 per 30000 of live births (Sillence et al. 1979a). It is inherited as an autosomal dominant trait with variable expressivity.

Different defects have been identified in patients, and they can be classified into two groups. In the first group, mutation of the gene coding the $\alpha1$(I) chain of procollagen produces a null allele. The second group is associated with the synthesis of structurally abnormal chains of type I collagen.

Synthesis of type I collagen in cultured fibroblasts isolated from these patients is reduced, but the collagen produced is structurally normal. These changes are caused by a null $\alpha1$(I) allele and result in decreased production of pro-$\alpha1$(I) chains as well as a reduced level of the proper mRNA (Barsh and Byers 1981; Barsh et al. 1985; Rowe et al. 1985). The decrease in pro-$\alpha1$(I) chains cannot be compensated for by elevated synthesis of pro-$\alpha2$(I) chains as procollagen molecules containing more than one pro-$\alpha2$(I) chains are unstable. The mechanism resulting in the null allele is unknown, although mutations of a promotor region of the gene or errors in the splicing, stability, or transport of the collagen mRNA have been proposed (Cole 1988).

Mutations in the genes coding the chains of type I procollagen result in the production of structurally abnormal collagen. Small deletions have been identified and are probably the cause of the abnormalities. The deletions were found in the $\alpha1$(I) and $\alpha2$(I) chains (Byers et al. 1983; Sippola et al. 1984; de Wet et al. 1986; Steinmann et al. 1986; Prockop 1985, 1988).

### 9.4.2 Osteogenesis Imperfecta Type II

Type II is the most severe form and is usually fatal at birth due to separation of the head and limbs from the trunk. It is heterogenous and has been classified into three subgroups. The type IIA subgroup is characterized by very short, broad bones and beaded ribs. Similar symptoms are found in the type IIB subgroup, but the rib changes are less severe. The type IIC subgroup is inherited as an autosomal recessive trait and is characterized by irregular, slender bones with acute angulation of the shafts and thin ribs. The incidence of osteogenesis imperfecta type II is on the order of 1 per 60000 births (Silence et al. 1979a).

It is associated with mutations in genes coding the chains of the type I procollagen. The reported defects are severe deletions in the $\alpha1$(I) and $\alpha2$(I) chains and point mutations (Bateman et al. 1986). Chu et al. (1983, 1985) reported the deletion of 643 base pairs from one allele for the pro-$\alpha1$(I) chain. Because the deletion was in phase, the synthesized chain had the correct amino acid sequence except that 82 amino acids were missing from the middle (Barsh et al. 1985). In fibroblast culture about one-half of the synthesized pro-$\alpha1$(I) chains is defective. The impaired chains have normal terminal peptides and are thus able to be included into procollagen molecules. The incorporation of a short, defective chain in the molecule leads to its degeneration either before secretion or shortly thereafter. This process is called "protein suicide" as only one-quarter of procollagen molecules do not contain defective chains. The severity of the disease is probably caused by this, and one can hypothesize that the ability of impaired procollagen chains to assemble into molecules is the basic mechanism. In the opposite situation, the impaired chains are degraded intracellularly and do not affect the structure of the secreted protein.

A deletion of 20 amino acid residues was found in the $\alpha$2(I) chain. This abnormality was caused by a defective exon 28 normally containing 54 base pairs and coding amino acids 448–465 of the chain (de Wet et al. 1983). Another deletion of 100 amino acids from the central position of the helical part of the $\alpha$2(I) chain has been reported (Byers and Bonadio 1985).

Point mutations are responsible for amino acid substitutions. Various substitutions have been described, including replacement of glycine by cysteine at residues 748, 904, and 988 of the $\alpha$1(I) chain (Cohn et al. 1988; Constantinou et al. 1987, cited in Cole 1988; Vogel et al. 1987). The especially sensitive point is probably at residue 988 of the $\alpha$1(I) chain. Model studies have suggested that substitution of the glycine there results in extremely unstable molecules. The substitutions also lead to overmodified lysine residues. The increased amount of hydroxylysyl glycoside may also interfere with the aggregation or stabilization of the fibers. A decreased synthesis of collagen was found in the cultured fibroblasts isolated from a baby with osteogenesis imperfecta type II. The glycine at residue 391 of the $\alpha$1(I) chain was substituted by arginine (Bateman et al 1987a, 1987b). The collagen type I which was secreted outside the cells was overmodified.

### 9.4.3 Osteogenesis Imperfecta Type III

Osteogenesis imperfecta type III is a moderately severe form of the syndrome. This syndrome is largely nonlethal at birth, although profound bone fragility leads to progressive deformation of the skeleton. The individuals present at birth with multiple fractures and blue sclerae, which become less blue with age. Fractures occur frequently during childhood. The skeleton is deformed, with kyphoscoliosis, very short stature, and short limbs. Hearing impairment has not been reported.

This type is inherited as an autosomal recessive trait (Beighton and Uersfeld 1985). The nature of the molecular defect, however, has not been definitely established. A few abnormalities have been noted, but there are not enough data to suggest that one of them is the common molecular basic mechanism. Pihlajaniemi et al. (1984) described a deletion of 4 kilo base pairs in the gene coding the pro-$\alpha$2(I) chain of procollagen. The deletion altered the sequence of the last 33 amino acid residues of the carboxyterminal propeptide and interfered with triple-helix formation. As a result, an $\alpha$1(I) trimer collagen is synthesized instead of type I collagen.

In autosomal dominant forms of osteogenesis imperfecta type III, de Vries and de Wet (1986) identified cysteine in the aminoterminal half of the helical region of the $\alpha 1$(I) chain. The sequence of amino acids in this chain was unaltered. The role of the abnormality remains unclear.

An abnormal mannose-rich oligosaccharide unit attached to the carboxyterminal propeptide of type I procollagen molecules was reported in cultured fibroblasts from a patient with osteogenesis imperfecta type III (Peltonen et al. 1980). The collagen was found to aggregate abnormally. Further characterization of the role of this abnormality in the pathogenesis of the disease has not been reported.

### 9.4.4 Osteogenesis Imperfecta Type IV

In most afflicted patients, the symptomatology of osteogenesis imperfecta type IV is limited to bone fragility. Skeletal symptoms similar to those of type I are observed. The bluish sclerae at birth become progressively less blue. There is no hearing impairment. The improvement about the time of puberty is more significant than in other types of the syndrome.

The causal biochemical defect is probably associated with a mutation in the gene coding the pro-$\alpha 2$(I) collagen chain (Tsipouras et al. 1986; Weinstrup et al. 1986a). Linkage has been established in some patients between polymorphism of the $\alpha 2$(I) locus and osteogenesis imperfecta type IV (Falk et al. 1986). This mutation caused impaired posttranslational modifications and reduced the secretion and stability of type I procollagen (Weinstrup et al. 1986b). A point mutation in the COL1A1 gene in a patient with osteogenesis imperfecta type IV has recently been reported (Marini et al. 1989).

## 9.5 Epidermolysis Bullosa

Epidermolysis bullosa is a group of disorders whose principal sign is the ability to form blisters after a minor degree of trauma. The blisters heal leaving scars. Due to the relationship of the blister formation to trauma, epidermolysis bullosa is classified as a mechanobullous disorder.

Different forms of the disease have been described, which vary in their severity, inheritance, and pathogenetic mechanisms. The blisters also range in severity: In a mild form they occur only on the palms and soles while in severe forms there may be such widespread blistering of the skin and mucosal surfaces that it causes death in utero or soon after delivery. The forms are commonly classified as simple, junctional, and dystrophic. In the simple form, the blister formation is associated with a split within the epidermis. In the junctional form, the split is formed through the lamina lucida of the epidermal basement membrane. The dystrophic form of the disease is distinguished by cleavage just beneath the lamina densa of the epidermal basement membrane (Eady et al. 1987). All three forms are inherited as autosomal traits.

The mechanism of the defect has not been elucidated, and in some forms of the disease, such as recessive dystrophic epidermolysis bullosa, the degradation of collagen is involved in the pathological phenomena. Early hints arose from ultrastructural investigations. Pearson (1962) showed that phagocytosis of collagen fibers took place in the area of blister formation. The activity of collagenase was a few times higher in these patients than in normal individuals (Eisen 1969; Lazarus 1972). Greater amounts of immunoreactive collagenase were found in the affected skin of patients with both recessive and dominant forms of dystrophic epidermolysis bullosa (Bauer et al. 1977a). The increase in the collagenase protein and enzyme activity was also detected in normal-looking skin of patients with generalized and localized recessive dystrophic epidermolysis bullosa (Bauer et al. 1977a; Eisen 1969). The last observation differentiates the recessive from the dominant form: In patients with the latter, the elevated amounts of immunoreactive collagenase are limited only to the area of the blisters (Bauer et al. 1977). Fibroblast cultures from patients with recessive dystrophic epidermolysis bullosa demonstrated an increased capacity to synthesize and secrete collagenase (Weinberg et al. 1989). This phenotypic trait appeared to distinguish this form from other genetically distinct forms (Bauer and Eisen 1978). The physicochemical properties of collagenase isolated from the cultured fibroblasts were determined. The procollagenase has been shown to be more thermolabile in the presence of low concentrations of calcium ions than that from controls. Interestingly, the collagenase from cultures of fibroblasts obtained from affected individuals was shown to have diminished catalytic efficiency, i. e., activity per unit of immunoreactive protein. The diminished catalytic efficiency was compensated for by overproduction of the enzyme, and thus the net activity in the skin was elevated. All these findings suggest that a defect in the regulation of the synthesis and secretion of collagenase is associated with pathogenesis of the recessive dystrophic form (Valle and Bauer 1980). The details, however, remain obscure.

Immunoultrastructural studies have demonstrated an absence of the anchoring fibrils in the skin of patients with recessive dystrophic epidermolysis bullosa (Goldsmith and Briggaman 1983; Tidman and Eady 1985). The fibrils are composed of type VII collagen. Recently, the carboxyterminal domain of type VII collagen was found in the basement membrane, which may be due to either production of abnormal type VII collagen or defective assembly of this protein (Rusenko et al. 1989). The role of impaired anchoring fibrils in the etiology of the disease remains unclear.

## 9.6 Chondrodysplasias

The chondrodysplasias form a heterogenous group of disorders characterized by an abnormal development of the skeletal system. Disproportionate shortness of stature, abnormal size and shape of the limbs, trunk, and skull are the most visible symptoms. More detailed descriptions can be found elsewhere (Sillence et al. 1979b; Rimoin and Lachman 1983; Revell 1986). The disorders are considered inherited defects of the cartilage growth plates. The role of collagen in the development of the chondrodysplasias is usually secondary and insufficiently recognized.

The abnormal processing of type II collagen has been described in Kniest dysplasia. Patients with this disease are characterized by short-limbed dwarfism with a short trunk, a flat nasal bridge, kyphoscoliosis, talipes equinovarus, a cleft palate, hearing defects, joint contractures and stiffness, myopia and vitroretinal degradation, and hernias (Siggers et al. 1974; Maumenee and Traboulsi 1985). Radiological features include, among others, platyspondyly, dysplastic femoral heads with wide acetabular margins, and an undermineralized skeleton. Histological studies show extensive degeneration of the extracellular matrix in the growth plate and adjacent resting epiphyseal cartilage (Horton and Rimoin 1970). The electron microscope investigations of Poole et al. (1988b) revealed that collagen fibrils in the growth plates of patients with Kniest dysplasia are thinner, have irregular shapes, and do not possess the banding pattern. These findings indicate abnormal fibril assembly. Interestingly, the carboxyterminal propeptide of type II procollagen was not detected within the extracellular space. This propeptide accumulated within large vascular dilatations in the chondrocytes, probably due to cleavage from procollagen within the cells. The intracellular accumulation of the C-propeptide was not associated with changes in the content of type II collagen in the cartilage. The mechanism of these phenomena remains unknown. It is possible that type II procollagen in Kniest dysplasia is secreted outside the cell, without the carboxyterminal propeptide. The abnormal fibril formation indicates that the carboxyterminal propeptide is required for fibrillogenesis. This carboxyterminal propeptide was found to be identical with chondrocalcin, the calcium-binding protein found in developing fetal cartilage matrix (Van der Rest et al. 1986). The role of chondrocalcin in Kniest dysplasia remains unknown.

There are only single reports on possible collagen involvement in other chondrodysplasias (Rimoin 1975). Increased amounts of collagen were found in cartilage from sufferers of thanatophoric dwarfism, a lethal dwarfism associated with micromelia (Svercar 1975). The amino acid composition of type II collagen was found to be normal although some differences in electrophoretic mobility of the collagen were reported (Hollister et al. 1975). There were no changes in collagen content and extractibility in cartilage from patients with achondrodysplasia and dystrophic dwarfism (Svercar 1975; Stanescu et al. 1976). Ultrastructural changes in collagen fibrils were observed in some forms of spondyloepiphyseal dysplasia (Byers et al. 1978), but their nature needs to be investigated.

Recently, achondrogenesis, an extremely rare skeletal dysplasia, has been found to be associated with inherited defects of type II collagen. The disease is a lethal, short-limbed dwarfism, and phenotypic heterogeneity has been reported. Two subgroups are described: type I known as Parenti-Fraccaro and type II, Langer-Saldino (Langer et al. 1969; Saldino 1971; Spranger et al.

1974). Eyre et al. (1986) reported that type II collagen was absent in the cartilage. Type I was predominant, and type XI made up about 10% of total collagen. This finding was confirmed by Murray et al. (1987). Microscopic investigations revealed hypercellular cartilage with decreased matrix traversed by numerous fibrous vascular canals. Immunohistochemical staining showed intracellular retention of type II collagen within vacuolar structures, probably the dilated rough endoplasmic reticulum of all chondrocytes. Deficient endochondral ossification was found (Godfrey et al. 1988; Godfrey and Hollister 1988). Type I collagen extracted from bone and types I and III collagens secreted by cultured dermal fibroblasts were normal. Hyaline cartilage was found to contain approximately equal amounts of types I and II collagens and decreased amount of type XI. Structural changes in type II collagen have been found on the basis of altered mobility. The abnormality of type II collagen was the cause of earlier descriptions of the absence of type II collagen in cartilage of patients with achondrogenesis (Godfrey and Hollister 1988). A heterogenous mutation of the COL2A1 gene is suggested as the genetic error in type II achondrogenesis. The nature of the molecular defect in type I achondrogenesis has not been discovered.

## 9.7 Congenital Dislocation of the Hip

A hereditary factor is suggested in some cases of congenital dislocation of the hip. The involvement of several members of the family is not unusual. The nature of inheritance of the defect or predisposition to dislocation remains unknown.

The dislocation is defined as displacement of the femoral head from its natural position within the acetabulum and in congenital cases is found at birth. Some reports suggest that joint hypermobility is a general feature in these children (Andren 1960; Carter and Wilkinson 1964; Wynne-Davies 1970; Felländer et al. 1970). The involvement of a connective tissue defect is possible. Fredensborg and Udén (1976) reported that total collagen and acid-soluble collagen were decreased in the umbilical cord of newborns with dislocation of the hip. This was not confirmed by Jensen et al. (1986). In the last study a slight decrease was reported, but only five cases were investigated. A decreased ratio of types III and I collagens was found in the umbilical cord. A similar decrease was described in the joint capsule by Skirving et al. (1984).

At present there are not enough data to suggest that congenital dislocation of the hip is related to the inherited defect of collagen. Dislocation is a common feature of other generalized disorders of collagen, and mild forms of these diseases might lead to the results described above. On the other hand, some forms of the dislocations may result from decreased collagen content in the hip only. More detailed studies on collagen structure and metabolism as well as investigations of a large number of affected children are needed.

## 9.8 The Nail-Patella Syndrome

The nail-patella syndrome, also known as hereditary osteo-onychodysplasia, is a disorder of the mesenchymal tissues, involving the collagen structure in the basement membranes. It is characterized by skeletal and ocular abnormalities and renal involvement. The most common irregularities include unilateral or bilateral hypoplasia or absence of the patella, bilateral accessory conical iliac horns, thickening of the scapula, and subluxation of the radial heads at the elbow. Changes in the fingernails are visible; the fingernails are atrophic or even absent, may be pitted or friable. Ocular manifestations have been reported in the majority of patients, including microcornea, congenital glaucoma, heterochromia iridis, and strabism (Bennett et al. 1973; Suki and Caskey 1979). Renal involvement occurs in 30%–40% of patients, and one-quarter of them develop renal failure (Simila et al. 1970). The syndrome is inherited as an autosomal dominant trait linked to the ABO blood group locus.

The morphological hallmark of the renal lesion is the presence of collagen-like fibrils in the glomerular basement membrane. The membrane is thick, and segmental glomerulosclerosis and cortical fibrosis may occur. Collagen bundles with their characteristic periodicity are found within the expanded lamina rara interna of the basement membrane (Ben-Bassat et al. 1971; Browning et al. 1988) (Fig. 9.6). The cause of the presence of cross-banded collagen within the basement membrane remains unknown. It has been suggested that a switch in the collagen phenotype can be responsible for the synthesis of other than type IV collagen. Another hypothesis puts forth that abnormal extracellular processing of type IV collagen may result in aggregation of this type of collagen in the form of cross-banded fibrils (Martinez-Hernandez and Amenta 1983). The nail-patella syndrome is a rare disease, and many clinical and biochemical aspects remain to be elucidated. Immunochemical studies of the basement membrane from the kidneys and other organs may provide more details about the collagen heterogeneity and generalization of the defect in patients with the syndrome.

## 9.9 Hereditary Progressive Glomerulopathy (Alport's Syndrome)

This disease called hereditary progressive glomerulopathy, hereditary chronic nephritis, or Alport's syndrome is a primary glomerular disorder occurring in families. It is characterized by renal involvement and ocular and auditory abnormalities. Renal involvement has its onset in early life as manifested by microscopic hematuria. Hematuria and proteinuria of modest degree become aggravated following infections of the upper respiratory tract or exercise. These changes lead to progressive renal insufficiency which develops in a large portion of patients in the age range 25–40 years (Grunfeld 1985). Other manifes-

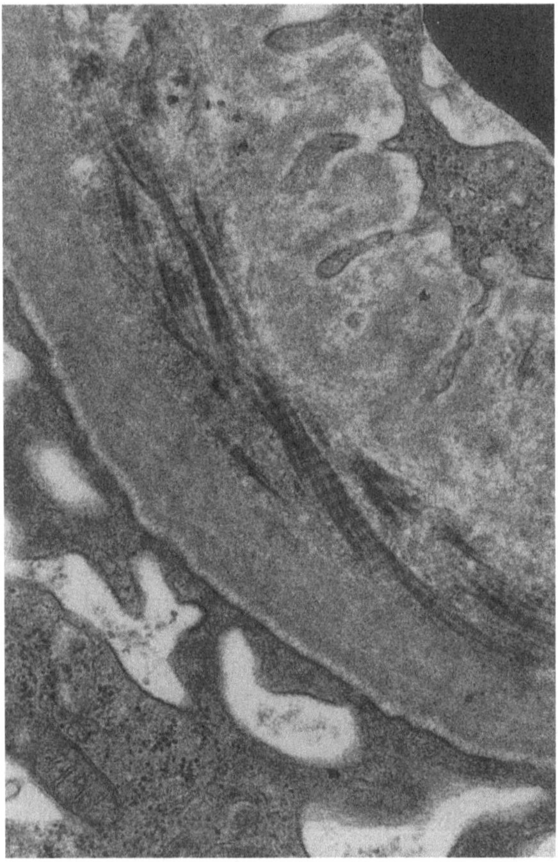

**Fig. 9.6.** Cross-banded collagen fibrils within the glomerular basement membrane of a patient with nail-patella syndrome. (From Browning et al. 1988 with permission)

tations include hearing deficits, which can progress to deafness, and ocular changes including cataract, retinitis pigmentosa, lenticonus, and macular lesions (Suki and Caskey 1979). The pattern of inheritance remains obscure but probably is an X-linked or an autosomal dominant trait, and a high rate of new mutations has been postulated (Feingold et al. 1985).

The defect is localized in the basement membrane and is probably generalized, involving various organs of the body. The majority of investigations have focused on the glomerulas basement membrane as the resulting renal insufficiency is fatal. The glomerular basement membrane was found to be thickened, and the lamina densa was distorted and transformed into a heterogenous network of strands enclosing clear electron-lucent areas (Hinglais et al. 1972; Schröder et al. 1986; Bernstein 1987).

Kleppel et al. (1987) investigated the type IV collagen isolated from glomerular basement membranes of patients with Alport's syndrome. They found after collagenase digestion the complete absence of the two 28 kDa monomers, the neutral and cationic ones. These monomers were present in all samples obtained from patients with other renal diseases. The other components, including the $\alpha 1$(IV) and $\alpha 2$(IV) chains, were present. These findings related to immunological studies which showed that the basement membrane did not bind antibodies from the serum of patients with Goodpasture's syndrome (Jeraj et al. 1984). Goodpasture's syndrome is associated with antibody-induced damage to the glomerular basement membrane (see Chap. 18). The absent monomers contain the antigenic determinants for the antibody. The mechanism of the defect leading to the absence of some parts of the collagen, whether related to a transcriptional, translational, or posttranslational abnormality, remains unknown.

## 9.10 Homocystinuria

Homocystinuria is an autosomal disease which is the second most common inborn error of amino acid metabolism (incidence 1 per 200000 live births). It is heterogenous as several metabolic defects lead to elevated excretion of homocystine. The most common one is deficiency of cystathionine synthetase. The enzyme catalyzes the formation of cystathionine from serine and homocysteine. The unused homocysteine leads to elevated levels in the plasma and excretion in the urine.

Clinical manifestations include ocular changes (ectopia lentis, acute glaucoma), skeletal malformations (osteoporosis, kyphoscoliosis, pectus carinatum or excavatum, vertebral collapse), as well as mental handicap. The majority of clinical symptoms are associated with the connective tissue and collagen in particular. Increased extractibility of skin collagen was the first collagen abnormality reported (Harris and Sjoerdsma 1966a; Griffiths et al. 1976). Kang and Trelstad (1973) found that homocysteine interfered with cross-link formation in vitro. A decrease in cross-links was found in collagen isolated from the skin of patients with homocystinuria. The mechanism of the impaired maturation of collagen has not been known with any certainty. It is suggested that homocysteine inhibits cross-link formation in another way from the lathyrogens. The impaired collagen maturation leads to similar clinical pictures of homocystinuria and osteolathyrism (see Chap. 24) (Griffiths et al. 1976; Gay and Miller 1978). The formation of stable thiazolidine rings through the reaction of the aldehyde precursors in collagen with homocysteine has been postulated as a molecular mechanism (Fig. 9.7). A similar mechanism has been proposed for the drug penicillamine (Chap. 24), which is a structural analogue of homocysteine. Other mechanisms may also be involved. A diminished activity of lysyl oxidase in cultured fibroblasts of patients with homocystinuria suggests enzymatic failure in collagen maturation (Lindberg et al. 1976). These findings were obtained in investigations in vitro. Detailed studies on the con-

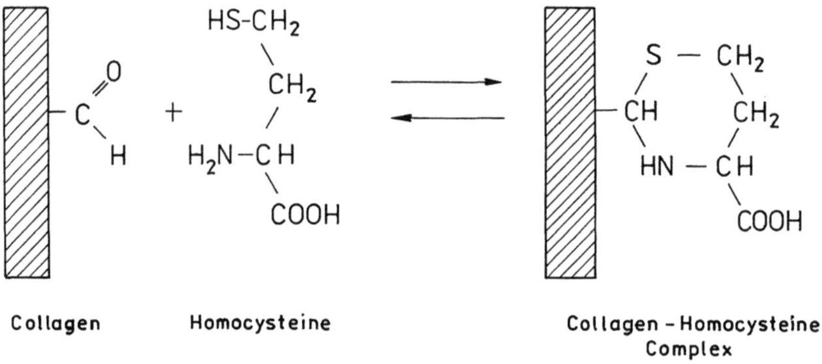


**Fig. 9.7.** Mechanism proposed for the mode by which homocysteine acts to interfere with collagen cross-linking. (From Kang and Trelstad 1973 with permission)

centration of active metabolites and activity of the involved enzymes are needed to analyze the possible causative mechanisms in vivo. The defect of the connective tissue in patients with homocystinuria may be the result of a complex of phenomena involving both collagen maturation and metabolism of other components of the extracellular matrix.

## 9.11 Alkaptonuria

Alkaptonuria or ochronosis is an autosomal recessive metabolic disease caused by a deficiency of the enzyme homogentisic oxidase. The deficiency results in increased urinary excretion and deposition in the connective tissue of homogentisic acid.

The cardinal feature is a blue-black color of the cartilages, pigmentation of the sclera of the eyes and pinna of the ears. The deposits produce arthritis, mostly involving the large joints. In urine, the homogentisic acid causes a dark coloration.

The mechanism of the cartilage damage has not been determined, but involvement of type II collagen is one possibility. It has been suggested that homogentisic acid reacts with collagen and accelerates its polymerisation (Deyl and Adam 1982). On the other hand, the inhibition of procollagen lysyl hydroxylase and subsequent impaired cross-link formation has been found in organ cultures (Murray et al. 1977). Secondary involvements of collagen resulting from the toxic changes in chondrocytes are also possible.

## 9.12 Aspartylglycosaminuria

Aspartylglycosaminuria is a recessive inherited lysosomal storage disease. The metabolic defect is caused by the deficiency of aspartylglucosaminidase, the enzyme which cleaves the *N*-glycosidic linkage between the asparagine-linked oligosaccharide side chains and the polypeptide core in many glycoproteins. Pathological accumulation of glycoproteins is one feature.

Clinical symptoms include psychomotor disturbances with mental handicap, hypermobile joints, and hernias. The typical coarse face with sagging skin is common. A variety of skeletal changes have been reported (Autio 1972; Aula et al. 1982).

Involvement of collagen in the pathogenesis has been suggested on the basis of clinical symptomatology. Nänto-Salonen and Penttinen (1982) found that fibroblasts cultured from these patients produce less collagen than those from controls. It was also reported that urinary hydroxyproline excretion was reduced in young patients as compared with age-matched controls. In adult patients, no significant difference in urinary output of collagen metabolites was detected (Nänto-Salonen et al. 1984a). Electron microscopic studies showed an abnormal size and shape of collagen fibrils in the skin (Nänto-Salonen et al. 1984b). The cause of the collagen defect remains unclear, but an asparagine-linked oligosaccharide occurs in procollagen and is removed with the terminal propeptides. Secondary changes are also possible. Proteoglycans are responsible for collagen fibrillogenesis. Significant changes in proteoglycan metabolism in patients with aspartylglycosaminuria have been reported. The increased urinary excretion of glycosaminoglycans is associated with a markedly increased proportion of excreted heparan sulfate (Nänto-Salonen et al. 1985). Cultured fibroblast production of proteoglycans is also altered (Nänto-Salonen et al. 1987). These changes are probably responsible for impaired fibrillogenesis and the reduced mechanical properties of the connective tissue.

## 9.13 Familial Cutaneous Collagenoma

Familial cutaneous collagenoma is a form of connective tissue nevi of the collagen type (Uitto et al. 1980a). The disease is inherited probably as an autosomal dominant trait (Uitto et al. 1979c). Afflicted patients have asymptomatic cutaneous nodules distributed symmetrically on the trunk and upper arms. Usually, the nodules appear at the age of 15–19 years. Their number increases during pregnancy (Henderson et al. 1968; Hegedus and Schorr 1972). Histopathologically, the nodules are characterized by an excessive accumulation of thick, coarse collagen fibers in the dermis. The lesions contain about ten times more collagen than unaffected skin in the patients.

The mechanism of an increased collagen accumulation is unknown. Fibroblasts obtained from the nodules showed a normal ratio of synthesis of types

I and III procollagens. There were no changes in the amount of collagenase, and the growth kinetics of the cells was normal. These findings suggest that local factor(s) are responsible for increased collagen synthesis. Tan et al. (1981) excluded an increased level of plasma growth factors as a causative mechanism. The aggravation of the disease during pregnancy suggests that hormonal or immune factors modulate the altered accumulation of collagen in the nodules (Bauer and Uitto 1982).

## 9.14 Focal Dermal Hypoplasia

The disease described as focal dermal hypoplasia is not limited to the skin and is associated with abnormalities in the skeleton (small stature, microcrania, kyphoscoliosis with abnormal vertebrae, polydactyly, rudimentary tail), eyes (subluxation of lens), and dental structures (microdentia, dysplasia of teeth, enamel defects). The defect is inherited probably as a X-linked dominant trait, and the clear predominance of female sufferers has been reported (Goltz et al. 1962, 1970).

Cutaneous defects appear as linear areas of thinning of the skin and herniations of adipose tissue in the form of yellowish papules. The decreased amount of collagen is a secondary phenomenon. However, it is very important in the pathogenesis of the disease. The basic cause is associated with significant impairment in fibroblast growth kinetics, probably due to intracellular defects (Uitto et al. 1980b). Fibroblasts obtained from the lesion had a large granular cytoplasm with vacuoles. In culture, they show a compromised growth potential with a mean duplication time twice that of controls. The final saturation density in cultures was one-fifth of the normal values (Uitto et al. 1979a). On the other hand, the ratio of collagen synthesis per cell was undisturbed, and the relative synthesis of types I and III procollagens was normal. The detailed nature of the fibroblast defect remains unknown.

## 9.15 Hyalinosis Cutis et Mucosae

Hyalinosis cutis et mucosae is known also as lipid proteinosis or Urbach-Wiethe disease. It is a rare disease that is transmitted as an autosomal recessive trait. It affects the skin, mucous membranes, and internal organs. The course of the disease is relatively benign (Hofer 1973; Grosfeld et al. 1965; Caplan 1967). In the skin, a hyaline-like material is found in the papillary dermis around the blood vessels. This material is composed of noncollagenous proteins, but accumulation of type IV collagen around capillaries has been seen. The collagen fibrils within the hyaline areas are altered; they are fine and arranged in a random fashion. Cultured fibroblasts from the patients have been shown to produce increased amounts of noncollagenous proteins at the expense of newly synthesized collagens. An about fivefold decrease in type I collagen synthesis and to a lesser extent of type III collagen synthesis were reported (Fleischmajer et al. 1984).

It is concluded that the collagen abnormalities in patients with hyalinosis cutis et mucosae are secondary phenomena to the primary overproduction and accumulation of noncollagenous proteins.

## 9.16 Fibrodysplasia Ossificans Progressiva

Fibrodysplasia ossificans progressiva, known also as myositis ossificans progressive, is a rare disorder of the connective tissue, inherited as an autosomal dominant trait. The main feature is progressive extraskeletal ossification, involving particularly the connective tissue of muscle. The first symptoms usually appear in early childhood and begin with fever and involvement of the neck muscles. The characteristic phalangeal abnormalities are detectable by X-radiographic examination. Further progression of the disease leads to stiffness of the trunk and main joints due to involvement of many muscles and ligaments (Tünte et al. 1967; Smith 1975).

Histopathologically, it starts as degeneration of the muscles and formation of new fibrous tissue which is deposited within the muscles. This new tissue is replaced by ectopic cartilage. The cartilage contains large amounts of type II collagen and subsequently becomes calcified, and bone is formed. Collagen production in the lesion usually culminates in ossification, and then type I collagen is predominant. Overproduction of proteoglycans has also been found in the affected muscles. Ultrastructural studies showed an abnormally shorter periodicty of the collagen fibrils (Maxwell et al. 1977). The nature of the defect and the cellular origin of the connective tissue matrix have not been elucidated. Altered phenotypic expression of the muscle cells or disturbed regulatory processes leading to bone formation are potential mechanism involved in the pathogenesis.

## 9.17 Hereditary Disorders of Collagen Metabolism in Animals

Hereditary disorders of collagen occur in many species of animals and are valuable took for research on abnormal connective tissue. Animal models of inborn defects in collagen metabolism offer many opportunities for experimental investigations that are not possible in human beings. They include prospective genetic studies, embryonic and fetal investigations, as well as wider access to tissue samples.

The first identified model was dermatosparaxis. It was discovered simultaneously in unrelated cattle in Belgium and the USA (Hanset and Ansay 1967; Hanset and Lapière 1974; Ansay et al. 1968; O'Hara et al. 1970). It is inherited as a recessive trait and is caused by the deficiency of procollagen N-proteinase, an enzyme required for the cleavage of the aminoterminal propeptide. Thus, the disease is an analogue of Ehlers-Danlos syndrome type VII. The major clinical feature of dermatosparaxis is fragility of the skin. In contrast to Ehlers-Danlos syndrome, dermatosparaxis is not associated with joint hypermobility. In affected animals, electron microscopy studies showed that the cutaneous collagen fibrils form an irregular network which resembles heiroglyphs (O'Hara et al. 1970). The disease has also been identified in sheep (Helle and Nesm 1972; Fjolstad and Helle 1974) and in a cat and a dog (Patterson and Minor 1977; Collier et al. 1980; Counts et al. 1980; Holbrook et al. 1980; Minor et al. 1987).

It is possible that, other than the deficiency of the procollagen N-proteinase, molecular defects of collagen lead to dermatosparaxis. An increased degradation of dermal collagen resulting in fragile skin in horses, minks and cats has been reported (Gunson et al. 1984; Minor et al. 1984a; Counts et al. 1977). This observation suggests the heterogeneity of the disease. Further studies will probably uncover various defects in the processing of collagen in the affected animals. Similar discoveries in humans led to a description of a variety of forms of the Ehlers-Danlos syndrome.

Autosomal dominant traits presenting as skin fragility and hyperextensibility are assumed to be clinical analogues of the Ehlers-Danlos syndrome type I. This syndrome has been described in dogs (Hegreberg 1982; Hegreberg et al. 1969, 1970a, b; Minor et al. 1983), mink (Patterson et al. 1982; Minor et al. 1987), and cats (Hegreberg 1982; Minor 1982). A defect in fibril formation resulting in the variability of fibril diameter has been found in these animals. The biochemical defect remains unknown. There are hints that the synthesis of proteoglycans is disturbed and that collagen fibrillogenesis is affected secondarily (Damle et al. 1985) (Fig. 9.8).

The analogue of the Menkes' (kinky) hair syndrome associated with copper deficiency was reported in mice (Menton and Hess 1980).

A genetic abnormality in collagen and elastin cross-link formation has been identified in mice. The defect is an X-linked trait. The affected mice have aneurysms of the aorta, weak skin, and bone deformities (Rowe et al. 1974).

A syndrome which resembles osteogenesis imperfecta has been identified in calves (Denholm and Cole 1983; Termine et al. 1984; Fisher et al. 1987). The prominent symptoms are brittle bones and teeth, fragile tendons, and hypermobility of joints. Biochemical investigations indicate a decrease in the amount of type I collagen in the tissues of affected animals. Changes in the molecular weight of all the chains of types I, V, and probably III collagens were reported (Minor et al. 1984b). This leads to a slowing in electrophoretic mobility; the cause is an increase in the posttranslational modifications of procollagen chains prior to triple-helix formation. The site of the disturbances remains partially unclear. It has been suggested that the differences are caused

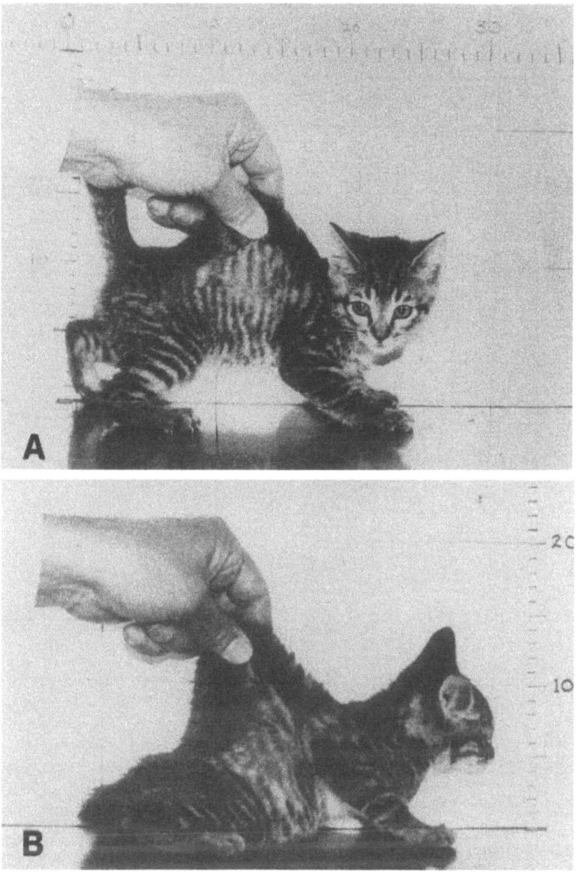

**Fig. 9.8. A** Sibling unaffected and **B** affected kittens from the Philadelphia cat analogue of the Ehlers-Danlos syndrome. (From Minor et al. 1987 with permission)

by a simple addition of further glucose residues to the galactosylhydroxylysine residues in the procollagen chains. The hydroxylation of proline and lysine is not affected (Minor et al. 1987).

There are reports on mice dwarfism associated with cartilage abnormalities which can be used as analogues of chondrodysplasias (Kimata et al. 1979; Brown et al. 1981; Seegmiller et al. 1981).

The molecular defects in animals with genetic disorders of the connective tissue may be identical with those found in affected humans. Clinical symptoms are usually slightly different due to the variety of other factors which affect the phenotypic expression of the genome. The majority of the defects in animals are probably undiscovered, and affected animals are eliminated from breeding. It is possible that future investigations will lead to the description of other animal models of the disorders of collagen.

# 10 Collagen Diseases

## 10.1 Introduction

The concept of diffuse collagen disease was introduced by Klemperer et al. (1942). They were the first to point out that some disorders are a nonorgan-directed, systemic involvement of certain tissues. This holistic viewpoint led to a new perception of the rheumatic diseases and allied conditions (Ehrlich 1984a). The original concept of Klemperer et al. (1942), who had recognized histopathological similarities between systemic lupus erythematosus and scleroderma, has been modified especially in the light of immunological discoveries, but these two diseases, along with dermatomyositis, periarteritis nodosa, various forms of vasculitis, and rheumatoid arthritis, are termed "collagen diseases". Currently this term is used as "a pars pro toto," i.e., it indicates that all these disorders develop in the connective tissue. These diseases, described by some modern investigators as "dyscollagenoses" or "connectivitis" are not, with the probable exception of scleroderma, disorders of collagen molecules.

## 10.2 Scleroderma and Related Conditions

Scleroderma or systemic sclerosis is a rare disease, and women are affected about three times more frequently than men. It is characterized by the deposition of dense, acellular connective tissue rich in collagen fibers in virtually any part of the body. This feature leads it to be considered as a "true" collagen disease (LeRoy 1981). In fact, disturbances in collagen metabolism are frequently linked to the pathogenesis, but despite numerous studies, the role of collagen in its etiology and natural history remains unclear.

Scleroderma often begins with Raynaud's phenomenon or as a chronic, nonpitting edema of the hands and fingers. As the disease progresses, the skin become tense and smooth; a typical face features tightening of the skin, a constricted mouth with radial furrows as well as sclerodactyly, and teleangiectases. Skin alterations are associated with systemic involvement of internal organs, including interstitial lung fibrosis, esophageal sclerosis, myositis, myocardial involvement, and changes in renal function (Jabłońska 1975; Kucharz and Jabłońska 1978; Hendel et al. 1984).

Histopathologically, early scleroderma displays a lymphocytic, inflammatory infiltrate in the deep dermis and subcutaneous fat. Later lesions shown significantly increased quantities of collagen in the dermis, which replaces the fat layer. In scleroderma the fibrotic reaction is closely associated with alterations in small arteries and in microvessels, but the precise relationship between these two phenomena remains obscure (Braun-Falco 1957; Hayes and Rodnan 1971).

The skin is the most intensively investigated tissue for collagen research in scleroderma. It is easily accessible, and cutaneous involvement is present in almost all patients. Earlier works on collagen content in sclerodermatous skin were controversial: Some authors reported increased collagen levels, while others found a normal or decreased content (Korting et al. 1964; Neldner et al. 1966; McConkey et al. 1967; Harris and Sjoerdsma 1966a; Fleischmajer 1964, 1983; Fleischmajer and Krol 1967; Black et al. 1970; Uitto et al. 1971a; Stachow 1976; Bauer and Uitto 1979). Changes in solubility and increased biosynthesis of collagen in skin samples were described (Laitinen et al. 1966b, 1969; Uitto et al. 1970b; Herbert et al. 1974). These discrepancies were clarified in the 1970s. Careful investigations showed that the "cutaneous fibrosis" in scleroderma was, in fact, the replacement of subcutaneous tissue by fibrous deposits (Fleischmajer et al. 1972). Rodnan et al. (1979) described an increase in the amount of collagen under a constant surface area of skin, and further electron microscopic investigations confirmed the accumulation of collagen in the subcutaneous tissue (Kobayasi and Asboe-Hansen 1972). The results of former studies depended on the depth of biopsy or separation of the epidermis, dermis, and subcutaneous tissue. The normal dermis contains about 85% collagen. Even in extensive fibrosis the margin for an increase in collagen is small. The results of solubility studies depend on the stage of the disease. Early inflammatory changes are characterized by an increase of soluble, newly-synthetized collagen while late fibrotic changes contain a high proportion of insoluble collagen (Uitto et al. 1971b). Interestingly, Laitinen et al. (1966b) were able to find an increased content of soluble collagen in the normal-looking skin of scleroderma patients.

Numerous works have concentrated on the rate of biosynthesis of collagen. Increased activity of procollagen proline hydroxylase was reported in most of them (Uitto et al. 1969, 1970a; Keiser et al. 1971; Bauer and Uitto 1982). The rise in synthesis is associated with the greater content of procollagen mRNA (Graves et al. 1983; Kähäri et al. 1984). The synthesis de novo of collagen was confirmed by electronmicroscopic investigations (Kobayasi and Asboe-Hansen 1972; Fleischmajer et al. 1978).

Contradictory results were obtained in studies on collagen types present in sclerodermatous skin. Immunochemical studies showed an increase in type III collagen while biochemical analyses showed the normal ratio of collagen types (Lovell et al. 1979; Fleischmajer et al. 1980; Gay et al. 1980; Seyer et al. 1981). Only a few studies referred to collagen degradation. There is no evidence for diminished degradation of collagen although it was an attractive hypothesis to explain the deposits of collagen due to lowered collagenase activity. The low collagenase activity was proposed on the basis of investigations of late fibrotic sclerodermatous changes with low cellularity. Uitto et al. (1979c) showed that sclerodermatous skin fibroblasts produced a normal amount of collagenase, and enhancement of collagenase activity has been suggested. The activation can be preceded by plasminogen activators liberated from the damaged endothelium (LeRoy 1981). Late fibrotic changes include a decreased level of collagenolytic enzymes.

Various indices of collagen metabolism have been determined in patients with scleroderma. Most reports have presented conflicting data, probably due to differences between stages of the disease. Increased urinary excretion of hydroxyproline was found in some patients while other studies were not able to see any difference as compared with healthy controls (Sjoerdsma et al. 1965; Rodnan and Luksick 1969; Emmrich 1970; Dreux et al. 1971). Serum indices seem to correlate closely with the fibrotic process. An increased hypoprotein (hydroxyproline-rich protein) level was reported in scleroderma patients (Rodnan and Luksick 1969). Recently, an increased serum level of type III procollagen aminoterminal propeptide was reported in about one-third of the patients (Shinkai 1984; Krieg et al. 1986; Hørslev-Petersen et al. 1988a; Zachariae et al. 1989). There was no correlation between the level of the peptide and the number of internal organs involved (Majewski et al. 1987).

The etiology of scleroderma remains unknown. It is generally accepted that the disease is caused by disturbed metabolic and/or cellular regulation, and various possible abnormalities have been suggested. LeRoy (1972, 1974) first discovered that scleroderma skin fibroblasts in culture produce several times more collagen per cell than normal ones. This has been confirmed by other investigators (Buckingham et al. 1978; Uitto et al. 1979b). The fibroblasts are also less sensitive to the mitogenic stimulus of serum than healthy ones (Buckingham et al. 1978). There are few possible mechanisms which explain overproduction of collagen by scleroderma skin fibroblasts: altered phenotypic expression, impaired feedback inhibition of collagen synthesis, selection in vivo of fibroblast clones that synthesize more collagen, and overstimulation of collagen production by a number of factors of various cellular origins.

Krieg et al. (1977, 1978, 1981) reported that this elevated collagen biosynthesis responded to procollagen aminoterminal propeptides, which were able to diminish collagen production by feedback inhibition (see Chap. 5). Perlish et al. (1985) reported that collagen synthesis in scleroderma fibroblasts was inhibited only 19% by the aminoterminal propeptide of type I procollagen while in control fibroblasts 42% reduction was achieved. Furthermore, scleroderma fibroblasts with elevated rates of collagen synthesis were inhibited to a lesser degree (about 10%). The diminished response to aminoterminal propeptide has been suggested as the defect leading to collagen accumulation, although the role of these regulatory circuits in vivo remains unclear (Peltonen et al. 1985).

In vivo fibroblast selection is the basis of a hypothetical model of fibrosis in scleroderma. Botstein et al. (1982) demonstrated that normal fibroblasts from a single, healthy subject are heterogenous with respect to the level of collagen synthesis, and threefold differences were reported between fibroblast clones. This heterogeneity of phenotypic expression was conserved over multiple population doublings in vitro. Serum obtained from patients with scleroderma contained a factor or factors which selectively stimulated growth of the fibroblasts producing more collagen and relatively suppressed the clones producing less collagen (Bashey and Jimenez 1977b). According to this hypothesis, the fibrotic changes in scleroderma are caused by the selective regulation of cell growth, favoring clones of fibroblasts programmed to produce increased amounts of collagen. These findings are consistent with reports on the variability of collagen synthesis by fibroblasts obtained from different layers of the skin. Increased production of collagen was described mostly in primary cultures of fibroblasts from reticular and subcutaneous layers (Fleischmajer et al. 1981a). Subsequent studies confirmed that cells obtained from deeper skin layers of patients with an early stage of the disease showed excessive overproduction of collagen (Krieg et al. 1985; Scharffetter et al. 1988). It is possible that the skin localization of fibroblasts is associated with susceptibility to regulation by obscure serum factors.

Various immune abnormalities have been reported in patients with scleroderma, although their role in the pathogenesis remains unknown. An immune reaction to collagen has been described, including both cell-mediated (Stuart et al. 1976) and humoral (Steffen 1969; Mackel et al. 1982) types. Antibodies to types I and IV collagens were found in the sera, and their levels correlated with the presence of pulmonary fibrosis as measured by abnormal capacity (Mackel et al. 1982). The association of enhanced autoimmunity to collagen and increased production of collagen moderated by cytokines released during immune reactions has been postulated, but these processes may represent changes secondary to causative phenomena (Johnson and Ziff 1976; Whiteside et al. 1985; Kähäri et al. 1988; Solis-Herruzo et al. 1988; Potter et al. 1985).

Disturbed receptor reactivity has been suggested as one possible mechanism of excessive activation of fibroblasts in scleroderma. Cell surface glycoproteins play an important role in numerous cellular functions, e.g., growth regulation, cell-matrix interactions, cell adhesion, and cell recognition (Olden et al. 1982; Maquart et al. 1985). Studies with cultured scleroderma fibroblasts showed a significant reduction in the amount of surface glycoprotein in the 120000 Da range (Sundar-Raj et al. 1984; Kähäri et al. 1987), and the decreased level was observed mostly in fibroblasts producing increased amounts of collagen. It has been proposed that the reduced amount of the 120000 Da glycoprotein reflects the decrease in receptor density on the cell surface, which may cause a lack of feedback down-regulation of collagen synthesis. The biological role of this glycoprotein remains unknown, and further investigations are needed to test the relevance of the hypothesis.

The role of vascular injury in the pathogenesis of scleroderma is the subject of numerous investigations. Vascular changes occur before the fibrotic stage of the disease and can be detected for diagnostic purposes. The presence of certain abnormalities of the nailfold capillaries can predict the development of the disease in one-third of patients with Raynaud's phenomenon (Maricq and LeRoy 1973; Campell and LeRoy 1975; Bachinger and LeRoy 1986).

Microvascular damage occurs in all of the major target organs of patients with scleroderma. This injury consists of swelling and disruption of endothelial cells, fraying of the basement membrane, and interstitial edema. Intimal arteriolar lesions include the loss of endothelial cells, matrix deposition, and development of an adventitial cuff of collagen. In some organs, up to 70% of vessels were shown to be obliterated and nonfunctioning (Fleischmajer et al. 1976; Norton et al. 1968; Norton and Nardo 1970).

One possible interaction between the vascular lesions and fibrosis is the serum endothelial cytotoxic activity which is probably associated with mitogenic activity against fibroblasts (Cohen et al. 1983b; Keyser et al. 1984). These two activities share the property of an inhibitory response to soybean trypsin inhibitor and chloromethyl ketone serine protease inhibitor (Kahaleh et al. 1979; Kahaleh and LeRoy 1983; LeRoy et al. 1983). It is possible that the stimulus for fibrosis originates from injured endothelium since the fibrosis is often perivascular in location (Fleischmajer and Perlish 1980). The nature of this process remains unknown, and probably fibrosis cannot be explained by the simple action of a single factor. The signal from the vascular compartment may be one of a group of stimuli responsible for fibrosis which develops with inflammation and multiple effects of vascular damage.

Localized scleroderma (morphea) exhibits circumscribed lesions of scleroderma in contrast to normal-appearing skin. The sclerotic plaque has an ivory-colored center and a surrounding violaceous halo. The violaceous border signifies that the disease is in an acute state. In linear scleroderma, the lesions appear in a bandlike distribution. Generalized morphea is a severe form of the local disease, characterized by widespread skin involvement with multiple plaques as well as other abnormalities. The disease affects only the skin and is not associated with systemic involvements. Histologically, changes are characterized by extensive collagen metabolism in localized scleroderma. Increased urinary output of collagen metabolites was found by Ammitzbøll and Serup (1984). An increase in collagen synthesis in cultured fibroblasts from patients with generalized morphea was reported by Krieg et al. (1983). Elevated serum levels of aminoterminal type III procollagen propeptide was reported in 4 of 5 investigated patients with widespread morphea (Zachariae et al. 1989).

An animal model of scleroderma is unknown, but disorders resembling it have been observed (for review see Sierakowski and Kucharz 1988). Gershwin et al. (1981) described a VDC 200 strain of chicken with inherited fibrotic changes of skin, pecten, and internal organs. Histological studies confirmed the similarity to scleroderma, showing perivascular deposition of collagen. No metabolic or structural studies of collagen in this strain of chicken were reported. It has been suggested that antibodies against certain components of the connective tissue are involved in the etiology of "the avian scleroderma" (Gershwin et al. 1984; Van de Water et al. 1984). An increased content of collagen in the skin has been found in the tight skin mouse, strain B, 10 D2 58N/SN, described for the first time by Green et al. (1976). Homozygotic mice of this strain die in an early fetal period while heterozygotic ones are born with cutaneous changes which become more severe during postnatal development.

Their skin is thick and unelastic and contains increased amounts of collagen fibers (Menton and Hess 1980; Menton et al. 1980). Similar to humans with scleroderma, the same area of skin contains two and half times more collagen than that of healthy mice, but the amount of collagen calculated per weight of the skin remains unchanged (Jimenez et al. 1984). There is no change in the DNA content in the skin, and an increase of synthesis of collagen is not associated with the proliferation of fibroblasts. Further investigations on the regulation of collagen biosynthesis in tight skin mice are needed to verify the similarity of the biochemical mechanism of this animal model to human scleroderma.

Scleroderma-like changes associated with morphological evidence of collagen deposition are found after prolonged exposure to some chemical agents, e.g., vinyl chloride (Haustein and Ziegler 1985; Kucharz and Sierakowski 1987). There are no published biochemical investigations on collagen either in exposed individuals or in experimental animals.

Bleomycin-induced fibrosis which involves the skin and internal organs is described in Chap. 24. There are some hopes that the application of bleomycin will enable us to produce animal models of a scleroderma-like disease, potentially useful in the testing of antifibrotic drugs.

An increased skin content and elevated synthesis of collagen by cultured fibroblasts was induced by fractions of glycosaminoglycans isolated from the urine of patients with scleroderma (Ishikawa et al. 1975, 1978, 1984). This process has not been further characterized, and another group was not able to induce scleroderma-like changes in mice in this way. This discrepancy may be explained by differences in the strain susceptibility of mice (Fox et al. 1982).

## 10.3 Other Collagen Diseases

Only single reports deal with collagen metabolism in such diseases as systemic lupus erythematosus or dermatomyositis. Changes in collagen metabolism in patients with rheumatoid arthritis are described in Chap. 11. It is generally accepted that impaired immune regulation leads to damage of the connective tissue, and that the changes in the structure and metabolism of collagen are probably secondary. Some alterations are possibly caused only by organ dysfunctions (renal insufficiency, hepatitis, etc.) or enhanced inflammation.

It has been known since early studies on hydroxyproline excretion that most of the patients with collagen diseases have a normal hydroxyproline output, and this index is without any diagnostic value, although it is noteworthy that in, e.g., dermatomyositis (Kibrick et al. 1964; Smith et al. 1965; Laitinen et al. 1966b; Uitto et al. 1970a), systemic lupus erythematosus (Laitinen et al. 1966; Kucharz and Drożdż 1975), and periarthritis nodosa (Smith et al. 1965), some patients have elevated hydroxyproline excretion. Usually such increases were found during the most active stages of the disease (Kivirikko 1970).

Laitinen et al. (1969) reported a decreased hydroxyproline content in skin collagen in systemic lupus erythematosus and dermatomyositis. This decrease was found in pathological lesions as well as in normal-looking skin. Similarly, an increase in soluble collagen content was shown in both affected and normal-looking skin in patients with these diseases. The in vitro incorporation of [$^{14}$C]proline into collagen in dermal samples was enhanced in periarthritis nodosa, systemic lupus erythematosus, and dermatomyositis. The labelled proline was incorporated into both soluble and insoluble collagen except in periarthritis nodosa; in these patients a lowered maturation of collagen in skin samples in vitro was confirmed by Uitto et al. (1970a), who measured prolyl hydroxylase activity. It is assumed that these changes reflect inflammatory involvement. It is also possible that some soluble mediators (e.g., histamine, interleukin 1, acute phase proteins) liberated during inflammation and/or disturbed immune reactions in collagen disease are partially responsible for the observed changes in collagen metabolism.

The serum activity of collagen peptidase, an enzyme cleaving collagen-like peptides containing the amino acid sequence Gly-Pro-Y, was found to be decreased in patients with systemic lupus erythematosus. Fujita et al. (1978) reported a decreased activity of X-Pro dipeptidyl-aminopeptidase (EC 3.4.14.1; dipeptidyl peptidase IV), and Iwase-Okade et al. (1985) described a significant decrease in collagen peptidase activity (PZ-peptidase) in systemic lupus erythematosus. The decrease did not show any correlation with the duration of the disease, as early cases also had a low enzyme activity. There was no decrease seen in patients with Sjögren's syndrome or a mixed connective tissue disease. The mechanism of diminished enzyme activity is unknown. It is suggested that a long-standing inflammatory process decreases enzyme generation in the liver; however, all investigated patients had normal liver function.

## 10.4 Drug-Induced Lupus-Like Syndrome

In some patients, the administration of several drugs produces symptoms closely resembling those of systemic lupus erythematosus. The syndrome has been described under various terms: iatrogenic lupus, drug-related lupus, or drug-induced collagen disease-like syndrome. The pathogenesis is unknown, and it is generally accepted that the syndrome is clinically compatible with systemic lupus erythematosus. It develops de novo during the application of some drugs to individuals previously free of autoimmune disease and disappears after discontinuation of the treatment. Clinical manifestations are indistinguishable from spontaneous systemic lupus erythematosus in the individual patient, but the statistical analysis have shown some differences in the incidence of organ involvement (Alarcón-Segovia 1969, 1975; Alarcón-Segovia and Kraus 1984; Alarcón-Segovia et al. 1967; Cush and Goldings 1985; Dorfmann et al. 1972a, b; Harmon and Portanovs 1982; Hess 1981, 1987; Kucharz 1976a, 1987a; Lee and Siegel 1968; Stratton 1985; Weinstein 1980).

More than thirty pharmacological agents have been reported to induce the syndrome resembling lupus erythematosus. The highest rate of induction is observed in patients treated with hydralazine, procainamide, isoniazid, and chlorpromazine. The application of these drugs to laboratory animals leads to the induction of certain signs of the lupus-like syndrome; this model was used for studies of collagen metabolism, including the effect of inductive drugs on collagen biosynthesis in cell or tissue cultures.

In 1972, Bhatnagar et al. described the inhibition of collagen synthesis caused by hydralazine. The drug decreased the activity of prolyl hydroxylase. This observation was confirmed by Chen et al. (1977) and Bashey et al. (1980). A decreased biosynthesis of collagen was also demonstrated in a 10-day-old embryonal chicken tibia incubated with hydralazine, procainamide, or chlorpromazine (Blumenkrantz and Asboe-Hansen 1974a). Hydralazine was found to inhibit the hydroxylation of proline and lysine residues as well as the glycosylation of hydroxylysine residues in procollagen. This phenomenon is attributed to the chelating properties of hydralazine, which may eliminate iron or manganese from the enzyme systems. Procainamide and chlorpromazine also decreased the hydroxylation of proline and lysine residues. The mechanism of their action is unknown; it is possible that it is caused by nonspecific blockage of nucleic acid synthesis (Blumenkrantz and Asboe-Hansen 1974a).

Kucharz and Drożdż (1978a) were the first to study collagen metabolism in the experimental lupus-like syndrome produced by long-term treatment of guinea pigs with hydralazine. They found a decreased maturation of collagen and suggested inhibition of lysyl oxidase as a cause of an increased soluble/insoluble collagen ratio. This inhibition was confirmed in vitro by Numata et al. (1981). The depression of collagen synthesis is the biochemical basis of skeletal defects induced by the drug. These defects resemble those observed in the experimentally induced manganese deficiency, which is in agreement with the hypothetical role of manganese deficiency in the drug-induced lupus-like syndrome (Comens 1961). Autoradiographic studies of Rapaka et al. (1977) showed that hydralazine blocked collagen secretion. The secretion was restored by $Fe^{2+}$ alone or with $Mn^{2+}$ but not by manganese alone. The results of this study did not confirm earlier reports of impaired glycosylation of hydroxylysine residues. Hydralazine inhibits glycosylation in vitro but not in vivo.

The influence of hydralazine on the extracellular stages of collagen biosynthesis causes a decrease in the mechanical properties of collagen fibers incubated in vitro in the presence of the drug (Kraemer et al. 1979). It was also found that hydralazine reduces the mechanical properties of the aorta in turkeys treated with the lathyrogen, β-aminoproprionitrile. About a 3.5-fold increase in the frequency of aortic rupture was observed in these birds (Simpson and Taylor 1982). Antifibrotic properties of hydralazine, based on its inhibitory activity on collagen synthesis, has been postulated in some experimental studies (Kucharz and Drożdż 1977a).

It was reported that long-term treatment of rats with hydralazine or its derivative, todralazine, led to changes in the reactivity of the connective tissue to catecholamines. Connective tissue binds catecholamines nonspecifically, a feature known as "silent receptors" (Powis 1973). Changes in the healing after isoprenaline-induced myocardial necrosis in rats with hydralazine-induced lupus-like syndrome were observed in biochemical and histological studies (Drożdż and Kucharz 1979; Drożdż et al. 1979b, 1981a). These changes were associated with a decreased amount of total collagen and a relative increase in a soluble fraction of collagen (Drożdż and Kucharz 1979). An increased activity of collagenolytic enzymes in leukocytes obtained from rats with hydralazine-induced lupus-like syndrome was also reported (Drożdż et al. 1983).

The role of the connective tissue changes in the pathogenesis remains obscure. Most of the investigators suggested immunological disturbances. Altered metabolism of collagen can be independent of the development of autoimmunity or at least partially associated with inflammation and immune reactions in tissues.

# 11 Musculoskeletal System

## 11.1 Introduction

The musculoskeletal system contains bones, structures that bind them together and striated muscles. The main function of the system is support, protection, motion, and maintenance of static skeletal and postural support. Although its mechanical functions are primarily important, there are several metabolic functions, e.g., those associated with calcium and phosphorus metabolism. The components of the system constitute a significant proportion of the body, and the greatest part is built up of connective and muscle tissues. It is estimated that more than half of the total collagen content of the body is present in the musculoskeletal system (Forbes et al. 1953, 1956), and the functional state of all parts of the system is directly related to collagen and other components of the extracellular matrix. The musculoskeletal system is classified into sections responsible for passive and active movements. The first one includes skeletal components which form the framework of the body, i.e., bones, cartilage, and various complex structures that bind them together (ligaments, joints, intervertebral discs). The active part of the system consists of striated, voluntary muscles which due to their ability to contract produce movements. All these components are affected in numerous disorders, including primary diseases of bone, cartilage, joints, or muscles. Besides the rheumatic diseases described in this chapter, the components of the musculoskeletal system are involved in disorders described in other chapters, including hereditary diseases of collagen (Chap. 9), collagen diseases (Chap. 10), and hormonal disturbances (Chap. 7).

## 11.2 Bone

### 11.2.1 Structure and Function

The bones are the principal organs of support and the passive instruments of locomotion. They form a framework of hard material which protects the vital organs and harbors the hemopoietic system of the bone marrow, and to which the muscles are attached. As anatomical structures they are complex, containing more than one type of component, forming a specialized connective tissue which is one of the hardest tissues of the human body with a great ability to withstand mechanical stress. The biological functions are associated with its composition, i.e., the predominately mineral extracellular matrix.

There are two main types of bone tissue, nonlamellar and lamellar. The former represents immature, primary tissue known also as "woven" or "coarse fibered" bone, seen in the skeleton of fetuses and young children, and appearing during the repair process. This primary bone is replaced in adults by lamellar bone tissue, except in a very few places (tooth sockets, insertions of some tendons, near sutures of the flat bones of the skull). Non-lamellar bone differs from the mature one in the arrangement of the collagen fibers. In primary bone tissue the collagen fibers form an irregular pattern, there are more cells, and the extracellular matrix contains less mineral than in secondary bone tissue. Lamellar, secondary, or mature bone contains the same structural components as primary bone, but the collagen bundles are arranged in parallel layers or sheets, i.e., lamellae. Lamellar bone is present in both structural types of adult bone, namely cortical (compact) and cancellous (spongy or trabecular). These two types are easily visible in cross-sections of bones where cortical bone is seen as areas without cavities, and spongy bone is located in the places interconnecting cavities (Lees 1987).

Three types of cells are distinguishable within bone, the osteoblasts, osteocytes, and osteoclasts (Martin and Nicholson 1988). Osteoblasts are responsible for the synthesis of the organic matrix. They are cuboid cells with a basophilic cytoplasm and are located exclusively at the surface of bone tissue, arranged in a pattern similar to that of single layer epithelium. Osteoblasts originate from connective tissue cells (Friedenstein 1976) and are formed from fibroblast-like precursors, pro-osteoblasts (Owen 1970, 1980). Osteocytes are small, mature cells completely surrounded by the mineralized bone matrix. They possess several long cytoplasmic processes which radiate through the matrix in canaliculi. In this way, the cells make contact with the interstitial fluid present between the walls of the canaliculi and the cytoplasmic processes. They are located in lacunae, small cavities in the bone matrix. Osteoclasts are very large, multinucleate cells with an acidophilic cytoplasm, usually occurring singly or in small groups on the inner surface of the bone. Their origin is not definitely established, but there is a lot of evidence that they are derived from the fusion of monocytes and originate, in contrast to osteoblasts and osteocytes, from a hemopoetic stem cell (Owen 1980).

The most distinctive feature of bone is the high content of inorganic matter. It is composed mainly of calcium and phosphorus, in the form of hydroxyapatite crystals. Bone apatite is smaller (10–40 nm) and more perfect in atomic arrangement and stoichiometry than the geological mineral, with the empirical formula $Ca_{10}(PO_4)_6(OH)_2$. Carbonate is a constant component of bone (4%–6% by weight), and small amounts of other ions (sodium, potassium, magnesium, fluoride, chloride) are also found.

### 11.2.2 Collagen in Bone

The organic matrix of bone contains predominantly type I collagen, which constitutes 85%–95% of the matrix (Rogers et al. 1952). The remaining, noncollagenous proteins are osteocalcin, osteonectin, bone sialoproteins, phosphoproteins, proteoglycans, and plasma-derived proteins (for review see Butler 1984; Mbuyi-Muamba et al. 1988). Type I collagen is probably exclusively found in bone (Scheven et al. 1988). Small amounts of types III and IV seen in bone are associated with vessels or adherent, unmineralized connective tissue and are not produced by osteoblasts (Giraud-Guille 1988; Mbuyi-Muamba et al. 1988). Bone collagen is a product of the same gene as type I collagen of the skin or other tissues. The significant differences are caused by posttranslational modifications. The best-known feature of bone collagen is its insolubility. It was found that only 0.5% of the total collagen can be extracted from bone tissue with neutral salt solution, and less than 1% is solubilized with acetic acid (Herring 1972). A special extensive procedure has been elaborated to increase the efficiency of extraction and allows the dissolution of about 5–10% (Glimcher and Katz 1965). This results from the cross-linking of bone collagen, the nature of which remains partially unknown. The insolubility can originate from both a greater amount of cross-links similar to those found in other tissues and the possible occurrence of cross-links unique to bone collagen (Kuboki et al. 1981a; Butler 1984). The last question remains to be investigated. There is also evidence that bone collagen differs in the level of glycosylation. It was found that the amount of hydroxylysine residues associated with carbohydrates in skin and bone collagen is the same, but the ratio of glucosylgalactosylhydroxylysine to galactosylhydroxylysine is five times lower in bone than in skin collagen (Pinnell et al. 1971; Royce and Barnes 1977). The biological role of this difference remains unknown, although it is possible that another glycosylation pattern is associated with cross-link formation and the uniquely high insolubility of collagen in bone.

There are few detailed studies on the total collagen content in human bone. Dequeker and Merlevede (1971) studied postmortem biopsies of the left iliac crest obtained under highly standardized conditions. They reported that the organic matrix represents 28% by weight of the fat-free dry bone, and collagen 23% i.e., approximately 80% of the organic matrix. Similar results were reported by Eastoe (1968). There was no significant difference between the sexes. Results referring to changes of collagen content with age are controversial. Cassuocio (1962) found an increasing content of collagen in human lumbar vertebrae (dry spongy bone), while Dequeker and Merlevede (1971) found no effect of age in iliac bone, and others reported a decreased amount of collagen in human femora and iliac bone with age (Rogers et al. 1952; Eastoe 1968). One possible explanation of this controversy is the decrease of total bone mass with age, especially significant in women (Dequeker et al. 1971). The extractability of bone collagen was found to decrease with age. In the bone tissue of individuals 20–30 years old, about 11% of collagen can be extracted with different salt solutions, while in samples obtained from subjects 70–89 years old, only 5% can be solubilized (Dequeker and Merlevede 1971).

Osteoblasts are the collagen-synthesizing cells within the bone tissue. The synthesis of collagen is highly regulated (Leblond 1989), modulated by the calcitropic hormones, prostaglandins, and other humoral factors. The role of hormones is reviewed later in this chapter.

Investigations on the role of prostaglandins in bone formation have provided controversial results. Most of these discrepancies arise from studies using mixed cell populations. It is believed that prostaglandins are multifunctional autocrine regulators of bone formation. In some systems, prostaglandins $E_1$, $E_2$, and $F_{2\alpha}$ stimulated collagen synthesis while in others, prostaglandins induced bone resorption (Chyun and Raisz 1984; Nefussi and Baron 1985; Nagai 1989).

There are few reports on the influence of other autocrine factors on bone collagen metabolism (for review, see Raisz 1988). For example, epidermal growth factor has been reported to inhibit collagen synthesis in osteoblast cells (Hata et al. 1984) while insulin-like growth factors I and II and platelet-derived growth factor stimulated collagen production (McCarthy et al. 1989; Centrella et al. 1989a, b).

The mechanism of collagen degradation in bone is only partially characterized. It is clear that the osteoclasts dissolve mineralized collagenous tissue, and two distinct metabolic pathways are commonly accepted. One of them involves dissolution of hydroxyapatite crystals with subsequent degradation of the collagenous fibrils. The other one starts with degradation of the matrix framework in which the mineral is embedded (Vaes 1980). In general, two collagenolytic systems are postulated to contribute the collagenases active at a neutral pH and collagenolytic cathepsins.

Neutral bone collagenase appears to be similar to most other tissue collagenases. It accumulates in the culture fluid after a lag period of a few days. It is released in an inactive form, called procollagenase (Vaes 1971, 1972, 1980; Eeckhout and Vaes 1977). Several reports provide evidence for the production of a collagenase inhibitor in bone tissue cultures. The inhibitor has been partially characterized (Sellers and Reynolds 1977; Cawston et al. 1983; Nagayama et al. 1984). The release of a neutral collagenase by explanted bones has been reported in numerous studies (Vaes et al. 1976; Vaes 1980). Its properties resemble those of the collagenases isolated from other tissues. Its molecular weight is about 40000 Da for the mouse form while lower values (about 28000 Da) have been reported for the rabbit form (Sakamoto et al. 1978; Sellers et al. 1977; Vaes et al. 1978).

Osteoclasts are capable of digesting calcified collagen in the acidic microenvironment created by the sealed zone of resorption. This process is mediated by lysosomal collagenolytic enzymes. Studies in vitro showed that degradation was most rapid near or below pH 4, and the rate of degradation was increased in the presence of calcium salts. It has been suggested that cathepsin N is the major lysosomal enzyme responsible (Etherington and Birkedahl-Hansen 1987). The activity of the enzymes involved in collagen degradation is regulated by several factors. Hormones controlling the calcium-phosphorus homeostasis are reviewed later in this chapter. Other humoral factors include tumor necrosis factor $\alpha$, epidermal growth factor, and prostaglandin $E_2$ (Shen et al. 1988).

The role of collagen in the mineralization of bone is not yet clearly defined. Various possible mechanisms have been postulated (reviewed by Glimcher 1976), and the process of calcification is a complex phenomenon involving a number of interrelated factors. The simple concept of collagen serving by itself as a nucleation catalyst for the deposition of mineral components (Glimcher and Krane 1968) has been replaced by a more complex mechanism, but the importance of collagen is still accepted (Revell 1986). It has been suggested that calcification begins in membrane-bound vesicles, known as matrix vesicles. They provide an enclosed environment for the accumulation of calcium and phosphate ions as well as being the source of enzymes (e.g., alkaline phosphatase, pyrophosphatase, and ATPase) and possible nucleating agents. Heterogenous nucleation is the basic mechanism for calcification. In this process, the ions aggregate on a substrate to form nuclei and subsequently crystals. Collagen is probably one of the nucleating agents in bone, a concept supported by reports indicating that almost all the mineral in lamellar bone is located within the collagen fibers. The crystals are initially deposited within the fibers and later between the fibrils. Structural features of collagen fibrils in bone have been postulated as a key mechanism in mineralization. According to a model proposed by Miller and Wray (1971) and Katz and Li (1973 a, b), of collagen bone differs from that of soft tissues in the packing arrangement of molecules within the fiber. The major variation is the average gap between molecules constituting the fiber, which in bone is about 0.6 nm while in tendon, it is 0.3 nm. These gaps, know as hole zones, are probably the place of initial calcification which continues until full mineralization of the fibril is complete (Revell 1986). The concept of collagen as a nucleating agent is not accepted by all investigators, or more frequently collagen is considered as a part of a complex mechanism of calcification. It has been postulated that collagen fibers are responsible for the general architecture of mineralization while nucleation depends mostly on some glycoproteins (osteocalcin, osteonectin) and is regulated by proteoglycans. On the other hand, these glycoproteins and proteoglycans occur in bone in association with collagen, and collagen fibers may be considered as a tissue structural scaffold for calcification.

Disturbed skeletal metabolism of collagen can be estimated by measuring specific indices in urine and serum. For more than two decades hydroxyproline output has been used for this purpose although soft-tissue collagen does influence urinary hydroxyproline excretion (Kivirikko 1970). Resorption of bone is associated with an elevated urinary excretion of glycosylated hydroxylysine (Segrest and Cunningham 1970; Askenasi 1974, 1975), and recently, urinary $\beta_1$-galactosyl-$O$-hydroxylysine was proposed as a reliable marker. It was found that the bone mineral content inversely correlated with the ratio of excreted $\beta_1$-galactosyl-$O$-hydroxylysine to creatinine. A high rate of bone mineral loss was associated with a high urinary excretion of $\beta_1$-galactosyl-$O$-hydroxylysine (Moro et al. 1988 b).

### 11.2.3 Collagen in Pathology of Bone

Modern pathology distinguishes a number of bone disorders. Many of them are directly or indirectly associated with collagen structure and metabolism in the bone tissue. Generalized defects of collagen synthesis and posttranslational modifications usually also involve collagen in bones. Changes occurring in hereditary disorders of collagen are described in Chap. 9. Such acquired disorders as scurvy or lathyrism has been described in separate chapters (Chaps. 22 and 24), and the present description focuses mostly on the so-called metabolic disorders.

#### 11.2.3.1 Disturbed Regulation of Calcium and Phosphorus Metabolism

The metabolism of calcium and phosphate is subject to precise regulation by a group of endocrine factors. Because of the crucial role of the mineral component in bone function, the effect of endocrine regulation of calcium and phosphorus metabolism is described in this chapter instead of in Chap. 7, which is devoted to the endocrine control of collagen metabolism.

Three hormones are mainly responsible here: parathyroid hormone (parathormone), calcitonin, and the active metabolite of vitamin $D_3$. Parathormone is a large polypeptide produced by the chief cells of the parathyroid glands. It exerts a number of effects on calcium and phosphorus homeostasis, including an increase in bone resorption with secondary compensatory partial elevation of bone formation, an increase in urinary phosphate output with a decrease in calcium excretion in urine, and an increase in the intestinal absorption of calcium. All these phenomena occur in cooperation with the remaining hormones, especially 1,25-dihydroxycholecalciferol (1,25-dihydroxy-vitamin $D_3$). The increase in secretion of parathormone results in an enhanced plasma calcium level. Calcitonin is a hormone secreted by the parafollicular cells (the C cells) of the thyroid gland. The production of calcitonin is independent of the mechanisms controlling the release of other thyroid hormones. An increased level of plasma calcium is the main stimulus for calcitonin release. An additional signal is mediated by gastrin in response to an increased dietary calcium intake. Calcitonin causes an enhanced deposition of calcium within the bones and increases urinary excretion of calcium. A reduction in the intestinal resorption of calcium is another mechanism leading to a decrease of the plasma calcium level. For details of the hormonal control of calcium and phosphorus homeostasis readers are referred to handbooks of endocrinology.

Parathormone stimulates osteoclast activity and collagen resorption. After long-term exposure of the skeleton to the hormone, the number of matrix-degrading cells also rises significantly. Increased bone matrix resorption is associated with an elevated urinary output of hydroxyproline, as found in patients with hyperparathyroidism. This excretion was shown to correlate with plasma alkaline phosphatase activity (Smith 1967; Piedra et al. 1987) and loss of bone mineral content (Hyldstrup et al. 1984). Removal of the parathyroid adenoma is associated with normalization of the urinary excretion of hydroxyproline (Laitinen 1977). The mechanism of the parathormone-induced collagen degradation is not fully understood. As mentioned earlier, collagenase and collagenolytic cathepsin are involved in bone collagen breakdown. In vitro, addition of parathyroid hormone to cultured bone tissue results in the release of collagenase. The activity of the enzyme has been shown to correlate with the extent of the bone resorption observed morphologically (Sakamoto et al. 1975). Further studies have revealed that parathyroid hormone leads to the release of latent collagenase and gelatinase from bone tissue (Jilka 1989). The activation of collagenolytic cathepsin as an alternative pathway of hormone-induced collagen degradation has also been suggested (Jilka and Hamilton 1985).

There are several reports on the role of 1,25-dihydroxy-vitamin $D_3$ in collagen metabolism. The majority of them have been done in vitro on isolated cells or tissue systems, and it is difficult to conclude that similar phenomena occur within the body. This reservation is especially important as in many in vivo systems a close cooperation of parathormone and 1,25-dihydroxy-vitamin $D_3$ has been shown. Vitamin D deficiency is well-known as a causative factor of the main form of rachitis (rickets). A rachitogenic diet applied to chicks is a common model for studies of the disease. Toole et al. (1972) reported an increase in lysine hydroxylation in both chains of type I collagen isolated from rachitic long bones. Similar results were found in rachitic rats (Barnes et al. 1973a). It is possible that this effect of vitamin D deficiency is mediated by low levels of calcium because calcium deficiency causes a similar abnormality (Barnes et al. 1973b). Vitamin D deficiency was also found to affect cross-link formation. It is possible that vitamin $D_3$ contributes to the activity of lysyl oxidase (Mechanic et al. 1972; Fujimoto et al. 1979). The urinary excretion of hydroxyproline has been shown to increase in most patients with rickets (Kivirikko 1970).

It is of interest that the urinary excretion levels of hydroxyproline and glycosylated hydroxylysine rise in children with vitamin D-resistent rachitis (Novikov et al. 1980). This suggests that calcium is the main modulator of collagen metabolism in bone.

Calcitonin is believed to express an opposite effect on collagen metabolism as compared with parathormone. There are only single reports on its influence: It lowers urinary hydroxyproline excretion (Aer 1968) and increases collagen synthesis in bones. Calcitonin treatment applied to patients with polyostotic fibrous dysplasia (McCune-Albright's syndrome), a disease characterized by progressive bone resorption, results in a significant decrease in hydroxyproline excretion in urine (Yamamoto et al. 1983).

1,25-Dihydroxy-vitamin $D_3$ has been shown to inhibit type I collagen synthesis in osteoblasts of fetal rat calvaria and cultured osteosarcoma cells (Genovese et al. 1984; Franceschi et al. 1988; Ishizuka et al. 1988). Regulation occurs at the level of DNA containing the $\alpha\,1$(I) promoter (Lichtler et al. 1989).

#### 11.2.3.2 Osteoporosis

Osteoporosis is a reduction in the amount of calcified bone mass per unit volume of skeletal tissue, a matter of decalcification or demineralization. It is the most frequent metabolic bone disorder. The most common form of this disease is postmenopausal osteoporosis (Stevenson 1988), which is characterized by a decrease in the bone collagen content (Burnell et al. 1982). These patients were shown to excrete higher amounts of galactosylhydroxylysine in urine (Moro et al. 1988a).

Chronic immobilization leads to a loss of bone mass as observed after muscular inactivity or reduced weight-bearing during space flights (Mack and LaChange 1967; Donaldson et al. 1970; Leach and Rambaut 1977). This type of osteoporosis is associated with intracortical striations, subperiosteal bone loss, surface erosion, and loss of definition of the endosteal margin. The mechanism is not fully understood. Increased osteoclastic resorption, depressed osteoblastic activity, altered parathyroid hormone secretion, as well as changes in piezoelectricity are suggested to contribute. A progressive increase of nonmineralized collagen in bone was found in monkeys immobilized for 6–28 weeks. During reambulation, the nonmineralized collagen content in the bone returned to normal values (Mechanic et al. 1986). The number of reducible cross-links in collagen markedly increases during immobilization and returns to control values after recovery (Yamauchi et al. 1988b). The cross-link concentration remains stable throughout immobilization. These findings suggest that synthesis of a new bone collagen occurs during immobilization and is a partial compensation for the elevated collagen breakdown due to osteoclast activity at the bone surface.

Osteoporosis resulting from the long-term use of heparin is probably associated with an increase in collagenase activity and enhanced bone resorption (Sakamoto and Sakamoto 1981; Sakamoto 1982).

#### 11.2.3.3 Paget's Disease

Paget's disease (osteitis deformans) is a common bone disorder of unknown etiology. It usually affects just a few bones and is characterized by a localized osteolytic process in which proliferation of osteoclasts is the dominant lession. It results in a mixed osteolytic and osteoblastic process of great intensity (Nagent de Deuxchaisnes and Krane 1964). There are no data indicating any definite abnormality of the structure of collagen although a rapid turnover of collagen in bone is well documented. The majority of studies are based on the increased hydroxyproline output, which achieves values 50 times greater than normal; this is the highest elevation observed in human pathology (Lee and Lloyd 1964; Goidanich et al. 1965; Kocher et al. 1965; Woodhouse 1972; Franck et al. 1974; Fabier et al. 1976). The increase correlates with disease activity and serum alkaline phosphatase levels (Nagent de Deuxchaisnes and Krane 1964). Elevated levels of free and total hydroxyproline in serum have also been reported in patients with Paget's disease (Gilbertson et al. 1983). This index normalized when patients were treated with calcitonin (Minisola et al. 1985), indicating the coupling between resorption and formation in bone tissue. This observation was confirmed by measurements of the levels of carboxyterminal propeptide of type I collagen and aminoterminal propeptide of type III collagen in serum. The elevated concentrations of both propeptides were found to be decreased after calcitonin treatment (Simon et al. 1984).

## 11.3 Cartilage

### 11.3.1 Structure and Function

Cartilage is a differentiated type of connective tissue which makes up much of the extracellular matrix. The matrix has a rigid consistency although is not calcified and is thus less resistant than bone. The main function of cartilage is to support soft tissues and, due to its smooth surface, to provide a sliding area for joints. It takes also part in the growth and formation of long bones.

Cartilage is devoid of blood vessels and is nourished by diffusion. Histologically, three types are distinguishable: elastic, fibrous and hyaline. They vary in the content of the matrix, but the differences are only quantitative. Elastic cartilage contains collagenous and numerous elastic fibers. Fibrous cartilage (fibrocartilage) contains a dense network of collagenous fibers, and the most common form, hyaline, consists of a moderate amount of collagenous fibers.

Hyaline cartilage is present in the articular surfaces of joints and forms some parts of the skeleton (e.g., the ventral ends of the ribs) as well as the rigid wall of the respiratory passages. Its physiological functions additionally include an important contribution to bone formation and joint movements (Roberts 1985). The cells of cartilage, chondrocytes, occupy small cavities, lacunae within the matrix. Chondrocytes are ovoid or spherical, and their surface is irregular, and they have short processes that extend into the matrix. The cells are usually arranged in groups known as isogenous groups representing the offspring of a single parent chondrocyte. Chondrocytes as whole cartilage are derived from the mesenchyme. Mesenchymal cells are transformed into chondroblasts, which synthesize and deposit the matrix. The differentiation of cartilage takes place from the center outwards. The growth of cartilage results from two processes: interstitial growth (proliferation of the preexisting chondrocytes) and appositional growth, due to differentiation of the peripheral cells. The newly formed cells in both processes produce a large proportion of the matrix components.

### 11.3.2 Collagens of the Cartilage

The matrix is a complex, precisely organized, extracellular system. Collagen constitutes about 40% of the dry weight of hyaline cartilage. Its fibers are embedded in the amorphous substance of the extracellular matrix, consisting predominantly of proteoglycans. All components of the matrix are produced by cartilaginous cells.

The major collagen of cartilage is type II. Cartilage is also almost the only tissue containing type II collagen, the other sites being the vitreous body of the eye and cartilagenous sclera. Type II collagen (see Chap. 2) is different from other interstitial collagens. The most significant variation is the high level of glycosylation of the hydroxylysine residues. Almost half of all hydroxylysine residues are bound to galactose or galactosyloglucose. The biological role of this high carbohydrate content remains obscure. It is speculated that glycosylation facilitates the interaction with proteoglycans and affects the steric alignment of the molecules and formation of different fibers than those of types I or III collagens or stabilization of cross-links formed in cartilage.

Type II collagen accounts for about 85%–90% of the total cartilage collagen. The remaining part is composed of a heterogenous mixture of so-called minor cartilage collagens. Recently, they have been isolated and characterized as types IX, X, XI, and XIII (Clark and Richards 1985; Németh-Csóka and Meszaros 1984; Slutskii 1984; Mayne and Irvin 1986; Yasui and Nimni 1988; Mayne 1989). The details of their structure and functions are not clear (see Chap. 2). It has been discovered that articular cartilage contains small amounts of type VI collagen (Ayad et al. 1984; Eyre et al. 1987b; Poole et al. 1988a), which is probably absent from other cartilages. Type VI collagen constitutes about 1%–2% of the total collagen. Its biological significance in articular cartilage remains unknown.

The phenotypic expression of collagen types synthesized by the chondrocytes has been investigated in various experimental systems. Chondrocyte differentiation proceeds through two steps. Dedifferentiated cells change into slightly differentiated cells, and this stage is characterized by transition of the synthetic ability from type I collagen to predominantly types II and IX collagens. The later stage is characterized by the secretion of mostly type X collagen (Benya et al. 1978; Benya and Brown 1986; Von der Mark 1986; Castagnola et al. 1988; Gerstenfeld et al. 1989).

Collagen metabolism in cartilage has been shown to be regulated by a number of factors, including mechanical loading (Vasan 1983; De Witt et al. 1984) and humoral factors (e.g., transforming growth factor; Redini et al. 1988).

Age-related changes of cartilage include a decrease in water content and extractability of glycosaminoglycans. The total glycosaminoglycan content appears to be constant, and their decreased solubility is related to altered interactions with collagen. The collagen content decreases when expressed as a percentage of dry weight. It has been suggested that the major age-related change in collagen is an increase in cross-links (Burgeson 1982; Nakano et al. 1984). It is, however, difficult to differentiate degradative changes from physiological aging.

## 11.4 Intervertebral Disc

### 11.4.1 Structure and Function

The intervertebral discs are complex structures that form the amphiarthroses, the slightly movable joints, between the bodies of two adjacent vertebrae. They are the strongest bond between the vertebral bodies. Apart from bone and articular cartilage, they are only structures which transmit the whole weight of the body and therefore they must combine mobility with great strength.

The basic anatomy of the 23 discs in the human body is the same; each one is biconvex and consists of a more fibrous outer ring (the annulus fibrosus) and an inner, semifluid nucleus pulposus. The two end plates of hyaline cartilage form an integral part of both the disc and the vertebral bodies (Coventry 1969; Naylor 1970).

The annulus fibrosus consists of a concentric ring of fibrous lamellae that encases the nucleus and unites the vertebral bodies by continuity of the fibrous structure with the margin of the vertebral segment and with the investing anterior and posterior ligaments. The collagen fibers in the annulus fibrosus are organized in parallel arrays to form circular sheets. At the periphery of the disc, the coarse fibers of the outer lamellae are anchored into the vertebral bone where they are called Sharpey's fibers. They merge into the cartilage of the end plates. The fibers within each lamella lie parallel to each other, at an angle of 120° to fibers in adjancent lamellae (Inoue 1981; Klein and Hukins 1982). Cells of the annulus fibrosus are similar to tendon cells and are probably fibrocytes (Knese 1978). Some investigators classify them into subsets based on the amount of basophilic material surrounding the cells (Butler and Fujioka 1972).

The nucleus pulposus occupies the central portion of the disc and consists of a viscid fluid structure, which histopathologically is sparsely cellular and consists principally of loose, delicate, fibrous strands embedded in a gelatinous matrix. It contains cells derived from the embryonic notochord that persist only in young humans and are not the cells which generate the matrix of the nucleus. The latter cells arise from the mesenchyma and are specialized chondrocytes. The amount of cells in the nucleus is maintained or even increases with age (Trout et al. 1982; Parke and Schiff 1971; Johnson et al. 1986; Vernon-Roberts 1987).

The disc is held superiorly and inferiorly by cartilagenous plates, which are firmly fixed to the bony end plates of the vertebral segments and differ little in structure from the hyaline articular cartilage seen in the joints, except that they have no collagenous "skin" or indeed any discrete superficial surface. The cartilage serves as an anchor for the fine filamentous fibers of the nucleus pulposus in its central position and the coarse fibrous plates of the annulus fibrosus peripherally. The cells of the end plates are similar to the chondrocytes of hyaline cartilage (Maroudas et al. 1975).

### 11.4.2 Collagen in the Intervertebral Disc

Collagen constitutes about 66%–68% of the dry weight of the human annulus fibrosus (Dickson et al. 1967; Herbert et al. 1985; Pedrini et al. 1973). Adams and coworkers (1977) described changes in the collagen content in different regions of the annulus fibrosus of various discs from a 44-year-old human being. It is unclear which of these changes are related to degenerative process and which reflect a natural diversity. Both types I and II collagens are clearly found. The annulus fibrosus of human lumbar discs as a whole contain a considerably higher proportion of type II collagen (50%–65% of total collagen) than does the pig annulus fibrosus (20%) (Eyre and Muir 1975, 1977). No significant variations with age were found in the relative proportions of types I and II collagens in individuals aged 5–66 years. Eyre and Muir (1977) described the distribution of each type of collagen in a radial section of the annulus of the T12/L1 disc from a 5 year-old spine. The proportion of type I to type II varied inversely and smoothly from being almost all type I collagen at the outer edge of the annulus to only type II on reaching the nucleus pulposus. A similar pattern was seen in the pig annulus fibrosus (Eyre and Muir 1975). It is possible that this collagen polymorphism is connected with mechanical properties of the annulus fibrosus.

Collagen constitutes about 25% of the dry weight of the nucleus pulposus and consists of mostly type II collagen (Eyre and Muir 1977). This type represents more than 85% of the total collagen in the nucleus, the remainder being type I (Eyre 1979). There are single reports indicating the occurrence of other collagens in the disc (Ayad et al. 1982; Wu et al. 1987).

Studies of the whale nucleus pulposus which, being so large, is easy to dissect into parts show that the collagen content at the periphery of the nucleus is higher than than in more central regions (Ludowieg et al. 1973). The structural orientation of collagen in the nucleus is unknown. It is suggested that collagen fibers are randomly oriented and closely associated with proteoglycans. Smith and Serafini-Fracassini (1968) described the aggregation of collagen into large-diameter fibers after the extraction of proteoglycans. The extractability of glycosaminoglycans in the nucleus pulposus is higher that in the cartilage, which indicates that proteoglycans are less strongly associated with the collagen network (Lyons et al. 1964; Rosenberg et al. 1967). It is believed that the fluid matrix of the nucleus pulposus derives from the association of proteoglycans with a deformable, fine, collagen network.

The presence of collagenolytic enzymes in the intervertebral disc was shown by Sedowofia et al. (1982). They found a significant activity of collagenase in extracts of both the annulus fibrosus and nucleus pulposus. The activity in the nucleus pulposus was higher as was the amount of latent collagenase, almost four times more than in the annulus fibrosus. The activity of gelatinase was the same in both parts of the disc. The cellular origin of collagenolytic enzymes and their possible role in the degenerative process are unknown.

### 11.4.3 Collagen in Pathology of the Intervertebral Disc

The involvement of collagen in the pathology of the intervertebral disc includes age-related degeneration and nonage-related changes like disc herniation and scoliosis. Degenerative changes in the disc resemble those occurring in the articular cartilage in osteoarthrosis.

### 11.4.3.1 Age-Related Degeneration

Age-related changes in the intervertebral disc are difficult to differentiate from generative lesions as pathologists report that after 40 years of age all human discs reveal signs of degeneration, particularly in the cervical and lumbal regions (DePalma and Rothman 1970; Maurice-Williams 1981; Vernon-Roberts and Pirie 1977; Beard and Stevens 1980).

It is believed that the overall collagen content and the ratio of types remain essentially the same after maturity. It is possible that some local changes develop, but most of them reflect pathological phenomena (Adam and Deyl 1984). The content of proteoglycans and their components (glycosaminoglycans) changes with age, predominantly in water content. These changes are usually secondary to proteoglycan alterations (Adams et al. 1977).

### 11.4.3.2 Chemonucleolysis

Enzymatic hydrolysis of a herniated nucleus pulposus in situ is therapy for the treatment of a prolapsed disc. Various enzymes have been used, including chymopapain and collagenase. The details of the chemonucleolysis are described by Brown (1983) and Weinstein et al. (1986). The present description is limited to a brief presentation of the application of collagenase to the management of disc disorders. Bacterial collagenase was recommended for "discolysis" by Sussman in 1968 on the basis of his studies in vitro. Years later, he employed collagenase in vivo to dissolve the nucleus pulposus in dogs (Sussman and Mann 1969) and obtained better dissolution effects than with chymopapain (Brown 1983). Preparations of collagenase have been found to have a sufficient margin of safety and were introduced into clinical trials in 1981 (Brown 1983). The detailed comparison of the effects of chymopapain and collagenase has been elaborated. Briefly, chymopapain splits the protein core of proteoglycans whereas collagenase depolymerizes native collagen. Both enzymes exhibit no cell toxicity; collagenase has a higher safety margin, and hypersensitivity to it is less common (Olmarker et al. 1987; Wintermantel et al. 1985; Hedtmann et al. 1987).

### 11.4.3.3 Scoliosis

Scoliosis or lateral curvature of the spine is a common deformity caused by various, in many cases obscure, factors (Nachemson and Sahlstrand 1977). A molecular defect(s) of the connective tissue components which make up the intervertebral disc has been suggested as the underlying mechanism of idiopathic scoliosis. In these patients no neurological, muscular, or skeletal abnormaly can explain the occurrence of scoliosis. Idiopathic scoliosis is associated with a familial predisposition, and biochemical findings suggest systemic metabolic abnormalities in the connective tissue. They include a raised glycosaminoglycan level in serum and urine (Kazmin and Merkurieva 1971; Chmiel et al. 1980) and increased urinary excretion of hydroxyproline (Zorab 1969, 1971; Zorab et al. 1971; Benson 1965, 1972). An elevated proteoglycan production was observed in cultured in vitro fibroblasts from patients with idiopathic scoliosis (Conen 1971; Nordwall and Waldeström 1976) as well as changes in skin collagen cross-linking (Francis et al. 1976b, 1977). Variations in the composition of the intervertebral discs have not been elucidated. Some reports indicate an altered content of glycosaminoglycans and collagen, but it is difficult to distinguish which changes are primary and which are formed as a response to overloading of the disc. A raised content of collagen in the annulus fibrosus and nucleus pulposus was reported by Pedrini et al. (1973) and Taylor et al. (1976a). An increased intermolecular cross-linking in the collagen of scoliotic discs has been suggested on the basis of diminished extractability with pepsin (Bushell et al. 1978). Collagen changes were associated with a decreased content of proteoglycans and other components of the amorphous matrix of the discs (Pedrini et al. 1973; Ghosh et al. 1980). Similar biochemical alterations were also described in the intervertebral discs of mice and chicken with inherited kyphoscoliosis (Mason and Palfrey 1977; Greve et al. 1988).

These findings have led to the hypothesis that an inborn, mild defect in the connective tissue produces hyperelasticity of the discs, which is responsible for the development of scoliosis. On the other hand, scoliosis often occurs secondarily to other changes in the musculoskeletal system (e.g., paralysis of some groups of muscles due to cerebral palsy). Variations in disc composition were also found in patients with secondary scoliosis, which suggests the important role unbalanced mechanical loads play in the development of disc abnormalities. This is consistent with biochemical analyses of discs obtained from animals with surgically produced models of scoliosis (Taylor et al. 1976b). It is possible that two mechanisms, a primary defect in the connective tissue and a tissue response to mechanical overload, occur simultaneously and are mutually related. The majority of the specimens available for investigation have been obtained from patients with advanced scoliosis; thus, the observed biochemical composition of the discs results from primary and secondary pathological phenomena. Further biochemical studies on the early stages and development of scoliosis are needed to elucidate the causative and associated pathophysiological phenomena.

## 11.5 Tendons and Related Structures

### 11.5.1 Structure and Function

A tendon is an elongated, cylindrical structure that attaches striated muscle to bone and through which the tension of a muscle is transmitted to the skeleton. They are the most common example of dense, regular connective tissue and consist almost entirely of parallel, closely packed bundles of fibers. The fibers

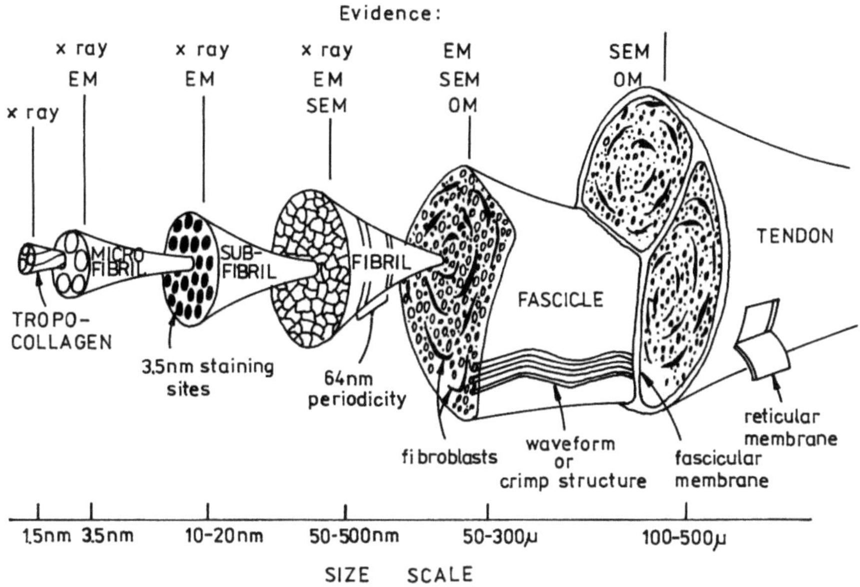


**Fig. 11.1.** Hierarchical structure of tendon. (*S*)*EM*, (scanning) electron microscopy; *OM*, optical microscopy. (From Kastelic et al. 1978 with permission)

are separated by a small amount of amorphous matrix (Rowe 1985a, b). Externally, the tendon is surrounded by a sheath of dense connective tissue, the paratenon. Under this layer another sheath is present, the epitenon, which continues on as the endotenon, separating the fascicles as well as carrying the blood supply. The structure and function of the mammalian tendon is reviewed by Elliot (1965), Davidson (1982), and Baer et al. (1988). Its typical feature is its hierarchical structure. Various terms have been used to describe the components, and the model of tendon hierarchy described by Kastelic et al. (1978) and employed in the present chapter is depicted in Fig. 11.1. According to this, the tendon consists of fascicles separated by endotenon. Each fascicle is built up of many fibrils, and among them fibroblasts are located. Fibrils are collagenous structures with a subordinate hierarchy including subfibrils and microfibrils. The most elementary component of all is the molecule of collagen. The structural hierarchies in collagen from the molecule through the fibril level are shown in Fig. 11.2. There is a large body of evidence for this structure although some aspects still remain controversial.

Tendon has been for years a classic subject for investigations of the biomechanical properties of connective tissue (for review, see Woo and Sites 1988; Oxlund 1986; Folkhard et al. 1987). Several factors affect the mechanical behavior. All collagenous structures change with maturation and aging; the stress-strain behavior of rat tail tendon as a function of age is shown in Fig. 11.3. With aging come also alterations in the hierarchical structure (Parry

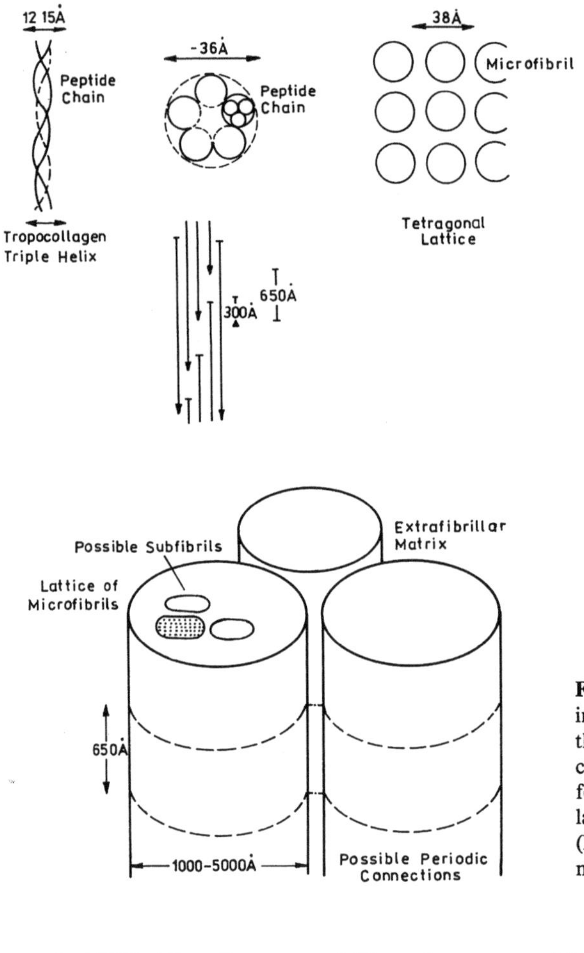


**Fig. 11.2.** Structural hierarchies in collagen from the molecule through the fibril level: tropocollagen, microfibril, lattice formed by microfibrils, and collagen fibril (perspective view). (From Baer et al. 1974 with permission)

et al. 1978). The diameter of fibers increases with age; in newborn rats, the mean diameter is 30 nm while at maturity, it is 450 nm. There is no further increase with old age. These molecular changes, including variation of extent of cross-linking, are not fully understood despite several investigations. It is clear that cross-linking contributes to the mechanical behavior of the tendon (Squier and Bausch 1984; Davison 1989).

Alterations in hydration of the extracellular matrix influence the mechanical properties. These changes probably reflect the relationship between proteoglycans and fibrillogenesis (Scott et al. 1981; Vogel et al. 1984).

### 11.5.2 Collagen in Tendinous Pathology

A number of pathological conditions are associated with tendons, ligaments, and fasciae, but their explanation at a molecular level is relatively poor. Inborn

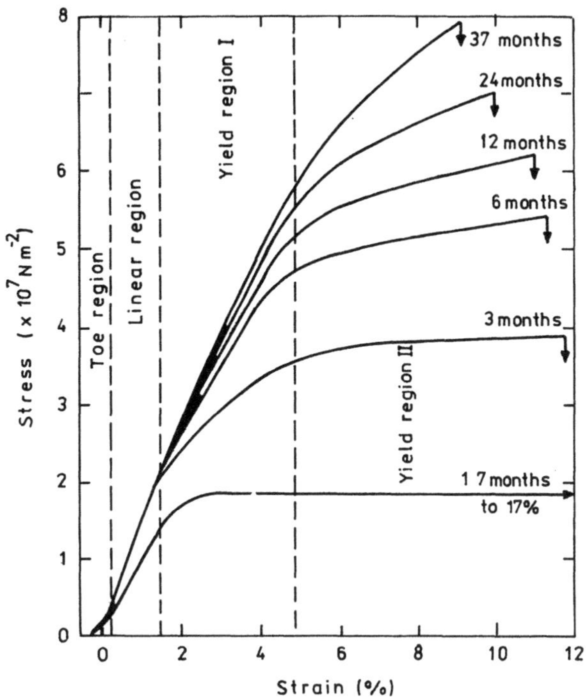


**Fig. 11.3.** Stress-strain behavior of rat tail tendon as a function of age. (From Kastelic and Baer 1980 with permission)

errors of collagen metabolism are commonly attributed to impaired properties of tendons and their ruptures. These diseases are reviewed in Chap. 8. Here, disorders of unknown cause, Dupuytren's contracture and carpal tunnel syndrome, are described.

### 11.5.2.1 Dupuytren's Contracture

Dupuytren's contracture is a relatively common disorder characterized by hyperplasia of the palmar fascia and related structures. It affects mostly the tendons of the palmar fascia, resulting in nodule formation and contracture of the aponeurosis. The cause is unknown.

Morphological investigations including electron microscopy studies have not shown abnormalities in collagen structure in patients with the disease (Jahnke 1960; Dahmen 1961). The nodules are composed of closely packed cells and dense fibrous connective tissue (Luck 1959). They contain higher amounts of collagen than normal aponeurosis does. Bailey et al. (1977) showed that the nodules, contractures, and apparently unaffected aponeurosis from these patients contained a significantly increased proportion of type III collagen than the aponeurosis of normal subjects. This is consistent with the view that the disease is not a local lesion of a specialized form of connective tissue as the changes are usually symmetrical, the feet are also involved, and an increased proportion of type III collagen has also been reported in the apparently unaffected aponeurosis. The cellular origin of this type III collagen remains obscure. Both fibroblasts and myofibroblasts have been suggested as the cells involved (Gabbiani and Majno 1972). Delbrück and Schroder (1983) showed that cultured in vitro fibroblasts from a Dupuytren's contracture synthesized more collagen and glycosaminoglycans per cell compared with the cells from normal human fascia. This finding leads to the conclusion that the former undergo changes in metabolic control which result in the impaired composition of the matrix of affected palmar fascia.

#### 11.5.2.2 Carpal Tunnel Syndrome

Carpal tunnel syndrome is a painful disorder defined as a compression neuropathy of the median nerve where it passes under the transverse carpal ligament at the wrist. Many elements can produce carpal tunnel syndrome (injuries of the wrist, rheumatoid arthritis, other causes of synovitis, a few systemic diseases) although some cases described as idiopathic occur without known cause. The idiopathic form is usually associated with a familial predisposition (Phalen 1966). Staubesand and Fisher (1980) described abnormal collagen fibrils in the transverse carpal ligament of patients with the syndrome. These fibrils had irregular outlines in cross-section, ranging from slight peripheral fraying to identical circumferences. This was confirmed by Stransky et al. (1987), who reported a changed distribution of fibers with various diameters in the cross-section of an affected transverse carpal ligament. In normal ligament the majority of fibers (more than 90%) have a diameter centered about 50 nm while in patients with idiopathic carpal tunnel syndrome only 35% of the fibers were within this range. Large fibers up to 200 nm in diameter as well as very small fibers (20 nm) have been observed. The same group (Weis et al. 1987) reported that in affected patients' ligaments the same area of cross-section is occupied by approximately half the number of fibers as compared with normal subjects. The fusion process of normal-sized fibers or a dysplastic process were proposed to explain the observed irregularities. The detailed mechanism remains unknown, and any molecular alterations in ligaments from patients with carpal tunnel syndrome have not been elucidated.

## 11.6 Joints

### 11.6.1 Role of Collagen in the Structure and Function

Joints are interruptions in the rigid bony skeleton and are classified into three groups: immovable (synarthroses), slightly movable (amphiarthroses), and movable (diarthroses). Synarthroses are generally found in the skull. An example of amphiarthroses is the intervertebral joints, discussed earlier in this chapter. This section emphasizes the diarthroses, i.e., the synovial joints. They comprise most of the body articulations and are characterized by a wide range of movement. The joint is a complex structure. The articulating body surface are part of the bone, known as the articular end plate. They are covered by hyaline cartilage. Between the gliding surfaces of the cartilage is the joint cavity. The stability of the joint is achieved due to the fibrous capsule and ligaments. Within the joint capsule is a specialized layer of connective tissue cells, the synoviocytes, which secrete the synovial fluid into the joint cavity. Other structures, such as fibrocartilaginous discoid forms (menisci), are present in same joints.

Collagen is present in all components of the joint. Ligaments have the some structure as tendons. The joint capsule is made up principally of parallel bunds of collagen, sparsely populated with fibrocytes. Collagen accounts for about 90% of the dry weight of the tissue. Mostly, type I collagen is found. Articular cartilage was discussed earlier under the general description of cartilaginous tissue.

The menisci of the knee joint are semilunar, weight-bearing structures located one in each of the medial and lateral compartments of the joint. The fibrocartilage of the meniscus is reinforced by collagen fibrils which are able to withstand tension but have low compressive, flexural, and torsional stiffness (Hukins 1982). Almost all (about 98%) is type I collagen (Eyre and Muir 1975; Eyre and Wu 1983). The remaining part consists of types III and V collagens. Type II collagen has been found in the inner one-third (Cheung 1987), a part which resembles hyaline cartilage more than the remaining portion of the fibrocartilage and contains less collagen than the outer portion.

Mechanical properties of the meniscus are related to the orientation of its collagen fibers. The fibrils are oriented circumferentially in the bulk tissue and radially in the surface region. Polarised light microscopy studies showed them to be aggregated into crimped fibers resembling those of tendon. Some of these fibers pass from the outer region of the meniscus into the bulk tissue (Bullough et al. 1970; Aspden et al. 1985).

## 11.7 Skeletal Muscle

### 11.7.1 Role of Collagen in the Structure and Function

Muscles are specialized organs for the production of movement. The cells responsible for this activity are known as muscle fibers. There are three types of muscles: striated voluntary or skeletal, smooth involuntary, and the heart. This chapter deals with the first group, the main tissue of the body flesh.

Skeletal muscles are structures built up of striated muscle and associated connective tissue. The whole is wrapped in an envelope of connective tissue, the epimysium, and a finer partition of it surrounding bundles or fascicles of the muscle fibers forms the perimysium. Individual fibers are surrounded by the endomysium, a fine connective tissue which extends from the perimysium.

The amount of collagen in the skeletal musculature is relatively high and is estimated at about half the total collagen content of the body. Collagen forms 1%–9% of the fat-free dry mass of the muscle (Lawrie 1979). Despite this high content, the majority of studies on the anatomy and physiopathology of striated muscle have focused on the muscular cells. Connective tissue rich in collagen is commonly recognized as a supportive structure only. There are a few reports concerning the types of collagen found. Duance et al. (1977) described a predominance of type I collagen in the epimysium; only traces of types III and IV collagens were noted. Type III collagen was the only form observed in the perimysium, and types III and IV collagens were detected in the endomysium. Type IV collagen was found together with fibronectin and laminin to delineate precisely each muscle fiber. This observation reflects the occurrence of basal lamina type structure in close contact with the sarcolemma (Hantaï et al. 1983). Light and Champion (1984) confirmed these results with biochemical methods. It was shown that both the epimysium and perimysium contained type I collagen as the major component and type III collagen as a minor component. Traces of type V collagen were found in the perimysium. A similar pattern was reported in the endomysium, where additionally type IV collagen was found. The determination of collagen in muscular tissue has been applied in the food industry to estimate the quality of meat (Jeremiah and Martin 1981). The collagen content is higher in the fast muscle than in the slow muscle. This difference is probably related to the role of collagen in the elastic properties of fast tetanic muscle compared with slow tonic muscle (Kovanen et al. 1980, 1984). Synthesis of muscle collagen is almost the same as that of noncollagenous proteins. The fractional synthesis rates of types I and III collagens are very similar (Palmer et al. 1980).

Collagenous structures have an important role in the biomechanics of muscle. They are responsible for the transmission of forces generated by the contraction of the fibers. The following problems are associated with the connective tissue: the cellular origin of the matrix proteins (including collagen), their role in the development of muscle, and their role in the pathogenesis of disorders affecting muscle. Physical activity affects the collagen content of muscle. An increase was shown in the experimental hypertrophy of muscles (Jablecki et al. 1973). On the other hand, immobilization results in a transient increase in collagen content due to a rapid breakdown of the noncollagenous proteins after decreased muscular activity. When continued, immobilization is associated with inhibited collagen synthesis (Jozsa et al. 1988; Kovanen and Suominen 1988; Kovanen 1989).

There is evidence that myoblasts and fibroblasts have a common progenitor cell (Miranda et al. 1978). The primary muscle clones are able to synthesize collagen and proteoglycans as well as myotubules (Lipton 1977). The mechanism which regulates the expression of the fibroblast-like phenotype in myoblasts is unknown. There are suggestions that collagen supports myogenesis. The type of collagen present affects the developmental pattern of myotubules of cultured myoblasts. It is possible that collagen determines the development of optimal (from the point of view of biomechanics) muscle cytoarchitecture. The ability to synthesize collagen was found also in myogenic continuous cell lines cultured in vitro. The lines were used for studies on factors affecting the differentiation of myoblasts (Abbott et al. 1974; Miranda et al. 1978). It was found that differentiation is associated with the suppression of collagen synthesis. A fraction of cells, developing from the common precursor cell, which do not undergo fusion can be separated from the myogenic cell cultures. These mononuclear cells showed many similarities to fibroblasts and were considered a major source of collagen in skeletal muscle. Postnatal aging is associated with a slow increase in the collagen content in muscles. Differences between trained and nontrained muscles as well as slow and fast muscles have been described (Kovanen et al. 1987, 1988).

### 11.7.2 Collagen in the Pathology

The data concerning the role of collagen in the pathology of striated muscle are limited to a few reports on muscular dystrophy. Muscular dystrophies are inherited myopathies, characterized by a progressive severe weakness of the muscles. Abnormalities in collagen have been reported in the Duchenne type. Pathological studies showed an increased amount of collagen in the perimysium and endomysium, and in final stages of the disease the muscle is almost entirely replaced by connective and fatty tissues. An increased deposition of fibrous tissue within the muscle is observed in preclinical forms of the disease. The cellular origin of this collagen is not known although there are suggestions that its accumulation is caused by overproduction either by dystrophic muscle cells or by mononuclear fibroblast-like cells (Ionesescu et al. 1971). In dystrophic muscles so-called satellite cells appear. It is possible that these cells undergo metaplastic transformation to fibroblasts. The quanitative changes in collagen are associated with an increased amount of types III and IV collagens (Duance et al. 1980a). Abnormalities in collagen content and composition are probably only secondary to the primary defect in the muscular tissue. The nature of this causative defect remains unknown. Alterations in the tendon collagen were also reported (Bartlett et al. 1973). In dystrophic chicken and mice models, an abnormal accumulation of collagen has been observed (DeMichelle and Brown 1984; Marshall et al. 1989). An increased collagen content and cross-linking have been shown to be a basis for the elevated resting tension and stiffness in the dystrophic muscles (Feit et al. 1989).

Altered collagen type pattern is seen in muscles from patients with polymyositis. Polymyositis is an inflammatory muscle disease of unknown etiology. Morphological studies demonstrated an increase in collagenous tissue in the perimysium and endomysium (Bohan and Peters 1975). Duance et al. (1980b) reported a significant enhancement in type III collagen and, to a lesser extent, types I, IV, and V collagens. The mechanism of these changes remains obscure and is suggested to be secondary to an initial, unknown cause of the disease. Udén et al. (1980) reported that structural changes in collagen isolated from the fascia of the longitudinal dorsal muscles from patients with congenital scoliosis resulted in a decreased ability of the collagen to aggregate platelets. Their nature has not been investigated.

## 11.8 Collagen in Rheumatic Disorders

### 11.8.1 Rheumatoid Arthritis

The term "rheumatic disorders" refers to a number of diseases which primarily involve joints. The cause of the majority of them remains obscure, but it is clear that their pathogenesis is associated with connective tissue abnormalities.

Rheumatoid arthritis is a form of chronic polyarthritis in which joint changes are accompanied by general systemic disturbances. The origin is unknown despite extensive efforts of investigators.

The main site of the pathological process is the more peripheral joints. Swelling and congestion of the synovial membrane are the earliest changes. They are followed by cell infiltration, effusion of synovial fluid into the joint space, and hypertrophy of the synovial membrane. Progressive destruction of the articular cartilage is associated with the formation of inflammatory granulation tissue, the pannus. These alterations lead to severe disability due to fibrous adhesion or bony ankylosis within joints. Articular changes are associated with systemic alterations, including subcutaneous nodules and muscle infiltration (Fassbender 1983).

Progressive inflammation in rheumatoid arthritis is associated with destruction of the cartilage and subchondrial bone (Harris et al. 1977). Therefore, a number of investigations have focused on collagenase, its synthesis and activation within the pannus and synovial fluid (for review, see Mainardi 1985). Collagenase is responsible for the breakdown of the cartilage matrix and is involved in the generation of chemotactic peptides and activators of the immune response. It has been suggested that the autoimmunity initiated by the action of collagenase may be one of the mechanisms of chronicity in some patients (Golds and Poole 1984).

The cellular origin of the collagenase in rheumatic joints remains to be clarified. Rheumatic synovial cells are able to produce collagenase, as found in several studies. The neutrophils also liberate the enzyme, and the synovial fluid contains at least two enzymes (Krane 1981; Cohen et al. 1983a). Collagenase activity has also been found in rheumatoid nodules (Hashimoto et al. 1973a).

Regulation of the action of collagenase may be a key mechanism for joint destruction (Sellers and Murphy 1981; Harris 1986). Synthesis and release of the enzyme are controlled by some products related to the immune phenomena

(Deshmukh-Phadke et al. 1978; Kerwar et al. 1984). Interleukin 1 has received the most attention and is thought to be involved in collagenase production and secretion (Killar and Dunn 1989). It is noteworthy that hydroxyapatite or other crystals release collagenase. This phenomenon may be responsible for so-called crystal-induced joint destruction (Cheung et al. 1981; Dieope et al. 1988). A role for the environment in the activation of collagenase in inflammed joints has been postulated.

Several reports showed the presence of active collagenase in synovial fluids from some rheumatoid patients (Evanson et al. 1967; Harris et al. 1969, 1970; Menzel and Steffen 1977). Latent collagenase was found in the majority of patients (Abe et al. 1973). Inhibitors and activators have been identified in rheumatoid synovial fluids. The presence of $\alpha_2$-macroglobulin and the tissue inhibitor of metalloproteinases have been discovered in the fluid (Shtacher et al. 1973). The presence of a heat labile and nondialyzable activator of latent leukocyte collagenase was reported in the synovial fluid obtained from the knee joint of patients with rheumatoid arthritis (Kruze and Wojtecka 1972; Kruze et al. 1978). An activator of procollagenase has been purified from synovial fibroblasts in culture (Vater et al. 1983). The mechanism of its action remains unknown.

There are only a few reports on collagen changes in rheumatic joints. An increased depolymerization of synovial collagen isolated from the affected joints was shown by Steven (1966). Similarly, Weiss et al. (1975a) showed an enhanced solubilization of collagen after pepsin treatment. These findings reflected the higher content of type III collagen (Weiss et al. 1975b). It is possible that these alterations are secondary to the inflammation.

The majority of markers of connective tissue metabolism are without value in the diagnosis of rheumatoid arthritis and degenerative rheumatic disorders. The urinary excretion of hydroxyproline is generally within the normal range, while an increase in hydroxyproline, hydroxylysine, and collagen-like protein levels in the sera of patients with rheumatoid arthritis was shown. There was no difference between patients with seropositive and seronegative forms of the disease (Kucharz 1986c). Collagen peptidase activity was reduced (Iwasc-Okada et al. 1985). A similar reduction of activity of dipeptidyl aminopeptidase IV, the enzyme which cleaves glycyl-glycyl-proline, was reported by Fujita et al. (1978).

An increase in the aminoterminal propeptide of type III procollagen was found in the sera and was shown to be associated with the inflammatory synovial mass. Normal levels were observed in most patients with degenerative diseases (Hørslev-Petersen et al. 1988b, c, d).

Synovial fluid hydroxyproline level correlated with the disease activity in rheumatoid arthritis (Wize et al. 1975; Manicourt et al. 1979, 1980, 1981; Kucharz 1986c). No changes in the aminoterminal propeptide concentration were found (Gressner and Neu 1984).

The possible involvement of a collagen autoimmunity in the pathogenesis of rheumatic disease is supported by a number of reports on the occurrence of collagen antibodies in the sera and synovial fluids of patients. Elevated levels of serum antibodies to type III, collagen have been demonstrated in patients with rheumatoid arthritis (Steffen and Timpl 1963; Michaeli and Fudenberg 1974; Steffen et al. 1974; Andriopoulos et al. 1976a, b, c; Rowley et al. 1987; Clague et al. 1980, 1981, 1983; Gioud et al. 1983; Trentham et al. 1981; Wooley et al. 1984; Meyer et al. 1983; Stockman et al. 1989). Charriere et al. (1988) noted a greater frequency of antibodies to type II collagen than type I collagen. The most frequent antibodies were those to types IX and XI collagens. Increased levels of collagen antibodies and collagen-anticollagen immune complexes in the synovial fluid of rheumatoid patients were reported (Menzel et al. 1976, 1978). Antibodies to the collagen-like region of the complement component C1q were found in the sera of patients with rheumatic disease (Wener et al. 1989). The diagnostic value of collagen antibodies remains low (for review, see Townes 1984; Verbruggen et al. 1986).

### 11.8.2 Osteoarthrosis

Osteoarthrosis is an abnormality of the synovial joints characterized by focal splitting and fragmentation of articular cartilage which is not directly attributable to an inflammatory process. Osteoarthrosis and intervertebral disc degeneration are degenerative joint diseases. The process of joint degeneration may be accompanied by subchondral bone sclerosis, subchondral bone cysts, narrowing of the joint space, and bone outgrowths at the joint margins in the form of osteophytes.

Osteoarthrosis resulting from an identifiable cause is called secondary; when the cause cannot be found the disease is said to be primary or idiopathic. Recently, a predisposition to familial primary osteoarthrosis was shown to be linked to the type II collagen gene (Palotie et al. 1989). Secondary osteoarthrosis is caused by the coexistence of a mechanical factor and an intrinsic predisposition. It is believed that any disorder which directly or indirectly alters articular geometry may induce or accelerate osteoarthrosis.

Osteoarthrosis is believed to be primarily a cartilage disease. The mechanism of the cartilage degradation is complex. Immune and enzymatic factors are suggested to be involved, although the disease is noninflammatory.

Nimni and Deshmukh (1973) showed that osteoarthritic cartilage synthesizes other collagen types than normal cartilage. These differences are associated with a lower thermal stability of collagen and other physicochemical alterations of the matrix (Herbage et al. 1972). Despite numerous investigations which provide evidence for variation between normal and osteoarthritic cartilage and subchondral bone, the cause is unknown (Adam and Deyl 1983; Adam et al. 1984; Ampe et al. 1986; Poole 1986). It is unclear whether the differences are the primary mechanism for the disease or result from degenerative changes in the diseased cartilage.

### 11.8.3 Animal Models of Rheumatic Disorders

Several animal models of rheumatoid arthritis have been described, but their relationship to the disease in humans remains obscure. The models include arthritis induced by antigens, adjuvants, mediators, and infectious agents (for review, see Cole et al. 1982). Collagen metabolism has been studied in a few models only, and these reports are reviewed here.

The most commonly used model is adjuvant arthritis. The induction of polyarthritis in rats was first demonstrated by Stoerk et al. (1954); it is produced by the injection of complete Freund's adjuvant in mineral oil. The adjuvant contains *Mycobacterium butyricum* or *M. tuberculosis.* The symptoms of arthritis develop in more than 80% of animals receiving the injection (DeLustro et al. 1984). An inflammatory swelling sets in at the site of injection on the same day. Some 2 or 3 weeks later, it has become very severe and thereafter remains without significant changes. On the contrary, inflammatory effects develop in the contralateral paw. These usual progress and become more severe in another 1 or 2 weeks. A fourfold increase in volume of the contralateral paw has been found (Kaibara et al. 1984), with changes including reddening, swelling, and functional impairment as well as osteoporosis, erosions, and periosteal reaction. The increase in weight of the paw has been attributed to an expansion in the amount of ligaments and cartilaginous tissue.

Local and systemic alterations in connective tissue metabolism were reported in rats with adjuvant arthritis. An increased breakdown of newly formed collagen and delayed maturation of collagen were noted (Trnavský and Trnavská 1973). The metabolism of the proteoglycans was found to be altered; in uninflamed dorsal skin their biosynthesis was diminished while in inflamed limbs the synthesis and turnover had increased (Exer et al. 1976).

The mechanism of pathogenesis of adjuvant arthrititis remains unclear. It is thought to result from a cell-mediated immune response to disseminated myobacterial antigen (Van Arman 1976; Billingham and Davies 1979). The role of cellular immunity in this disease is supported by the ability to transfer the arthritis with spleen cells or lymph nodes but not with serum (Waksman and Wennersten 1963; Pearson and Wood 1964). Humoral immunity is also suggested to be involved in the induction of arthritis (MacKenzie et al. 1978). Trentham et al. (1980) found both humoral and cell-mediated immunity to type II collagen in rats with adjuvant arthritis. This finding has not been confirmed in the studies of DeLustro et al. (1984).

This animal model of arthritis seems to be related to immune phenomena. The disturbed collagen metabolism is probably a secondary reaction resulting from the inflammatory changes. Adjuvant arthritis has been used in the experimental evaluation of several antirheumatic agents.

An experimental model of inflammatory disease that closely mimics arthritis in humans can be induced in rodents by the intradermal injection of native type II collagen. The model was reported for the first time by Trentham et al. (1977). The collagen from human, chick, or rat cartilage induced polyarthritis in rats. The collagen preparations were emulsified in incomplete or complete Freund's adjuvant, but there was evidence that the adjuvant was only a supportive agent that increased the immune response to collagen. Later, several studies showed that arthritogenicity was a specific and intrinsic property of type II collagen unrelated to adjuvanticity. Injections with types I or III collagens or denatured type II collagen did not produce any effect.

The major histologic abnormalities seen in rats with collagen-induced arthritis include synovial hyperplasia with infiltration of the subsynovial tissue by mononuclear and polymorphonuclear cells and exudation of cells into the joint space. Bone and cartilage changes include fraying or fragmentation of the cartilage surface, marginal erosion, and periostitis (Stuart et al. 1979). The earlier changes were seen on the 5th day after immunization (Caulfield et al. 1982). The early deposition of fibrin was substituted by synovial hyperplasia on the 12th day. Later, on the 19th day, cellular infiltrations were reported. Detectable cartilage alterations were present when cellular infiltrations were found. Structural abnormalities in chondrocytes were associated with a loss of proteoglycans in the matrix (DeSimone et al. 1983). Extraarticular manifestations of arthritis have been described although they were found within the ears only. The external ear lesions were visible later than the joint changes. Destruction of the cartilage plate with focal regenèration and subperichondral infiltrations were reported (Cremer et al. 1981; McCune et al. 1982). The changes resemble those of relapsing polychondritis in humans. Alterations found in the inner ear similar to otospongiosis observed in otosclerosis and Menière's disease are detailed in Chap. 17.

Collagen-induced arthritis is without doubt caused by an immune response to type II collagen although the exact mechanism remains unknown. Immunization of animals with native type II collagen is associated with the induction of cellular and humoral immunity against it. In rats, the ability to respond to type II collagen and susceptibility to arthritis are linked to the RT1 major histocompatibility locus, and rats with $RTH1^{u}$, $RT1'$ and $RT1^{a}$ type develop arthritis (Griffiths and DeWitt 1981; Griffiths et al. 1981). In mice, the immune response towards collagen was found to be major histocompatibility complex I-region-restricted (Wooley et al. 1981, 1983). Arthritis can be passively transferred from an immunized rat to syngeneic, nonimmunized recipients with cells but not with serum (Trentham et al. 1978). However, the transfusion of large amounts of concentrated antibodies leads to the induction of arthritis (Stuart et al. 1982, 1983; Stuart and Dixon 1983). The mechanism of the transferred disease seems to be somewhat different than that in immunized animals. The role of collagen-specific antibodies in the induction of arthritis has also been suggested in other than passive transfer experiments. The development of collagen-induced arthritis is complement-dependent (Kerwar et al. 1982, 1983), and the formation of a localzed antigen-antibody complex is required for the induction of polyarthritis (Kerwar et al. 1983). Moreover, the arthritis can be suppressed by the prior intravenous administration of either type II collagen coupled to syngeneic spleen cells or type II collagen in solution (Englert et al. 1984).

Generally, collagen-induced arthritis is a model of autoimmunity arthritis employing collagen as an antigen. It is used in several investigations. However, the collagen metabolism remains poorly recognized.

The role of collagenase in the pathogenesis of rheumatoid arthritis prompted the induction of joint changes by injection of enzyme preparations. The subsequent arthritis is limited to those joints that have been injected with collagenase, but the changes exhibit an inflammatory nature and are different from those found in experimental models of osteoarthrosis. The collagenase-induced arthritis model is not used for testing antirheumatic drugs but is of importance in studies on the pathogenesis of articular manifestations of rheumatoid arthritis.

Purified rheumatoid synovial collagenase was injected intraarticularly into the knee of rabbits. Three injections were given at 2-day intervals. Arthritis-like changes were observed. One day after the last injection, a cell-rich exudate was found. One week later, signs of proliferative synovitis were evident, and 3 weeks later, there was no exudation and fewer signs of inflammation, but distinct fibrotic changes of the synovium were seen. No pathological alteration of the cartilage was observed. There were no changes in control animals injected with trypsin (Steffen et al. 1979). It was also noted that the intraarticular injection of collagenase $\alpha_2$-macroglobulin complexes was more arthritogenic than collagenase alone. Such complexes were found within the synovial fluid of patients with rheumatoid arthritis. This suggests that the causative role of the proteolytic enzymes cannot be limited to the simple breakdown of the articular cartilage. It is possible that degradation triggers several secondary phenomena of a metabolic and immune nature.

# 12 Cardiovascular System

## 12.1 Introduction

The circulatory system is composed of two divisions, the cardiovascular and the lymph-vascular. The former consists of the heart, which is a muscular pumping device, and a closed system of vessels called arteries, veins, and capillaries. The lymph-vascular division consists of a network of lymph capillaries, lymphatic vessels, and lymph nodes. Collagen of the lymph nodes is described in Chap. 20.

The main function of the circulatory system is the transportation of vital materials between the external environment and the internal fluid of the body. The transported materials include water, oxygen, carbon dioxide, inorganic and organic nutrients, hormones, immunoglobulins, and waste products of catabolism. Its effective functioning depends on differences of pressure which form one of the forces moving blood in the cardiovascular system. The mechanical properties of the system are associated with proper resistance and elasticity. Collagen is present in almost all structures of the system and takes part in maintaining the mechanical features of the heart and vessels.

## 12.2 The Heart

### 12.2.1 General Structural Features

The heart is a highly specialized portion of the circulatory system. It is a muscular pump that maintains the flow of blood from early embryonic development until death. Functionally, it is a pair of muscle pumps, consisting of four chambers, a right atrium and ventricle (the right pump) and a left atrium and ventricle (the left pump). Both pumps are linked to each other by the pulmonary and the systemic circulations. The heart, like all parts of the cardiovascular system, is composed of three layers, the endocardium, myocardium, and epicardium. The greater part of the wall of the heart is made up of cardiac muscle fibers and is called the myocardium. It is covered externally by the inner layer of the serous pericardium, and this, together with a thin subserous layer of connective tissue, forms the epicardium. The chambers of the heart are linked by the endocardium.

The heart is a complex organ of important vital function. Although human and animal hearts contain only a small amount of collagen and other connective tissue components, there are several pathophysiological conditions associated with alterations in their collagen (Robinson et al. 1983, 1984). The present section summarizes the available results of investigations on collagen in the normal and diseased heart. The role of collagen is described in normal heart function and development as well as a myocardial hypertrophy and infarction. Valvular collagen is, at least in part, responsible for the unique properties of the valves.

### 12.2.2 Collagen in the Normal Heart Muscle

Connective tissue components, and collagen in particular, form a network in the heart muscle. This network, described for the first time by Holmgren in 1907 and extensively investigated under the scanning electron microscope, consists of three components: a complicated weave of collagen bundles that surrounds the myocytes and sequesters them into groups, collagen struts that interconnect the myocytes within groups, and bundles that join the myocytes to the capillaries (Caulfield and Borg 1979). In this system, myocytes are centrally placed and directly connected to all components. It is believed that the weave contributes to the viscous and elastic properties of the heart. Myocyteto-myocyte struts can prevent slippage of adjacent cells during the cardiac cycle and support the stretch of adjacent cells during diastole. Capillary patency in the early phases of systole is maintained by myocyte-to-capillary bundles of collagen. Details of the morphological structure of the heart extracellular matrix have been described in recently published papers (Borg and Caulfield 1979, 1981; Robinson et al. 1983; Caulfield et al. 1984; Weber 1989).

Biochemical and histological techniques were applied to the measurement of collagen in heart muscle. Early measurements showed that the collagen content was 0.61% – 1.35% of the wet weight (Blumgart et al. 1940) or 4.01% – 4.76% of the dry weight (Montfort and Pérez-Tamayo 1962). Introduction of hydroxyproline measurements led to more precise estimates. The collagen content was reported to be 41.5 – 52.5 mg/g dry weight (Caspari et al. 1977) or 3.33 – 5.42 μg of hydroxyproline/g wet tissue (Wegelius and Von Knorring 1964). Similar values were reported by Frederiksen et al. (1978), who found 6.1 mg of collagen per gram of wet tissue.

There are discrepancies in the literature concerning the effect of aging on the collagen content of the myocardium. Using histological techniques, no change with age in the human ventricular collagen content was found by Dogliotti (1931). An increase was reported by Ehrenberg et al. (1954) and Lenkiewicz et al. (1972). Using biochemical methods, Clausen (1962) found a progressive increase in with age. Oken and Boucek (1957) showed the ventricular concentration of collagen to be higher in the first decade than subsequently. Some investigators were, however, not able to detect any variation with age in the human

**Table 12.1.** Collagen content in the heart muscle of rat

| Age (months) | Collagen content | | |
|---|---|---|---|
| | Whole heart (a) (mg of Hyp/g wet tissue) | Left ventricle (b) (mg/g of protein) | Right ventricle (b) (mg/g of protein) |
| 0.1 | ND | $2.91 \pm 0.55$ | $3.27 \pm 0.08$ |
| 0.5 | $2.71 \pm 0.63$ | $3.47 \pm 0.15$ | $4.51 \pm 0.24$ |
| 1 | $2.91 \pm 0.25$ | ND | ND |
| 2 | $3.68 \pm 0.13$ | $3.71 \pm 0.08$ | $5.65 \pm 0.11$ |
| 6 | $7.15 \pm 0.69$ | ND | ND |
| 8 | ND | $4.12 \pm 0.24$ | $6.37 \pm 0.36$ |
| 12 | $5.98 \pm 0.26$ | ND | ND |
| 18 | ND | $5.97 \pm 0.10$ | $8.50 \pm 0.09$ |
| 24 | $10.37 \pm 0.57$ | $5.92 \pm 0.23$ | $7.72 \pm 0.36$ |

Data from (a) Mays et al. (1988) and (b) Cappelli et al. (1984).
ND, not determined; Hyp, hydroxyproline

myocardium (Blumgart et al. 1940; Montfort and Pérez-Tamayo 1962; Wegelius and Von Knorring 1964; Caspari et al. 1977).

More detailed studies were done in animals. A progressive increase in collagen content in the rat heart was shown by several investigators (Schaub 1964/65; Capelli et al. 1984; Mays et al. 1988). The results are summarized in Table 12.1. It is interesting that the collagen changes are related to diastolic stiffness of the rat ventricular myocardium. During maturation, i.e., from birth to 2 months, the increase in collagen content was paralleled by an increase in static elasticity and by a reduction of plasticity of the myocardium. Later, in maturity and aging, the collagen concentration still increased but was accompanied by a reduction of the static components and an elevation in the dynamic components of myocardial stiffness.

Changes in the total collagen content were found to be associated with altered solubility. A decrease in the soluble collagen level in the rat myocardium with age was reported by Von Knorring (1970). The role of this in the variations of mechanical properties remains unknown.

A progressive increase in collagen concentration in rabbit heart ventricles was reported in the 28-day-old fetus to animals 6 months of age (Caspari et al. 1975).

In the normal human and animal heart, the concentration of collagen is higher in the right ventricle than in the left one, and higher in the epicardial part of the muscle than in the endocardial part. Only single reports compare the ventrical and atrial collagen levels: that in the atria is estimated to be two- to threefold higher. There was no significant difference between the right and left atria (Caspari et al. 1975).

### 12.2.3 Collagen in Myocardial Hypertrophy

Cardiac hypertrophy has been a subject of several investigations. Despite numerous studies, there are conflicting results in the literature concerning the effect of hypertrophy on the collagen content in the heart. Such discrepancies arise due to the differing techniques employed for collagen determination (biochemical or histological). On the other and, cardiac hypertrophy is caused by a range of factors, several experimental models have been used, criteria varied for the selection of necropsy material in human heart studies.

The increase in collagen content in the human hypertrophied heart was described for the first time using a gravimetric method by Blumgart et al. (1940). It was greater in 9 of 24 hypertrophied human hearts, but 5 of the 9 had associated arteriosclerosis. When chemical methods for collagen determination were applied, no variations in content were found (Oken and Boucek 1957; Montfort and Pérez-Tamayo 1962; Shekhonin et al. 1988). This was partially confirmed by Caspari et al. (1977), who estimated the collagen concentration in human ventricles from measurement of the hydroxyproline level. The content of collagen in the right or left ventricle of a heart without valvular abnormality was unchanged. There was an increase in the total amount of collagen due to the larger size of the ventricles. In patients with aortic stenosis the concentration of collagen in the left ventricle was significantly higher than normal (controls $41.5 \pm 2.0$ mg/g dry weight; aortic stenosis $56.1 \pm 6.0$ mg/g dry weight). In four patients with aortic regurgitation no changes in collagen content in the left ventricle were found.

An interesting question of the role of the collagen which accumulated during pressure-induced myocardial hypertrophy in the recovery of left ventricular function after reversal of the hypertension was investigated by Hess et al. (1984). They performed myocardial biopsies in patients with hypertrophy caused by aortic stenosis before and after surgical valve replacement and demonstrated persistent signs of fibrosis associated with myocardial stiffness. There are no other studies in humans although several animal models have been investigated as discussed later. It is, however, clear that the kinetics of the recovery of abnormalities in contractile and noncontractile proteins, including collagen, in the posthypertrophied myocardium is different.

A rise of the collagen level of the heart ventricles was observed in animals with experimentally induced hypertrophy. Bartošova et al. (1969) determined the increase in collagen and the muscular parts of the heart in rats in various models of hypertrophy. Expansion of the muscle was not accompanied by collagen accumulation in cardiomegaly induced by sideropenia or thyroxine treatment. The most pronounced elevation of collagen content was found in hypertrophy stimulated by hypoxia or isoprenaline administration. An increase in resistance flow and long-term adaptation to physical stress stimulated growth of both the collagenous and muscular parts of the heart. In these experiments, injection of isoprenaline led probably to cardiomyopathy with necrotic changes, and accumulation of collagen reflected healing of the necrotic foci (see Sect. 12.2.4).

Surgical constriction of the ascending aorta was used to produce myocardial hypertrophy in animals. Skosey et al. (1972) reported a progressive increase in total heart collagen. A substantial elevation was observed on the 7th day after banding of the aorta. Mean total collagen content on day 11 was twice that of controls. Its concentration measured as collagen content per 100 mg heart tissue fell during the early period of cardiac hypertrophy and was enhanced after the 5th day following aortic ligation. The early fall was caused by a significant increase of heart weight immediately after banding of the aorta.

An increase in collagen content in the right ventricle of cats with hypertrophy produced by chronic constriction of the pulmonary artery was shown by Buccino et al. (1969). The rise was greater in the epimyocardium than in the endomyocardium. The myocardial collagen concentration in cats with hypertrophy was significantly higher in right ventricular than in left ventricular specimens. Interestingly, constriction of the pulmonary artery caused an elevated collagen content in the nonstressed left ventricle. The mechanism of this increase has not been elucidated, and functional interactions in the pulmonary and systemic circulation have been suggested to be a causative factor.

Bonnin et al. (1981) used the fractioned synthesis rate of collagen, i.e., the percentage of total collagen synthesized in 1 day, as an index of collagen formation in the right ventricle of dogs with pulmonary artery stenosis. The measurements were done on days 5, 12, and 28 after banding of the artery. The fractional synthesis rate of collagen in the right ventricle of intact animals was slow (0.56%/day) and increased significantly under the conditions of the experiment. The highest values were observed on the 5th day after banding (4.8%/day). After 12 and 28 days, the synthesis rate was 2.6% and 1.3%/day, respectively. The increase in noncollagenous proteins was found only on the 5th day of the experiment. The collagen content in the right ventricle expressed as a mass fraction fell over the first 12 days of hypertrophy and was restored on the 28th day. The initial decrease in collagen content was caused by the rapid growth of muscle protein, despite elevated collagen synthesis.

Renovascular hypertension has been used as a model of myocardial hypertrophy. Structural remodeling of the collagen matrix of the heart was investigated in long-tailed macaque (*Macaca fascicularis*). Systemic hypertension of gradual onset was induced by unilateral perinephric cellophane insertion. The macaques were sacrificed in three phases. These sequential studies revealed that the matrix was remodeled with the development of the hypertrophy. At the evolutionary phase of hypertension only a slight thickening and increase in density of the collagenous weave were reported. The early compensatory phase was characterized by thickening of the collagen strands to pillarlike structures. In the late compensatory phase some myocytes had become encased in collagen. These structural changes have been suggested to play a significant role in deleteriously affecting the ventricular function of the pressure-overloaded heart muscle. They were accompanied by quantitative collagen alterations. In the evolutionary phase an increase in collagen and a greater proportion of type III collagen were shown. The total collagen content grew in the next phases of hypertrophy while the proportion of types I and III collagens was opposited as compared with controls. Necrosis of myocytes was evident in the late compensatory phase, and the proportion of collagen types reflected scar formation. The investigators concluded that the left ventricular hypertrophy was followed by reparative fibrosis. Hemodynamic changes may arise at least partially from the variation in force generation in the hypertrophic ventricle (Weber et al. 1988; Abrahams et al. 1987).

Morphometric analysis of the collagen network in the hypertrophied myocardium of rats was given by Michel et al. (1986). They compared two models of myocardial hypertrophy, renovascular hypertension and aortocaval fistula. Renovascular hypertension represented the pressure overload state, and the fistula resulted in the volume overload state. In renovascular hypertension, the growth of the cardiac mass was concentric, due to a significant elevation in ventricular wall thickness, and was associated with a marked fibrosis. Volume overload resulted in a significant increase in the surface area of the left ventricular cavity higher than an increase in the wall thickness. The pattern of fibrosis was visualized by staining with sirius red. Renovascular hypertension was associated with a rise in collagen density. There was a positive correlation between the left wall thickness and collagen density. The mean fiber thickness enlarged in the renovascular hypertensive group but did not change with the aortocaval fistula.

Morphometric studies were confirmed by biochemical investigations and indicated that collagen accumulation is associated with a rearrangement of the collagen structure of the myocardium (Thiedemann et al. 1983). Augmentation of the collagen content in the left ventricle of rats with hypertension due to aortic banding or renovascular obliteration was inversely related to the myosin concentration. It has been postulated that changes in myosin and collagen are independent of the mechanism of pressure overload but are correlated with the severity of hypertrophy (Kozlovskis et al. 1987).

It is clear from a number of studies that pressure overload leads to decreased cardiac distensibility (Holubarsch et al. 1983). An increase in collagen in pressure-overload hypertrophy is partially adaptive in that it enhances the tensile strength and three-dimensional delivery of force by the myocardium. This is accomplished at the expense of reducing distensibility (Jalil et al. 1989). Several factors are probably involved in this process, including changes in the capillary bed as well as the hypothetical release of stimulatory peptides. The detailed molecular mechanism remains unknown. As mentioned earlier, regression of hypertensive hypertrophy through blood pressure control also involves the left ventricular collagen. Antihypertensive treatment in spontaneously hypertensive rats led to a decrease in collagen concentration in the left ventricle as compared with untreated hypertensive controls (Motz and Strauer 1989).

### 12.2.4 Collagen in Myocardial Infarction

Myocardial infarction represents the most frequent lethal eventuality of coronary artery insufficiency. True myocardial infarction, that is, ischemic death or necrosis of myocardial cells, has been investigated mostly in reference to the early stages of cellular injury. On the contrary, the healing of myocardial infarction is largely unexplored. The process of connective tissue replacement is generally thought to resemble the healing of other tissues which have no ability to regenerate parenchymal cells. As a result of proliferation of the connective tissue, a scar is formed which interfers with the overall functioning of the heart. The most explored area is the so-called evolution of myocardial infarction, described in numerous morphological studies. After early reliable evidence of irreversible damage detectable under the electron microscope and followed by various changes observed under the light microscopy (e.g., interstitial edema), coagulation necrosis develops. The necrotic myocardium is removed by polymorphonuclear neutrophils; infiltrations of these cells occurs in the first 24–72 h. The cells mediate early degradation of collagen in the infarct zone (Cannon et al. 1983).

During the 2nd week monocytic infiltration predominates, and phagocytosis is brisk. It is postulated that macrophages also stimulate the proliferation of fibroblasts and capillary buds. The process of phagocytosis is accompanied by the simultaneous growth of fibroblasts and the subsequent deposition of connective tissue components. The mechanism of fibroblast and capillary invasion remains unclear. It is believed that the fibroblasts may migrate along a scaffold of fibrin stimulated and attracted by some specific growth factors produced by activated macrophages. A large accumulation of glycosaminoglycans in the infarction area as discovered with histochemical and biochemical methods is an early sign of fibroblast proliferation, but its significance is not clear. It is possible that proteoglycans are responsible for the space orientation of collagen fibers and affect their size. These factors are of importance in the mechanical properties of the myocardial scar.

The next stage of the healing process includes an increase in collagen production and deposition of the newly formed fibers. The number of capillaries in the formerly necrotic area decreases, and fibroblasts are also less active. The collagen scar replaces the granulation tissue usually between the 3rd and 6th week. Morphologically, the scar contains collagen bundles with few resting fibroblasts and rare capillaries (Kischer and Shetlar 1979; Lautsch 1979).

Most of the studies on the healing were done in animals with experimentally induced infarction. Two types of general categories of these animal models have been described: The first one involves mechanical occlusion of the coronary artery, and in the second one disseminated ischemia following necrosis of the myocardium is produced by pharmacological agents (e.g., isoproterenol) (Porter 1896; Smith 1918; Seyle 1970; Rona et al. 1958; Hardforth 1962). An interesting study on connective tissue in myocardial infarction was done by Judd and Wexler (1969). They used isoproterenol injected subcutaneously into rats to produce necrotic foci in the heart muscle. A rise in the collagen content expressed as an increased amount of hydroxyproline was observed at 5 and 7 days after injection. This increase was preceded by glycoprotein and glycosaminoglycan accumulation (Judd and Wexler 1970). There were no comparable changes in other organs (liver, kidneys); however, necrotic foci were found in organs other than in the heart. Similar results were reported by Shetlar et al. (1979), who used ligation of the circumflex coronary artery of dogs to produce infarction. The early phase of the healing was accompanied by an increase in proteoglycan content.

Drożdż and Kucharz (1979) noted a significant increase in the collagen content in the heart muscle of rats with isoproterenol-induced myocardial necrosis. The rise in insoluble collagen concentration was greater than that of soluble form in a late stage of the healing process. Soluble collagen predominated in the early stages of scar formation. An elevation of the serum level of hydroxyproline and urinary output of collagen metabolites accompanied healing of the myocardial necrosis.

Recently, collagen-induced changes in the electrocardiogram (ECG) of experimental animals have been reported. Intravenous injection of type III collagen produces abnormal ECG patterns after 40 s. The changes mainly consist of an ST fall in lead I, an ST fall or elevation in lead II, and an ST elevation in lead III. The changes peak at 3–5 min and disappear after 15 min. About one-fifth of the animals die due to atrioventricular block. Arrhythmia and cardiac arrest are probably related to a collagen-induced thromboxane $A_2$ release (Hori et al. 1989).

Only a few clinical studies on collagen metabolism in patients with myocardial infarction have been published. Takahashi (1979) reported an increase in serum hydroxyproline content in patients with myocardial infarction. Kucharz et al. (1982a) investigated serum hydroxyproline, hydroxylysine, and hypoproteins (so-called collagenlike protein) levels and noted a decrease of all these indices in the first few days after infarction and an increase after 10–25 days. A negative correlation between indices of collagen metabolism and transaminase activity suggests that the increase was not produced by necrotic changes in the myocardium. There was no significant alteration in the activity of collagen peptidase in serum (Kucharz et al. 1987). The relationship of the level of collagen metabolites in serum to the extent of necrosis has not been investigated. An increase in hydroxyproline excretion in patients with myocardial infarction was reported by Widomska-Czekajska et al. (1968). It is possible that factors other than the healing process affect the output of collagen metabolites.

### 12.2.5 Collagen of the Valves in Health and Disease

The heart valves are mechanical devices that permit the flow of blood in one direction only. They are thin, translucent leaflets of connective tissue which are under continuous mechanical stress. There are four cardiac valves in the mammalian heart, the aortic and mitral valves on the left side and the pulmonary and tricuspid valves on the right. All but the mitral valves have three cusps. The mitral valves consist of two leaflets only and are thought to have more a complex structure than the others. Valves are composed of a dense, avascular, connective tissue core. Both sides of this core are covered by endocardium. The valves are bound to the connective tissue skeleton of the heart and are extensions of the annuli fibrosi. On the ventricular side, the atrioventricular valves are anchored additionally to the heart wall by chorda tendineae.

Collagen makes up approximately half of the dry weight of the valves (Bashey et al. 1967). Porcine atrioventricular heart valves contain 18% collagen per unit of wet weight and 70% per unit of dry weight, based on hydroxyproline estimation (Jimenez and Bashey 1978). There are no significant differences in the collagen content between the various valves. Valvular collagen was identified as a mixture of types I, III, and V collagens (Mannschott et al. 1976; Denníng and Pinnell 1979; Morris and McClain 1972). Its most important feature is its extreme insolubility. Due to this, for a long time very little information existed concerning the biochemical characteristics of valvular connective tissue. Collins et al. (1977) found valvular collagen to be insoluble in salt solutions, acids, or denaturing solvents. The reason for this unusual feature remains unclear. Valvular collagen is highly cross-linked, preponderately one derived from two hydroxylysine residues (dihydroxylysinonorleucine). This type of cross-link is usually found in hard tissue collagens and may be associated with the mechanical function of the valves. A hydroxylysine-rich type I collagen has been isolated from pepsin-digested porcine heart valves. The increased hydroxylysine content appeared to be distributed throughout the length of the $\alpha$-chain as revealed by analysis of cyanogen bromide peptides. Only 20.7% and 35.7% of the $\alpha_1$- and $\alpha_2$-chains, respectively, contain galactose or glucose residues. The site of glycosylation is hydroxylysine residues, but the carbohydrate content has not been found to be greater than in other tissues. Elaboration of special procedures, combining pepsin digestion with buffer extraction and reduction of disulfide bonds, allowed for isolation of undenatured collagen from the valves (Jimenez and Bashey 1978). All three types of collagen identified in the valves contain a higher hydroxylysine content than similar ones from other tissues (Bashey et al. 1978).

The collagen content in the valves changes with age. Bashey et al. (1967) reported a significant age-related decrease in individuals between 20 and over 60 years old. In the 20- to 30-year old age group, collagen represented about 60% of the dry weight of the tissue while in the over 60-year-old age group it accounted for less than 40%. According to Bashey and Jimenez (1988), the newborns and children have a low (about 30% of dry weight) content. The collagen level rapidly reaches a peak at the age of 20–30 years and then decreases slowly. These findings reflect only the relative content. Morphological studies indicate that aging is associated with increased valvular thickness and size, and more fibrous changes appear within the valves (Nakao et al. 1966; Oka and Angrist 1961). Changes in the water content may also influence the relationship of collagen content to size and weight of the valves in situ.

Incubation in vitro of bovine valves with radiolabelled proline was used in a few studies to measure the biosynthesis of collagen. The highest turnover was found in the mitral valves (Mori et al. 1967). The highest rate of biosynthesis was noted in the distal areas of the valves rich in fibroblasts as compared with the basal area containing predominantly chondrocyte-like cells. The ability of the valvular connective tissue to synthesize collagen in vitro was also shown in chickens (Mitomo et al. 1968, 1969). Incorporation of proline into all valves decreased progressively with age (Mori et al. 1967).

The change under pathological conditions is obscure. The most common disorder affecting the valves is rheumatic fever. Despite the administration of antibiotics, it is still responsible for the development of valvular deformities in a significant number of patients. The valvular damage is caused by the immune response to an antigen that the streptococci share with human heart valves. The natural response to the streptococcal infection is followed by immune-mediated development of fibrotic changes, thickening and calcification of the cusps, and adhesions of the chordae (Goldstein et al. 1967). Detailed mechanisms of this long-term valvular fibrosis remain unknown.

Rheumatic valves were found to contain twice as much collagen as normal ones, and the rate of collagen biosynthesis was higher than in controls. The synthesis of noncollagenous proteins was unaltered. Incorporation of proline in vitro into the sclerotic regions of the mitral valve was significantly elevated as compared with normal-looking tissue (Bashey and Jimenez 1977a; Henney et al. 1982).

Degeneration of the valves and the chorda occurs in several diseases and leads to so-called floppy valves. The valves are rich in mucinous material. Accumulation of this mucinous material is either a primary metabolic defect or a secondary phenomenon caused by impaired functioning of the prolapsed cusps. An increase in collagen content and synthesis was shown in human floppy mitral valves (Henney et al. 1982) and was associated with an altered collagen type pattern: Type I collagen was decreased and type III increased (Cole et al. 1984). Ultrastructural changes were revealed in the collagenous network of the ruptured chorda tendineae (Lee et al. 1983). An absence of collagen bundles was paralleled by increased extensibility and reduced resistance to breaking (Lim et al. 1983).

Mitral valve prolapse is a common feature of the inherited disorders of collagen and is an another sign of a systemic defect of the connective tissue (see Chap. 9; Child et al. 1986; Handler et al. 1985).

Induction of experimental endocarditis in some species of animals was shown to produce morphological damage in the valves. There are some suggestions that a metabolic defect of the valvular connective tissue facilitates the development of endocarditis. An increase in valvular collagen synthesis was reported in chickens inoculated intravenously with *Streptococcus faecalis* that was followed by the development of endocarditis (Oka et al. 1966). It seems that an increased solubility of collagen may accompany or precede the development of endocarditis in humans (Drożdż et al. 1989).

## 12.3 Vessels

### 12.3.1 General Characteristics

The mammalian vascular system consists of blood and lymphatic vessels. The blood vascular system is composed of the following structures: the arteries whose function is to carry the blood and, with it, nutrients and oxygen to the tissues; the capillaries, a diffuse network of thin tubules whose function is the interchange between blood and tissues; and the veins, the afferent vessels of the heart.

All blood vessels except small capillaries are composed of three basic layers: the intima (internal tunic), the media (middle tunic), and the adventitia, which is the most external layer. The intima consists of a layer of endothelial cells lining the interior surface of the entire circulatory system. These endothelial cells are located on a thin layer of loose connective tissue, the subendothelial layer. The media is a muscular layer. In arteries, it is separated from the intima by an internal elastic lamina, a fenestrated layer composed of elastin. In larger vessels, a similar external elastic lamina separates the media from the adventitia. The media consists of smooth muscle cells and a significant amount of the extracellular matrix. The adventitia is composed chiefly of collagen and elastin. There is a continuous connection between the tissue of the organs through which the vessel runs. These three layers differ in various parts of the vascular system.

Arteries are classified into three sorts: arterioles, muscular arteries, and elastic arteries. Arterioles are thin and have a typical endothelium, with no subendothelial layer. The media and adventitia are thin. The thick and well-developed media is characteristic of muscular arteries. Large elastic arteries have a relatively thick intima, the media is rich in elastic structures, and the adventitia is without an external limiting membrane and is also rich in collagen structures.

Veins are classified according to their size into venules and small, medium, and large veins. They are composed of the same layers as the arteries. The muscular layer of the media is usually very thin, and adventitia is developed only in the large veins. Veins contain valves which are especially numerous in the limbs.

Capillaries are the simplest part of the vascular network. They are composed of a single layer of endothelial cells in the form of a tube. The cells rest on the basal lamina, a product of the endothelium. The main role of capillaries is mediating the exchange between the blood and tissues. Depending on the structure of the endothelium, capillaries are classified into the following groups: the continuous one whose wall has no gaps, the fenestrated type whose wall contains pores, and the sinusoidal sort with large spaces between the endothelial cells.

### 12.3.2 Collagen in the Normal Vessel Wall

Collagen is an important component of all blood vessels. Its role is not limited to a biomechanical function; it provides anchorage for the cellular elements and plays a central role in hemostasis (see Chap. 20). Collagen comprises a significant part of the dry weight of the vascular tissue; it is estimated that arteries contain about 20%–50% collagen (Grant 1967; McCloskey and Cleary 1974; Bartos and Ledvina 1979; Barnes 1988; Mayne 1986, 1987).

Different collagen types have been found in the vessel wall. Type I collagen predominates, tends to localize in proximity to the smooth muscle cells of the media, and is abundant in the adventitia (Gay et al. 1975; Farquharson and Robins 1989). Type III collagen was shown in the subendothelium. In the media, type III collagen is thought to form a network interconnecting the elastic lamellae (McCullagh et al. 1980; Madri et al. 1980; Bartholomew and Anderson 1983). Types I and III collagens account for about 80%–90% of the total collagen of the vessel wall (Mayne 1986). Type III collagen is estimated to make up 30%–44% of the total collagen, depending on the artery type: in the vena cava, it was found to constitute 21% of the total collagen (Hanson and Bentley 1983). Type I collagen is thought to contribute to the tensile strength of the wall, and type III collagen takes part in the elastic properties of the vessel although the nature of this remains unknown. The association of collagen type I with the mechanical resistance of vessels was confirmed by Harbers et al. (1984) and Löhler et al. (1984). They found that in mice in which a retrovirus (the Moloney murine leukemia virus) was integrated into the first intron of the $\alpha 1$ (I) chain, causing complete inhibition of type I collagen synthesis, all embryos died from rupture of a major blood vessel.

Type IV collagen was detected in the basement membranes located beneath the endothelium and that surround the smooth muscle cells (Wick et al. 1979; Madri et al. 1980; Fitch et al. 1982; Palotie et al. 1983).

Types V and VI collagens are noted in the vessel wall. Their location remains controversial. Type V collagen was reported in the vicinity of the basement membrane and also in association with bundles of interstitial collagen fibrils (Fitch et al. 1984). The possible coaggregation of types I and V collagens similar to that shown in the cornea (see Chap. 17) was considered. Type VI collagen was found to be distributed throughout the vessel wall (Von der Mark et al. 1984). It is known as "intima" collagen since it was first demonstrated by Chung et al. (1976) in the human aortic intima. Type VIII collagen, so-called endothelial cell collagen, was originally isolated from bovine aortic endothelial cells (Sage et al. 1980). The presence of type VIII collagen in sheep aorta and carotic artery but not in the jugular vein was described by Kittelberger et al. (1989). This collagen was predominant in the subendothelial intimal region. There are some reports indicating the presence of other types of collagen in the vessel wall (Barnes 1988).

The cellular origin of collagen of the vessel wall is based mostly on studies of in vitro cell cultures. There is clear evidence that several factors can modulate the phenotypic expression of the cultured cells, and it is very difficult to conclude from the in vitro studies the nature of the collagen-producing cells in situ.

The ability to synthesize types I, III, and V collagens by intact chick embryo aorta was reported (Morton and Barnes 1983). It is also of interest that development of the vessel wall is associated with increased type III collagen synthesis and reduced type I collagen production. This finding has been confirmed in other experiments (Scott et al. 1977; Rauterberg et al. 1977). This is a phenomenon opposite to the developmental changes in other tissues, where type I collagen predominates with development and aging.

Several reports showed that smooth muscle cells in culture synthesize collagen, and an excess of type III over type I collagen was observed (Rhodes 1982). There are, however, differences between species, and a higher rate of type III collagen synthesis was found in monkeys and guinea pigs while in humans and pigs, the type pattern was different, and in some studies the predominance of type I collagen was suggested. Monkey and guinea pig smooth muscle cells also produce type V collagen (Mayne et al. 1978).

The technique of in vitro culture of endothelial cells has been used for investigating the production of matrix components by these cells. Howard et al. (1976) reported the ability of the aortic endothelial cells to synthesize type IV collagen. This was confirmed in several reports (Sage and Bornstein 1982; Kramer et al. 1985; Bartlet et al. 1985). Further studies showed synthesis of types I and III collagens by these cells as well (Rhodes 1982). It is also clear that synthesis of interstitial collagens depends on the conditions of culture. It is possible that types I and III collagens of the subendothelium are produced by the endothelial cells.

It is generally accepted that collagen of the adventitia is produced by the fibroblasts present there.

### 12.3.3 Atherosclerosis

Atherosclerosis is a common cardiovascular disease contributing significantly to mortality, especially in Western societies. Despite extensive investigations, its etiopathogenesis is poorly understood.

The primary site of the disease is the intimal layer of the arterial wall. The changes reduce the blood flow, cause thrombosis, and in this way impair the function of several organs, especially those sensitive to ischemia.

Pathologically, early changes are associated with endothelial damage and are known as fatty streaks. The lesion exhibits the accumulation of lipids within so-called foam cells. The foam cells are partially composed of transformed smooth muscle cells and blood monocytes. The natural history of the fatty streaks remains unknown. It is believed that the early changes precede the definitive atherosclerotic feature, the fibrous plaques, which are irregular, grey nodules that protrude into the lumen of the artery. They are made up of smooth muscle cells, lipids, and matrix components. Progression of the fibrous plaques results in advanced, complicated lesions. This stage of the disease is characterized by cell necrosis and calcification and is commonly associated with thrombosis (Ross and Glomset 1976).

The process of atherosclerosis is a highly complex phenomenon. Several factors have been attributed to the development and progression of the disease; however, they are associated only on the basis of epidemiological studies, and the etiology of the disease remains unknown.

Collagen is the major component of the atherosclerotic plaque and comprises about 30% (Anastassiades et al. 1972; Barnes 1985a; Ribeiro et al. 1983; Levene and Poole 1962; Kadar et al. 1975; Bihari-Varga 1986). The type pattern remains controversial. Some studies claim that the type I collagen proportion increases compared with the normal arterial wall (McCullagh and Baliam 1975; Labat-Robert et al. 1985). This finding was based on the investigation of advanced changes. In an earlier stage, type III collagen was found to predominate and was located in the subendothelium. The presence of type III collagen was suggested to stimulate platelet aggregation and release of factors that attract smooth muscle cells and cause their proliferation (Bihari-Varga 1986). The relation of the stage of the disease as estimated morphologically to collagen content was described by Murata et al. (1986). They investigated the collagen content in human aortic wall obtained at autopsy. An increase in the total collagen level of the aortic intima correlated with the grade of atherosclerotic changes. It was accompanied by elevated levels of types I and V collagens. A slight reduction in type III collagen was seen in the intima with the most advanced grade of atherosclerosis. A huge rise in extractibility of type IV collagen was shown in the intima of atherosclerotic human aortas. Some 81/μg of type IV collagen per gram of dry defatted weight was isolated from the normal tissue while 473 μg/g was extracted from the aortas with the most advanced disease (Murata et al. 1986).

Comparative localization of types I, III, IV, and V collagens in the intima of atherosclerotic lesions at different stages was carried out by Shekhonin et al. (1987) with immunohistochemical methods. In the extracelllular matrix of the fatty streaks and in some areas of the fibrous plaques containing large amounts of subendothelial cells, types I and III collagens were revealed, and an increased content of types IV and V collagens was shown. The fibrous plaques consisted mostly of interstitial collagen types I and III and contained moderate amounts of type V collagen. Type IV collagen and fibronectin were absent in the plaques. The authors pointed out the similarity in the matrix composition between the advanced atherosclerotic lesion and granulation tissue.

Structural alterations of collagen isolated from atherosclerotic plaques have been reported. Increased stability resulted from elevated cross-linking and/or interaction with lipids (Bihari-Varga et al. 1969; Claire et al. 1976). The plaque collagen was shown to be more antigenic than that isolated from healthy aortas (Jacotot 1983), but the nature of the structural alterations involved was not elucidated.

Several studies focused on the collagen content and synthesis in the arterial wall in animals fed with an atherogenic diet. An increase in the collagen content in the aortic wall was reported in rabbits and rats fed with a fat-rich diet (Modrak and Langner 1980). The synthesis of aortic collagen was elevated under these conditions (Ehrhart and Holderbaum 1977, 1980; Langner and Modrak 1977; Fischer et al. 1980; McCullagh and Ehrhart 1974). An increase in activity of aortic lysyl oxidase in response to the feeding of rabbits with an atherogenic peanut oil-cholesterol diet was reported by Kagan et al. (1981). The rise occurred initially in the aortic arch, with much less of an increase appearing subsequently in the abdominal aortic wall. An increase in the arch was evident after 60 days of feeding, and values measured on the 90th day were about three times higher than controls. The differences in anatomical distribution of the elevated lysyl oxidase activity are consistent with the development of lesions primarily in the arch in diet-induced atherosclerosis. The changes in lysyl oxidase activity did not seem to be an immediate response to the serum cholesterol level since the enzyme activity was only slightly elevated in rabbits fed cholesterol for 30 days, although the cholesterol concentration in serum was high at this time (Kagan et al. 1981). An increase in aortic prolyl hydroxylase activity was found in animals with experimental atherosclerosis. Some authors suggested that this was correlated with an elevated cholesterol level in the serum (St. Clair et al. 1975; Modrak and Langner 1980). Synthesis of collagen was preferentially stimulated in comparison with noncollagenous protein synthesis (Kramsck and Chan 1976; Langner and Modrak 1977).

A number of studies have investigated the ability of the cells derived from the artery wall to synthesize collagen in in vitro culture. Enhanced synthesis was found in smooth muscle cells from the atherosclerotic aortas of animals receiving special diets (Scott et al. 1977; Pietilä and Nikkari 1980). Several authors described the effect of increased serum lipids on collagen synthesis by the artery wall, but the results are contradictory. Some of them were able to stimulate collagen synthesis by aortic explants with various lipids (McCullagh and Ehrhart 1974, 1977), while others found that hyperlipoproteinemic serum did not (Holderbaum et al. 1975). Augmentation of collagen synthesis by liver homogenates from hypercholesterolemic rats but not by hyperlipidemic serum was reported by Rönnemaa and Doherty (1977). These findings suggest that the high fat serum affects the metabolism of the arterial wall indirectly, through factors generated in organs other than the vessel wall. The possible involvement of the platelet-derived collagen-stimulating factors has also been hypothesized (Burke and Ross 1977).

The majority of these studies were concerned with total collagen production, and type pattern was not elucidated.

Aumailley and Bricaud (1981) using pulse-label experiments in organ culture reported a higher level of collagen synthesis in aortic samples obtained from rabbits on cholesterol-enriched diets and from mini-pigs in which atherosclerosis occurs spontaneously. A significant prevalence of type III collagen synthesis in atherosclerotic tissue was observed.

Animal models of atherosclerosis have been used to evaluate various methods of the prevention or treatment of the disease. Mestranol and norethynodrel (contraceptive steroids) were reported to diminish collagen accumulation and synthesis in the aortas of rabbits fed on an atherosclerotic diet (Fischer et al. 1981). The hypothetical relationship of this finding with a lower incidence of atherosclerosis in women before the menopause needs further studies.

Serum and urine levels of indices of collagen metabolism are not significantly altered in patients with atherosclerosis. Recently, Bonnet et al. (1987, 1988) found an increased serum aminoterminal procollagen type III peptide level in patients with coronary artery disease. The elevation appeared to be unrelated to risk factors, and there was no relationship between the peptide concentration and the severity or extent of stenosis of the coronary arteries.

It is very difficult to estimate the role of matrix changes in the pathogenesis of atherosclerosis. It is possible that increased collagen synthesis and accumulation are secondary phenomena, but in turn they initiate or augment further pathological events, such as platelet aggregation and thrombosis. Feedback regulation of collagen synthesis by platelet-derived substances may be involved in the progressive accumulation of matrix within atherosclerotic plaques. It can be hypothesized that collagen deposition is related to the stage of the disease. In some stages collagen synthesis may be involved in the pathogenic phenomena, and in other stages collagen accumulation may reflect nonspecific scar formation.

### 12.3.4 Hypertension

Hypertension is a syndrome with a complex etiology. All causative factors result in an increased pressure of blood. This affects the vascular wall, and in turn the vascular wall contributes to the regulation of blood pressure.

Several investigations have provided evidence that collagen metabolism is altered in animals with hypertension and probably also in patients with elevated blood pressure. An increase in collagen content associated with enhanced synthesis in the arterial wall was found in rats with spontaneous and induced long-term hypertension (Fischer and Llaurado 1967; Foidart et al. 1978; Wolinsky 1972; Ooshima et al. 1975; Fischer 1976). These alterations were reversed by antihypertensive drugs. It seems that the accumulation of collagen was induced by mechanical stress of the vessels. This suggestion was drawn from the results of an artificial increase of blood pressure in veins of animals with arteriovenous anastomoses. The veins responded with a significant increase in collagen content and synthesis (Smith et al. 1976; Todorovich-Hunter et al. 1988). However, the collagen content in the veins of hypertensive animals did not increase, and the elevated collagen production was limited to the arteries, i.e., the vessels that were under increased pressure (Iwatsuki et al. 1977).

Bashey et al. (1989) studied collagen type pattern in the aortas of spontaneously hypertensive rats. In these animals there was no increase in collagen content in the aortas. In comparison with the age-matched controls, a decrease in type I collagen content was associated with a marked increase in type V collagen. There was no change in the relative level of type III collagen. Investigation of the passive mechanical properties of the tissue showed that hypertensive rat aortas were stiffer that those from controls, and the difference increased with the age. There was no correlation between mechanical properties and collagen content.

An interesting feedback relationship between blood pressure and vascular collagen content was shown in hypertensive rats treated with the lathyrogen β-aminopropionitrile. Inhibition of collagen maturation which reduced the mechanical properties of the arterial wall resulted in normalization or at least a decrease in high blood pressure (Nissen et al. 1978). This finding correlated with the beneficial effects of some antihypertensive drugs. Hydralazine was found to decrease collagen synthesis and to impair maturation (see Chap. 10). This may be an additional mechanism of antihypertensive action of the drug. On the other hand, drugs that reduce blood pressure by other than direct vascular effect actions (e.g., via influence upon the central nervous system) have also been reported to modify collagen metabolism in the vasculature. Their effect on collagen is probably secondary to the reduced mechanical stress of the wall. The pathophysiological role of the increased accumulation of collagen is not fully understood. It can be assumed that the collagen metabolism response leads to enhanced mechanical resistance of the vessels and occurs as a homeostatic mechanism of the cardiovascular system.

There are two distinct compartments of the connective tissue in the pulmonary parenchyma: the extra-alveolar interstitium (the loose tissue that surrounds arteries, veins, and bronchioles) and the alveolar interstitium, the space between the alveolar basement membranes (Amenta et al. 1988). The compartments have different structures and functional properties. The extra-alveolar tissue is very distensible and is made up of collagen and elastic fibers. The distensibility of the alveolar interstitium is limited (Gil 1982).

The total amount of collagen varies under pathophysiological conditions. It is estimated at 5%–20% of the dry weight of adult human and animal lungs (Johnson and Andrews 1970; Wright et al. 1960; Métivier et al. 1978).

The amount of collagen decreases in the lungs of rats fed a protein-deficient diet in the neonatal period (Kalenga and Eeckhout 1989).

The distribution of collagen throughout the pulmonary structures is known only partially. Vessels and the tracheobronchial tree are rich in collagen. According to Rennard and Crystal (1982) the total mass of parenchyma is much larger than that of the vessels and bronchial structures, and most of the collagen is present in the alveolar structures.

Lung collagen is highly insoluble; only a small fraction can be solubilized (Laurent 1986). The details of the maturation of collagen within the lungs remain unknown.

The contribution of collagen to the various compartments of the lungs is associated with certain patterns of collagen types. Type I collagen is the most ubiquitous. Types I and III collagens predominate in the parenchyma. Type I fibers form small bundles or occur individually in the alveolar septa. The distribution of types I, III and IV is similar. These types predominate in the alveolar septa, interstitium of the major airways, and vasculature. Type III collagen assembles into filaments 15–20 nm in diameter. Fine filaments of 5–10 nm diameter are made up of type VI collagen. Fibers derived from type I collagen are thick (30–35 nm in diameter) and are associated with thinner fibers.

Type II collagen is limited to the cartilage of the tracheobronchial tree. Type IV collagen is present in all basement membranes of the epithelial and endothelial cells.

Aging is associated with an increase in the collagen content of the lungs and a decrease in its solubility (Schaub 1963; Drożdż et al. 1979c; Burri et al. 1974). Mays et al. (1989) showed a progressive increase from 1.94 mg per lung in 2-week-old rats to 56.47 mg per lung in 24-month-old animals, while only a threefold increase in the lung weight was observed. The gross collagen synthesis rate decreased very significantly. In 1-month-old rats 13.5% of total collagen is synthesized daily while in 2-year-old animals only 0.97% of the total collagen is produced per day. Fractional rates of total collagen degradation, calculated from the differences between rates of synthesis and deposition, decreased twentyfold from 1 month and to 2 years of age. Degradation of newly synthesized collagen increased from 27.6% at 1 month to a maximum of 82.3% at 15 months. It is clear that lung collagen metabolism is active throughout life. A significant proportion of the collagen produced is degraded, and only a small fraction is deposited. In young animals, the major pathway

# 13 Respiratory System

## 13.1 General Structural Features

The respiratory system consists of the lungs and the airways. The lung is the organ of external respiration whereby oxygen and carbon dioxide are exchanged between the body and the surrounding air. The respiratory airways are classified into the upper part, the conducting zone, and the respiratory zone. The upper respiratory tract consists of the nose and sinuses, the pharynx, and the larynx. Connective tissue contributes to the structure of all the parts of the upper respiratory tract, mainly as cartilages which form the scaffold and connective tissue which supports the mucous membrane.

The conducting zone of the airways consists of a dichotomous system of tubes. It begins with the trachea and continues in the bronchi and bronchioles. The airways are linked by a pseudostratified columnar ciliated epithelium containing a variety of cell types. The trachea and the bronchi are prevented from collapse by cartilage. The bronchioles are airways without cartilage and have mucous glands in their wall. The terminal bronchioles supply the respiratory unit of the lung known as the acinus. In humans there are approximately 25000 acini. The terminal bronchioles branch in the same dichotomous manner and form two respiratory bronchioles. There are usually two or three orders of respiratory bronchioles, which finally divide into alveolar ducts. These ducts are entirely surrounded by alveoli. The alveolar sacs and terminal alveoli arise from the alveolar ducts. The alveolus is composed of a variety of cells, mainly the types I and II pneumocytes which form the cellular lining. Other cells including alveolar macrophages are present. The well-developed network of vessels is an integrated part of the respiratory system.

## 13.2 Collagen Content in the Lungs

Collagen contributes to the structure of all sorts of the pulmonary connective tissue. The tissue forms a continuum that begins at the hilum and accompanies the bronchi and vessels to the alveolar structures. Cartilage of the tracheobronchial tree also contains collagen. Visceral pleura is another collagen-rich structure, and it contains fibers that penetrate to the parenchyma.

of breakdown appears to be extracellular, whereas in older rats it is intracellular. Degradation is believed to be the principal mechanism regulating the amount of collagens deposited (Bienkowski 1984). The mechanism responsible for these phenomena as well as their relation to collagen cross-linking remain unknown (Last et al. 1989).

Collagen degradation in the lungs is an active process. At maturity, the collagen content is constant despite active synthesis by the parenchymal cells. Fibroblasts and alveolar macrophages are known to produce collagenase. Phagocytosis of collagen fibers by activated fibroblasts and macrophages in the lung tissue has been observed. This alternative pathway of collagen breakdown occurs under conditions of activation of the connective tissue, for example by inflammation.

Unlilateral pneumonectomy in many species causes compensatory growth of the remaining lung. This growth is associated with a rapid synthesis of collagen; about a threefold elevation has been shown in rats. Collagen degradation also increases in the remaining lung, but this rise does not last as long as the enhancement of synthesis. These changes resulted in a 75% increase in collagen content by 28 days (McAnulty et al. 1988).

## 13.3 Cells Involved in Collagen Production

Cells participating in collagen production in the lungs have been identified only partially. The parenchyma contains fibroblasts, types I and II pneumocytes, and endothelial cells. Additionally, macrophages and lymphocytes are abundant, especially in inflammatory lesions.

It is difficult to discover the cellular origin of collagen in situ. The total amount of synthesis by the adult pulmonary parenchyma is estimated to be $2-4 \times 10^4$ collagen chains per cell per hour (Bradley et al. 1974, 1975; Collins and Jones 1978). Fibroblasts are believed to contribute significantly to this production (Raghu et al. 1989); it is estimated that they synthesize $4.0-7.5 \times 10^5$ collagen chains per cell per hour (Breul et al. 1980). On the other hand, type II pneumocytes in vitro produce more collagen than pulmonary fibroblasts (Fulmer et al. 1977; Leheup et al. 1989). Whether other cells are capable of synthesizing collagen is not clear. The endothelial cells are able to secrete collagen, and it is likely that they contribute to collagen production within the lungs. Type I pneumocytes probably do not synthesize collagen.

The synthesis of collagen in the vasculature and tracheobronchial tree has not been quantitatively investigated. Mesothelial cells of the pleura are known to be active in collagen production.

The type pattern of collagen produced by different cells in situ remains unknown. The parenchymal cells produce predominantly types I and III collagens. The same types predominate in the vasculature. The cells of the tracheobronchial tree secrete type II collagen.

## 13.4 Collagen and Lung Disease

### 13.4.1 Introduction

The respiratory system is affected by a variety of disorders. Some of them have been found to be associated with alterations of the connective tissue. In general, two major groups of noninfectious diseases are distinguished, the diffuse interstitial fibrotic disorders and destructive lung diseases, and are reviewed in this chapter with emphasis on the collagen involvement.

Pulmonary manifestations in systemic diseases are associated with collagen abnormalities. The inherited disorders of collagen molecule are reviewed in Chap. 9. Collagen diseases, especially systemic sclerosis, are discussed in Chap. 10. Association of neoplasms with collagen pathology is described in Chap. 21.

### 13.4.2 Fibrotic Lung Disorders

Diffuse interstitial pulmonary fibrosis is characterized by fibrosis of the alveolar structures. The disease forms a heterogenous and complex group of pulmonary conditions, and more than a hundred causative factors have been listed. It may be considered a nonspecific reaction of connective tissue to various noxious agents. The disease decreases lung compliance and leads to a reduction of lung volume and in consequence of gas exchange.

The known factors that are able to induce pulmonary fibrosis are classified into inhaled dusts, inhaled gases, chemical substances which are introduced parenterally or orally, and radiation. The classification is temporary and does not reflect the mechanism of the fibrosis induction. For example, immune disturbances are either a primary mechanism (as in graft-versus-host disease) or a secondary complication as suggested in some forms of drug-induced pulmonary fibrosis. There is also a large group of disorders of unknown etiology associated with fibrosis. The most common among them are sarcoidosis, alveolar proteinosis, and so-called idiopathic pulmonary fibrosis.

In the present chapter, a short review of the most explored forms of pulmonary fibrosis is given. Radiation-induced fibrosis is described in Chap. 23, and bleomycin-induced fibrosis is reviewed in Chap. 24.

Inhaled inorganic dust (e.g., silica, asbestos) is known to produce severe pulmonary dysfunctions in humans and has been used to induce fibrosis in animals. An increase in the pulmonary collagen content was reported in animals so treated with silica (Chvapil et al. 1979b; Lugano et al. 1982; Lehtinen et al. 1983; Reiser et al. 1983; Callis et al. 1985; Poole 1987). In rats, a single intratracheal instillation of silica is sufficient for the development of fibrosis associated with collagen accumulation. Long-term treatment with silica has been also studied. Vuorio et al. (1989) showed that the collagen content in the whole lung of rats increased significantly after 4 months of exposure. Twelve months of treatment with silica led to almost twice the increase in total collagen content in the lungs. This increase was associated with a time-dependent increase in lung weight. At the same time the body weight of animals remained practically constant. The increase in the lung mass could be explained by an elevation of the cell number as shown by an enhanced content of DNA. An increase in types I and III procollagen mRNA was shown. The lungs responded to silica with a greater increase in type III than type I procollagen mRNA.

Ozone is a pneumotoxin that has been shown to induce pulmonary fibrosis (Last et al. 1979, 1984; Reiser et al. 1987a; Reiser and Last 1981). It has been suggested that ozone-induced fibrosis results from damage to the alveolar epithelium. The epithelium is not able to heal, and fibroblasts proliferate through the normally intact epithelial lining layer (Haschek and Witschi 1979; Fulmer and Crystal 1976; Hance and Crystal 1975). An increase in total collagen content was reported in animals exposed to various concentrations of ozone (Sun et al. 1988). Microscopic evidence of fibrosis was also found.

Organic compounds such as butylated hydroxytoluene, carbon tetrachloride, or paraquat are well-known pneumotoxic agents. A rise in collagen accumulation and synthesis has been investigated in various animal models employing these agents. The pulmonary fibrosis is a result of acute and chronic damage of the lungs and is used for studies on the pharmacological control of fibrogenesis (Hollinger and Chvapil 1977; Hollinger et al. 1978; Smith et al. 1974; Kehrer 1985; Pääkkä et al. 1989).

The mechanism of the fibrotic reaction of the pulmonary parenchyma is not known. The contribution of several factors affecting collagen metabolism has been suggested. Aalto et al. (1976) showed that treatment with silica released from the macrophages a soluble factor which stimulated the synthesis of collagen. The role of the immune mechanism and cytokine release in fibrosis development has also been described (Clark et al. 1989). The repair process after damage may be another hypothetical mechanism of the induction of fibrosis. It is of interest that a single exposure to silica may bring about fibrosis development, suggesting that the mechanism once initiated is a self-stimulating phenomenon and acts much longer than the original irritating exposure. It is possible that the extreme insolubility of silica and its resistance to elimination are responsible for its potentiation effect.

Propranolol, a $\beta$-adrenergic receptor-blocking agent, is known to be able to produce fibrosis and to stimulate collagen synthesis. The role of $\beta$-adrenergic receptor blockade in the control of collagen production by fibroblasts has been postulated as a mechanism of fibrogenesis (Smith and Sommers Smiths 1988; Sommer Smiths and Smith 1989). Most of the observations on collagen-stimulating factors have been done in vitro, and it is difficult to estimate the role of these phenomena in vivo.

Sarcoidosis, a granulomatous disease of unknown etiology, commonly invades the lungs. Pulmonary manifestations develop into interstitial fibrosis. Urinary excretion of hydroxyproline was found to be increased in patients with sarcoidosis (Massaro et al. 1966; Pawelec 1973). An increase in serum galactosylhydroxylsyl-glucosyltransferase activity was found in 80% of patients, but there was no correlation with the clinical activity of the disease. A similar relationship was shown for the aminoterminal propeptide of type III procollagen (Poole et al. 1989).

Cryptogenic fibrosing alveolitis is a relatively rare fibrotic lung disorder of unknown etiology (Turner-Warwick et al. 1980). Histological studies have provided evidence of extensive fibrosis, but biochemical measurements did not confirm an increased accumulation of collagen in the lung tissue (Fulmer et al. 1980). Kirk et al. (1984) showed that the mean percentage of type III collagen compared with type I was significantly lower in postmortem lung samples from patients with this form of fibrosis. In those patients who responded to treatment the percentage of type III collagen was similar to that found in controls. There was a positive correlation between the percentage of type III collagen and the response to treatment at 6 months as shown by improvement in the results of functional pulmonary tests. Estimation of the type I/type III collagens ratio has been suggested to be a useful marker for staging the disease and to predict a response to treatment.

Fibrotic lung disease is associated with changes in serum and urine collagen marker levels (Pawelec 1975; Kucharz and Stawiarska 1981). An increase in serum and bronchial lavage fluid levels of the aminoterminal propeptide of type III procollagen was shown in patients with pulmonary fibrosis due to asbestos inhalation (Pascalis et al. 1986; Cavalleri et al. 1988). Similar changes were reported in other forms of pulmonary fibrosis (Okazaki et al. 1983; Okuno et al. 1985).

The active fibrous form of tuberculosis is associated with elevated urinary hydroxyproline output (Pawelec 1972a, b) while in patients with inactive pulmonary tuberculosis, unchanged urinary hydroxyproline levels were found (Pawelec 1974).

### 13.4.3 Destructive Lung Disorders

The destructive lung disorders are also known as emphysematous diseases because emphysema is the major abnormality that occurs. The term "emphysema" is defined pathologically as a condition characterized by enlargement of the air spaces distal to the terminal bronchiole, accompanied by destruction of their walls. Its development arises from a variety of external factors together with some intrinsic susceptibility. The etiological factors include smoking, atmospheric pollution, and infections as well as an inherited low activity of antiprotease, $\alpha_1$-antitrypsin (Soskel and Sandberg 1986).

Although the primary focus in studies has been on elastic tissue, several lines of evidence implicate an involvement of collagen.

Ultrastructural studies on human lung tissue samples obtained from patients with emphysema showed a focal accumulation of abnormal collagen fibers (Martin and Boatman 1965; Belton et al. 1977; Yu et al. 1977). The reorganization of collagen within the lung during the development of emphysema has been suggested to cause a distortion of the alveolar epithelium and capillaries including detachment of the epithelial cells from the basement membranes (Huang 1978). It may be an additional factor stimulating the loss of tissue integrity and development of emphysema.

One of the forms of destructive lung disease involves a hereditary deficiency of $\alpha_1$-antitrypsin. In patients affected by this disorder, an early onset of emphysema is the most prominent clinical complication. It was found that neutrophil collagenase is present in the lower respiratory tract (Gadek et al. 1984). It is also clear that an elevated activity of neutrophil elastase may contribute to the enhanced collagen degradation as the enzyme is capable of cleaving certain types of collagen (see Chap. 4). On the other hand, intratracheal instillation of collagenase is unsufficient to induce emphysema in animals. This finding suggests that the collagen abnormality is probably secondary to the primary destruction of the elastic tissue of the lungs.

A number of studies have been devoted to the effects of nitrogen dioxide on lung connective tissue. Nitrogen dioxide is one of the most important toxic components of motor and industrial gases and tobacco smoke. Chronic exposure even to low concentrations leads to the development of lung emphysema (Freeman et al. 1968ab, 1969; Nakajima et al. 1980). Most biochemical studies on nitrogen dioxide-induced emphysema have been carried out in animal models. A decrease in total collagen content in the lungs of guinea pigs exposed chronically was shown by Drożdż et al. (1977). The changes were accompanied by increased serum and urinary hydroxyproline levels (Kucharz et al. 1972). Similar changes were reported in guinea pigs exposed to industrial exhaust gases (Drożdż et al. 1976). Increased breakdown of pulmonary collagen induced by inhalation of nitrogen dioxide was reported by Kleinerman (1979) and Hatton et al. (1977). An enhanced urinary hydroxyproline level was found to correlate with nitrogen dioxide exposure in human beings and has been suggested as a marker of environmental exposure (Yanagisawa et al. 1986). It is of interest that in the skin of animals exposed to nitrogen dioxide an increase in collagen content was found (Drożdż et al. 1977; Drożdż and Kucharz 1976). The similar phenomenon in humans has not been investigated.

Several studies have been done on experimental, protease-induced emphysema in animals (Soskel and Sandberg 1986). Proteases which are able to destroy elastin are the only enzymes that produce emphysema. An increase in total and insoluble collagen was found in the lungs of hamsters with emphysema induced by intratracheal instillation of elastase. This elevation resulted from enhanced synthesis and is preceded by a transient increase in salt-soluble collagen (6–20 days after elastase treatment) (Yu and Keller 1978). Completely the opposite findings have been reported in rats exposed to tobacco smoke. Tobacco smoke is another factor which induces emphysema and produces depression in collagen synthesis within rat lungs (Joyce and Garrett 1978). The relationship of collagen production to the development of emphysema in various animal models remains unknown.

Cystic fibrosis, a systemic disease of the exocrine glands, is associated with a high incidence of lower respiratory tract infections which lead to degradation of the tissue and a progressive loss of pulmonary function. Urinary excretion of hydroxyproline, proline, and hydroxylysine was found to be increased in patients with cystic fibrosis. This rise was correlated with impaired pulmonary function. The collagen markers have been suggested to serve as indicators of the severity of pulmonary involvement in patients with cystic fibrosis (Ammitzbøll et al. 1988a).

# 14 The Liver

## 14.1 Introduction

The liver is the largest gland of the body. It is an accessory digestive organ, although secretion of bile is only one of its numerous functions. Briefly, it produces constituents of blood plasma, detoxifies lipid-soluble factors, stores glycogen, synthesizes urea, and takes part in the water and electrolyte balance of the body.

It is a highly cellular organ with cells constituting four-fifths of the organ volume. The hepatocytes, the main parenchymal cells, are predominant and account for 92% of the total cell volume. The remaining, nonparenchymal cells are endothelial cells, Kupffer cells, and fat-storing cells.

The connective tissue of the liver makes up just a small part of the weight of the organ. Four compartments are considered: the capsule, the fibrous tissue of the portal triads, the intralobular reticulum fibers, and the connective tissue of the vessel wall. The capsule (Glisson's capsule) contains mainly collagen. It sends thin septa into the liver parenchyma. The vessels enter the liver at the hilum. After divisions, together with the intrahepatic bile ducts they form the socalled hepatic triads. The triads are surrounded by connective tissue. Fine collagenous structures can be demonstrated as reticulin within the lobules. Intrahepatic vessels and bile ducts are built of certain connective tissue parts, including basement membranes. The unique structure is found in the sinusoids, the blood vessels lined by endothelial and phagocytic (Kupffer) cells which do not make contact and have no basement membrane. The structure of the sinusoids allows free access of blood plasma to the hepatocytes. The space between the sinusoids and hepatocytes is called the space of Disse, in which delicate collagen fibrils can be recognized.

## 14.2 Collagen Content in the Liver

Normal human liver contains collagen in the amount of about 5.5 mg/g of fresh tissue. The same value for the rat is about 0.91 mg/g. The origin of this variation remains unknown. The distribution of different genetic types of collagen has been recently investigated although the insolubility of hepatic collagen has been considered a main methodological problem of the interpretation of the results. In humans, the main collagens present in the liver are types I and III. They occur in approximately equal amounts and account for about 72% of the total collagen. Type V collagen is estimated as 16.3%. The amount of type IV collagen is 9.0%. A small fraction of type I trimer was found in the liver, making up 0.9%.

The relative amounts of different types of collagen in the rat liver are similar, but types I and III collagens predominate more significantly than in humans, i.e., types I and III collagens constitute about 88% of the total collagen content.

Using immunohistochemical localization of some collagen types, it has been shown that types I and III collagens are present in the portal tract, around the central veins, and along sinusoids (Wick et al. 1978). Type IV, as expected, occurs in the basement membranes. It was found in blood vessels, lymphatic vessels, nerves, and bile ducts. There was no type IV collagen in the interstitial matrix of the normal portal tract, but it is found in the space of Disse (Hahn et al. 1980). This suggests that in the space of Disse in which the morphologically distinct basement membranes are absent, some components of the membranes exist. Type V collagen has been located along the hepatic sinusoids (Biempica et al. 1977).

## 14.3 Collagen Synthesizing Cells

The cells responsible for collagen production in the normal and fibrotic liver have not been fully identified. The ability to produce collagen has been studied in cell cultures in vitro. Under these conditions various factors different from those in vivo can affect the phenotypic expression, including collagen synthesis.

The classic concept that fibroblasts are the cells responsible for collagen production has not been rejected, although the ability to produce collagen has been attributed to other cells as well. The role of the fibroblast in the development of hepatic fibrosis is probably limited to some forms, for example, that of the portal tract is attributed to elevated collagen synthesis by fibroblasts.

The role of hepatocytes in collagen synthesis has been a subject of numerous investigations. Isolated hepatocytes contain prolyl hydroxylase and, when cultured in vitro, produce collagen (Ohuzi and Tsurufuji 1972; Sakakibara et al. 1978; Berman and Foidart 1980; Foidart et al. 1980; David 1987; Clement et al. 1988a, b). The role of hepatocytes in collagen accumulation in vivo remains controversial. The ability to synthesize collagen under conditions of cell culture may arise from an altered phenotypic expression in vitro, or collagen may be necessary for cell adherence to the culture substrate, and thus its production is a form of adaptation facilitating cell survival in vitro.

The role of hepatocytes in collagen production in vivo was investigated in the model reported by Hata and Ninomya (1984). Protein synthesis by parenchymal and nonparenchymal cells isolated from the liver of rats injected with radiolabelled proline was determined. It was found that 0.1%–0.2% of the labeled protein created was collagen. Considering the population of hepatocytes in the liver (70% of total cell number) about 80% of the total collagen production of the liver can be attributed to them.

There are various, more or less direct indications that perisinusoidal cells produce collagen in the injured liver (Irving et al. 1984). Newly formed collagen fibers are present in the space of Disse, the cells have well-developed endoplasmic reticulum and Golgi apparatus, and autoradiographic studies indicate that they are a location of collagen synthesis in the liver (McGee and Patrick 1972; Groniowski 1975). McGee (1982) presented the hypothesis that the fat-storage cells are the ones with a fibroblastic function in various organs. The perisinusoidal cells contain numerous droplets of lipids, are rich in vitamin A, and are able to produce collagen in culture (Kawase et al. 1986; Maher et al. 1988; Masuoka et al. 1989). Similar cells were described in the lungs, kidney, adrenal gland, and gut. The role of vitamin A in the deposition of connective tissue components has not been elucidated. Studies on the cultured fat-storing cells showed that differentiation of these cells to myofibroblasts was paralleled by an increase in collagen synthesis. The main collagen type secreted by the myofibroblast-like cells was type I. Collagen types III and IV were present in lesser amounts (Geerts et al. 1989).

## 14.4 Fibrosis as a Response of the Liver Bioecosystem

The normal functioning of the liver is highly dependent on the cytoarchitectural organization and precise cell-to-cell and cell-to-matrix interactions. This three-dimensional structure reacts to various changes in the internal and external environment of the body but the structure of the liver remains constant with certain limits. All its components including cells, extracellular spaces, and matrix are distributed in a highly ordered manner. This stable structural and functional homeostasis provides continuous adaptation of the hepatic function to the needs of the whole body. Destruction of hepatocytes is the most common initial response to several harmful factors. The death of these cells usually triggers a sequence of disturbances caused by alterations in the homeostatic mechanisms of the bioecosystem. Usually, destruction is associated with proliferation of nonparenchymal cells, expression of new phenotypic features, and the appearance of cells absent in normal liver. Regardless of the nature of the harmful agent, fibrosis is the response of the liver. It results from the overaccumulation of extracellular matrix components, especially collagen, and is associated with a profound distortion of the cytoarchitectonic structure. These changes lead to separation of the hepatocytes from vessels and isolation of them from nutrients. It also produces altered interactions between cells and with the matrix.

On the basis of these findings, Rojkind and Mourelle (1988) presented a hypothetical explanation of liver fibrosis as a complex response of its bioecosystem. It is beyond the scope of this book to review all the mechanisms involved or thought to be involved in hepatotoxicity. Accumulation of collagen is believed to play a key role, and this chapter focuses on the mechanisms of collagen deposition within the liver (Mauch and Krieg 1986).

## 14.5 Collagen Content in the Fibrotic Liver

An increase in fibrous tissue in the cirrhotic liver was reported for the first time by Morrione (1947, 1949) and confirmed by Fels (1958) and Kent et al. (1959). The early studies were followed by a number of reports which showed an increased content of collagen in human and animal fibrotic liver (Grimaud et al. 1980; Gascon-Barré et al. 1989; Popper and Udenfriend 1970; Martinez-Hernandez 1985; Chojkier and Brenner 1988; Bazin et al. 1975; Ohtani 1988; Rojkind and Pérez-Tamayo 1983; Hassanein et al 1983).

It is estimated that the total collagen content increases five-to-tenfold in the fibrosis as compared with controls (Barrows et al. 1980; Perier et al. 1984). The augmentation is associated with a higher proportion of insoluble collagen (Bazin et al. 1976; Frey and Bayle 1978). Changes in the quantitative composition of the different types of collagen have also been found. There are, however, discrepanies between the results of various studies. Seyer (1980) and Rojkind et al. (1979) found a decrease in the relative content of type III collagen in human cirrhotic liver tissue. Type III collagen accounted for 36% of the total collagen of normal tissue, while in the fibrotic hepatic samples it reached only 26%. The ratio type III/types I + III collagens ranged from 47.3% to 54.2% and from 37.5% to 39.6% in the normal and cirrhotic liver, respectively. According to the same investigators, there were only slight changes in the relative content of types I and III collagens in the liver of rats treated with carbon tetrachloride. Type I made up 43.9% and 40.0% of the total collagen in the normal and fibrotic liver, and type III constituted 43.9% and 42.5%, respectively.

Studies on the kinetics of collagen accumulation in the liver of rats treated with carbon tetrachloride showed a progressive increase in collagen content. The curves of the collagen accumulation measured during 0–18 weeks of the experiment were described by the exponential equation $y = \mathrm{a}e^{\mathrm{b}t}$, where $y$ is the collagen content after time $t$, and a, b are coefficients. A typical pattern is shown in Fig. 14.1. The increase in type III collagen was found to be more rapid than that of type I collagen. Relatively speaking, a reduction of collagen solubility was shown during the induction of fibrosis of the liver (Kucharz 1987b).

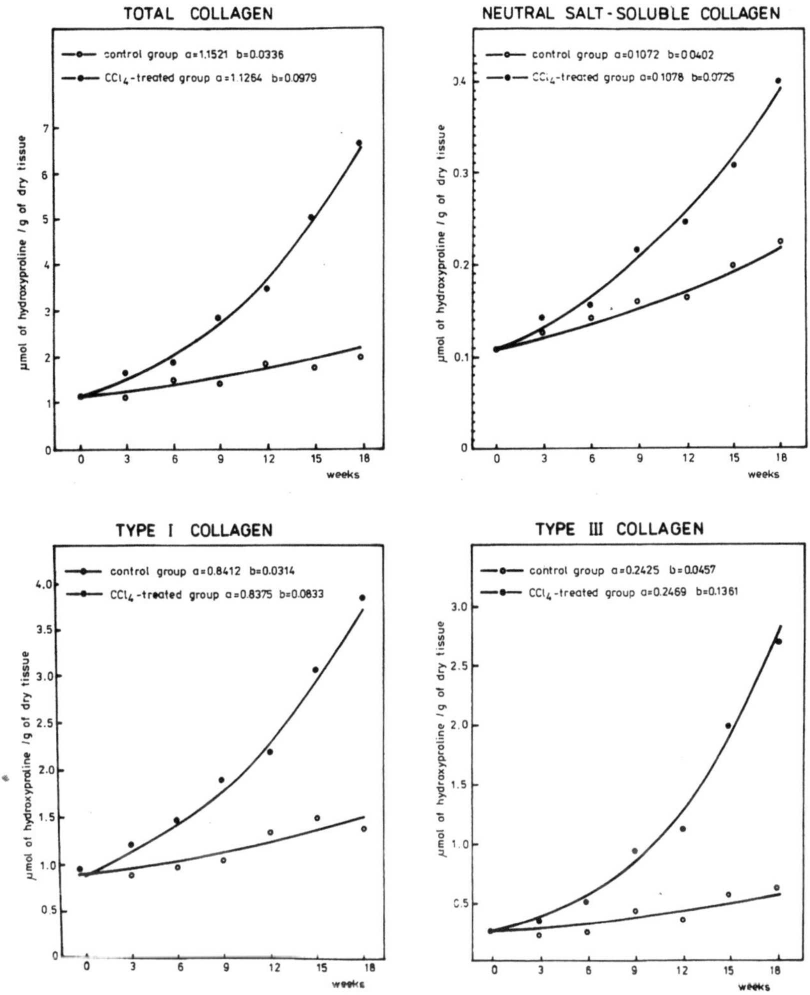


**Fig. 14.1.** Dynamics of collagen accumulation in fibrotic and normal rat liver. The curve of collagen accumulation fits the exponential equation $y = ae^{bt}$. Coefficients of the equation are shown in the figure. (From Kucharz 1987c with permission.)

## 4.6 Metabolism of Collagen in the Liver

A characteristic feature of the fibrotic liver is a very significant stimulation of collagen synthesis. In normal rats, the amount of synthesized collagen is very small (about 1.32 μmol of hydroxyproline/g of tissue per day) while in the cirrhotic liver daily synthesis increases to about 20.8 μmol of hydroxyproline/g (Rojkind and Mourelle 1988).

Increased synthesis of collagen in carbon tetrachloride-induced fibrosis is associated with a progressive rise in transcription of the procollagen $\alpha 1$ (I) gene despite changes in other gene transcription (Panduro et al. 1988).

Increased synthesis is associated with an enhanced activity of enzymes involved in collagen metabolism. A rise in prolyl hydroxylase activity was found in liver samples obtained from patients with various forms of cirrhosis, steatosis, and acute viral hepatitis (Fuller et al. 1976; Tuderman et al. 1977b; Jain et al. 1978). Similar results were reported in various animal models of hepatic fibrosis (Takeuchi and Prockop 1969; Bańkowski et al. 1986; Pawlicka et al. 1988; Bańkowski and Pawlicka 1989). The mechanism, however, remains unclear. Risteli and coworkers (Risteli 1977a; Risteli and Kivirikko 1974, 1976; Risteli et al. 1976, 1978) showed that the enhanced activity of prolyl hydroxylase was not accompanied by an increase in the amount of the immunoreactive enzyme protein, rather the ratio between active and inactive enzyme in the liver was altered. Under normal conditions only 4% of the enzyme remains in the active form, while in fibrotic liver tissue 16% is able to catalyze hydroxylation. The mechanism of activation has not been elucidated.

A stimulation of other enzymes of collagen biosynthesis has also been reported (Risteli et al. 1976, 1978; Risteli and Kivirikko 1974, 1976). An increase in the level of insoluble collagen content is consistent with an enhanced activity of lysyl oxidase (Siegel et al. 1978; Shiota et al. 1987).

Breakdown of collagen in the fibrotic liver has been investigated as a possible mechanism of overaccumulation. Collagenolytic properties of hepatic tissue were suggested for the first time by Frankland and Wynn (1962a). On the basis of histological studies, they described a rapid reduction of collagen fibers in the liver after discontinuation of treatment with hepatotoxins. The ability of the hepatic tissue to degrade collagen was observed in direct measurements by the same authors (Frankland and Wynn 1962b). Their pioneer reports were followed by a number of studies which were mostly carried out on animals with carbon tetrachloride-induced hepatic fibrosis. An augmentation of collagenase was found (Morrione and Levine 1967; Maruyama et al. 1981; Hutterer et al. 1964; Carter et al. 1980; Pérez-Tamayo et al. 1987). The increase of enzyme activity was great after discontinuation of the treatment with fibrosis-inducing agents. Changes in collagenase activity were found in the liver tissue after partial hepatectomy (Okazaki and Maruyama 1974).

An elevated activity of collagenolytic cathepsin was reported in the liver of animals with carbon tetrachloride-induced fibrosis (Hirayama et al. 1969; Murawaki and Hirayama 1980; Kucharz 1987c) and after chronic ingestion of ethanol in rats (Kucharz 1980). An increase in collagen peptidase activity in the liver tissue of rats with fibrosis was shown by Kucharz (1987c).

The cellular origin of collagenase and other enzymes capable of degrading collagen in the liver remains unclear. Collagenase activity was shown in the Kupffer cells (Fujiwara et al. 1973), hepatocytes (Maruyama et al. 1983), and fibroblasts, isolated from the damaged liver. The mechanisms regulating collagenase production in situ are unknown, and a role of cell-to-cell interactions has been suggested. In cell cultures, it was found that the hepatic fibroblasts are able to synthesize and secrete collagenase only when cocultered with hepatocytes and when the medium contains phorbol myristate. The contribution by phorbol esters, compounds with tumor-promoting properties, suggests that collagenase secretion is associated with an alteration in the phenotypic expression. It is possible that the rapid regeneration which occurs in the early stages of fibrosis is a natural equivalent for the phorbol ester stimulation of the cells. The role of hepatocytes in collagenase production by fibroblasts remains unclear. The cell-to-cell interaction may be related to the secretion of soluble collagenase-stimulating factors. When the hepatocytes that are cocultured with fibroblasts started to degenerate, the production of collagenase was reduced.

Immunohistochemical studies showed the presence of collagenase bound to normal collagen fibers in the liver (see Chap. 4). This suggests that the biochemical measurement of collagenase activity may be affected by different procedures of preparation of the sample, leading to release and/or activation of the enzyme bound to the fibers.

The half-life of collagen in the fibrotic liver is shortened. Using pulse labeling with radioactive proline, it was shown that the half-life of type I collagen decreases from 30–35 days in normal rats to 17–20 days in animals with liver fibrosis. The same data for type III collagen were 15–17 days and 10–12 days, respectively (Rojkind and Mourelle 1988). These results depict collagen metabolism in the whole liver. It has recently been hypothesized that different compartments of the hepatic connective tissue are characterized by a significantly different turnover of collagen. At least two metabolic pools of collagen are thought to occur within the liver (Rojkind et al. 1982).

Accumulation of collagen in cirrhotic liver is associated with abnormalities in the metabolism of components of the matrix other than collagen (Kucharz 1986d; Rojkind and Mourelle 1988). Impaired interactions of collagen and glycosaminoglycans isolated from fibrotic liver have been reported. In vitro, glycosaminoglycans obtained from fibrotic rat liver stimulated faster collagen fibrillogenesis than an analogue preparation from controls. These studies were done on unfractionated preparations of glycosaminoglycans. It is difficult to determine whether alterations in collagen or glycosaminoglycan pattern and/or structure cause the increase in the fibril formation. It is also unclear whether similar phenomena occur in vivo (Kucharz 1984, and unpublished results).

## 14.7 Regulation of Collagen Metabolism and Development of Fibrosis

Numerous mechanisms have been postulated to take part in the regulation of collagen metabolism in the fibrotic liver. It is, however, very difficult to prove the real role of these processes in the development of fibrosis. It is clear that damage to the liver produces a response from various cells, distorts the cytoarchitectural structure of the liver, and alters cell-to-cell and cell-to-matrix interactions. Many forms of fibrosis are associated with inflammation and the accumulation of lymphocytes, monocytes, and granulocytes. All these phenomena are responsible for an abnormal regulation of collagen metabolism, and it is impossible to point out a single mechanism which results in increased collagen accumulation (Kucharz 1987b).

Certain investigations have focused on the role of the intracellular pool of free proline. Expansion of this pool is associated with elevated collagen synthesis. The increase of free intracellular proline was shown to enhance collagen synthesis in the liver slices cultured in vitro (Rojkind and Diaz de Leon 1970; Rojkind and Kersenobich 1975). Under experimental conditions, it was found that 7 weeks of treatment with carbon tetrachloride led to a doubling of the proline pool in the liver of rats. This was caused by an augmented synthesis of proline de novo, although increased transport of this imino acid to the cells was also postulated (Rojkind and Dunn 1979). Mata et al. (1975) reported a rise in the level of free proline in the serum of patients with cirrhosis. Moreover, Tyopponen et al. (1980) found an increased content of free proline in liver samples obtained from patients with cirrhosis.

The role of the intracellular pool of free proline in the regulation of collagen metabolism is described in Chap. 5. The relationship of this process in the liver to other mechanisms that have been suggested to be involved in fibrogenesis remains unknown. It is of interest that enhancement of the free proline content has been thought to be a major causative mechanism of the bile duct fibrosis in fascioliasis (see below).

The relationship between collagen biosynthesis and degradation during induction of hepatic fibrosis was investigated in rats treated with carbon tetrachloride for 3–18 weeks (Kucharz 1987c). It was found that the collagen concentration increased progressively, and the general trend of collagen accumulation could be described by an exponential equation. The accumulation was accompanied by an increase in the activity of collagenase, collagenolytic cathepsin, and collagen peptidase. The activity of these enzymes, especially collagenase, rose rapidly during the first weeks of treatment. Further induction of the fibrosis was accompanied by only a slight enhancement of collagenolytic enzymes. On the basis of these findings the following hypothesis was postulated: An increase in collagen breakdown is a self-defense mechanism against fibrosis. When this mechanism is exhausted due to a decreased number or functional capacity of the cells, the rise in collagen content becomes more rapid. This hypothesis is consistent with earlier observations of the production of collagenase by fibroblasts cocultured with hepatocytes. The advance of fibrosis and decrease in the number of parenchymal cells are possible mechanisms for the slowing of the further elevation of the collagen-degrading enzyme production. The presence in the mixed culture of hepatocytes had no effect upon collagen production; thus, a decrease in collagenase secretion facilitates overaccumulation of newly-synthesized collagen in the liver. The described hypothesis may be an explanation of the transition of reversible fibrosis into the irreversible stage of the disease: The loss of the ability of cells to secrete adequate amounts of collagenase and other collagenolytic enzymes may be a turning point.

Various local factors that are able to stimulate collagen synthesis have been described in the damaged liver. McGee et al. (1973) found that the liver of mice receiving a single dose of carbon tetrachloride contained a factor or factors which stimulated an increase in the activity of prolyl hydroxylase. The factor was called collagen-stimulating factor and was shown to be a group of peptides. It was suggested that these peptides were released from the necrotic hepatocytes.

Hatahara and Seyer (1982) isolated a factor from fibrotic rat liver which stimulated collagen synthesis in cultured fibroblasts without affecting their rate of proliferation. It appeared to be a complex, phospholipid-containing polypeptide. The factor caused a specific rise in the rate of transcription of the COL1A1 and COL1A2 genes. An increased accumulation of type I collagen mRNAs was found. The rate of intracellular degradation was not altered (Raghow et al. 1984). The relation of this finding to the earlier described collagen-stimulating factor and their possible role in the development of cirrhosis remain obscure, and several fundamental aspects of the mechanism of action is still to be investigated.

Factors regulating fibroblast proliferation have been shown to be secreted by cultured Kupffer cells. Those from normal rat liver produce two factors (25 000 Da and 5 000 Da) that inhibit fibroblast proliferation. On the other hand, the conditioned media of Kupffer and mononuclear macrophagic cells obtained from rats treated with carbon tetrachloride secrete a factor that stimulates fibroblast proliferation. This factor also stimulated proliferation of cultured fat-storing cells from rat liver. Its molecular weight is 17 000 Da, and the preliminary characterization reveals that it is not interleukin 1 and that its production is not inhibited by indomethacin. This suggests that the factor is not prostaglandin (Armedáriz-Borunda et al. 1989). Earlier, Shiratori et al. (1986) showed that Kupffer cells from rats receiving carbon tetrachloride produce a growth factor for culture liver fat-storing cells. These findings are consistent with the concept of the liver as a bioecosystem because the factors are local phenomena regulating the internal milieu of the liver. The role of these factors in human pathology remains unknown.

A difference in the response to prostaglandin $E_2$ between normal and fibrotic liver slices in culture has been shown by Peters et al. (1989). 16,16-Dimethyl prostaglandin $E_2$ diminished collagen production by slices of liver from rats with nutritional cirrhosis, while in samples of normal liver it had no effect.

Prostaglandin $E_2$ did not change the proline pool. A decrease in collagen formation has been proposed as the hypothetical explanation of prostaglandin-induced inhibition of fibrogeneis.

The role of cytokines (e.g., interleukin 1, interferons) in the stimulation of collagen production has been established as described in Chap. 8. Inflammation is an almost constant phenomenon in several forms of hepatic fibrogenesis and could be a mechanism responsible for the release of the cytokines. All the studies on these phenomena were, however, done in cell cultures, and it is difficult to conclude that similar processes occur in situ under conditions of complex cell-to-cell and cell-to-matrix interactions (McGee 1982; Rojkind and Mourelle 1988).

Infection with the helminth *Schistosoma mansoni* produces hepatic fibrosis (ElMeneza et al. 1989; Grimaud et al. 1987; Dunn and Kamel 1981; Emonard and Grimaud 1989). The worm bears eggs into the small mesenteric vessels of the host. Some eggs become trapped in the presinusoidal areas of the liver and induce a delayed hypersensitivity reaction. T-cell-mediated granulation and inflammation result. Several secreted factors stimulating collagen synthesis have been described (Wyler 1983; Wyler and Rosenwasser 1982; Wyler et al. 1984). These factors are probably secreted by the cells involved in the immune response.

The role of ethanol and its metabolites in the stimulation of collagen synthesis is not clear (Henley et al. 1977; Pawlicka et al. 1988). It has been shown that ethanol accelerated proline uptake by cultured mouse liver cells (Mendenhall et al. 1984). An increased proline uptake by hepatocytes from patients with chronic active liver disease has also been reported (Hassanein et al. 1989).

Liver fibrosis is enhanced by acute phase proteins (Van Gool et al. 1986a, b). An increase in prolyl 4-hydroxylase has been found to correlate with enhancement of haptoglobin, ceruloplasmin, and $\alpha_2$-macroferroprotein. A decrease in hepatic collagenase activity was noted.

Iron overload leads to hepatic fibrosis. Liver damage has been reported in patients with familial hemochromatosis. The deposition of iron in familial hemochromatosis appears to be greatest in the periportal areas, which may explain the formation of early periportal fibrosis. The mechanism of the collagen accumulation is poorly understood. Experimental studies have suggested that in vivo accumulation of iron in the hepatocytes may stimulate them to excess collagen synthesis. An increase in prolyl hydroxylase activity in the liver has been found (Weintraub et al. 1988).

Liver fibrosis is a useful model for the investigation of pharmacological agents affecting collagen metabolism, with special emphasis on the inhibitors of fibrosis. These studies are described in Chap. 24.

Interesting findings on the possible role of collagenase inhibitors in the regulation of collagen breakdown in the fibrotic liver have been reported. It was found that the increase in total collagenase activity in the rat liver during carbon tetrachloride-induced fibrogenesis was associated with the relative increase in the active fraction of the enzyme. Indirectly, this could indicate that the quantity of collagenase inhibitors in the liver is diminished during the induction of fibrosis. The decrease in inhibitors may result from hepatocyte damage as these cells are known to be a source of the inhibitors of proteolytic enzymes, including those of collagenase (Kucharz 1987c). This finding is consistent with the report of Kucharz (1985) who has shown changes in the inhibitory activity of serum in patients with cirrhosis (see below).

## 14.8 Collagen in Liver Regeneration

Surgical removal of two-thirds of the liver in rats is widely used as a model of liver regeneration (Higgins and Anderson 1931). It has been found that connective tissue is formed in the regenerating liver in response to cell growth in order to maintain a normal parenchyma-to-matrix ratio. Hepatectomy is followed by an initial increase in the volume and number of hepatocytes. This results in a reduction in collagen content as related to the weight of the tissue. Some 2 or 3 weeks posthepatectomy, the level of collagen reaches 70% of normal values (Harkness 1952). Morphological studies showed that fine reticulin fibers inside the liver lobule regenerated faster that the thick collagen bundles of the large vessels (Harkness 1957).

Biochemical studies revealed that the collagen content 5 days after hepatectomy was about half that of normal. When regeneration was complete (11 days posthepatectomy), liver collagen content had returned to normal values. Collagen synthesis in the liver measured 5 days posthepatectomy was more than four times higher than in controls. The rate of synthesis returned to normal values at 11 days posthepatectomy.

Regeneration was associated with an altered pattern of collagen types. A significant decrease in the relative proportion of type III collagen was seen. Partial normalization of types I, III, and V collagens was reported after restoration of the hepatic mass. The synthesis rate of types I, III, and V collagens in the liver of hepatectomised rates followed the same trend as changes in collagen type pattern (Rojkind et al. 1983). The biochemical findings are partially consistent with earlier morphological studies. Type V collagen predominates in the sinusoids (Rojkind and Ponce-Noyola 1982), and regeneration of the sinusoids precedes the formation of large vessels and portal tracts, which are rich in type I collagen (Harkness 1957). The differences in the selective properties of collagens synthesized and deposited in the matrix of regenerating or cirrhotic liver are probably related to the synthesis of collagen carried out by different cell populations. Other regulatory mechanisms have also been suggested (Rojkind et al. 1983).

## 14.9 Collagen Metabolism in Diseases of the Bile Ducts

Disorders affecting the bile ducts are common. Inflammation, infection, or stones in the biliary tract lead to cholestasis, which damages the liver. Collagen abnormalities are usually secondary to the liver dysfunction. Primary collagen accumulation within the bile ducts accompanying infection with *Fasciola hepatica* is a main feature of this parasite infestation (Elwy 1967).

Interesting studies on collagen metabolism in the bile duct have been carried out in experimental fascioliasis. Infection with the fluke, *F. hepatica*, and certain other trematodes leads to significant enlargement of the bile duct. The duct which harbors the worm is about 20 times thicker than a normal one. The increase is associated with the deposition of fibrous material, mostly collagen. A rise in total collagen content in the bile duct was reported. Both types I and III collagens were shown to increase in infected animals, but type I increased less than type III (Mark and Isseroff 1983). It has been found that the flukes release large quantities of proline (Kurelec and Rijavec 1966; Ertel and Isseroff 1974, 1976). Free proline is known as one of the stimulatory factors responsible for an increase in collagen synthesis (see Chap. 5). This role of the large amounts of proline in the bile fluid secreted by the worms has been confirmed by an experimental induction of fibrosis with the infusion of proline. Vacanti and Folkman (1979) and Wolf-Spengler and Isseroff (1983) infused proline solution into the bile duct of mice and rats and found hyperplasia and fibrosis similar to that observed in animals with implanted flukes. These observations have been confirmed by subsequent studies (Modavi and Isseroff 1984). Intraperitoneal infusion of proline in rats produced a hyperplasia similar to that following implantation of *F. hepatica* in the peritoneal cavity (Isseroff et al. 1977). Further confirmation of the role of the excess of proline was obtained by the inhibition of fibrosis. Proline analogues (azetidine or 3,4-dehydroproline) were able to diminish fibrosis in rats with implanted flukes or receiving proline infusion. As described in Chap. 24, other mechanisms of antifibrotic activity of proline analogues are also possible, and it is difficult to interprete the above-described experiment.

Ligation of the biliary duct without infection of flukes has been found to increase the proline content and uptake by hepatocytes but also by the duct cell fraction. The glycine conent is not affected (Chen et al. 1983). The role of this mechanism in the fasciola-induced biliary obturation is unknown.

## 14.10 Indices of Collagen Metabolism in Hepatic Disorders

The liver biopsy is still the most reliable procedure to establish the stage of fibrosis. This method has limitations and cannot be used frequently. Several attempts have been made to find biochemical indices for the measurement and follow-up of liver fibrosis, but their diagnostic value remain controversial (Hahn and Martini 1980a; Hahn and Schuppan 1982; Hahn 1984).

Increased urinary excretion of hydroxyproline was reported in patients with cirrhosis of the liver (Emmrich et al. 1967; Kratzsch 1969a, b; Kaznachev et al. 1972; Komarov et al. 1977; van Hussen et al. 1971). It was observed in patients with cirrhosis regardless of the etiology of the cirrhosis. Several authors investigated hydroxyproline excretion in patients with alcoholic liver disease, including precirrhotic steatosis (Resnick et al. 1973; Wu and Levi 1975). It was found that withdrawal from alcohol consumption in patients with chronic alcoholic liver disease was associated with a rapid increase in hydroxyproline excretion. The mechanism of this phenomenon remains unclear. It has been suggested that alcohol inhibits collagen degradation in the liver.

An elevated urinary output of hydroxyproline was also reported in patients with viral hepatitis (Resnick et al. 1973). Experimental studies with animals treated with various hepatotoxins confirmed that the enhanced excretion of hydroxyproline and hydroxylysine is associated with damage to the liver (Kucharz 1980).

An increased level of collagen metabolites (hydroxyproline, hydroxylysine, collagenlike protein) were shown in patients with chronic liver disorders (Iber et al. 1957; Iob et al. 1966; Hirayama et al. 1972; Mata et al. 1975; Myara and Cosson 1988; Komarov et al. 1977; Gressner 1982). A relationship between the severity of the liver damage and level of serum hydroxyproline was found in patients with viral hepatitis. These indices were normalized during recovery from the disease (Drożdż and Kucharz 1978). Although certain correlations between liver function and the serum level of collagen metabolites have been reported, the measurement of serum or urinary hydroxyproline has no diagnostic value or application to clinical practice.

In recent years, many groups reported the results of the determination of amino- or carboxyterminal propeptides in serum or urine. Attention has been paid to the aminoterminal propeptide of type III procollagen. There is a positive correlation between its serum level and extent of liver fibrosis as measured histologically in tissue samples (Rohde et al. 1978; Wick et al. 1978; Rohde et al. 1979; Frei et al. 1984; Weigand et al. 1984; Bolarin et al. 1984; Ackermann et al. 1981; Gabrielli et al. 1989; Savolainen et al. 1984, 1988). A detailed study on the serum content of the aminoterminal propeptide of type III procollagen in patients with alcoholic liver disease was given by Niemelä et al. (1983). The patients were divided into four groups on the basis of liver histology, those with normal light microscopy, fatty liver, alcoholic cirrhosis with hepatitis, and inactive cirrhosis. All subjects with alcoholic cirrhosis with hepatitis had elevated levels of the propeptide. Patients with a normal liver histology as seen under the light microscope had normal levels of the propeptide, and those with fatty liver and inactive cirrhosis exhibited only slightly elevated levels. In a follow-up study, it was found that the peptide concentration decreased slowly during recovery from alcoholic hepatitis and increased rapidly after a new drinking bout. Serum levels of the aminoterminal propeptide of type III procollagen were shown to be useful in the developing diagnosis in patients suffering from viral hepatitis. In a prospective longitudinal study, it was found that the propeptide level was elevated at the onset of viral hepatitis and increased in patients with chronic hepatitis and active cirrhosis, while in those who recovered a reduction was seen (Bentsen et al. 1987).

Further clinical studies are needed to estimate the practical value of the determination of the aminoterminal propeptide of type III procollagen (Surrenti et al. 1987). It seems that this index has no value in the differential diagnosis of chronic hepatitis. On the other hand, it could be useful in monitoring the progress of primary biliary cirrhosis (Hahn and Martini 1980a; Hahn 1984; Rojkind 1984; Babbs et al. 1988).

An elevated serum level of antigen related to type IV collagen (7-S collagen) was reported in patients with alcoholic liver disease. The highest values were observed in patients with hepatitis and inactive cirrhosis. It has been suggested tht this index reflects the formation of real basement membranes in the perisinusoidal space, a process known as capillarization of the sinusoids, which is found during the development of liver cirrhosis (Niemelä et al. 1985).

Enzymes of collagen biosynthesis have been determined in the sera of patients with hepatic disorders. Tuderman et al. (1977b) reported an increased level of immunoreactive prolyl hydroxylase in patients with chronic active hepatitis and primary biliary cirrhosis. The highest values were shown in patients with this type of cirrhosis. There was no correlation between the serum level of immunoreactive prolyl hydroxylase and other indices of liver function. In patients with primary biliary cirrhosis a negative correlation between the hydroxylase and $\gamma$-glutamyl transpeptidase was noted (Kuutti-Savolainen et al. 1979a). The same group carried out further studies with a larger group of patients with primary biliary cirrhosis and confirmed that 77% of patients had an elevated serum level of immunoreactive prolyl hydroxylase. Contrary to their previous results, enhanced activities of the enzyme were found in five of seven patients with alcoholic cirrhosis. Three of four patients with viral hepatitis as well as all patients with cancer metastases in the liver had an increased level of the enzyme. There were no changes in the serum level of prolyl hydroxylase in individuals with fatty liver and disorders of the biliary tract. A progressive increase of the enzyme level was shown in patients with biliary cirrhosis. Treatment of these patients with d-penicillamine diminished the serum enzyme level. A similar decrease to normal values was observed in patients who recovered from viral hepatitis (Savolainen 1979; Savolainen et al. 1983).

The serum level of immunoreactive prolyl hydroxylase paralleled the activity of the enzyme in the liver, but no correlation has been found with collagen content. The only association with liver function was shown in patients with primary biliary cirrhosis; the prolyl hydroxylase level in the serum correlated with the activity of alkaline phosphatase (Kuutti-Savolainen et al. 1979a; Bolarin et al. 1984, 1987).

Only single reports deal with enzymes of collagen synthesis other than prolyl hydroxylase. An increase of serum galactosylhydroxylysyl-glucosyltransferase was reported in patients with primary biliary cirrhosis, acute viral hepatitis, and hepatocellular carcinoma (Anttinnen 1977; Kuutti-Savolainen et al. 1979b; Bolarin et al. 1982, 1984).

Changes in serum inhibitors of collagenolysis in patients with chronic liver disorders were reported by Kucharz (1985). He found, using rat hepatic slices cultured in vitro, that the serum of healthy individuals inhibited collagenolysis in 57.17% while in patients with cirrhosis inhibition was very low (28.82%). Interestingly, in patients with chronic persistent hepatitis and chronic active hepatitis inhibition was 65.69% and 71.22%, respectively. A possible diagnostic value of this measurement has been suggested.

Collagenase activity in peripheral blood granulocytes obtained from patients with various liver diseases was determined by Ikeda et al. (1983). The total collagenase activity was found to be enhanced in patients with chronic active hepatitis and liver cirrhosis and correlated with the extent of fibrosis of the liver. Active collagenase, i.e., the enzyme activity without pretreatment with organomercurial compounds, was increased in patients with chronic active hepatitis but not in other investigated disorders. The diagnostic value of the measurement of granulocyte collagenase activity needs further study.

A few reports suggesting the diagnostic usefulness of serum collagen peptidase level in patients with chronic diseases of the liver have been published. An enhancement of its activity in patients with cirrhosis was described by Grasedyck et al. (1971) and von Hahn et al. (1974). Kucharz et al. (1981) found a significant elevation of the enzyme activity in all investigated patients with cirrhosis ($n = 30$) and in only one-quarter of patients with chronic hepatitis ($n = 18$). The enzyme activity correlated with the bilirubin level and alanine transaminase activity. A negative correlation with serum albumin level was found.

Prolidase (proline dipeptidase) is an enzyme that cleaves dipeptides with proline or hydroxyproline as their carboxyterminal amino acid. It is possible that prolidase takes part in the degradation of collagen metabolites. Two forms were found in rat liver, and the levels of both increased after carbon tetrachloride treatment (Miech et al. 1988). A rise in plasma prolidase activity was reported in patients with chronic liver diseases. There was no correlation between prolidase and liver function and the stage of fibrosis. This enzyme is also seen in ascites fluid (Myara et al. 1984).

# 15 The Teeth and Periodontal Tissues

## 15.1 Structure and Collagen Content in Normal Teeth

The teeth form a highly specialized body part located in the oral cavity that take part in the separation and mastication of food. In some animals, the teeth are important weapons. The human dentition consists of two sets of teeth: the deciduous (baby) teeth and secondary (permanent, succedaneous) teeth. The morphology of the two sets is similar. The tooth consists of a crown which is visible above the gum and a root or roots that fit into a socket (alveolus) in the bone. Mostly it consists of dentin covered by enamel or cementum in the crown or root, respectively. The root is attached to the bone by the periodontal ligament. The central part of the tooth is made up of loose connective tissue filling the pulp chamber. The pulp chamber continues into the root canal where it communicates with the periodontal structures via the apical foramen.

The major feature of the tooth and the periodontium is the great resistance to mechanical stress. This property is closely related to the function of the dentition and is associated with a high content of collagenous tissues. Three calcified tissues that form the hard parts of the tooth have been distinguished on the basis of their structure and embryological development: dentin, cementum, and enamel.

### 15.1.1 Dentin

Dentin makes up the majority of the tooth. In terms of chemical composition, it is similar to bone tissue. It consists of about 80% hydroxyapatite crystals and 20% organic material. Its structure, however, is quite different from that of bone. Dentin is produced by odontoblasts, the cells which line the pulp cavity, and formation continues throughout life, although the rate of production decreases with age. Thus, the dentin layer thickens at the expense of the pulp. The odontoblast serves as a dormant pool of cells. In case of trauma or dental caries the odontoblasts are activated and involved in the rapid formation of reparative dentin.

Structurally, dentin is composed of long dentinal tubules. They are numerous (about 35000 per $mm^3$ of dentin), and they run at right angles to the pulp. Each tubule houses the odontoblastic process, which is located one third of the way along the pulp side (Thomas 1979). The tubules are formed during the migration of the odontoblasts during dentinogenesis. Collagen constitutes almost all the organic matrix of dentin. Morphologically, the demineralized dentin is composed of tightly packed collagen fibers without any distinctive pattern. The fibers found in predentin are smaller than those in dentin (20–70 nm) (Jessen 1967; Johannesen and Bang 1972). The major biochemical feature of dentin collagen is its insolubility; it cannot be dissolved by neutral or acid solutions even after denaturation (Carmichael et al. 1974; Munksgaard 1979). Small portions of total collagen can be obtained after pepsin digestion (Carmichael et al. 1977). This finding indicates the occurrence of the highly cross-linked collagen type. The nature of the cross-links seems to differ from those found in other tissues as concluded from experiments with lathyrogenic animals. In animals which became lathyric due to feeding with a diet supplemented with $\beta$-aminopropionitrile, a powerful lathyrogenic agent, only a small quantity of collagen can be extracted from the dentin (Carmichael et al. 1974; Wohllebe and Carmichael 1979).

Collagen of the dentin has been identified as type I. The other types have not been found in normal dentin although organ cultures showed that odontoblasts synthesize and secrete type III collagen and type I trimer (Volpin and Veis 1973; Dodd and Carmichael 1979; Munskgaard et al. 1978). The solubility of predentin is slightly higher than that of dentin, and small amounts of collagen may be extracted from the odontoblast layer. Traces of type I trimer have been found in rat or bovine predentin (Wohllebe and Carmichael 1978; Munksgaard 1979).

Collagen cross-links of predentin and dentin have recently been investigated with regard to the extreme insolubility as well as the role of the cross-link pattern in the mineralization of the tissue. The major reducible cross-link of dentin was reported to be dihydroxylysinonorleucine (Mechanic et al. 1971). Pyridinoline, a trifunctional nonreducible cross-link, is known to be a major cross-link of dentin (Eyre and Oguchi 1980; Kuboki et al. 1981b). Deoxypyridinoline, a minor cross-link, seems to be unique to bone and dentin (Eyre et al. 1984a; Robins and Duncan 1988). There are some controversies on the amount of reducible cross-links in dentin. Walters and Eyre (1983) showed that the content of pyridinium cross-links in mineralized dentin increased with age, concomitantly with reducible cross-links. Aging of dentin is associated with mineralization, and some investigators found that pyridinoline cross-links in bone were present solely in unmineralized collagen (Banes et al. 1983; Mechanic et al. 1985). This has been confirmed in recent studies of Linde and Robins (1988) who reported that bovine predentin, the unmineralized precursor of dentin, contained twice as many reducible cross-links as dentin. Only minute amounts of nonreducible cross-links were found in predentin, whereas both pyridinoline and deoxypyridoline were present in collagen from mineralized dentin.

Differences were also found between rat and bovine dentin. Bovine dentin was shown to contain about one reducible cross-link per collagen molecule. Rat dentin was less mature and contained twice as many reducible cross-links. In contrast, bovine dentin contained twice as many pyridinium cross-links.

These observed differences in cross-linking between species may reflect some physiological differences in their dentition. The role of collagen maturation in the mineralization of teeth is suggested on the basis of cross-linking changes in predentin and dentin.

### 15.1.2 Cementum

Cementum covers the root dentin and forms the dental side of the tooth suspension apparatus. Its structure and composition resemble those of bone. Cementum is formed by cementoblasts, mostly located close to the apex of the tooth, an area known as the cellular cementum. Cementoblasts secrete the matrix and, like osteoblasts, are trapped in lacunae. These cells are known as cementocytes. Cementum as investigated in cow teeth is composed 95% of type I collagen, the remainder being type III collagen (Birkedal-Hansen et al. 1977). Immunofluorescence studies showed that type III collagen is located mostly in structures related to periodontium, while the other areas of cementum contain type I only (Wang et al. 1980). In human teeth no traces of type III collagen in cementum have been found (Christner et al. 1977).

### 15.1.3 Enamel

Enamel is the hardest substance of the body and almost entirely (99.5%) consists of inorganic substances. It does not contain collagen. Silness et al. (1976) found that in the fissures of human teeth, a cementum-like substance covers the enamel. A similar covering has been reported on the crown of teeth in certain animals (Glimcher et al. 1964; Ainamo 1970; Lustgarten and Shapiro 1974). It is thought that this cover is produced by mesenchymal cells which penetrate the enamel surface after enamel formation is complete.

### 15.1.4 Dental Pulp

Dental pulp fills the dental chamber. It consists of loose connective tissue rich in vessels and nerves. Its main function is to supply the dental tissues with nutrients. The size of the chamber decreases with age, and the pulp becomes more fibrous. The loose connective tissue consists of types I and III collagens. Type IV collagen has been found in association with the blood vessels within the pulp (Lesot et al. 1978; Lesot and Ruchl 1979; Rao et al. 1979; Thesleff et al. 1979; Cournil et al. 1979; Shuttleworth et al. 1980). The content of collagen in the dental pulp is estimated as low, i.e., 16.8% (Orlowski 1974), but the turnover of collagen is relatively high (Orlowski 1976; Orlowski and Doyle 1976). Kishi et al. (1979) found a high content of collagenase in the culture medium of bovine dental pulp. All pulp enzyme released was in a latent form. The molecular weight was estimated at 45000 Da. The presence of collagenase in the pulp seems to be an explanation of the low collagen content and a high synthesis of collagen within the dental pulp. A specific inhibitor of collagenase (glycoproteid, molecular weight 36000 Da) has been isolated and purified from bovine dental pulp (Kishi and Hayakawa 1984; Kishi et al. 1984).

### 15.1.5 Periodontium

Periodontium is a complex structure responsible for the attachment of the tooth to the bone and for compensation of the forces exerted by mastication. Periodontal ligaments and the gingiva are the fibrous components of the periodontium. Collagen accounts for about 50% of the soft connective tissue of the periodontium (Ballard and Butler 1974; Page et al. 1980; Birkedal-Hansen et al. 1977; Shuttleworth and Smalley 1983; Robertson 1976), forming fibrils assembled in bundles of Sharpey's fibers that bind the tooth as well as free gingival fibers. It was found that collagen in the periodontal membrane is synthesized by fibroblasts. The bundles are formed by migrating fibroblasts and represent the tracts of their migration (Garant and Cho 1979a, b; Trelstad and Hayashi 1979). This phenomenon can explain the mechanism of the development of an oriented and genetically controlled collagenous network. Gingival and periodontal membrane contains a majority of types I and III collagens (Butler et al. 1975; Dabbous et al. 1979; Wang et al. 1980). Type III collagen is estimated at 10%–25% of type I (Limeback and Sodek 1979; Sodek and Limeback 1979; Duncan et al. 1984). Type I trimer was also detected (Narayan and Page 1976). Gingival collagen can be extracted with salt and acid solutions to 10%–30% (Schneir et al. 1975). The major cross-link of insoluble collagen was identified as dehydro-$(OH)_2$-Lys-Nor-Leu. The rate of collagen turnover in gingiva is high; its half-life is estimated as 2 days (Sodek 1977). Many studies have reported the presence of collagenase in the gingiva (Robertson and Simpson 1976). Christner (1980) was the first who described collagenase found in human periodontium ligament. It has properties typical of mammalian collagenase. Most of the interest in periodontal collagenolytic enzymes has been associated with their role in periodontal diseases, and these aspects of collagen degradation under pathological conditions are reviewed in the next part of this chapter.

The synthesis of collagen and formation of bundles in the periodontium are associated with their insertion into the hard tissues. This process remains unclear. It has been suggested that periodontal fibroblasts differentiate at the hard-tissue surface and then migrate across the periodontal membrane and anchor the fibers to the opposing hard tissue. The migration is suggested to take place from the alveolar bone toward the cementum (Garant and Cho 1979a, b). Periodontal ligaments were found to be able to regenerate after tooth reimplantation (Proye and Polson 1982).

The gingiva is a tissue complex whose external part is composed of epithelium which covers a connective tissue portion (Alvarez-Mojuro and Alvarez-Mojuro 1978). The gingival collagen forms fasciculi that are distributed throughout the entire gingiva. The abundant blood vessels are surrounded by collagen.

Mechanical injury to the gingiva may lead to repair with retention of severed fibrils or, alternatively, to a complete replacement of fibrils by inflammatory cells followed by root resorption (Grevstad 1987). The progress in restoration is influenced by the initial tissue reaction to injury. Gingival wound repair in rats has been described by Grevstad (1988).

## 15.2 The Role of Collagen in Development of Teeth

The formation of a tooth, odontogenesis, is a complex phenomenon relatively well explored and described. The tooth develops from the mesenchyme and the ectoderm. The latter's role is in the form of proliferation of the oral epithelium. The epithelium invaginates into the mesenchyme, and through a few stages of histo- and cytodifferentiation the tooth is formed. The detailed description of these phenomena is reviewed in textbooks of histology. Here, the role of collagen in this process is emphasized. Certain types of collagen appear at various stages of development. It has been shown that collagenous matrices play an important part. The basement membrane is an indispensable component of the early phases. The epithelial and mesenchymal parts of the tooth germ are separated from the basement membrane, a structure rich in type IV collagen. Experimental elimination of the membrane showed that the odontoblasts failed to differentiate (Thesleff et al. 1978). The basement membrane disappears when predentin starts to mineralize. The dental papilla produces types I and III collagens (Thesleff et al. 1979). The amount of type III collagen in mineralization remains unclear.

There are various experimental data and clinical observations indicating that agents which impair collagen synthesis produce defective tooth development. For example, lathyrogens, ascorbic acid deficiency, proline analogues, and tetracyclines are well-known to affect tooth development and to lead to defective dentition.

## 15.3 Collagen and Dental and Periodontal Pathology

Alterations in collagen occur in a number of diseases of the teeth and supporting structures. Many of them are part of the generalized defect of the connective tissue. Among them, osteogenesis imperfecta is the best known hereditable disorder involving teeth. Scurvy (vitamin C deficiency) is an example of an acquired abnormality of collagen metabolism manifested as profound changes in the oral cavity. Hormonal disturbances, especially those affecting calcium and phosphorus metabolism, lead to visible dental defects. These systemic diseases have been reviewed in other chapters. The present description is restricted to disorders with specific localization to the dental tissues. Among them, the most extensively investigated is periodontal disease. Hydantoin-induced gingival hyperplasia is the other one limited to the gingival tissue. It is probably associated with metabolic alterations.

### 15.3.1 Periodontal Disease

Periodontal disease is a chronic inflammation of the periodontal supporting structures. The disease leads to progressive destruction of the periodontium and loosening of the teeth. Its pathogenesis is understood only partially. The causative agent seems to be complex. In general, bacteria which accumulate close to the gingival margin and elevated local activity of collagenase are suggested to be responsible for the breakdown of the periodontium. The lesions include separation of the junctional epithelium and formation of a crevice close to the enamel surface. The crevice is invaded by bacteria which augment further degradation of the collagenous structures. Gradual progression of the disease in an apical direction is associated with breakdown of the periodontal membrane and alveolar bone and impaired synthesis of collagen (Larjava et al. 1989). This process leads to the destruction of the binding structures of the teeth within the jaw. Increased degradation of the periodontal collagenous structures is attributed to collagenase. Increased collagenase activity has been found in the gingival crevicular fluid in numerous studies (Uitto and Raeste 1978; Golub et al. 1979; Kowashi et al. 1979; Villela and Birkedal-Hansen 1985; Kryschtalskyj et al. 1986), but its origin has not been finally elucidated. It is possible that the collagenase originates from various cells. An early hypothesis attributed collagenase production to bacteria, which make up a constant component of dental plaque. Various species occurring in the oral cavity have been examined for the ability to produce collagenase. Evidence has been provided that *Bacteroides gingivalis* is a potent manufacturer (Uitto et al. 1986).

The role of other bacterial species is more controversial. It was found that the black-pigmented *Bacteroides* as well as *Actinobacillus actinomycetemcomitans* were also able to produce collagenolytic enzymes (Robertson et al. 1982; Rozanis and Slots 1982), but this is not accepted by all investigators (Uitto et al. 1986). Involvement of bacteria in the pathogenesis of periodontal disease has not been directly proved. Enzymes of bacterial origin have never been demonstrated in tissues or fluids obtained from the lesions. The mode of cleavage of collagen molecules by bacterial collagenase is different from that by vertebrate enzymes, and degradation products originating from bacterial enzymes have not been detected in the periodontal tissue. Production of tissue collagenase has been shown in cultures of clinically healthy gingiva (Uitto et al. 1978, 1984; Narayanan and Page 1983). Moreover, inflamed gingiva produces more collagenase than normal tissue (Fullmer and Gibson 1966; Beutner et al. 1966; Bennick and Hunt 1967; Heath et al. 1982). Collagenase has been extracted from the gingiva, and a positive correlation was found between the severity of inflammation and the activity of the extracted collagenase (Uitto et al. 1981). The presence of vertebrate collagenase in the inflamed gingiva and crevicular fluid has been shown using specific antibodies (Woolley and Davies 1981). The cellular origin of the collagenase remains unclear. The polymorphonuclear leukocytes which are abundant in the inflamed tissue are suggested to be the cells responsible for collagenase production. Macrophages are also possible sources of collagenase, and recently the epithelial cells were suggested to be involved in collagen degradation. The last hypothesis is supported by morphological studies which revealed that breakdown appears to begin perivascularly in infiltrated tissue of the subepithelial space (Thilander 1968).

The ability of normal gingival tissue to degrade collagen is associated with a rapid turnover of collagen under normal conditions. This finding indicates that healthy gingiva contains a potent proteolytic system, including collagenase (Kowashi et al. 1980). Various mechanisms of activation of collagenase have been postulated to be involved in the pathogenesis of periodontal disease. Most of them are similar to those observed in the inflammation of other tissues. The increased activity of collagenase may be a result of the following factors: interleukin 1, prostaglandins, bacterial toxins, proteases, and lysosomal enzymes released during phagocytosis (Wang et al. 1983).

Interesting suggestions arose from the discovery of the epithelial-mesenchymal cell interaction. Coculturing with epithelial cells led to increased secretion of collagenase by fibroblasts (Johnson-Mueller and Gross 1978). The nature of the active substance responsible for collagenase stimulation remains unknown.

The constant presence of bacteria in the vicinity of the inflamed gingiva suggests that bacterial products (e.g., endotoxins) can be directly or indirectly involved in the increased collagenase production and/or activation, although the mechanisms of these phenomena are still obscure.

Increased collagenase activity in saliva obtained from patients with periodontal disease was reported by Iijima et al. (1983). A correlation was found between total collagenase activity in saliva and that in crevicular fluid. The electrophoretic pattern of collagen breakdown products generated by the enzyme indicated that the collagenase originated from the tissues. In normal humans, the active enzyme constitutes about 45% of the total activity of collagenase. The same value in patients with periodontal disease was about 25%. The nature of the latent enzyme in saliva remains unknown. The latency may be caused by an increased release of the procollagenase or elevated formation of the enzyme-inhibitor complex. The role of the salivary glands in this is unclear. Kato et al. (1980) showed an increase in collagen peptidase in developing rat salivary glands. The activity increased rapidly from 1 to 10 days and then decreased until 4 weeks of age. The changes were associated with an accumulation of collagen in the gland. A natural inhibitor of collagen peptidase appeared at 5 weeks of age and subsequently increased in activity. After subtracting the inhibitory effect, the activity of collagen peptidase was shown to increase gradually with maturation. The occurrence of collagen peptidase in human salivary glands is not known. Hino et al. (1975) found no enzyme activity in human submandibular glands, but his finding could have resulted from a high content of the inhibitor. A physiological role of collagen peptidase in the salivary gland and its possible secretion into the saliva are unclear. Collagenase has not been investigated in the salivary glands. It is, however, possible that collagen peptidase is accompanied by collagenase.

An enhanced cellular immune response to type I collagen was reported in patients with periodontal disease (Mammo et al. 1982). An increase in the response correlated with the severity of the disease. The process was suggested to be secondary to inflammation but could be of importance in maintaining the chronicity of the disease, even after removal of the initiating factor or factors.

### 15.3.2 Hydantoin-Induced Gingival Hyperplasia

The involvement of collagen metabolism in gingival hyperplasia due to long-term treatment with hydantoin remains obscure despite numerous investigations. It is well-known that a large percentage of patients treated with hydantoin derivatives for epilepsy suffer massive gingival overgrowth. The hyperplastic gingiva contains collagenous tissue, but the cellular-extracellular matrix content stays unchanged (Birkedal-Hansen 1982). The ratio of type I/type III collagen has also been found to be normal in the altered tissue (Ballard and Butler 1974; Schneir et al. 1978). Hydantoin has been found to be an inhibitor of prolyl hydroxylase when added to the fibroblast culture in vitro (Liu and Bhatnagar 1973; Blumenkrantz and Asboe-Hansen 1974b). This observation has not been confirmed in vivo as the hydroxyproline content in the chains of collagen molecules isolated from the hyperplastic gingiva was normal (Schneir et al. 1978). It is possible that hydantoin-induced gingival hyperplasia is not associated with altered collagen metabolism but rather with uncontrolled overgrowth of the gingival cells and overproduction of collagenous structures. This latter possibility may reflect the normal rate of production of the extracellular matrix by an enhanced number of cells. On the other hand, fibroblasts obtained from the altered tissue, when cultured in vitro, synthesize at a significantly faster rate than those isolated from the normal gingiva (Hassell et al. 1978). One can speculate that these findings result from the stimulation of fibroblasts by bacterial products. The fibroblasts prior to stimulation are "sensitized" to bacterial products by hydantoin. This effect of the bacterial plaque is obviously opposite to that observed in periodontal disease, and there are not enough data to support this hypothesis. Thus, the pathogenesis of hydantoin-induced gingival hyperplasia remains unknown.

# 16 The Skin and Wound Healing

## 16.1 Collagen in the Skin

The skin is the largest organ system of the body. It is an organ of protection, forming a tough but pliable covering that is the major barrier between a person and the environment. It plays a role in temperature regulation and sensory reception and is involved in immunological surveillance.

It consists of the epidermis which lies above a complex of structures of the basement membrane zone. Under this zone, the dermis and hypodermis are found. The dermis is composed of an outer or superficial papillary layer and an inner or deep reticular layer. The former is characterized by a loose arrangement of relatively thin fibers embedded in a considerable amount of ground substance; the latter is built up of a dense pattern of thick fibers. Beneath the dermis is a loose connective tissue layer, known as the hypodermis. Functionally, the hypodermis provides support for the overlying skin and attachment to deeper tissues.

Collagen represents about 70% of the dry weight of skin. It contributes to the formation of the structural network underlying the dermis. Type I collagen predominates, accounting for 80%–90% of the total collagen content. The remaining 10%–15% consists of type III collagen. Other types of collagens are found in small quantities (IV, V, and I trimer). The ratio of type I/type III collagens is related to age; in fetal tissue the amount of type III collagen is almost equal to that of type I (see Chap. 6). Fibroblasts are the cells responsible for collagen production in the skin. The possible contribution of other cell types has been suggested (Bauer and Uitto 1982; Burgeson 1987a).

## 16.2 Collagen in Cutaneous Diseases

Collagen is also involved in cutaneous pathology. Abnormalities in the structure and metabolism of collagen have been shown in several hereditary disorders (Chap. 9) and socalled collagen diseases (Chap. 10). The remaining disorders have been the subject of only a few investigations. This chapter summarizes the collagen alterations in patients with various dermatological diseases.

An increase in prolyl hydroxylase activity in skin biopsies from patients with psoriasis has been noted. It was accompanied by an elevated level of immunoreactive prolyl hydroxylase and galactosylhydroxylysyl glucosyltransferase. There was no correlation between the enzymes in serum and the skin although in some patients a high activity of the serum enzymes was found (Kuutti-Savolainen 1979; Kuutti-Savolainen and Kero 1979; Fleckman et al. 1973; Uitto et al. 1970a; Keiser et al. 1971).

One report indicating an increase in prolyl hydroxylase activity in the skin of patients with lichen ruber planus was published (Kuutti-Savolainen and Kero 1979). Oikarinen et al. (1989) showed an increased solubility of collagen from the lesioned skin of patients with granuloma annulare, as compared with the skin of controls or unaffected skin from the patients. Granuloma annulare is characterized by a focal degeneration of collagen associated with inflammation and fibrosis in the surrounding area. The collagen content in the skin was low but the prolyl hydroxylase activity was elevated. Surprisingly, this rise in the enzyme activity was not associated with an increased collagen synthesis level per cell. Type I/type III collagen ratio was unaltered. The increased degradation of collagen in the skin of patients with this inflammatory disorder has been suggested as well as other mechanisms (e.g., cell selection).

Similar findings were seen in patients with skin sarcoidosis, a systemic disease affecting the lungs. The available data on collagen involvement in the pulmonary form of sarcoidosis are reviewed in Chap. 13.

In a patient with scleroderma adultorum Buschke, a disorder of unknown etiology characterized by progressive swelling and nonpitting induration of the skin, an increased level of hydroxyproline in the deep layer of the skin and an elevated urinary output of hydroxyproline were reported (Roupe et al. 1987). This single case is unsufficient to estimate the role of disturbed collagen metabolism in this disease.

Changes of the epidermal basement membrane were described in patients with dermatitis herpetiformis. These findings were based on immunohistochemical localization of type IV collagen and were not associated with biochemical measurements (Karttunen et al. 1984).

Ultrastructural and biochemical studies revealed a loss of collagen in the affected skin from patients with necrobiosis lipoidica, a chronic disease characterized by yellowish, irregularly demarcated lesions on the skins. The disease is often associated with diabetes mellitus. There was no change in the type I/type III collagens ratio. Cultured fibroblasts from affected skin produced less collagen than those from normal-looking skin. The decreased synthesis was caused by a reduced level of mRNA for type I procollagen. The secretion of collagenase was not increased (Oikarinen et al. 1987).

Antibodies to collagen were shown in the sera of patients with leprosy (McAdam et al. 1978). It is well-known that *Mycobacterium leprae* causes antigenic stimulation, and it is possible that it acts as an adjuvant in immune reactions. Degradation of collagen in the affected skin has been suggested. On the other hand, it is unclear whether collagen-specific antibodies are involved in the pathogenesis of any complications of leprosy.

Chronic ultraviolet radiation is one of the common causes of cutaneous damage. Its effect on collagen has been investigated in various models. Histochemical studies have shown that both animal and human skin collagen undergoes changes in staining characteristics after chronic irradiation. An increase in "reticulin" fibers and a decrease in the amount of collagen fibers have been found (Knox et al. 1962; Kligman et al. 1982, 1983). Biochemical characterization of these changes have revealed a loss of insoluble collagen and an elevated soluble collagen fraction content (Smith et al. 1962; Johnson et al. 1984). Plastow et al. (1987) noted that hairless albina mice receiving chronic ultraviolet radiation show an increase in type III collagen in the irradiated skin. This finding is consistent with an elevated amount of "reticulin" found in microscopic studies (Schwartz et al. 1989). Application of a sunscreen protected against these changes (Plastow et al. 1988). An increase in total skin collagen in mice irradiated from 4 to 6 weeks has been reported. At week 20 the total collagen level returned to control values (Kligman et al. 1989), and it was concluded collagen synthesis was stimulated until late in the course of irradiation. In vitro, ultraviolet light was found to inhibit fibril formation. Differences between light of various wavelengths have been described (Fujimori 1985).

## 16.3 Collagen in Keloids

Keloids are dermal nodules, located in the majority of patients on the anterior chest, shoulders, and upper back. The development of keloids is usually preceded by a skin injury at the site of the lesion.

Histopathologically, keloids are characterized by an extensive proliferation of connective tissue of the skin. Fibroblasts and dense deposits of collagenous fibers are seen. Biochemical studies have confirmed the morphological investigations and indicate that collagen is the predominant component. The amount of collagen expressed per mass of dry weight of tissue is in keloids similar to that in normal skin, but due to a significant increase in thickness of the dermis, there is a large increase in total collagen per square unit of skin.

The type pattern of collagen in keloid tissue remains controversial. An increased amount of type III was reported (Bailey et al. 1975), while other studies showed normal type I/type III collagens ratio (Clore et al. 1979). Abergel et al. (1985) presented a detailed study on keloid tissue: they reported that keloids contained more type I collagen and slightly less type III collagen than normal skin. The amount of type V collagen was unaltered. The above-mentioned discrepancies were probably caused by differences in the examined specimens. Type III collagen predominates in hypertrophic scars, and some of these scars could be mistaken for keloids (Abergel and Uitto 1987).

The biosynthesis of collagen in fibroblast lines isolated from the keloid tissue increases (Abergel et al. 1985). Elevated synthesis is associated with a raised type I protocollagen mRNA level. The mechanism of this phenomenon remains unknown. Abergel and Uitto (1987) excluded gene amplification as an explanation for altered collagen gene expression in keloid cell cultures. The other possible mechanisms include enhanced rate of transcription or reduced degradation of procollagen type I mRNA.

The activity of collagenase in keloids hase been shown to be close to normal values (Abergel et al. 1985). It is of interest that the amount of intracellular degradation of newly synthesized collagen is relatively low in cultured keloid fibroblasts (Abergel and Uitto 1987).

The available data of collagen metabolism in keloids lead to the conclusion that overaccumulation of collagen is associated with enhanced production. The mechanism of the stimulated synthesis remains unclear. An increased level of specific mRNA and low intracellular collagen degradation rate have been suggested. Other possibilities must be taken into consideration. Altered maturation of collagen may lead to reduced degradation. The selection of fibroblast populations active in collagen production is also potentially one of the mechanisms. A similar phenomenon has been suggested to be involved in the etiology of systemic sclerosis (see Chap. 10). It is noteworthy that keloids make useful models for studies on the pharmacological control of collagen metabolism (see Chap. 24).

## 16.4 Wound Healing

Wound healing is not limited to the skin, although dermal wounds constitute the most investigated model of the healing process. Tissue injury initiates a complex series of events leading to repair and restoration of normal structure and function or scar formation. These processes represent a very precisely regulated cascade of steps, including cellular activation, matrix deposition, and tissue remodeling. The subject has been comprehensively reviewed by Clark and Henson (1988). Here, a short overview is presented followed by comments on the contribution of collagen.

Wound repair is a dynamic process, and classification of the phenomena into a sequence of events is an oversimplification. It is clear that the phases of healing overlap each other. On the other hand, in the description of wound repair, separation into phases is very useful and is applied for the purpose of reviewing. According to Clark (1988), the whole comprises three phases: inflammation, granulation, and matrix formation with tissue remodeling. The inflammatory one is divided into early and late subphases.

Disruption of blood vessels causes extravasation of blood. This initiates a number of events, including platelet aggregation and blood clotting. These processes are associated with the generation of several humoral factors. Platelets contain factors responsible for the stimulation of collagen synthesis (Bańkowski et al. 1980). Generation of kinins, histamine, leukotrienes, and prostaglandins and initiation of the classic complement cascade lead to the production of substances that attract neutrophils and monocytes and mediate vasodilatation (Dvorak et al. 1988). Neutrophils infiltrating the inflammatory area adhere to the blood vessel endothelium and migrate outside the vessels. Their functions include destruction of bacteria in the wound (Tonnesen et al. 1988). Neutrophil infiltration does not last long in the area of inflammation. The late inflammatory phase is characterized by monocyte infiltration. The vascular response that facilitates the migration of cells is uniform despite the nature of the harmful agent. Immediately after injury, the vessels constrict. In a short time this vasoconstriction is followed by an active vasodilatation. Vasodilatation facilitates fluid leakage from venules and diapedesis of cells. The lymphatic drainage is usually impaired. All these mechanisms produce classic signs of inflammation – redness, swelling, and heat. Pain arises from the release of chemical mediators and from volume changes, i.e., swelling (Alvarez 1987).

Accumulated monocytes become activated and convert to tissue macrophages. They are able to phagocytise not only bacteria but also tissue debris. Phagocytosis is associated with the release of substance responsible for the recruitment of additional cells and modulation of the cellular activity in order to form the granulation tissue. Elimination of the destroyed or dead tissue fragments by phagocytes facilitates formation of the granulation tissue. This process includes ingrowth of fibroblasts with concomitant deposition of loose connective tissue and growth of new blood vessels. Fibroplasia and angiogenesis are the predominant features of formation of the tissue. Like all events of repair, these phenomena are modulated by several humoral factors (Jackson 1979; Wahl 1986).

The early event of wound healing, which is seen in some cases before granulation tissue formation, is re-epithelialization. Although this is a very early event, it is considered as the second phase of wound healing because it represents new tissue formation. Such humoral factors as platelet-derived growth factor, epidermal growth factor, transforming growth factor $\alpha$, transforming growth factor $\beta$, and fibroblast growth factor-like peptides are involved in this process (Clark 1988; Sporn and Roberts 1986). The epithelial cell proliferation begins from the free edge of the tissue across the defect. The growing cells undergo phenotypic alterations. They achieve the ability to migrate. The signals that terminate proliferation and produce reversion of the phenotype at the completion of epithelialization remain unknown (Grotendorst and Martin 1986).

The early stage of granulation tissue formation includes production of fibronectin and hyaluronic acid. The fibronectin network provides a provisional substratum for the migration and growth of cells and serves as a framework for fibrillogenesis. Hyaluronic acid is involved in the growth of parenchyma cells. Remodeling of the matrix is associated with the disappearance of fibronectin and hyaluronic acid and the formation of collagen fibers embedded in the ground substance (McDonald 1988; Overall et al. 1989).

The activation and migration of fibroblasts are the main features of fibroplasia. The cells resemble mesenchymal cells and produce a significant amount of loose connective tissue rich in fibronectin and hyaluronic acid. Angiogenesis arises simultaneously with fibroplasia. The regulatory mechanisms of the formation of new blood vessels are not clear. A role for chemotactic and mitogenic stimuli has been postulated (Schwartz et al. 1982). The endothelial cell phenotype alters during angiogenesis.

The last stage of wound repair consists of matrix formation and remodeling. Matrix formation is initiated concurrently with the development of granulation tissue, but the process of remodeling continues after the dissolution of granulation tissue.

Involvement of collagen in dermal wound repair has been recently reviewed (Jackson 1982; McPherson and Piez 1988). In the inflammatory phase, collagen contributes to platelet aggregation (see Chap. 20). Fragments of the collagen molecule are chemotactic for fibroblasts (Postlethwaite et al. 1978). The roles played by histamine and cytokines in collagen metabolism are discussed in other chapters (Chaps. 7 and 8). Plasma proteins (socalled acute phase reactant proteins) have been shown to increase collagen synthesis in dermal slices, but the mechanism is unknown (Borel et al. 1976).

Activation of fibroblasts in the second phase of repair is associated with elevated collagen synthesis (Ross and Benditt 1961).

Humoral factors increase collagen production either by stimulation of net synthesis or by induction of fibroblast proliferation. It is, however, difficult to elucidate the regulatory phenomena in vivo as most of the studies have been carried out on isolated cell systems in vitro (Jalkanen and Penttinen 1982; Laato et al. 1987; McPherson and Piez 1988). The majority of the collagenous proteins synthesized during granulation tissue deposition are types I and III collagens. Significantly more type III collagen is produced as compared with normal skin. As the matrix matures the type I/type III collagen ratio returns to normal values (Gay et al. 1978). The granulation tissue contains collagen with a low level of cross-linking, that is easily degraded by cathepsins and collagenases. This susceptibility to breakdown facilitates dissolution of the tissue (Bailey et al. 1973; Bazin et al. 1976).

Collagen plays an important role in the contraction of the wound (Ehrlich 1984b; Chvapil and Koopman 1984). The myofibroblasts are contractile cells anchored to the collagenous network, and in this way they reduce the volume of the wound. Wound contraction is a different phenomenon than scar contraction. The latter is associated with matrix remodeling. Collagen is also important in re-epithelialization as it provides support for epidermal cell migration and proliferation (Clark et al. 1982; Woolley et al. 1985).

Granulation carrageenan tissue induced by implantation of polyvinyl sponge or carrageenan injection is a model widely used in studies on collagen metabolism. The mechanical properties of the healing wound are investigated as another model for studies on the pharmacological modulation of tissue repair (Viljanto 1964; Doillon et al. 1988). The development of carrageenan-induced granuloma is associated with changes in systemic collagen metabolism, including elevation of serum prolyl hydroxylase (Matsumoto et al. 1988).

# 17 The Eye and The Ear

## 17.1 The Eye

### 17.1.1 General Structural Features

The human eye is a complex and intrinsic mechanism. It is an important organ of sense responsible for input of a large amount of information from the environment to the brain. The eye is a roughly spherical structure; the eyeball is situated in the bony orbit supported by the accessory structures, including the conjunctiva, eyelids, and lacrimal apparatus.

The eye consists of three concentric layers: an external one composed of the sclera and cornea; a middle one, the vascular layer or the uvea, consisting of the choroid, ciliary body, and iris; and an inner one, the retina. Several compartments are located within the eyeball, including the anterior and posterior chambers filled by the aqueous humor and the space behind the lens filled by the vitreous body. The retina is the photosensitive layer. Light passing through the cornea is refracted by a series of different media, including the lens that focuses and inverts the image, before it reaches the retina. Ocular structures that are not located on the light path are lightproof.

Almost all structures of the eye are made up of collagen. The function of collagen is not limited to increasing the mechanical properties; those parts that serve as transparent and refracting media contain a high proportion of specialized collagenous structures.

Embryologically, the eye develops from the neural ectoderm, ectoderm, and mesenchyma. Cells of various origins contribute to collagen production within the embryonal and adult eye.

### 17.1.2 The Cornea and the Sclera

The external layer of the eye is called the tunica fibrosa. It consists of the sclera and the anterior transparent cornea. The cornea is a highly specialized form of connective tissue. It is colorless and transparent and is the front "window" of the eye. The cornea provides the structural strength to the front part of the eye. It is also responsible for most of the dioptric power. The corneal curvature provides 42 diopters and the lens only 20 diopters.

The cornea is composed of the following layers, from outermost to inward: a stratified, squamous, nonkeratinizing epithelium; an epithelial basement membrane; Bowman's membrane; the stroma; Descemet's membrane; and an endothelium. The epithelium consists of four to five cellular layers and is approximately 50–70 μm thick. The external surface is smooth and stabilizes the precorneal tear film. The presence of the tear film prevents the cornea from desiccation and provides a very smooth refracting surface. The epithelial cells rest on a basement membrane. The presence of this basement membrane is not accepted by all investigators. Some studies claim that the subsequent layer, Bowman's membrane, is a mixture of the basement membrane and stromal components (Friend 1988). Bowman's membrane is an acellular, densely packed layer. It is, at least partially, a modified anterior layer of the stroma and consists of short collagen fibrils that are randomly arranged. It occurs in some species (e.g., humans, primates, and birds) but not in others (e.g., rabbits). The stroma represents 90% of the corneal thickness; it is approximately 500 μm thick. It is responsible for the shape, resistance, and transparency of the cornea. Under the corneal stroma is Descemet's membrane, an acellular layer 6–10 μm thick that is the basement membrane of the corneal endothelium.

Collagen is abundant in the cornea and makes up about 70% of its dry weight (Maurice 1969; Kay 1988). The functions of the cornea, including its transparency, are highly related to the presence of collagen fibrils. Epithelial cells of the cornea have been shown to be able to synthesize collagen in vitro. Most studies have been carried out on embryonic epithelium, and types I and II collagens are the major ones produced in culture (Linsenmayer et al. 1977; Trelstad et al. 1974). Type V collagen does not appear to be present in the cornea, and type III collagen has been observed only transiently (von der Mark et al. 1977). Type IV collagen was identified in rabbit corneal epithelial cells in culture, located in the subepithelial structures (Sundar-Raj et al. 1980).

The supramolecular arrangement of collagen fibrils within the stroma received the attention of several investigators. Corneal collagen is organized into striated fibrils uniform in diameter (about 25 nm in diameter) which form orthogonal lamellae or layers. Every layer of fibrils runs parallel to the surface of the cornea and at different angles to each other in successive lamellae. The lamellae are embedded in a ground substance rich in proteoglycans. Corneal proteoglycans regulate collagen fibrillogenesis and affect the strict morphological arrangement associated with transparency (Birk and Lande 1981). According to Smelser et al. (1965) turnover of the stromal collagen is very slow.

There are several conflicting data about type patterns of collagens that make up the corneal stroma, arising from the application of different techniques, as well as the determination of collagen in situ or measurement of collagen production in cell cultures. Another source of discrepancy is the investigation of other animal species. Studies have been carried out on embryonal and adult tissues, and these differences may add more variability to the results.

Type I collagen is the major constituent of the corneal stroma and accounts for about 90% of the total collagen content. The corneal type I collagen has been shown to differ from type I collagen isolated from other tissues (Church 1980; Kao and Foreman 1980; Marchini et al. 1986). The major difference is the high level of hydroxylysine glycosylation (Friend 1988). This high glycoside content may be related to the control of fibrillogenesis.

The presence of collagens other than type I in the corneal stroma remains controversial. Some biochemical analyses have shown the presence of type III collagen in bovine corneal stroma and claimed that type III collagen constituted about 10% of the total collagen content (Schmut 1977, 1978; Praus et al. 1979). Others reported type III collagen in fetal bovine corneas as a minor component and were not able to detect it in adult tissue (Tseng et al. 1982; Lee and Davison 1984). Freeman (1978, 1982) did not confirm the presence of type III collagen in any significant quality in rabbit, bovine, or human tissue. On the contrary, Newsome et al. (1982) and Nakayasu et al. (1986) found type III collagen in adult human corneas. Some investigators reported type III collagen only in the early developmental stages and were not able to detect it in adult tissue (Ben-Zvi et al. 1986; Von der Mark et al. 1977; Conrad et al. 1980; Cintron et al. 1988).

Type V collagen constitutes about 10% of the normal stromal collagen of the cornea (Cintron et al. 1981; Lee and Davison 1984). This proportion is higher than in other tissues. Type V collagen is suggested to be involved in the morphogenesis of the corneal stroma.

There are only single reports indicating the presence of type II collagen in the stroma. This type probably occurs only in some developmental stages (Harnisch et al. 1978). Type VI collagen has been located in the corneal stroma; it may be associated with fine filaments (Cintron and Hong 1988).

An interesting concept of the supramolecular structure of collagen fibrils in the developed avian cornea has been presented by Linsenmayer et al. (1985). They have shown that the primary and secondary corneal stromas are heterotypic structures composed of at least two genetically different types of collagen. The primary stroma is made up of morphologically indistinguishable fibrils composed of types I and II collagens. In the secondary stroma, the fibrils are made up of types I and V collagens. The collagens are co-assembled into uniform fibrils. The antigenic domains of type V collagen are masked under normal conditions, thus explaining the false identification of collagen types distribution in the corneal stroma based on immunohistochemical investigations. After pretreatment with acids the stroma becomes reactive to the antibody against type V collagen. It has been hypothesized that the acid pretreatment facilitates antibody access to type V collagen by producing alterations such as swelling. Studies on the temperature-altered molecular packing of collagen molecules showed that the inaccessibility of the epitopes to which the type V collagen-specific antibodies bind was due to their packing within the fibrils (Fitch et al. 1984). These results suggest that several studies on the collagen composition of the corneal stroma based on immunohistochemical staining are difficult to interpret. Embryonic and adult stromas are built up of homopolymeric fibrils that are heterogenous in collagen type composition. It remains unclear whether or not the phenomenon is widespread in other tissues.

Descemet's membrane is the basement membrane of the corneal endothelium. Morphologically, it consists of an anterior layer of banded organization and the posterior layer, which is amorphous and granular. Type IV collagen, the specific form in basement membranes, is found here. Type IV collagen is also predominantly synthesized by the corneal endothelial cells in culture. Other types of collagen are only minor products found in culture.

The sclera is the tough, fibrous coat of the eye. It is continuous with the cornea and accounts for approximately four-fifths of the outer layer. It is built up of dense connective tissue containing collagen bundles intersecting in various directions. Its external surface is connected with the episclera and Tenon's capsule, also made up of connective tissue rich in collagen.

The collagen fibrils of the sclera have a pattern similar to that of tendon. Their diameter ranges from 30 to 300 µm. The sclera is rich in collagen and accounts for 80% of the tissue dry weight (Freeman 1982). Type I collagen predominates, in human sclera accounting for 94% of the total collagen content. The remaining 6% consists of type III collagen. No other collagen types have been found in the human sclera (Freeman 1982). In avian eyes, some cartilaginous components of the sclera (the scleral support ring) have been demonstrated; they are composed of type II collagen (Trelstad and Kang 1974).

The zone of transition between the cornea and sclera is called the limbus. Beneath the epithelium of the limbus is a trabecular meshwork that is continuous with Descemet's membrane. The aqueous humor drained from the anterior chamber by the trabecular meshwork is removed by Schlemm's canal. Several reports noted the occurrence of special, cross-striated collagen fiber bundles, called "curly collagen", within the trabecular meshwork (Garron and Feeney 1959; Leeson and Speakman 1961; Spelsberg and Chapman 1962; Rohen 1962; Holmberg 1965; Vegge and Ringvold 1971). The quantity of these bundles increases with age, forming part of the so-called plaque material underneath the inner wall of Schlemm's canal (Lütjen-Drecoll et al. 1986). Immunohistological staining for type IV collagen showed a distinct line underneath the endothelium of the trabecular lamellae near the basement membrane. Increased staining of type VI collagen was seen within the trabecular lamellae where the elasticlike fibers were located (Lütjen-Drecoll et al. 1989). The role of type VI collagen in the formation of the aggregates called "curly collagen" or long-spacing collagen has been proposed.

The evidence for the presence of a latent collagenase in human aqueous humor has been provided recently. Its properties are similar to those of tissue collagenases. It is interesting that the molecular weight of the collagenase is about 40000 Da (Vadilo-Ortega et al. 1989).

### 17.1.3 The Uvea

The tunica vasculosa uvea consists of the choroid, ciliary body, and iris. Its main functions are the elimination of light and provision of vessels. The choroid is highly pigmented and is a vascularized loose connective tissue lining the back of the eye. The ciliary body extends from the edge of the retina to the root of the iris. The iris is shaped like a disc with a central aperture, the pupil. The iris consists of a stroma of loose connective tissue. Eye color is related to the number of pigment cells of the iridal stroma.

Collagen accounts for about 41% – 49% of the dry weight of the uvea. This collagen is a mixture of types I and III. Most of the collagen is associated with the vasculature (Schmut 1978). It is of interest that the zonular fibers, i.e., suspensory apparatus of the lens, a part of the ciliary body, does not contain collagen (Freeman 1982).

### 17.1.4 The Lens

The lens is a transparent, biconvex body suspended by zonular fibers originating from the ciliary body. It has considerable flexibility that allows it to undergo the process of accommodation. The lens comprises the capsule, anterior epithelium, and lens substance. The capsule completely envelops the cells and is built up of the basement membrane. Cells lie only at the inner anterior surface in the form of a low cuboidal epithelium. Towards the equator of the lens, the cells of the anterior epithelium become columnar and are transformed into the lens fibers, which are laid down concentrically.

The lens capsule contains collagen. Type IV collagen is a predominant form of the protein of the capsule as in all basement membranes (Olsen et al. 1973a; Heathcote et al. 1980; Fitch et al. 1983; Schwartz and Veis 1978, 1980). In culture, the epithelial cells change their phenotypic expression and synthesize type I collagen as well as small amounts of type III collagen.

### 17.1.5 The Vitreous Body

The vitreous body is a transparent, jelly-like connective tissue. It is composed predominantly of an extracellular matrix and contains a high proportion of water. The vitreous body fills the posterior compartment of the eye and makes up about 80% of the total volume of the eye. It reflects light and protects the retina from shock and vibration.

The vitreous consists of a continuous network of fine collagenous fibrils embedded in a highly hydrated, amorphous matrix. The diameter of the fibrils ranges from 10 to 20 nm (Gross et al. 1955; Olsen 1965; Reeser and Aaberg 1979; Swann and Sotman 1980). The collagen is very similar to type II collagen of cartilage. Vitreous type II collagen, however, differs in alanine content, intermolecular cross-links, degree of hydroxylation, and mode of fibril growth (Snowden and Swann 1980; Snowden et al. 1982; Ayad and Weiss 1984; Liang and Chakrabarti 1981; Schmut et al. 1979, 1984). It is not clear whether the variations are caused by posttranslational modifications or whether they originate from different primary structures. This may indicate that vitreous and cartilage collagens are different types or the alterations are artifacts caused by the preparation techniques. Traces of other types of collagen similar to socalled minor cartilage collagens have been reported in the vitreous body (Kay 1986).

### 17.1.6 The Retina

The retina is a thin, complex membrane consisting of photoreceptor cells, rods and cones, which sense light intensity and color. The retina is composed of layers. The photoreceptor cells are the most external layer, and light must transverse all layers before arriving at the rod and cone outer segments.

Embryonically, the retina is formed from an invagination of the diencephalon called the optic vesicle. The vesicle proliferates outwards and subsequently forms the optic cup. The cup contains two layers of neuronal ectoderma. The outer layer develops into the pigment epithelium while the inner layer develops into the neural retina.

The retina rests on a single layer of pigmented epithelial cells. The epithelial cells lie on a thin basement membrane, called Bruch's membrane. Bruch's membrane consists of two basement membranes and the space between them, which is filled with loosely arranged collagen fibers. Bruch's membrane provides mechanical support for the photoreceptor cells and is a selective barrier for transporting nutrients from the choriocapillaries to the retina. The pigment epithelial cells are responsible, at least partially, for the production of Bruch's membrane. Embryonic chick pigment epithelial cells in culture synthesize several types of collagen (Newsome and Kenyon 1973). Interstitial collagen production was seen under the electron microscope only when basement membranes were present (Kennedy et al. 1986). The nature of this regulatory process of phenotypic expression of collagen genes remains unknown. The fibril diameter increased in correlation to the embryonic age (Newsome and Kenyon 1973). The ability of pigment epithelial cells to synthesize type IV collagen has been shown in cultures of embryonic cells obtained from other animals (Li et al. 1984; Shen et al. 1985).

### 17.1.7 Collagen in Ocular Pathology

Ocular manifestations are almost constant features of several inherited and acquired disorders of the connective tissue. Biochemical aspects of the Ehlers-Danlos syndrome, Marfan syndrome, homocystinuria, alkaptonuria, and osteogenesis imperfecta are described in Chap. 9. Nonspecific changes in the eyes are noted in rheumatoid arthritis, systemic lupus erythematosus, and many other disorders involving connective tissue. The present chapter focuses on abnormalities that are associated with altered collagen content and metabolism and are limited to the eyes.

A number of studies have been carried out on abnormalities of the cornea. A disease or trauma of the cornea frequently produces ulceration that heals with scarring, i.e., various forms of corneal fibrosis. Scarring is among the most frequent events leading to the loss of vision. Several other causes can result in ulceration: viral or bacterial infection, chemical trauma, or nutritional deficiency.

Ulceration can also exist in association with systemic autoimmune diseases. The basic pathogenic mechanism of corneal ulceration is a rapid destruction of the extracellular matrix due to a significant increase in hydrolases operative at physiologic pH. Collagenase has been found to be one of the most important enzymes responsible for corneal ulceration. The presence of enhanced amounts of active collagenase in tissue samples or corneal epithelial cells in cultures obtained from patients with corneal ulceration has been reported by many investigators (Gnädinger et al. 1969; Itoi 1969; Itoi et al. 1969; Brown et al. 1969; Berman 1975; Berman et al. 1971, 1973b). The enzyme was found to degrade type I collagen and type I trimer (Davison and Berman 1973). Enzymes capable of degrading other types of collagen are probably also secreted. The corneal collagenases require zinc for their activity (Berman and Manabe 1973).

The molecular characteristics of corneal collagenases remain unclear. A high molecular weight fraction (about 725000 Da) and a light fraction (about 40000 Da) have been reported. It is possible that the heavy fraction is a complex of the enzyme with $\alpha_2$-macroglobulin.

A few cell types have been postulated as a source of collagenases. It was suggested that the enzyme was released from the marginal epithelial cells of the initial epithelial defect (Brown and Weller 1970). Polymorphonuclear leukocytes have been shown to play an important role in the degradation of the corneal matrix. The cells predominate in some infections of the cornea and are observed after burns. The leukocytes migrate from blood vessels or may pass into tears and then move to the damaged area. Several factors are considered as attractants for leukocytes. The substances released by the injured epithelium are chemotactic (Weimar 1959), especially plasminogen activator from the epithelium. Plasminogen activators lead to cleave of the third component of complement. This component is a strong chemotactic agent (Ward 1967). Direct activation of latent collagenase by plasminogen activator is an additional important mechanism for matrix degradation. The activator was found to work on the 40000 Da molecular weight latent collagenase from the ulcerating cornea (Berman 1980). The cellular origin of this collagenase remains unknown.

Fibroblasts are also thought to be involved in collagenase production. They may secrete a soluble collagenase that diffuses into areas close to the cornea. The collagenase is produced in response to some phagocytic stimuli. A mutual interaction between the cells contributing to collagenase production has been hypothesized as the mechanism of the enhanced synthesis.

Experimental and clinical studies on inhibitors of collagenase (L-cysteine, *N*-acetyl-L-cysteine, calcium ethylenediaminetetraacetic acid) showed accelerated healing and no further ulceration in a significantly higher percentage of patients than in controls receiving only vehicle (Slansky et al. 1970, 1971; Berman and Dohlman 1975). An increased level of natural antiprotease, $\alpha_1$-antitrypsin, was found in tears which bathe ulcerating human corneas and was found to correlate with the severity of ulceration. An elevated level of $\alpha_2$-macroglobulin was also shown in tears. The role of the antiproteases in the inhibition of collagenase has been controversial, as $\alpha_1$-antitrypsin does not inhibit corneal collagenase isolated from keratoplasty tissues. The second antiprotease investigated, $\alpha_2$-macroglobulin, inhibits human corneal collagenase. The raised level of antiproteases may reflect greater permeability of the inflamed vessels and leak of serum proteins (Berman et al. 1973a). The presence of complexes of the corneal collagenase with $\alpha_2$-macroglobulin was shown in the tears of patients with corneal ulcers (Berman et al. 1975; Berman 1976). Current attempts to synthesize specific inhibitors of collagenase which can be used in the treatment of corneal ulcerations are reviewed in Chap. 24.

The healing process of corneal wounds or ulcers is a complex phenomenon. It includes replacement of the damaged epithelium, stromal healing, as well as changes in Descemet's membrane that may lead to the formation of a socalled retrocorneal fibrous membrane. The epithelium has been shown to play an important role. Its presence is necessary to transform cells into fibroblasts at the edge of the wound. Removal of the epithelium significantly decreases collagen synthesis in the healing stroma. The epithelium cells also contribute directly to collagen synthesis. An elevated incorporation of proline has been found in healing corneal wounds (Gnädinger et al. 1971).

Healing of the corneal stroma is associated with the formation of collagen fibrils that differ in their organizational pattern from the normal one. This variation is a cause of a marked loss of elasticity and transparency of the stromal scar (Gasset and Dohlman 1968; Swarz and Graf-Keyserlingk 1969). The altered structural architecture of the collagenous fibrils in the scar results from a change in the type of collagen synthesized. The most dramatic one was noted in the minor collagen components of the cornea. A switch from type V collagen to type I trimer was reported by Freeman (1980). Other changes included a different glycosylation of the hydroxylysine residues and cross-link formation (Cintron 1974; Cintron et al. 1978; Cannon and Cintron 1975). Matrix variations, especially alterations in glycosaminoglycan composition, have been postulated to affect collagen fibrillogenesis and produce thick, opaque, scar tissue.

The corneal endothelium and Descemet's membrane contribute to scar formation. They respond to several harmful stimuli by the production of a retrocorneal fibrous membrane known also as the posterior collagenous layer. This abnormal tissue appears between Descemet's membrane and the corneal endothelium. It is composed of matrix components rich in collagen and fibroblastlike cells. Its cells synthesize predominantly type I collagen in culture (Kay et al. 1982). The hypothetical explanation of its formation mechanism has been described by Kay et al. (1985). They found that the fibrosis is preceded by an accumulation of polymorphonuclear leukocytes. Polymorphonuclear leukocytes produce humoral factors which are able to transform endothelial cells into fibroblastlike cells. This transformation is associated with a change in phenotype expression and the synthesis of type I collagen. The ability to produce type IV collagen drops dramatically (Kay 1989). In the late stage of fibrosis, types I and III collagens are synthesized.

Noninflammatory changes in the cornea, including its thinning and dilatation of the central part, is called keratoconus. The disease may be unilateral or bilateral and is probably caused by a group of nonspecific factors. The decrease in total collagen content and impaired collagen maturation in the cornea have been postulated as the mechanism (Cannon and Foster 1978). It is possible that some forms of keratoconus are the symptoms of systemic hereditable disorders of collagen molecules (see Chap. 9).

Involvement of collagen in the pathology of parts of the eye other than the cornea is poorly elucidated. Only single reports have been published, although it is possible that several ocular disorders are associated with abnormalities in collagen within various parts of the eye.

Glaucoma is a condition which develops when intraocular pressure increases. Impaired or blocked drainage in Schlemm's canal is one of the major causes. Lütjen-Drecoll et al. (1989) reported an increase in "curry collagen," that was made up at least partially of type VI collagen within the trabecular meshwork of glaucomatous eyes.

Anterior capsular cataract is associated with the formation of a white-grey opacity appearing under the anterior lens capsule. Accumulation of fibrillar collagenous structures that are normally absent in the lens has been shown. The collagen is produced by myofibroblastlike cells which develop from the lenticular epithelium. The mechanism of this process resembles that of scar formation (Novotny and Pau 1984; Novotny et al. 1989).

Some forms of degeneration of the retina are associated with an increased accumulation of collagen (Witschel 1981). There are some suggestions that a type II collagen similar to that of the vitreous body is involved (Laqua 1981). Collagen accumulation may be a primary manifestation of the disease or may result from other unknown conditions as a secondary phenomenon associated with retinal degeneration. These aspects of collagen pathology remain to be elucidated.

## 17.2 The Ear

The ear is a special organ of sense, responsible for hearing and detection of changes in acceleration as well as orientation of the head relative to gravity. It is a complex structure, consisting of an external, middle, and internal part. Connective tissue (bone, cartilage, fibrous tissue) make up almost all parts of the ear (Yoo and Tomoda 1988). The specific role of connective tissue in its functions has not been investigated. The only exception is otosclerosis, a disorder which leads to deafness due to changes in the otic capsule, fixation of the stapes, and involvement of the cochlea and other parts of the labyrinth. The disorder occurs bilaterally and is common: It is reported in about 8% of white Americans (Lindsay 1980). The earliest changes are found in the bone of the otic capsule. Resorption of the bone leads to "otospongiosis" with increased vascularity. Histiocytes are present in the lesions, and the stapediovestibular joints are involved. It has been hypothesized to be an autoimmune disease. Autoimmunity to native type II collagen of cartilage nests associated with globular ossei in the endochondrial layer of the otic capsule has been suggested as a pathogenetic mechanism. Some changes can be explained by the activity of the inflammatory cells recruited by the local immune reaction.

There are clinical and experimental data supporting the autoimmune hypothesis of otosclerosis. Patients with otosclerosis have significantly higher serum levels of antibodies against type II collagen. There was no difference in the level of antibodies against types I and IV collagens as compared with controls (Yoo et al. 1982). Experiments with rats showed that immunization with type II bovine collagen produced hearing loss (Yoo et al. 1983). Spongiotic changes were observed in the anterior border of the external meatus bone near the tympanic annulus. The changes were characterized by the presence of osteocytes, osteoblasts, and wider vascular spaces associated with fibrosis as well as bone resorption, and new bone formation. The histological picture resembled that found in humans in the early stages of otosclerosis. There was no change in the oval window. This was probably due to the short period of immunization. A strong deposit of immunoglobulins was shown in the enchondral layer of the otic capsule. Immunoglobulins were also found around the lacunae and maxillar joints of the collagen-immunized rats. The role of cellular immunity and inflammatory cells in the pathogenesis has been postulated (Yoo 1984a). It is possible that the genetic predisposition to the autoimmune disorder is associated with some human leukocyte antigens and results from their linkage to immune response genes.

The hypothesis of the collagen autoimmune etiology of otosclerosis has not been accepted by all researchers. Harris et al. (1986) were not able to repeat the induction of middle and inner ear morphological abnormalities in rats immunized with type II collagen. There was no otosclerosislike lesion in rats immunized with collagen and with collagen-induced arthritis in experiments reported by Bretlau et al. (1987). Recently, the same group did not confirm elevation of type II collagen-specific antibody levels in the sera of patients with otosclerosis. There was also no sign of otosclerotic change or abnormality of the otic capsule or organ of Corti in MRL/1 mice with spontaneous type II collagen autoreactivity (Sølvsten-Sørensen et al. 1988).

It is difficult to explain these discrepancies. It is possible that an autoimmune etiology is limited to some subsets of the disease. A similar proposal has been put forward to explain the etiopathogenesis of some forms of Menière's disease, a disorder of the inner ear associated with increased pressure of the endolymph (Yoo 1984b). Further studies will evaluate these hypotheses.

Enhanced activity of collagenolytic proteases has been postulated to cause severe mucosal damage in otitis media with effusion. Inflammation of the middle ear is associated with infiltration of macrophages and neutrophils. These cells are thought to liberate the proteolytic enzymes. Higher activities of collagenolytic cathepsin and collagenase were reported in patients with acute otitis media than in those with a chronic form of the disease (Hamaguchi et al. 1987). Estimation of the level of collagenolytic enzymes in effusions obtained from the ear has no diagnostic value.

# 18 Urinary and Reproductive Systems

## 18.1 Introduction

The urinary and reproductive systems have different functions but a common embryological development, and they are known as the genitourinary system. In all vertebrates, these two systems originate from the mesoderm and in humans as well as in higher mammals undergo complex developmental changes. The common development produces several interactions of the systems, especially noted in the pathology. In the male, the systems are not fully separated (semen passes through the urethra), while in the female, they are morphologically separated. The urinary system is relatively similar in both sexes, while the reproductive system is obviously different.

Connective tissue contributes to the structure of the systems, but the role of collagen is not limited to mechanical strength only. In general, the present knowledge on collagen in the urinary and reproductive systems is scant and is insufficient for a comprehensive analysis of the role of collagen in health and disease.

## 18.2 The Urinary System

The primary function of the urinary system is to maintain homeostasis. This regulation includes control of the volume and composition of blood plasma by means of excretion or retention of water and electrolytes. The kidneys maintain the acid-base balance and excrete a number of metabolic waste products. The formation of urine is the way of eliminating these substances as well as controlling body fluid volume. The kidneys also take part in the control of blood pressure and are organs of endocrine secretion. The urinary system is composed of the paired kidneys and their excretory ducts, ureters, urinary bladder, and urethra.

The kidney contains only a small amount of collagen, but collagen plays an important physiological function; it constitutes the glomerular basement membranes that are part of the filtration barrier involved in urine formation (Heathcote 1982). Connective tissue makes up the tough capsule of the kidney and the renal interstitium. The basic unit of the kidney is the nephron, comprised of the renal corpuscle and tubules. Collagenous structures form the basement membrane of the glomerulus and the tubules as well as contribute to the structure of the Bowman's capsule. Arteries, veins, and ducts that collect urine also contain collagen. It has been estimated that collagen accounts for about 2% of the dry weight of the renal cortex in rats (Chvapil 1967). A similar amount has been suggested for the whole kidney (Deyl et al. 1972a). In humans, the kidney cortex contains about 0.8 and 12.5 µmol of 3-hydroxyproline and 4-hydroxyproline per gram of wet tissue, respectively (Man and Adams 1975).

Heterogeneity of the kidney collagen has been partially elucidated. Type I collagen is located mostly in the large vessels, renal capsule, and pelvis (Remberger et al. 1976; Roll et al. 1980). In the interstitium, type III collagen has been shown to be associated with type I collagen fibers and bundles. Fine filaments reacting to antibody against type VI collagen have been identified in the interstitium of the rat kidney (Karkavelas et al. 1988). Type IV collagen is specific for the basement membrane and is estimated to be a main component of the glomerular and tubular membranes. Type III collagen is probably a major constituent of the renal interstitium. Glomerular mesangial cells have been shown to produce type III collagen (Sterzel et al. 1986).

A continuous basement membrane is present all along the nephron. Immunohistochemical techniques have revealed the occurrence of type IV collagen in all basement membranes (glomerular and tubular). The most intensive labeling was found in Bowman's capsule, the mesangial matrix, and the proximal tubule basement membrane. Less intensive staining was shown in the glomerular and the distant tubular basement membranes (Desjardins and Bendayan 1989).

There is strong evidence of renal morphological changes arising from disorders of the kidneys. In contrast, there are only single reports on biochemical alterations, including collagen abnormalities, in these patients. The main difficulty in these investigations is the preparation of pure components, especially native basement membranes. The role of collagen in renal pathology may be considered in the following aspects: alterations in the basement membranes and changes in collagen content in the interstitium and urinary ducts.

A significant decrease in the 3-hydroxyproline content in the glomerular basement membrane isolated from patients with congenital nephrotic syndrome of the Finnish type was reported by Tryggvason et al. (1978). The activity of enzymes of collagen biosynthesis was not significantly altered in the kidney cortex of these patients.

The Goodpasture syndrome is a severe, often fatal disorder of unknown etiology. It affects mostly young men and involves the kidneys and lungs. It usually develops following viral infection of the upper respiratory tract. The lung involvement is manifested mostly by hemoptysis. In the kidneys, progressive glomerulonephritis leads to the nephrotic syndrome. Most of the patients die due to kidney insufficiency in the first 2 years after appearance of the early symptoms of the disorder.

Goodpasture syndrome is an autoimmune disorder. The antigen is located within the glomerular basement membrane; the antigenic epitope is a noncollagenous (collagenase-resistant) part of the type IV collagen molecule (Thorner et al. 1989), and its chain composition remains unknown. After collagenase digestion the epitope is found in the so-called M2* fragment (Wieslander et al. 1984; Butkowski et al. 1987).

In mice with glomerulonephritis produced by injection of purified human glomerular basement membranes, the main antigenic component was found to be laminin (Matsuo et al. 1986). Type IV collagen is probably insufficient to induce glomerulonephritis. The collagenolytic system is involved in Masugi type glomerulonephritis in rats, and an increase in activity of collagenase has been found in the kidneys (Lubec 1977).

Keller et al. (1988) reported an elevated serum level of the aminoterminal propeptide of type III procollagen in a group of 100 patients with acute and chronic renal dysfunction. A weak correlation was established between the propeptide and creatinine levels. A rise in the propeptide serum level did not result from the impaired renal elimination, and the renal clearance of the propeptide remained unaltered in these patients. There was also no correlation between the propeptide level and proteinuria. The cause is unknown. There was no difference between acute and chronic renal failure, glomerulonephritis, and intestinal nephritis. This suggests that an increase is caused by nonspecific changes in the matrix metabolism due to decreased renal function. It is possible that bone tissue or skin contribute to the enhanced propeptide level in the serum of patients with disturbed renal function.

Deyl et al. (1972a) studied the effect of unilateral nephrectomy on the collagen content in the opposite kidney. Almost twice as much collagen was found in the hypertrophied kidney 4 weeks after nephrectomy. The increase was observed regardless of the age of the animals. The solubility of collagen from the hypertrophied kidneys was higher than that from the controls, with the exception of very old rats.

Unilateral obstruction in the rabbit kidney was reported to lead to fibroblast proliferation. A raised content of interstitial collagen fibers first became apparent on the 7th day of the experiment and increased progressively. Changes in the structure of the basement membranes were observed (Cuppage et al. 1967; Nagle and Bulger 1978; Nagle et al. 1973, 1976). The biochemical studies of Man et al. (1978) showed that obstructive nephropathy was associated with an increase in collagen content in the renal cortex. Only a slight increase in type IV collagen was reported.

In the urether and bladder, collagen contributes to the mechanical strength of the walls. An increase in the collagen content in the narrowed pelviureteric segment has been suggested as a cause of idiopathic hydronephrosis (Notley 1968, 1970, 1972). A similar anomaly has been proposed as the underlying mechanism responsible for primary obstructive megaloureter (Notley 1971; Gosling and Dixon 1978).

The urinary bladder is a tough, muscular, storage sac located in the pelvic cavity. It serves as a reservoir for urine. Its wall is composed of four coats, from the inside, the mucous, submucous, muscular, and serous. The muscular layer contains the detrusor muscle. This muscle consists mainly of smooth muscle cells and collagen fibrils. Detrusor collagen determines to a great extent the passive mechanical properties of the bladder wall (Kondo and Susset 1974). The amount of collagen is not constant but varies under pathological conditions. Neurogenic bladder dysfunction has been shown to increase the detrusor collagen content in humans (Mayo et al. 1973), dogs (Kondo and Susset 1973), and rats (Uvelius and Mattiasson 1986). Denervation of the rat bladder by removal of the pelvic ganglia leads to both hyperplasia and hypertrophy of the smooth muscle cells. Such denervated detrusor has more total collagen than the controls, but the concentration is lower due to overgrowth of the muscle cells. Similar changes were reported in rats with urinary outlet obstruction (Uvelius and Mattiasson 1984).

Peyronie's disease is a pathological fibrosis in the tunica albuginea of the penis. At the onset of the disease, inflammatory cell infiltrations have been shown. They are followed by the deposition of collagen. Peyronie's disease is a common cause of penile curvature (Smith 1966). An increase of type III collagen was found in a Peyronie's disease plaque (Somers et al. 1989). The mechanism of collagen accumulation remains unknown.

Collagen contributes to the generation of urethral pressure. It has been found that urethral pressure correlates with the skin collagen content in postmenopausal women (Versie et al. 1988). Both measurements depend upon the estrogen level, and a generalized decrease in collange content in various organs has been suggested in postmenopausal women. Collagen implants has been used for the treatment of urinary incontinence (Dairiki-Shortliffe et al. 1989).

## 18.3 The Ovary

Histologically, the ovary is made up of thousands of ovum-containing follicles embedded in connective tissue. The follicle consists of an ovum surrounded by a layer of epithelial cells. The ovum matures during a menstrual cycle under the control of hormones. A mature ovum is released and, if fertilized, will implant in the uterine endometrium. For the detailed structure of the ovary, the reader is referred to textbooks on histology. This chapter focuses on the collagenous components of the ovary only.

Our knowledge of the collagen of the ovary is based mostly on morphological studies. The ovary is covered by a simple cuboidal epithelium that rests on a layer of dense connective tissue known as the tunica albuginea. The ovary is divided into a central medulla and an outer cortex. The medulla is built up of loose connective tissue and is highly vascularized. The cortex contains the follicles embedded in the stroma. Collagenous structures of the ovary may be divided into collagen associated with the vessels, a framework of the medulla, a stromal network of fibers forming baskets around the follicles, as well as the basement membranes. Ultrastructural investigations of surgical biopsies of the ovary obtained from women with the Stein-Leventhal syndrome have revealed that the collagen fibrils are made up of filaments. The filaments are complex structures composed of subfilaments 3.0–4.5 nm in diameter. The fibrils are connected by interfibrillar bridges. The fibroblasts that secrete collagen have been shown to have incomplete cell membranes with some "pores" leading to direct contact of the cytoplasm with the extracellular matrix (Petkov 1978a, b).

The follicle consists of an ovum surrounded by several coats that develop during maturation of the follicle. Connective tissue contributes to the formation of the layers. The mature follicle, i.e., the Graafian follicle, is covered by the external theca that interconnects with the tunica albuginea and by the internal theca that rests on the basement membrane. All these structures are made up of connective tissue and are rich in collagen.

Ovulation is a process of release of the ovum from the follicle. It is associated with penetration of the coats and degradation of the connective tissue. The nature of this degradation remains unclear. The participation of proteolytic enzymes, including those with collagenolytic properties, has been suggested by a number of investigators (Lipner 1973; Cajander and Bjersing 1976). The role of proteolytic enzymes in ovulation was first suggested by Schöcket (1916). Espey (1967) found that the tensile strength of the follicle wall diminishes prior to follicular rupture. At 1–2 h before ovulation, the total disappearace of collagen fibrils and the emergence of multivesicular structures in the fibroblasts of the rabbit ovary were reported by Okamura et al. (1980, 1985). These multivesicular structures are believed to carry the degrading enzymes. Collagenolytic activity of cultured samples obtained from the apex of rabbit follicles was noted by Espey and Coons (1976). The same investigators, however, were not able to detect changes in collagenolytic activity in samples obtained at different time points prior to ovulation. These results were confirmed by Morales et al. (1978a, b). The enzyme was found to be collagenase capable of digesting type I collagen (Morales et al. 1983). Injection of gonadotropin to immature rats produced changes in the collagen content in the follicles: It increased threefold from 35 to 56 h after gonadotropin injection but decreased by about one-quarter from 61 to 66 h. Accumulation of collagen was associated with an increase in the activity of procollagen prolyl hydroxylase. An enhanced activity was shown at 10 h after gonadotropin injection and reached a peak at 13 h (Okamura et al. 1985). During the period of follicle growth the activity of collagenase activated by trypsin was found to increase in a parallel fashion with the elevated collagen content. This activity was shown to decrease close to ovulation. In the same period, the activity of collagenase activated by aminophenylmercuric acetate in the insoluble pellet of the follicles was reported to be high. This was explained hypothetically by an increase in inhibitors of collagenase induced by the surge of luteinizing hormone (Morales et al. 1983). Activity of collagen peptidase in the follicles was also noted (Espey and Rondell 1967; Okamura et al. 1985). Other enzymes, like cathepsin, were found to increase just prior to ovulation, and it is possible that they cooperate in the process of follicular rupture (Motohashi et al. 1977). In summary, it can be hypothesized that collagenase (or a collagenolytic enzyme system) is an ovulatory enzyme, although certain additional mechanisms of the enzyme activation and inhibition have been postulated to modify its effective activity during ovulation.

The cellular origin of collagenase in the ovary remains unclear. Granuloma cells release collagenase in vitro (Too et al. 1984). They also produce plasminogen activator when stimulated by follicle-stimulating hormone, luteinizing hormone, or relaxin (Strickland and Beers 1976). The direct and/or indirect involvement of plasminogen activator in collagen degradation has been suggested. Granulosa cells in vitro also synthesize proteoglycanase, an enzyme that degrades amorphous matrix (Too et al. 1984). The role of prostaglandins in follicular rupture has been clarified in various experiments. Gonadotropin-induced ovulation is blocked by indomethacin injection, and infusion of prostaglandin $F_{2\alpha}$ reversed this blockage (Okamura et al. 1985). Prostaglandins are known to stimulate fibroblasts to secrete collagenase, plasminogen activator, and other proteinases. However, they have no effect on collagen synthesis as measured by procollagen prolyl hydroxylase activity (Okamura et al. 1985). The involvement of prostaglandins in ovulation was one of the suggested mechanisms that relate follicular rupture to uterine involution. Both phenomena are regulated by hormones and associated with a rise in collagenase activity (Espey 1980).

Morphological changes in collagen fibrils of the ovary were reported in the formation of the corpus luteum and follicular atresia. The formation of the corpus luteum resembles the repaire process after inflammation (Woessner 1982). Fibroblasts migrating into the follicle probably play the main role in collagen deposition in the corpus luteum (Van Blerkom and Motta 1978).

## 18.4 The Uterus and Pregnancy

The uterus is a pear-shaped, thick-walled, muscular organ, whose main function is to house the embryo and fetus. It consists of a broad body, a narrow, cylindrical cervix, and a rounded part that passes above the openings of the fallopian tubes and is known as the fundus. The walls are formed of the following layers, from the outside: the serosa, smooth muscular coat, myometrium, endometrium, and mucous membrane that lines the uterus and undergoes cyclic changes under the control of ovarian hormones.

Our main interest here is related to the extremely dynamic state of collagen under conditions of pregnancy, parturition, and in the postpartum period. The overgrowth of the uterus during pregnancy is associated with an alteration of the collagenous framework. These changes provide the mechanical resistance indispensable for protecting the organ during labor. Cervical collagen participates in forming a firm closure until the end of pregnancy. At term, the cervix opens within a very short time. After labor, the collagen of the uterus is degraded extremely rapidly, and the enlarged organ returns to its previous size.

The mature, nonpregnant human uterus contains 1000–1500 mg of collagen, i.e., about 22 mg per gram of wet tissue (Cretius 1959). Similar values have been found in the rat uterus (20–30 mg per gram of wet tissue; Kao and

McGavack 1959). Type I collagen predominates and has been estimated to account for four-fifths of the total collagen content. The remaining one-fifth consists of type III collagen and a small amount of type IV collagen (Kao and Leslie 1977; Chung et al. 1976). The turnover of uterine collagen has been reported to be relatively fast. In young rats, the half-life was shown to be 12 days, while in older animals this lengthened to more than 1 month (Kao et al. 1961 a, b). The occurrence of active enzymes of collagen biosynthesis in the uterus has been reported (Salvador and Tsai 1973; Spiro and Spiro 1971).

Most uterine collagen is associated with the myometrium. Collagenous fibers and amorphous matrix form a network supporting the muscular fibers. The smooth muscle cells are surrounded by a basement membrane. Fibroblasts are located in the spaces between the muscles. The cellular origin of myometrial collagen is not clear. Fibroblasts are obviously suspected to be involved, and the smooth muscle cells are also capable of secreting collagen. According to Ross and Klebanoff (1971), the myometrial smooth muscle cells are a major source of collagen production.

Endometrial changes are the subject of extensive morphological studies in relation to the menstrual cycle, pregnancy, or hormonal stimulation. Biochemical investigations are scarce. Endometrial collagen accounts only for 2% of the total uterine collagen. Collagen alterations during the menstrual cycle remains controversial. While the myometrial collagen is thought to be unaltered, investigations of the endometrial collagen have provided contradictory results. In rats, the highest level of collagen was shown at proestrus and the lowest at metestrus (Yochim and Blahna 1976). A similar trend was found in the activity of prolyl hydroxylase (Kao et al. 1964). An enhanced degradation of collagen has been suggested at metestrus. Alterations in the collagen content in the human endometrium are unknown. Agrawal and Fox (1972) studied 300 human endometrial biopsies and found a subendothelial layer of collagen in 6.6%. It was not possible to relate the presence of collagen to any clinical or pathological findings.

Pregnancy is associated with profound changes of the uterus, including the endometrium and myometrium. Endometrial changes begin with transformation of the stromal cells into decidual cells. This process is regulated by hormones of the fertilized egg. The fertilized ovum, following early divisions that occur in the tubes, is translocated to the uterus. There, the trophoblast penetrates the endometrium. The penetration is associated with destruction of several connective tissue structures, including the basement membrane of the epithelial cells, the stroma, and the wall of the vessels. This indicates that the trophoblast in fact penetrates through collagenous tissue. The enzymes involved in implantation remain unknown. It is possible that collagenase is involved in transmission of the trophoblast.

The growth of the uterus in pregnancy is one of the most extensive physiological changes in the amount of connective tissue within a single organ. The volume of the human uterus (including the conceptus) increases about hundredfiftyfold, while the surface area increases thirtyfold (Zimmer 1965). The wet weight of the organ increases ten- to sixteenfold, depending on whether

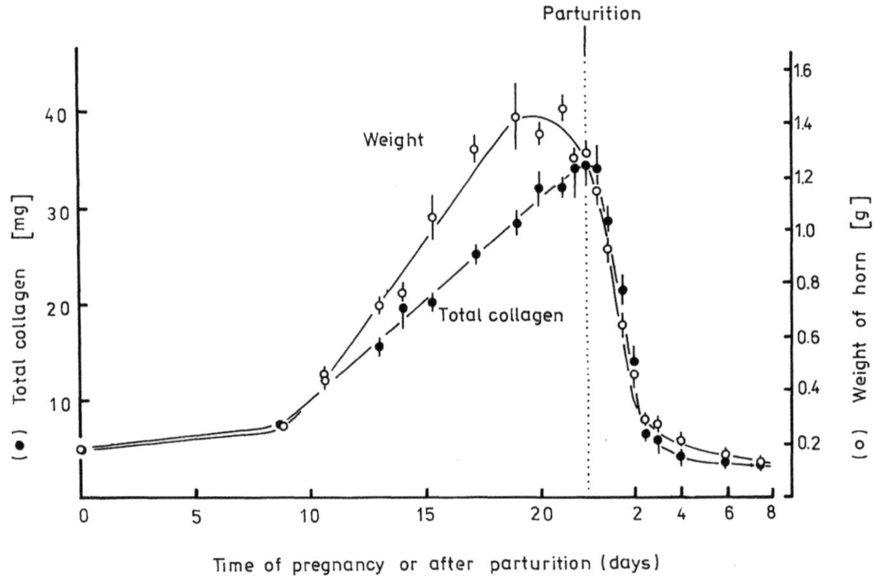


**Fig. 18.1.** Changes in weight and collagen content of the horn of the rat uterus in pregnancy and after parturition. (From Harkness 1964 with permission.)

there were previous pregnancies, and the collagen content increases seven- to eightfold (Woessner 1982). The kinetics of accumulation of collagen in the uterus during pregnancy is shown in Fig. 18.1.

There are no significant changes in collagen solubility. It has been suggested that the resistance of collagen to collagenase digestion is reduced, in particular, degradation of newly synthesized collagen that is deficient in cross-linking.

Postpartum involution of the mammalian uterus is characterized by an unusually fast removal of about 90% of the collagen content. In rats, at 3 days postpartum the collagen content is reduced to one-fifth of that at term, and by 6 days postpartum it has returned to the nongravid level (Harkness and Harkness 1954, 1956; Harkness and Morales 1956). Collagen degrading enzymes present in the involuting uterus were reported for the first time by Woessner (1962), who described collagenolytic cathepsin, and Jeffrey and Gross (1970), who found uterine collagenase. These reports have been followed by a number of studies which provide biochemical characterization of the enzymes (Halme et al. 1980; Woessner 1977, 1979; Woessner and Ryan 1973). Uterine collagenase is under hormonal control as described later in this chapter. Other collagenolytic enzymes are also involved. Lysosomal digestion of collagen resembles that which occurs in inflammation (Etherington 1973).

The mechanism of uterine growth in pregnancy is regulated by mechanical and hormonal factors. The role of a mechanical factor has been shown in the studies of Cullen and Harkness (1960), using ovariectomized mature rats. A pregnancylike state was induced by injection of hormones, and significant growth of the uterus was obtained after injection of paraffin wax to the uterine horn. The wax mimicked the fetus. Hormones alone were insufficient to stimulate growth of the uterus (Takács and Verzár 1968). Mechanical stimulation of the endometrial lining at the progestational state in rats has been shown to produce a growth of cells resembling the decidua. This process is known as formation of the deciduoma. An increase in collagen content of the uterus accompanied formation of the deciduoma. Resorption of the deciduoma was similar to postpartum involution (Jeffrey et al. 1975).

The uterine cervix differs from the remainder of the uterus. The corpus consists mostly of myometrium, while more than 90% of the cervix is made up of fibrous connective tissue (Rorie and Newton 1967). Collagen accounts for 50% of its dry weight (Fitzpatrick and Dobson 1979, 1981). Types I and III collagens constitute almost the total collagen content; type III collagen has been found to occur in a proportion of one-fifth (Kleissl et al. 1978). Traces of type IV collagen are associated with the basement membranes of the cervix. Smooth muscle makes up only 10%–15% of the organ mass, and small amounts of elastin and proteoglycans (about 1% each) have been reported (Fitzpatrick 1977; Fitzpatrick and Dobson 1979).

The cervix remains closed until the end of pregnancy. Its properties vary with the physiological state, from being hard "as cartilage" in the nonpregnant state to softening as the organ ripens to become easily distensible by labor. The softening is very rapid and, in final stage, occurs within a few hours of parturition (Fitzpatrick 1977; Stys et al. 1978). This transformation of the cervix involves significant changes in the collagenous structures, whose nature is poorly understood, despite numerous investigations. The key phenomenon is the extreme alteration in mechanical properties. Harkness and Harkness (1959) investigated the resistance of a ring cut from the rat cervix and measured its internal circumference. It was shown that the resistance, determined as the weight needed to rupture the tissue ring, was very high up to the 12th day of pregnancy. Before labor, the resistance decreased extremely. The inner circumference was small during pregnancy, while close to labor it increased significantly. The breaking strength of the cervix recovered within 24 h of parturition. The inner circumference also returned to its previous size within this time. Altered mechanical properties of the cervix can result from a decrease in collagen content, impaired supramolecular structure of collagenous fibrils, and/or changes in collagen-proteoglycan interactions. Early biochemical investigations focused on the collagen content in the cervix. The cervix increases during pregnancy, although this growth is relatively small compared with the enlargement of the uterus.

Changes in total collagen content are opposite to those of collagen concentration due to a rapid hydration of the cervix. According to Cretius (1959) the human cervix at term contains 82.6% water while in nongravid women, water accounts for 78%. The collagen content has been found to be 24.4 mg and 35.5 mg per gram of wet tissue at the time of labor and in the nongravid cervix, respectively. More detailed results have been obtained in animal experiments. Woessner (1982) summarized a few reports on collagen content in the cervix of pregnant rats. Collagen level at term increased twofold while wet weight increased fivefold. This "dilution" of collagen is, at least partially, an explanation of the change in mechanical properties of the cervix. There is no variation in type pattern of collagen, but it is generally accepted that at the time of labor the solubility of collagen increases. The extent of this increase remains controversial. Danforth and Buckingham (1964) reported that at term about 25% of the cervical collagen was salt-soluble as compared with 15% in controls. Values given by other authors were lower. Von Maillot and Zimmermann (1976) were able to solubilize 11% of the total collagen at labor and 3% and 6% in early pregnancy and early parturition period, respectively. Lower percentages of soluble collagen were found by Cretius (1965), although the same trend of changes was revealed. An increase in solubility can result from a reduction in cross-linking or degradation of the fibers.

Increased collagenolytic activity has been described at the end of pregnancy in the human cervix (Ito et al. 1977, 1979; Kleissl et al. 1978; Kitamura et al. 1979; Uldbjerg et al. 1983) and in animals (Fitzpatrick and Dobson 1979). Recently, Raynes et al. (1988) measured collagenase activity in the culture medium of ovine cervical explants and saw no significant difference in total enzyme activity produced by explants from nonpregnant, early pregnant, or late pregnant animals when expressed as units per milligram of wet tissue. When the enzyme activity was expressed relative to DNA content and hence cell number, there was an augmentation in the enzyme production per cell. This finding does not support the concept of the role of collagenase in cervical softening in the ewe.

A few reports have been published on the urinary excretion of hydroxyproline in pregnant women. It was found to be lower in patients in early pregnancy than in age-matched, nonpregnant controls. From the 24th week of pregnancy, the urinary hydroxyproline level rose above that in nonpregnant women (Piukovitch and Morvay 1973; Klein and Yen 1970). Lower values were reported in women with premature delivery (Piukovitch and Morvay 1973).

Kucharz (1986a) reported a progressive increase in total and free hydroxyproline and collagenlike protein levels in the serum of pregnant women. The values in the third trimester of pregnancy were about 2.5-fold greater than those in nonpregmant women. Significantly higher values were noted also on the first and the third days after labor. This finding remains contradictory to reports on urinary excretion of hydroxyproline in the postpartum period. Meilman et al. (1963) and Klein and Yen (1970) showed a low level of hydroxyproline in urine after labor, i.e., in the period of extensive collagen breakdown in the uterus. It is possible that alterations in kidney function affect the urinary excretion of amino acids.

The elevation in serum activity of collagen peptidase (PZ-peptidase) in pregnancy and the postpartum period has been described by Rajabi and Woessner (1984), thereby suggesting increased degradation of collagen under these conditions.

The concentration of the aminoterminal propeptide of type III collagen in the serum of women with a normal or complicated pregnancy was reported by

Risteli et al. (1987). In the former, the concentration was shown to be relatively constant until the third trimester, then increased towards term. In most cases, the concentration began to rise at week 32 or 33, and values were highest at term. There was no significant change during normal labor and 2 days of the postpartum period. The concentration of the aminoterminal propeptide of type III collagen was elevated in the twin pregnancies without complications. Very high levels of the propeptide were seen in patients with preeclampsia at 28–37 weeks of pregnancy. High values were also reported in women with hypertension and intrahepatic cholestasis of pregnancy. The propeptide concentrations in the serum of patients with diabetes or signs of preterm labor did not differ significantly from those in normal pregnancies.

The serum level of the immunoreactive noncollagenous domain of type IV collagen rose during the second trimester of pregnancy. The concentration dropped with the onset of the third trimester, but an elevation was shown towards the end of pregnancy. The values in patients with preeclamptic symptoms were slightly higher than in healthy pregnant women (Bieglemayer and Hofer 1989).

Indices of collagen metabolism have no value in obstetrical diagnosis. Their changes are nonspecific and are under the influence of many factors, such as hormones, placental function, fetal growth, and certain systemic alterations, including the above-mentioned changes in renal function. Attempts to estimate the placenta and fetus state by hydroxyproline determination measurements are also without clinical value (Eachempati et al. 1975).

Hormones form the main factor controlling collagen metabolism of the uterus. The ovary releases two kinds of steroid substances, estrogens and progesterone. Estrogens are secreted by the cells of the Graafian follicle in response to stimulation by the follicle-stimulating hormone released from the anterior pituitary. Estradiol and estrone are the most active estrogens. Estrogens are responsible for the growth of the sex organs and of the endometrium after menstruation. They also bring about the development of the secondary sex characteristics. Progesterone is released by the corpus luteum, the follicular cells that remain in the ovary after ovulation and expulsion of the ovum. Secretion of progesterone occurs in response to stimulation by the pituitary luteinizing hormone. The main function of progesterone is development of the endometrium to maintain the implanted embryo. Thus, it supports the progress of pregnancy. In the menstrual cycle, progesterone is secreted after ovulation, and if the ovum fails to be fertilized, the paucity of pituitary stimulation causes the cessation of progesterone secretion. In pregnancy, progesterone secretion is stimulated by chorionic gonadotropin.

An influence of estrogens and/or progesterone has been investigated in various systems, including the uterus of intact animals, uterine slices cultured in vitro, and tissues derived from other systems. The effect of estrogens on collagen metabolism in the uterus of rat has been investigated in many laboratories, but the reported results are ambiguous. Most revealed that estrogens increase collagen content and synthesis in the uterus (Kao et al. 1964; Smith and Allison 1966; Dyer et al. 1980). This is in agreement with the decrease in collagen content noted in ovariectomized animals (Harkness et al. 1957). Moreover, the effect of ovariectomy can be reversed by estrogen treatment (Morgan 1963; Smith and Kaltreider 1963).

Hormones are found to control the activity of collagenase in the uterus and the cervix. Cultured uterine slices do not produce collagenase if progesterone is added to the medium (Jeffrey et al. 1971). Progesterone has been found to reduce the level of active collagenase, whereas the total enzyme level (active and latent) remains constant (Tyree et al. 1980). Estradiol does not change collagenase production in vitro. In the postpartum uterus, estradiol inhibits collagenase activity (Woessner 1982).

Humoral regulation of cervical collagenase has been described. Prostaglandins are known to be a potent stimulus for collagenase production. Goshowaki et al. (1988) showed that prostaglandins $E_2$ and $F_{2\alpha}$ significantly stimulated the synthesis of the enzyme by rabbit uterine fibroblasts. Prostaglandin $I_2$ was inactive in this experimental model. Elevated collagenolytic activity was found in human cervical biopsies obtained from patients with prostaglandin $E_2$-induced labor as compared with samples from women wo spontaneously delivered. In patients with a prompt clinical response, the increase was nearly twofold. No differences were found in the concentration of collagen or glycosaminoglycans or the activity of elastase. The role of augmented collagenase activity in the induction of labor with prostaglandin $E_2$ has been postulated, although the mechanism in the light of other studies is not clear (Ekman et al. 1983; Norström et al. 1985).

Only single reports deal with hormonal control of cervical collagenase. Dehydroepiandrosterone was shown to induce collagenase production by cultured explants of rabbit uterine cervix. The main metabolite in vivo of this hormone, 17$\beta$-estradiol, was found to depress enzyme secretion (Ito et al. 1984). Infusion of estradiol prior to parturition did not change collagenase activity in the ovine cervix (Rayness et al. 1988).

The influence of estrogen on collagen metabolism in tissue other than the uterus has been studied, and reduction of collagen content was found in most of them. In ovariectomized rats, a single dose of 17$\beta$-estradiol decreased collagen synthesis in the molar periodontal ligament (Dyer et al. 1980). A reduction of collagen synthesis was observed in cultured aortic smooth muscle cells (Beldekas et al. 1981). Increased degradation of collagen in the aorta wall in vivo was reported in animals treated with estradiol (Fischer and Swain 1978). There was no change in collagen synthesis in the liver of such animals (Amma et al. 1978). Treatment with 17$\beta$-estradiol led to a rise in the serum hydroxyproline level (Amma et al. 1978). There are also suggestions that estrogens control collagen metabolism in bone and skin (Ernst et al. 1988). Reduced cutaneous collagen has been described in postmenopausal women (Versi et al. 1988). In mammary fibroblasts, estradiol increased collagen production (Sheffield and Anderson 1984). It may be concluded that organs associated with reproduction (uterus, mammary gland) respond to estrogens with an increase in collagen production, while in other tissues a reduction is seen.

## 18.5 Placenta

The placenta is an organ whose main function is to maintain the exchange of oxygen, carbon dioxide, and nutrients between mother and fetus. It also produces several hormones as well as stores some nutrients for the fetus. The exchange of substances via the placenta is not simple because the placenta has the capability of selective transportation, thus protecting the fetus against certain toxic compounds.

Rodents and primates, including man, produce a so-called hemochorial placenta. The uterine epithelium, subepithelial stroma, and blood vessels of the endometrium are lysed by the chorion after implantation of the embryo. The outer layer of the chorion, i.e., the placental villi, are bathed in maternal blood. In some mammals, the maternal blood is separated from the fetal tissues as the placental villi contact the endothelium of the maternal vessels. This type of placenta occurs in cows, horses, and sheep and is known as the epitheliochondrial placenta.

The human placenta is a disc-shaped organ. At term, it reaches about 20 cm in diameter. The fetal body is connected to the placenta by the umbilical cord. The placenta consists of the maternal part consisting of the endometrium, transformed into the decidua basalis, and the fetal part which is produced by the fetus. The fetal part is made up of the chorion frontalis, which in the form of the placental villi invades the uterine wall. At term, the decidua occupies the entire surface of the endometrium, and that portion underlying the fetal chorion frondosum is known as the decidua basalis. The decidua basalis anchors the placenta on the uterine wall. It is composed of cells derived from the endometrial stroma and contains certain amounts of the extracellular matrix of the connective tissue.

The placenta undergoes rapid growth and remodeling during gestation. The chorionic villi of the mature placenta form a treelike network. They contain syncytiotrophoblasts, i.e., a layer formed by the breakdown of intercellular membranes of trophoblast and single cytotrophoblastic (Langerhans') cells. The trophoblastic basement membrane is a supportive structure for the syncytiotrophoblast and separates it from the stroma. The interstitium of the villi contains fibroblasts and Hofbauer cells. The latter decrease in number in the mature placenta and are probably transformed macrophages.

Morphologically, collagenous structures are found in the trophoblastic basement membrane and the interstitium of the chorionic villi. The endothelial basement membrane and the wall of placental vessels are also composed of collagen.

The major collagen found in the placenta is type III, which constitutes about 60% of the total collagen content. The reason for the large quantity of type III collagen remains unknown. It is possible that its predominance reflects the characteristic of all fetal tissues, which contain a greater proportion of type III collagen than those of adults. The high content of vasculature in the placenta can explain this as type III collagen is associated with blood vessels. Under microscopictudy, type III collagen was localized in thin, nonbanded fibers of 10–15 n diameter that were abundant in the interstitium. The fibers formed dense bundles or aggregates (Amenta et al. 1986). There are no detailed reports of localization of type III collagen within the placental vessels.

About 30% of the total collagen content in the placenta is composed of type I collagen. Types I and III collagens are probably responsible for the structural function, although in the placenta the only supportive use of the collagenous stroma is maintaining the shape of the placenta. Using type I collagen-specificantibodies, the occurrence of type I collagen was found to be restricted to the interstitium of the chorionic villi. The positively stained fibers were cross-banded, with a diameter of 30–35 nm, and the banding interval was 64 nm (Amenta et al. 1986).

Collagens other than types I and III account for less than 10% of the total placental collagen. The placenta is considered as a good source of types IV, V, and VI collagens. Type VI collagen is localized in the perivascular regions of the interstitium of the chorionic villi. Electron microscopic studies revealed that it appeared to coat the surface of collagen fibers of all diameters (Amenta et al. 1986). The nature of this interaction remains unknown. It is possible that type VI collagen is a component of the heterogenous fibers similar to those reported in the cornea (see Chap. 17).

Type V collagen is also limited to the interstitium and forms thin filaments about 10 nm in diameter. According to Amenta et al. (1986), type I collagen was often observed to be embedded within a dense matrix of type V collagen. There was no type V collagen in the basement membranes of the placenta. The cellular source of types V and VI collagens is unknown. It can be speculated that fibroblasts of the stroma produce them.

The placenta is relatively rich in type IV collagen; it is one of the few readily available human tissues. Type IV collagen occurs in the basement membranes. It was found in both laminae but preferentially in the lamina densa. The collagen is thought to be produced by the cytotrophoblasts, the precursor cells of the syncytiotrophoblast. The endothelial cells are considered as the producers of the basement membranes of the villous capillaries. These cells in culture were able to synthesize type IV collagen. In culture, however, other types of collagen were also noted, but this could reflect phenotypic changes under conditions of cell culturing. The role of the endothelial cells in the production of type III collagen in the placenta remains unknown. The trophoblastic basement membrane is responsible, at least partially, for selective transportation between the fetus and mother. Overproduction of the basement membrane, seen as its thickening, is consdered a sign of cytotrophoblastic hyperplasia The nature of the basement membrane alterations and their role in placental pathology are obscure. The basement membrane of the capillaries forms a scaffold for the vessels and protects blood from direct contact with the collagenous stroma. Such contact may lead to platelet aggregation and thrombus formation.

Our knowledge of placental pathology has enlarged in recent years very rapidly. Most studies are based on ultrastructural investigations, but the relationship of the observed morphological changes to functional disturbances are relatively unclear.

As described earlier, two sites of collagenous structures have been seen within the placenta, the interstitial collagens and the basement membrane collagen. The pathology of interstitial collagenous structures is fibrosis of the villous stroma, which is associated with congenital syphilis, maternal diabetes, and preeclampsia. The development of fibrosis is probably ascribable to several causative mechanisms, and its relation to the function of the placenta is unknown, although is has been suggested that fibrosis impairs blood flow and gas exchange. Changes in the basement membranes have been detected in some forms of placental pathology. Thickening was shown in maternal diabetes, preeclampsia, and rhesus incompatibility (Brown 1982). Histologically, the thickening was limited to the trophoblastic membranes, and the endothelial ones were not affected. The role of these changes in the functional state of the placenta remains unknown.

It can be hypothesized that several forms of placental pathology are associated with abnormalities in the stucture and metabolism of collagen. At present, mot of our knowledge of this pathology originates from microscopic studies, and further biochemical investigations are needed to provide more data about the possible role of collagen in disorders of the placenta.

Little is known about the collagen content in the placental membranes. The main function of the collagenous structures which occur in the chorioamnion unit is its mechanical resistance. Premature rupture of fetal membranes is known to be a major contributory factor in prematurity and perinatal mortality.

Changes in collagen type pattern in fetal membranes during gestation were reported by Al-Zaid et al. (1988). Collagen type I accounted for about half of the total collagen content in the chorioamniotic membrane regardless of the gestational stage. The quanty of type III collagen significantly decreased from 38.8% at the 26th gestational week to 17.3% at term. Opposite effects were found for type V collagen, which constituted only 10% at the 26th gestational week, while it accounted for 30.6% of the total collagen at the 40th gestational week. The biological role of these alterations remains unclear. There are also differences in collagen content at varying sites of the membranes. A linear decrease in the collagen content of the chorion distal to the placenta was found (Al-Zair et al. 1980b). The type pattern at various distances from the placenta was not investigated.

Al-Zair et al. (1980a, b) were not able to show any significant changes in the total collagen content of preterm and full-term ruptured membranes. On the other hand, there are suggestions that a profound decrease in type III collagen is associated with the premature rupture of fetal membranes (Skinner et al. 1981). The reduction in level of one type of collagen can be responsible for disrupted architecture of the collagenous network at the site proximal to the rupture point, as has been shown by Bou-Resli et al. (1981). It is clear that premature rupture is a condition of multifactorial etiology, and the changes in collagen responsible for the decreased mechanical properties of the membranes are to be considered either as a primary or secondary mechanism.

The ability of amniotic fluid cells to synthesize collagen was reported (Priest et al. 1977). Two types of cells were cultured, fibroblasts and amniotic-fluid-type cells (Hoehn et al. 1974). The amniotic-fluid-type cells are from a different origin, either endothelial or epithelial (Priest et al. 1977; Megaw et al. 1977). They produce collagen similar to the basement membrane collagen, i.e., type IV. Fibroblasts synthesized collagen indistinguishable from that made by cultured fetal dermal fibroblasts. There are suggestions that the cultured cells from amniotic fluid may be useful in the diagnosis of the inherited disorders of collagen.

The presence of hydroxyproline in the amniotic fluid was reported by Heinrich et al. (1977). They found a relatively constant level in patients with a normal pregnancy and an elevation in those with EPH-gestose.

It is of interest that a potent inhibitor of collagenase has been isolated from the amniotic fluid (Murphy et al. 1981). The role of this inhibitor in collagen metabolism in the placenta and the membranes is unknown (Aggeler et al. 1981).

# 19 Collagen in the Nervous System

The nervous system is the most highly developed of the organ systems. It comprises the body's control center (the central nervous system) and the communication network (the peripheral nervous system). Nervous tissue is made up of nerve cells, the neurons, which are the structural units of the system. Neurons are highly differentiated cells, consisting of a body and several protoplasmic extensions by which neurons relate to one another and to other structures. The network of neurons and their extensions is surrounded by supportive, protective, and nutritive cells. Until the past decade, interest in the role of collagen in the nervous system was minimal, as it is almost absent in the brain and spinal cord.

In the central nervous system, collagen is present only in the meninges and vasculature. On the contrary, the peripheral nervous system contains substantial amounts of collagen which effect the function of the nerve (Mei Liu 1988).

The central nervous system, brain, and spinal cord are completely enclosed by connective tissue membranes, the meninges. There are three meninges: dura mater, arachnoid, and pia mater. The dura mater is the strongest, thickest, and outermost membrane. It is chiefly made up of thick collagen fibers. The arachnoid is thin, netlike structure. The pia mater is a thin connective tissue net closely adherent to the surface of the brain and spinal cord. The arachnoid has no blood vessels. In contrast, the pia mater contains a large number of vessels. Both meninges, together called the leptomeninges, consist of interlacing collagenous bundles surrounded by fine elastic networks (Shellswell et al. 1979). The main cellular elements are fibroblasts and macrophages.

Although the brain parenchyma contains no collagen, the presence of glycosaminoglycans in the central nervous system has been documented by histochemical and biochemical methods (Robinson and Green 1962; Singh and Bachhwat 1965; Castejón 1970; Margolis et al. 1976; Branford White and Hudson 1977; Vitello et al. 1978; Kiang et al. 1978; Margolis and Margolis 1979; Bertolotto and Margassi 1984). Their physiological role has not been uncovered. It is suggested that proteoglycans take part in the regulation of the ionic environment and fluid volume in the brain as well as in the storage and release of some neurotransmitters or enzymes of their metabolism. Glycosaminoglycans in the brain are produced by the glial cells (Glimelius et al. 1978; Dorfman and Ho 1970; Norling et al. 1978). This is an interesting phenomenon, providing evidence for the expression of genes for the synthesis of connective tissue macromolecules in the cells of the central nervous system (Pycock et al. 1975).

The peripheral nerve contains a number of nerve fibers which are associated in fascicles and held together by connective tissue. Collagen is the major component (Sunderland 1965); the outer layer of the nerve, the epineurium, is composed of type I collagen. The collagen fibers are mainly arranged longitudinally. Each of the smaller fascicles is enclosed in concentric layers of connective tissue, which form the perineurium. This is made up of concentrically arranged cellular layers which together with fibrous sheets divide the nerve into fascicles. The nature of the cells remains unclear. Their morphological structure resembles that of fibroblasts. The cells, however, are covered on all sides by the basal lamina which is a feature of Schwann cells, muscle cells, or vascular pericytes but not of fibroblasts (Mei Liu 1988). The perineurium is a mixture of type I collagen when found close to the epineurium and type III collagen when located near the endoneurium. The endoneurium is a structure surrounding the individual nerve fibers and is composed of thin fibers of type III collagen (Shellswell et al. 1979).

The connective tissue sheets of the peripheral nerve provide mechanical support to the nervous cell extensions. They protect the nerve trunk against compression and permit longitudinal changes during movements of the body. The perineurium is the most important supporting structure. In addition to mechanical properties, the connective tissue serves as the regulatory barrier for fluids and electrolytes. An interesting functional property of collagen here is associated with signal transduction in the motor end plate and probably in other cholinergic synapses. The basal lamina of the synaptic cleft is different from that which covers other parts of the muscle cell. The variations include distribution of type IV collagen and the occurrence of type V collagen (Sanes 1982). The collagen of the synaptic cleft is thought to be involved in the anchorage of acetylcholinesterase, via the part of the molecule which has a collagenlike structure (see Chap. 2). This collagenlike tail is probably responsible for binding the enzyme to the basement membrane.

Collagen in the peripheral nerve is synthesized by the cells of the perineurium and the Schwann cells. The Schwann cells are of ectodermal origin and represent elements similar to neuroglial cells of the central nervous system. The myelin sheath and the sheath of Schwann are parts of these cells. Schwann cells in tissue culture produce connective tissue macromolecules, including types I, III, IV, and V collagens (Bunge et al. 1980; Carey et al. 1983).

The collagen produced by the Schwann cells and fibroblasts takes part in nerve regeneration. The regeneration of a cut peripheral nerve is preceded by degeneration of the distal segment. The degenerated nerve is removed by phagocytes, but the basal lamina and the endoneural collagen remain intact and form the tube for the regenerating axon. The matrix molecules, including collagens, are neurotrophic and probably are responsible for the activity of the hypothetical neurotrophic factor, postulated since the early years of experimental neurohistology (Ramon y Cajal 1968). Use of collagen grafts is proposed to hinder the regenerating axons from going astray when the nerve ends are separated by a wide gap (Millesi 1981).

Only single reports dealing with collagen in the pathology of the nervous system have been published. Deficiency of basement membranes in dystrophic spinal roots and peripheral nerves was reported (Madrid et al. 1975; Jaros and Bradley 1979). The nerves and the roots contained less collagen than normal, but type V collagen was present (John and Purdom 1984). Deficiency in type V collagen may be caused by impaired cooperation between Schwann cells and fibroblasts. Schwann cells from dystrophic nerves freed of fibroblasts did not produce type V collagen in cultures. When dystrophic nerve cells and Schwann cells were recombined with fibroblasts derived from normal skin or nerves, the basement membrane defect was corrected (Bunge et al. 1982; Cornbrooks et al. 1983). A decreased content of types I and III collagens in the roots and nerves of muscular dystrophic mice was also reported (John and Purdom 1984). A decrease in the amount of type I collagen in the dystrophic roots may explain why many axons are not myelinated.

Collagens contribute to healing of open penetrant lesions of the rat cerebrum. Types I and III collagens have been identified in the cicatrix. Types IV and V collagens have been detected in the basement membrane of the glia limitans. The glia limitans is formed between the neurophil and the cicatrix. Healing in neonatal animals has been shown to be delayed (Maxwell et al. 1984).

# 20 Collagen in Hematology

## 20.1 Collagen and Hemostasis

Hemostasis is a complex process which protects against bleeding and insures fluidity of the blood. Hemostasis involves platelets, plasma constituents, and the vessel wall. Collagens in the wall and perivascular space play an important role, and the most explored hemostatic function is collagen-induced platelet aggregation. The role of collagen in direct activation of the intrinsic pathway of coagulation has also been postulated in several studies.

Following injury to the blood vessel wall, the extracellular matrix of the wall is brought into contact with the circulating blood. Platelets, the cell-derived morphological elements of blood, adhere to the exposed subendothelial collagenous structures. Adhesion is followed by the release of several biologically active substances and platelet aggregation. A cascade of biological phenomena is initiated by these early events of hemostasis. They include secondary aggregation and a complex process of coagulation, resulting in the formation of a hemostatic plug basically composed of stabilized fibrin. Intrinsic and extrinsic pathways of coagulation represent a precisely regulated process whose final stage is the transformation of fibrinogen into fibrin. The hemostatic plug protects against bleeding and facilitates healing of the vessel wall. After repair of the wall, the fibrin is degraded and solubilized.

This chapter is not intended to review all aspects of hemostatic mechanisms. The role of collagen and its structural features in platelet aggregation receive the main emphasis. The collagen-induced activation of coagulation is also briefly considered.

Early morphological observations of the adhesion of platelets to subendothelial and perivascular tissue and in vitro aggregation of platelets following addition of collagen fibers led to the recognition that collagen is involved in hemostasis (Bounameaux 1959; Hugues 1960, 1962; Hovig 1963; Zucker and Borelli 1962). Adhesion of platelets to collagenous structures is a preceding stage of aggregation. Under normal conditions, the intact endothelium constitutes a nonthrombogenic surface. Any damage to the endothelium leads to exposure of the underlying matrix and creates conditions for the interaction of platelets with collagenous components of the vessel wall. The ability of platelets to adhere to the subendothelium has been reported in a number of experiments. The endothelial cells were removed mechanically with a balloon catheter. Although the adhesion process has been clearly documented, mostly by morphological methods, the nature of the matrix component responsible for the adhesion remains obscure, and these studies suffer from a lack of specificity (Stemerman et al. 1971; Stemerman 1974; Beachey et al. 1979). Elimination of certain components of the subendothelial tissue by enzymatic digestion showed that platelets adhere more avidly to exposed collagen fibrils than to the undigested subendothelial surface (Baumgartner 1977). Adherence has been considered as a mechanism that partially reverses the interrupted integrity of the vessel wall. Its molecular nature is not fully understood. A few mechanisms have been proposed, including formation of an enzyme-substrate complex between collagen and glucosyltransferase of the platelet membrane or mediation of adhesion by fibronectin, but both of these have been rejected (Barnes 1982; Gastpar et al. 1978). Recently, several reports suggested that the interaction between collagen and platelets is mediated through a specific receptor or receptors. Chiang and Kang (1982) isolated and purified a platelet membrane constituent that is responsible for binding to the $\alpha 1$(I) chain of chick skin collagen. The purified receptor was a glycoprotein with a molecular weight of 65000 Da. Adhesion and aggregation of platelets were inhibited by an antibody against the receptor (Chiang et al. 1984). Two other platelet membrane proteins were found, with molecular weights of 90000 Da and 58000 Da, which also reacted with the antibody raised against the 60000 Da receptor. This indicates that the three proteins are immunologically cross-reactive. Furthermore, they are functionally related to platelet glycoproteins IIb and IIIa as monoclonal antibodies against the glycoproteins react with the isolated receptor proteins (Chiang et al. 1989). The biological role of this relationship remains unknown.

Adhesion is almost concomitant with aggregation. It is followed by a sharp change of platelet form; the platelets disperse and become flat, lying on the collagen surface. Initially, adherent platelets release a number of active substances that stimulate the attachment of platelets to each other and recruit nonadherent platelets from the vicinity. Such substances as ADP, serotonin, platelet-derived growth factor, and coagulation factors are released. The synthesis of thromboxane $A_2$, a derivate of arachidonic acid, is an another event associated with aggregation.

Several investigations were carried out to determine the role of the molecular and supramolecular structures of collagen in platelet adhesion and aggregation. Aggregation is the final stage of a series of events initiated by the contact of collagenous structures with platelets. It can be visualized and measured. The most frequently used method is estimation of light transmittance through a platelet suspension incubated in the presence of various substances affecting aggregation.

Aggregation depends upon some level of supramolecular structure of collagen. Monomeric collagen in solution is unable to aggregate platelets; this occurs only when the collagen fibrils appear (Brass and Bensusan 1974; Jaffe and Deykin 1974; Simons et al. 1975; Sylvester et al. 1989). There is no doubt that collagen must reach some level of quaternary structure to be able to aggregate platelets, but the nature of the structural requirements remains unclear.

Several groups investigated the relation of the type of collagen to its ability to aggregate platelets. Type III collagen is much more active than type I collagen (Hugues et al. 1976; Lapière et al. 1978; Balleisen et al. 1975). Type II collagen is a relatively weak inductor. Type IV collagen has no or very little aggregatory activity, and neither in solution nor in the intact basement membranes is it able to aggregate platelets (Barnes and MacIntyre 1979; Barnes et al. 1980). The explanation of the lack of activity is probably related to its inability to form fibrils. Several reports indicate that the fibrillar structures are necessary for the aggregatory activity. Type V collagen is considered a protein with low aggregatory activity. This conclusion was probably drawn from experiments using monomeric or denatured type V collagen because when fibrils of the native collagen are formed, its aggregatory activity is similar to that of type I collagen (Barnes et al. 1980).

Different requirements have been shown for platelet adhesion. Adhesion may be mediated by collagens and certain noncollagenous compounds of the matrix. This explains the adhesion of platelets to intact basement membranes, but this adhesion is not followed by aggregation (Fauvel et al. 1978; Barnes 1982). Adhesion appears to be independent of collagen conformation. This is consistent with earlier findings indicating that platelets adhere to the monomers of type I collagen. Adhesion to denatured collagen has been reported to be as good as that to native molecules. Types I, III, V, and VI collagens are able to support adhesion, but the adhesion activity of type V collagen is significantly lower (Parsons et al. 1983).

Recent research has provided more details on the adhesion and aggregation sites in the collagen molecule. A number of sites on both chains of type I collagen have been established (Fitzsimmons and Barnes 1985; Fitzsimmons et al. 1986), distributed along the length of the molecule. The sites varied in their affinity, but most of them exhibit a moderate ability to aggregate platelets. In contrast, type III collagen possesses a single, highly reactive platelet-aggregatory site, whose primary structure has been described by Morton et al. (1987). Further studies of this group provided evidence for separate aggregation and adhesion sites (Morton et al. 1989). In types I and III collagen molecules, the sites that have almost no aggregatory activity are able to adhere to platelets. In general, adhesion is in accord with the known platelet-aggregatory activity of collagen-derived fragments.

Further studies on the molecular mechanisms associated with platelet adhesion and aggregation may be a key for the elaboration of drugs that will be able to inhibit these phenomena specifically and to prevent thrombosis.

In addition to its influence on platelets, collagen has been suggested to contribute directly to coagulation. Niewiarowski et al. (1966) and Wilner et al. (1968) reported that collagen initiated the intrinsic pathway of coagulation by activation of factor XII. Other points of coagulation initiation have been also postulated: activation of factor XI (Walsh 1972) and an increase in activity of platelet-associated factor V (Osterud et al. 1977). These findings, however, remain controversial (Barnes 1982).

Recently, Völkl (1989) described a new test for platelet function in vivo based on intravenously injected clostridial collagenase in rats. A transient decrease in the platelet concentration reflects the platelet-vessel wall interactions.

## 20.2 Collagen in the Bone Marrow

Bone marrow contains cells which are involved in hematopoiesis and osteoinduction. In culture, several types of cells develop from the bone marrow, including an adherent stromal layer comprised of adipocytes, endothelial cells, fibroblasts, reticular cells, and macrophages (Dexter et al. 1977; Boman and Balian 1989). Adherent stromal cells have been found to be responsible for the synthesis and deposition of the extracellular matrix and factors regulating the growth and proliferation of hematopoietic cells (Song and Quesenberry 1984). The normal bone marrow stroma is composed of distinct collagen types. Using immunofluorescence techniques, types I, III, IV, and V collagens were found within the stroma (Bentley et al. 1981). Collagen together with other components of the extracellular matrix is important in the support of the nonadherent hematopoietic cells. Collagen has been shown to modulate cell growth and differentiation (Zuckerman and Wicha 1982, 1983).

In culture, bone marrow cells produce types I, III, IV, and V collagens (Shvelidze and Tsagarelli 1978, 1980; Bentley and Foidart 1980; Bentley 1982; Zipori et al. 1985). Adherent stromal cells also produce a 17000 Da protein sensitive to digestion by bacterial collagenase (Waterhouse et al. 1986; Boman and Balian 1989). The protein was not a degradation product of collagen, and its nature and function remain unknown. It has been hypothesized that this protein is a minor collagen (like types IX or X collagens) and contributes to the mineralization of bone. The bone marrow stromal cells have been found to contain cells with osteogenic potential, and the protein could be a mediator of this activity (Owen 1985).

The term myelofibrosis refers to a condition in which there is replacement of normal bone marrow stroma by fibrous connective tissue. The condition is probably caused by multiple agents; idiopathic myelofibrosis is found in agnogenic myeloid metaplasia, and myelofibrosis occurs in such diseases as polycythemia vera or chronic myelogenous leukemia (for review, see Laszlo 1975; Gilbert 1980). Historically, two types of bone marrow fibrosis were recognized: reticulin fibrosis and collagen fibrosis. The former is believed to be a precursor to the severer latter type. The early phase of myelofibrosis is characterized by an increased accumulation of fibers stainable by silver impregnation which are made up of type III collagen. In the advanced stage, the content of both types I and III collagens appears to be significantly elevated. A dominance of type I collagen was found in the late stage (Ward and Block 1971; Charron et al. 1979). Myelofibrosis is associated with a raised level of collagen biosynthesis. An increase in serum aminoterminal type III procollagen propeptide was found in patients with myeloproliferative diseases. The highest values were reported in patients with agnogenic myeloid metaplasia, and a moderate augmentation was found in patients with polycythemia vera with or without myeloid metaplasia (Hochweiss et al. 1983). The serum level of aminoterminal type III procollagen propeptide correlated with the fibrosis of the bone marrow measured as the reticulin grade. These findings were confirmed in other studies (Vellenga et al. 1983; Arrago et al. 1984; Hasselbalch et al. 1985, 1986). The rise in serum level of antigens related to type IV collagen (serum 7S collagen) was reported in patients with idiopathic myelofibrosis and chronic myelogenous leukemia. The highest level of serum 7S collagen was noted in patients with the grade 4 bone marrow fibrosis (range 0–5). This is consistent with the suggestion that type IV collagen in the late stage of fibrosis is replaced by type I collagen. A positive correlation between the serum level of aminoterminal type III procollagen propeptide and serum 7S collagen was also described (Hasselbalch et al. 1986).

## 20.3 Collagen in the Lymph Nodes

Lymph nodes are small, encapsulated, peripheral organs through which lymph flows and are part of the system responsible for the immune response to foreign antigens. They consist of lymphoid tissue and stroma made up of loose connective tissue. The stroma is composed of a delicate network of reticular fibers. Only a few studies have been published about collagen in the lymph nodes. Immunohistochemical methods at the light microscopic level revealed the presence of type IV collagen (Karttunen et al. 1986; McCurley et al. 1986). Ultrastructurally, the reticular fiber contains collagenous fibers with amorphous material between them (Bairati et al. 1964). The presence of types I, III, and V collagens has been noted (Konomi et al. 1981). Type IV collagen has been seen in the basement membranes of the walls of sinuses. It is of interest that the aminoterminal propeptide of type III procollagen has been found in the reticular fibers and walls of blood vessels and sinuses. Therefore, it has been concluded that a significant number of molecules of type III collagen are not completely cleaved in the conversion of procollagen to collagen. Type IV collagen has also been shown to be a genuine component of these reticular fibers (Karttunen et al. 1989).

# 21 Collagen and Neoplasia

## 21.1 Introduction

It is difficult to define neoplasia clearly. It can be described as a progressive accumulation of pathological tissue. The overgrowth of the neoplastic tissue is not limited by the normal regulatory processes controlling tissue regeneration and occurs also when the unknown stimulating factor is removed. Traditionally, neoplasms are classified into malignant and benign tumors. The growth of the benign form is limited to one location as it does not metastasize. Malignant tumors grow very fast and produce distant foci of neoplasia, i.e., metastases (Wieczorek 1981).

The term "neoplasm" in practice is substituted by the term "tumor." In the strict sense, tumor is a morphological description and does not indicate the nature of the overgrowth. Nonneoplastic tumors caused by inflammation or parasites are common. Malignant tumors consisting of cells of epithelial origin are known as cancers. Cancers constitute 90% of all malignant neoplasia in humans. Neoplasms of mesenchymal origin are called sarcomas. It is common practice to use the term "cancer" to describe all forms of malignant neoplasia.

There are several points of mutual relationship between the tumor growth and collagen (Fig. 21.1). The tumor consists of neoplastic cells and stroma that consists of extracellular matrix including numerous structures. The stroma is produced by the tumor cells or can originate from the host cells. The production of the stroma by the host cells is modulated by the obscure humoral factors released by the tumor.

The relationship of the stroma with tumor cells has a number of possible consequences. Many functions of the extracellular matrix have been suggested to modulate neoplastic growth. One of them is the presence of specific receptors for matrix components on the surface of neoplastic cells. Access of nutrients to the tumor is, at least partially, controlled by the connective tissue structures of the stroma. In addition, the proliferation and invasion of the neoplasm are associated with the destruction of stromal barriers including basement membranes. This process occurs in the primary location of the cancer and in the distant sites of metastases.

The present chapter focuses on the collagenous structures in the stroma of neoplasms and the role of collagen breakdown in tumor invasion.

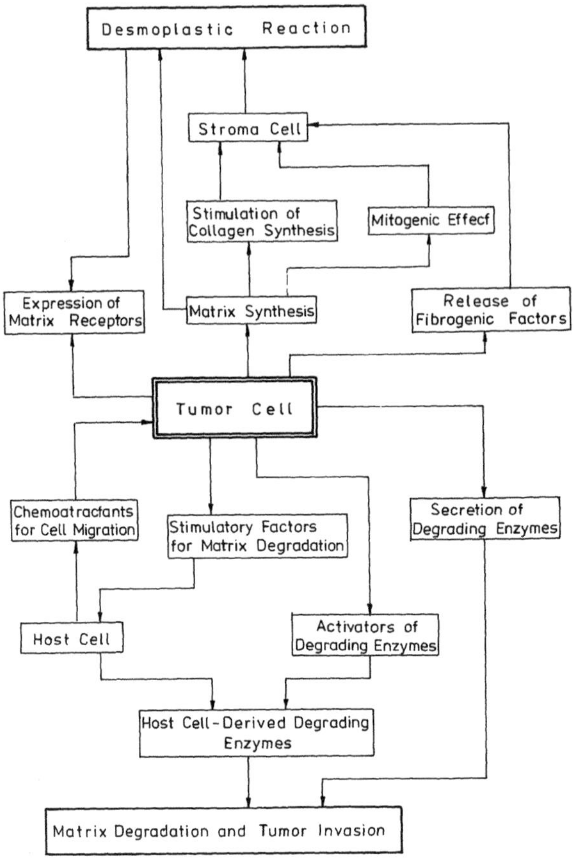


**Fig. 21.1.** Relationship between tumor growth and the connective tissue matrix

## 21.2 Collagen in the Stroma of Neoplasms

It is well-known that many neoplasms contain a prominent stroma built up of several elements including cells (fibroblasts, histocytes, inflammatory cells), blood vessels, and extracellular matrix. Collagen is a constant component of the matrix. The presence of collagen in the stroma of neoplasms poses several questions. The stroma may originate from tumor cells or host cells. Tumor cells can produce collagen, and this characteristic may also apply to the non-neoplastic cells of the tissue which is the origin of the neoplasm. The ability to secrete collagen can be a new feature of the tumor cells acquired during the metaplasia and neoplasm formation. Collagen structures originating from host cells my accumulate as a so-called pseudocapsule. The pseudocapsule is formed by compaction and displacement of the preexisting stroma by the expanding tumor. Production of collagen de novo is also a common feature of host cells. The secretion of matrix compounds including collagen is regulated by several factors, and stimulation of the host cells by products derived from the tumor cells and inflammatory cells is considered one of the main causes of enhanced collagen production in the stroma (van den Hoof 1983; Martinez-Hernandez 1988).

Quantitative and qualitative changes in the collagen of tumors have been reported. These changes are associated with carcinogenesis, and some of them may precede the development of advanced neoplasia. Many tumors have an abundant amount of extracellular matrix that is dense and fibrotic. This fibrotic reaction of the tumor is called the desmoplastic or scirrhous response. These tumors have received special attention from investigators (Alitalo and Vaheri 1982). The fibrotic reaction is most profound in human breast cancer and in some forms of gastric cancer.

Among breast cancers, the ductal infiltrating carcinoma is one of the most invasive. The tumor is characterized by the deposition of an abnormal extracellular matrix in the proximity of tumor nests and cords (Barsky et al. 1982; Al-Zuhair et al. 1986; Kao et al. 1986; Tremblay 1976). Ultrastructural and biochemical studies show that there is an accumulation of significant amounts of type I trimer collagen. This form of collagen is almost absent in normal adult breast tissues or in nonneoplastic mammary dysplasia (Luparello 1987). Type I trimer collagen is probably synthesized by the tumor cells. The malignant epithelial cells were shown to contain prolyl hydroxylase, an enzyme of collagen synthesis (Al-Adnani et al. 1975). In ex vivo fragments of the scirrhous stroma, type I trimer collagen was abundant, while in cultures of tumor cells this type of collagen was the main, if not the only, one produced (Minafra et al. 1988). The ability of neoplastic cells to synthesize collagen remains stable through sequential passages of the cultures. The role of type I trimer collagen in tumor cell behavior is not known.

Several studies showed that all normal mammary tissues contain well-organized basement membranes. The degree of disorganization of the basement membrane correlates with the level of differentiation of carcinoma; well-differentiated carcinomas contain defective basement membranes whereas poorly differentiated carcinomas have little or no basement membrane (Albrechtsen et al. 1981; Gusterson et al. 1982; Barsky et al. 1983; Wewer et al. 1986).

The presence of type IV collagen in normal human breast tissue and in benign and malignant lesions has been investigated for its possible application in differential diagnosis. In fibroadenoma, phyllodes tumor, and mastopathy as well as in normal breast tissue, type IV collagen was found linearly placed around epithelial cell nests. There was no type IV collagen revealed by immunohistochemical methods in scirrhous, mucinous, apocrine, and tubular carcinomas. Invasive ductal carcinoma was negative or discontinuous when stained for type IV collagen, and invasive lobular carcinoma reacted positively only while in situ (Tsubura et al. 1988). The varying distribution of basement membranes is useful in the differential diagnosis in surgical pathology of the breast and the early detection of micrometastases (Liotta et al. 1979). The biological role of these differences remains unknown.

It is believed that the disorganization of the basement membrane is a common property of poorly differentiated carcinomas. Although most reports have focused on breast cancers, single studies on other neoplasms have appeared. Forster et al. (1984) noted that patients with colorectal carcinoma which exhibited more complete basement membrane organization had more favorable survival times and a lower incidence of metastasis than those patients with carcinomas exhibiting less organized basement membranes. Immunostaining of type IV collagen was found to have prognostic value in colorectal cancer (Martinez-Hernandez and Catalano 1980; Havenith et al. 1988; Pignatelli and Bodmer 1988). Similar findings have been described for endometrial adenocarcinomas (Stenbäck et al. 1985; Liotta et al. 1986) and schwannomas (Oda et al. 1988).

Scirrhous carcinoma of the stomach is characterized by an especially rich fibrous stroma in which exist isolated carcinoma cells. Morphological studies showed that both the submucosal and mucosal layers were thickened by the deposition of a large mass of collagen fibers. The collagen content per unit transverse section of the tumor was two to four times higher than normal. A dense fibrous tissue was composed of types I and III collagens. Type IV collagen was diffusely distributed throughout the tumor stroma, submucosa, and fragmented regions of the muscle layer. Isolated stroma cells produced significantly more collagen than the undifferentiated or highly differentiated carcinoma cells. Addition of the culture medium of carcinoma cells resulted in an increase in the production of collagen by the stroma cells. This indicates that scirrhous carcinoma of the stomach is caused by the desmoplastic reaction of stroma cells induced by infiltrating malignant epithelium (Nagai et al. 1985). A stimulatory effect of the scirrhous carcinoma cells on normal fibroblasts cultured in vitro was also found by Naito et al. (1984).

Patterns of collagen types synthesized by the tumor have been used in the analysis of the cellular origin of these neoplasms. Yamagata and Yamagata (1984) investigated a virus-induced osteosarcoma of mice known as Finkel-Biskis-Jinkins osteosarcoma. Injections of the specific virus into young mice leads to the development of a sarcoma containing an extensive extracellular matrix. The tumor contains types I, III, V, and I trimer collagens. Contrary to its appearance, collagen type analysis indicates that this osteosarcoma is induced from osteogenic cells, and cartilage cells are not involved in the neoplastic process. Similarly, the bronchogenic carcinoma of the lung which is composed of small, oat-shaped cells was found to produce enhanced amounts of type II collagen. This type of collagen, specific for cartilage, is a constant component of the tracheobronchial tree (Svojtková et al. 1977).

The ability to synthesize collagen with unusual properties was shown in vitro in cultured mouse teratocarcinoma $TSD_4$ cells. The isolated collagen was similar to type I trimer (Little et al. 1977).

Transformation of the human fibroblast line with oncogenic virus SV40 was reported to reduce the ability to produce collagen (Bańkowski et al. 1977). The mechanism of this change is probably related to hypermethylation at certain cytosine residues within the procollagen type I genes. An additional mechanism is implicated because demethylation was not sufficient to reverse the effect (Parker and Gevers 1984).

Systemic changes of indices of collagen metabolism are of minor value in the diagnostic evaluation of patients with neoplasms. An increase in urinary excretion of hydroxyproline is associated with bone metastases and hormonal alterations affecting the bone tissue. An increase in hydroxyproline output was reported in patients with bone metastases of breast cancer (Cuschieri and Felgate 1972; Gielen et al. 1976). Similar results were found for metastatic prostate cancer (Mooppan et al. 1980; Gasser et al. 1981; Hopkins et al. 1984). Total and nondialyzable hydroxyproline excretion levels were shown to be more reliable indices than the free urinary hydroxyproline or serum hydroxyproline content.

## 21.3 Collagenolysis and Tumor Invasion

The ability to invade tissues other than the tissue of origin is an essential property of neoplastic cells. The invasion can be limited to surrounding tissues in benign tumors or is associated with active migration of the tumor cells into tissues of different types located in various parts of the body. The latter process is known as the formation of metastases and is a specific feature of malignant tumors.

During invasion tumor cells must transverse several barriers made of connective tissue matrix. They include matrix components of the primary tumor, basement membrane of the epithelium (in the case of carcinoma), interstitial stroma, and basement membranes of lymphatic or blood vessels. After intravasation, neoplastic cells migrate to distant places, and initiation of metastases is associated with crossing the vascular wall during extravasation as well as the perivascular interstitial stroma before entering the parenchyma of the target organ. It is clear that dissemination must be associated with active degradation of the connective tissue matrix, including the collagenous barriers to cell migration. The process is a very complex phenomenon, and its basic mechanism still remains unclear, although several single steps have been described. The major difficulty is to investigate in vivo where local and systemic factors orginating from the host and the tumor have an influence.

According to Liotta et al. (1982b) invasion of the matrix is a three-step process. First, tumor cells attach to components of the matrix. This step depends on the attachment properties of the cells as well as on the ability of the matrix to bind the cells. Thus, tumor and host factors play a role in the early part of the invasion. Second, the matrix is degraded by hydrolytic enzymes. Degradation of the collagenous network of the interstitial stroma or basement membranes is necessary for local permeability to cells. Normally, the matrix contains collagen fibers that form a dense network with spacing at least two orders of magnitude less than the diameter of a cell. The basement membranes are even less permeable structures. Breakdown is brought about by active collagenolytic enzymes in cooperation with other hydrolases. The enzymes are secreted by the tumor cells or by the host cells under the influence of stimuli originating from the tumor cells (see later). Collagenolysis is controlled by a complex system of activators and inhibitors, and the regulatory mechanisms are still poorly understood. The third step of invasion is cell locomotion into the area of the degraded matrix. The direction of the movement can be effected by chemotactic factors, some of which are generated during the breakdown of the matrix or accompanying inflammatory reaction of the host (Liotta and Rao 1985).

The increased adherence of tumor cells to matrix components was found in in vitro models. Tumor cells were inoculated onto deposits of matrix components previously produced by the cells cultured in vitro. Adhesion is followed by degradation of the matrix. Special attention was paid to the production of collagenase in the tumors. The role of proteolytic enzymes in the invasive growth was postulated in the pioneering work of Sylven (1949, 1958). Collagen degradation associated with invading tumor cells was shown under electron microscopy for the first time by Birbeck and Wheatley (1965). This was succeeded by several biochemical investigations into the presence and characteristics of collagenolytic enzymes in the tumors (Dresden et al. 1972; Woolley 1984a, b).

Tumor-derived collagenases have been found to be similar to vertebrate collagenases. Their molecular weights are within the same range as those isolated from nonneoplastic tissues. For example, collagenase from rat prostate carcinoma and rabbit V2 carcinoma have a molecular weight of 71000 and 33000–35000 Da, respectively (Huang et al. 1979; McCroskery et al. 1975). The cross-reactivity of antibodies against human rheumatoid synovial collagenase with the enzyme from human carcinomas and melanomas provides further proof of their relatedness (Woolley et al. 1980). Neoplastic tissues contain collagenases capable of degrading various types of collagen, including types IV and V (Nethery and O'Grady 1989). The many similarities raise the question of the cellular origin of the enzymes. Practically all tumors are grown in vivo as a mixture of tumor cells and host cells (e.g., fibroblasts, macrophages, granulocytes, and endothelial cells). The ability to secrete collagenase has been shown in several experiments using cultured tumor cells free of cellular contamination. This capacity in vitro does not exclude the possibility of a contribution by the host cells in vivo to the production of enzymes responsible for the breakdown of the matrix. An enhanced activity of collagenase in various human and animal tumors and cultured tumor cells has been noted by numerous authors. For example, collagenase was identified in skin cancer (Hashimoto et al. 1972, 1973b; Yamanishi et al. 1972; Abramson et al. 1975; Ohyama and Hashimoto 1977; Bauer et al. 1977b). Wirl (1977) described an increase in collagenase activity associated with chemical carcinogenesis of mouse skin. Some tumors also contain a collagenase which degrades basement membrane collagen (Liotta et al. 1977, 1979; Salo et al. 1983; Garbisa et al. 1988). The tumorous type IV collagen-degrading metalloproteinase has a molecular weight of about 60000 Da (Liotta et al. 1982a). Type V collagen-degrading enzyme activity was reported in several tumors, e.g., invasive hepatocellular carcinoma (Kobayashi et al. 1987).

Elevated active collagenase levels were shown in tumor homogenates of human malignant melanoma (Yamanishi et al. 1973). Specific collagenase activity was reported in pure populations of the cultured human malignant melanoma cells (Tane et al. 1978). The activity was found in the cell homogenates but was absent in the culture medium. Early cultures exhibited a high level of collagenase activity, and as they were successively subcultured, the activity diminished. It has been suggested that the presence of collagen in the vicinity of the cells is a stimulus for collagenase production.

Regulation of collagenolysis includes the action of various more or less specific collagenase inhibitors. Natural inhibitors are present both in the serum and extracellular matrix. The serum inhibitory activity mostly results from the presence of $\alpha_2$-macroglobulin and, in a small way, of $\beta_1$-anticollagenase. The role of inhibitors in the regulation of tumor growth is insufficiently explored. There are several reports on serum antiproteases in patients with various neoplasms, but their serum level responds to many factors, including inflammation (Bernacka et al. 1988).

Interestingly, it was shown that collagenase isolated from human gastric carcinoma is resistant to inhibition by serum antiproteases (Woolley et al. 1980). The mechanism of this resistance is unknown; it is possible that the binding affinity of the inhibitor to the enzyme is altered.

The synthesis of inhibitors by tumors has not been fully elucidated. It has been suggested that a local decrease in inhibitor levels may be responsible for an elevated collagenase activity. Despite the noted resistance of tumor collagenase to inhibitors, enzymes of tumor origin can be inhibited when an excess of natural inhibitors is added. It was found that natural inhibitors, when provided in excess, diminished or blocked tumor cell invasion in vitro (Thorgeisson et al. 1982; Pauli et al. 1981). This finding is strong evidence for the role of matrix degradation in tumor invasion. The ability of the cartilage-derived proteinase inhibitor to inhibit collagenase from carcinoma cells could explain why the cartilage is rarely invaded by neoplasms (Kuettner et al. 1977).

The cellular source of tumorous collagenase has not been determined. It is clear that the many kinds of cells making up the tumor are able to secrete active or latent collagenase. Synthesis of the enzyme by pure tumor cell cultures is not enough evidence for the same ability in situ. Studies of Woolley (1982) and Biswas (1982) showed that collagenase was produced both by tumor and host cells. According to Biswas (1982), about 60% – 80% of collagenase in rabbit V2 carcinoma transplanted into nude mice was derived from the tumor cells. It is impossible to generalize this finding to all kinds of tumors.

Detailed investigations on interstitial collagenase procduced by chemically induced mammary tumors in rats were reviewed by Wirl (1984). The ability to produce collagenase in vivo and in vitro was attributed to different cells, stroma cells in vivo and cuboidal cells, epithelial cells, and macrophages in vitro. This alteration was related to the inability of some cells to interact with fibrillar collagen due to the morphological arrangement of the tumor (Biswas 1984, 1985). In vitro, fibrillar collagen was found to be a potent collagenase stimulator. Caution in the interpretation of cultured tumor cells' ability to secrete collagenase is suggested.

Another mechanism of collagenase influence upon tumor invasiveness was reported by Maslow (1987). Pretreatment of various neoplastic cell lines with collagenase increased cell motility significantly as measured by the penetration of filters in vitro. The locomotory behavior was found to be related to the colonization potential (Goldroses and Maslow 1985; Carr et al. 1985; Gehlsen and Hendrix 1986). The mechanism remains unknown.

There are several suggestions that collagenase activity correlates with tumor aggressiveness. The activity in malignant tumors is significantly higher than in corresponding benign neoplasms, both in human and animal tumors (Tarin et al. 1982; Liotta et al. 1982b). The ability to form metastases is related to the activity of collagenase capable of degrading type IV collagen (Liotta et al. 1980; Turpeenniemi-Hujanen et al. 1984, 1985). All highly metastatic tumor cells contain elevated levels of collagenase that facilitates crossing of the basement membrane. Leukemia cells were found to have greater amounts of collagenase than corresponding normal blood cells (Kucharz and Drożdż 1978c). On the other hand, it is too easy to explain all aspects of invasion and metastasis by an enhanced collagenase activity. It is clear that active collagenase takes part in invasion, but several additional factors are responsible for the metastatic behavior of tumor cells.

# 22 Effect of Nutrition

## 22.1 Malnutrition and Overnutrition

The role of proper nutrition on collagen metabolism and content in tissues has been described in a number of reports. The term "nutrition" refers to the quality, quantity, and mutual proportions of certain alimentary components, and it is very difficult to elucidate the mechanism that interrelates nutrition with collagen disturbances. Most of the data in this field were obtained in animal models, i.e., different species of animals were subjected to food restrictions or were fed with special diets. In these experiments, the diet modifications were applied for a period which corresponded to half an average lifespan. Such a long-lasting influence of dietary factors could result in several secondary metabolic alterations which indirectly affect collagen. Thus, it is extremely difficult to interpret collagen changes under these conditions.

Malnutrition and especially impaired protein intake were shown to decrease the total collagen content and to raise the level of insoluble collagen in tissues (Čabak et al. 1963). The relationship between the collagen and noncollagenous protein content in tissues of starving animals remains unclear. In humans, malnutrition or kwashiorkor increases the urinary output of hydroxyproline (Satwekar and Radhakrishnan 1964; Picou et al. 1965; Whitehead 1969).

An interesting hypothesis on the role of collagen metabolism in aging in relation to food restrictions has been presented by Deyl and Adam (1982). It is well-known that food restrictions increase the lifespan of several species of animals (McCay 1947). These early observations have been confirmed by a number of experiments, and differences in the lifespan of undernourished and control (receiving normal diet) animals are very clear. Deyl et al. (1971) supported previous observations of the increase in collagen content with aging (see Chap. 6) and found a distinct elevation in the collagen level in the lungs, liver, and kidneys of rats around 14 months of age. This increase was not organ-specific. In animals with food restrictions, the same rise was significantly delayed and appeared at 25 months of age. The elevation of tissue collagen content was associated with an increased mortality rate, and the significant shift of collagen accumulation between the two groups of rats was the same as the change in their average lifespan. There is no unequivocal explanation of the role of collagen in aging. According to Deyl and Adam (1977) the food restriction led to a delayed shift in the collagen type pattern in the tissues. This change may be related to longer survival, although details of the relationship remain unknown.

Overnutrition is almost always associated with an increase in the consumption of fat-rich food. There are some suggestions that lipid metabolites act as cross-linking agents. Modification of the protein structure with the lipid metabolite malondialdehyde in vitro has been noted by several investigators (Menzel 1967; Davídková et al. 1975). A similar phenomenon was reported in vivo in fat-fed animals (Davídková et al. 1975). A high-fat diet in vivo changed the contraction-relaxation properties of rat tail tendon. Some lipids were also found to be bound to insoluble collagen. Collagen from the tissues of animals fed on a diet rich in fat was more resistant to pepsin cleavage than that of controls (Deyl and Adam 1982). The detailed mechanism is obscure.

Only single reports have focused on the role of quality alterations of diet in collagen metabolism, with the exception of ascorbic acid deficiency (see the next section). Vitamin $B_6$ deficiency was shown to impair cross-link formation in collagen (Fujimoto et al. 1979; see Chap. 24). Vitamin D affects collagen metabolism in association with other abnormalities in calcium and phosphorus metabolism (see Chap. 11). Small and unspecific changes of collagen metabolism were reported in animals fed on a folate-deficient diet (Barnes and Kodicek 1972). An elevated supply of iron can produce liver fibrosis (see Chap. 14). Other quality changes in diet have not been investigated with respect to collagen metabolism.

## 22.2 Scurvy

Ascorbic acid deficiency leads to a disease state known as scurvy. The disease is characterized by fragile vessels, bleeding, delayed wound healing, and impaired structure of the connective tissue (e.g., periodontal ligaments). It has been recognized since antiquity, and in the 1920s it was found to be caused by the deficiency of vitamin C.

The relationship between ascorbic acid and collagen metabolism has finally been proved and is reviewed by Gould (1960), Barnes and Kodicek (1972), and Barnes (1969, 1985b). The association of ascorbic acid deficiency with failure in collagen production was shown for the first time in experiments on wound healing and the tensile strength of wounds (Lanmam and Ingalls 1937; Taffel and Harvey 1938; Bartlett et al. 1942). A reduced level of insoluble collagen in scorbutic guinea pigs was demonstrated by Hunt (1941). A decrease in collagen production under conditions of ascorbic acid deficiency was shown in carrageenan granulomas. The basic mechanisms of impaired collagen production in scurvy is a decrease in the hydroxylation of proline and lysine residues (see Chap. 3), as found in whole-cell preparations from scorbutic animals, cell and tissue cultures, and direct investigations of prolyl hydroxylase preparations (Stone and Meister 1962; Bates et al. 1972; Jeffrey and Martin 1966ab, Myllylä et al. 1978). A dose-dependent increase in collagen synthesis was reported in human skin fibroblasts cultured in the presence of various levels of ascorbic acid (Pinnell 1985). However, the alterations observed in vivo are more complex and cannot be explained by defective hydroxylation only (Bates 1979).

Early attempts to isolate or to identify hydroxyproline-deficient collagen from tissues of scorbutic guinea pigs were unsuccessful (Gould 1960). The rapid degradation of underhydroxylated collagen precursors is a possible explanation. Impaired hydroxylation was not associated with a reduced urinary excretion of hydroxyproline; on the contrary, hydroxyproline output was normal or slightly elevated despite the low synthesis of collagen (Barnes et al. 1969; Barnes and Kodicek 1972). This finding is relevant to the raised excretion level of hydroxyproline shown in humans with scurvy (Bates 1977; Windsor and Williams 1970). Additional defects in the metabolism of collagen have been suggested. In guinea pigs, lack of ascorbic acid led to a fall in the incorporation of proline into skin collagen. The proline/hydroxyproline ratio was only slightly raised. Impaired synthesis of precursor forms of collagen was concluded from these findings and confirmed in further studies (Tajima and Pinnell 1982; Lyons and Schwarz 1984; Takehara et al. 1986; Pinnell et al. 1987). An effect of ascorbic acid on collagen maturation has also been postulated (Grinnell et al. 1989). The intracellular degradation of collagen was not altered by the level of ascorbic acid (Gessin et al. 1986). Ascorbic acid stimulated the conversion of type I procollagen to collagen and the subsequent deposition of type I collagen into the extracellular matrix. Secretion of the molecules from the cells was probably augmented. Secretion of the molecules from the cells was probably augmented (Pacifici and Iozzo 1988). The effect of ascorbic acid on the secretion of other types of collagen remains unknown (Heino et al. 1989).

# 23 Effect of Ionizing Radiation

Connective tissue is one of the most radiation-resistant components of the body. Since the discovery of X-rays and radioactivity, many investigations have been undertaken, and the effect of ionizing radiation on living matter has been considered at many levels, ranging from the complex living body down to simple molecules and water itself. The radiobiological effects on collagen have been investigated both in vitro and in vivo; the early findings in this field are reviewed by Bailey (1968a).

Changes in the physicochemical properties of collagen and the structure of the fibers were studied as early as 1950. Perron and Wright (1950) found that irradiation produced alterations in the intensity of the low angle diffraction pattern of dry kangaroo tendon. Variation in the organization of the fibrils seen under electron microscopy confirmed earlier findings (Bailey and Tromans 1964; Stoianova et al. 1960). Destruction of the banding pattern to give an amorphous appearance occurred when the fiber was irradiated while wet. In the dry state, the lateral adhesion of the fibrils was affected, but the fragments retained their banding pattern. The physical properties of irradiated fibrils are affected. A progressive, dose-dependent decrease in the shrinkage temperature of collagen, which reflects the integrity of its structure, was described (Cassel 1959; Bailey et al. 1962). This phenomenon is attributed to disorganization of the interchain hydrogen bonds with or without rupture of the peptide backbone. Similarly, Braams (1961, 1963) and Bowes and Moss (1962) reported a marked decrease in the tensile strength of collagen fibers irradiated under both wet and dry conditions; the decrease was more significant in the latter case. Radiation's influence upon solubility appears to be the result of two competing reactions: Formation of cross-links by an indirect mechanism dependent on the presence of water and peptide chain scission. The results of investigations were contradictory, and both increased and decreased solubility of irradiated fibers has been reported (Bowes and Moss 1962; Kuntz and White 1961; Kotek et al. 1964). It has been shown that the effect of ionizing radiation is dose-dependent; while at low doses polymerization prevails, at high doses polypeptide breakdown occurs and leads to increased solubility.

The increase in thermal aggregation of tropocollagen solutions irradiated with high doses (1–30 krad) was reported by Jeleńska and Dancewicz (1969). They found that irradiation induced the formation of aldehyde groups, and this is probably the mechanism responsible for the raised agglutination of molecules. Conformational alterations in tropocollagen irradiated in vitro are associated with the formation of cross-links involving $\varepsilon$-amino groups (Jeleńska and Dancewicz 1972). Other amino acid variations were also reported, such as the formation of dityrosine in irradiated collagen solutions (Majewska and Dancewicz 1976; Boguta and Dancewicz 1983). All of these findings arose after application of high doses of radiation to collagen in vitro.

The mechanism of irradiation-induced effects in vivo is more complex and consists of direct absorption of radiation energy by the collagen molecules and physicochemical changes induced by free radicals as well as indirect effects caused by altered metabolic factors.

Early investigations showed resistance of the connective tissue to irradiation. Verzar (1964) was not able to observe changes in the physical properties of the tail tendon of rats receiving a high dose of 900 rad to the whole body. Local irradiation of the tail with a very high dose was not sufficient to produce changes in elasticity. It is estimated that doses greater than 1500 rad lead to morphological alterations in the connective tissue of the skin. These doses are at least three times higher than the mean lethal dose for whole body irradiation. On the other hand, variations in the metabolism of collagen were reported after irradiation with low doses. An increased hydroxyproline output was found in rats irradiated with 700 rad (Vlădescu 1975). Elevated excretion began after 24 h, and maximal values were observed at 48 and 96 h. This phenomenon was investigated using various doses of radiation. It was shown that urinary hydroxyproline excretion decreased in animals irradiated with 100 or 250 rad and rose in those receiving 500 or 700 rad. It is suggested that the reduced solubility of collagen after irradiation with low doses was responsible for the diminished hydroxyproline output while higher doses led to the breakdown of cross-links in collagen and thus increased the urinary hydroxyproline level. Aggravated changes in urinary hydroxyproline output were observed in adrenalectomized rats receiving low or high doses of radiation. This trend was similar to that found in irradiated, intact animals (Vlădescu et al. 1975). Elevated hydroxyprolinuria was also reported in accidental irradiation in men (Katz and Hasterlik 1955; Gerber et al. 1961; Nesterova et al. 1980). Increased urinary excretion of pyrol-2-carboxylic acid, a specific derivative of hydroxyproline, was noted in rats irradiated with 750 rad (Konno et al. 1984) and in men after accidental exposure (Gerber et al. 1961).

The content and solubility of collagen in tissues of irradiated animals have been investigated, but the results are insufficient for a general description of the in vivo, radiation-induced alteration in collagen metabolism. Vlădescu et al. (1976) studied the influence of radiation on the collagen content in the liver of rats. Low doses (100 or 250 rad) increased the insoluble collagen level while higher doses decreased the content of mature collagen. The effects of long-term low-dose irradiation on the insoluble collagen content in the tissues of dogs were reported by Vinogradova (1979). She found that the insoluble collagen level decreased over 6 years of daily irradiation with a low dose (up to a total dose of 735 rad) in the lungs and tendons. No significant changes were found in the aorta wall and cartilage. In the skin and liver, a decline in the insoluble collagen content was seen only after long-term exposure. Interestingly, in the tendon, short-term exposure (up to 245 rad) produced an increase in the insoluble collagen content while prolonged irradiation led to a decrease. A reduction in the total collagen content in the skin of rats exposed to gamma-radiation (500 rad) was reported by Drożdż et al. (1982). It was accompanied by a decrease in the neutral-salt-soluble and acid-soluble collagen fractions in the skin. An elevation in the insoluble collagen content in the liver was described as well. No changes in the collagen of the lungs, kidney, and heart muscle were observed.

Studies on the effect of ionizing radiation on collagen structure in vitro require very high doses due to its relatively high resistance and have been useful in understanding radiation-induced macromolecular alterations. However, these results cannot be applied to in vivo systems. The high doses resulted in changes which were caused by numerous physicochemical phenomena, and most of them are secondary. Radiation-induced variations were proposed as a model of physiological aging of the connective tissue. Although some similarities have been reported, there is no rational indication to assume that the mechanism of these two processes is related. Moreover, detailed studies uncovered numerous differences in their molecular mechanisms, and the hypothesis of the similarity of aging and postirradiational changes has been rejected.

# 24 Pharmacological Control of Collagen Metabolism

## 24.1 Introduction

The major part of this book reviews alterations in the structure, heterogeneity, and metabolism of collagen under physiological and pathological conditions. Therefore, the logical consequence are efforts to modify collagen metabolism, i.e., to correct with pharmacotherapy pathological phenomena associated with or caused by collagen abnormalities.

There are numerous observations on the effect on collagen metabolism of various drugs or other exogenous substances with possible pharmacological applications as well as toxins. For the purpose of review, these compounds have been classified into groups according to their mechanism of action. The first one contains those whose basic mechanism is the relatively specific inhibition of the synthesis or maturation of collagen. These substances are potential specific antifibrotic agents. Members of the second group affect collagen degradation. These compounds are diverse. Some promote collagen breakdown (more or less specific stimulation of synthesis, secretion, or activation of collagenase) and in this way exert antifibrotic activity. The opposite phenomenon, an inhibition of collagenase, is also included. These substances are of possible use in the treatment of corneal ulcerations caused by overproduction of collagenase or other disorders associated with an elevated degradation of collagen (e.g., rheumatoid arthritis). The third class of the present classification involves a variety of drugs or chemical agents which affect collagen metabolism indirectly. These substances have an influence upon processes associated with inflammation and thus are known as nonspecific antifibrotic agents. Several drugs and environmental toxins are included here which have been reported to influence collagen metabolism but whose mechanism of action remains unclear and is believed to be indirect. Among them, some antibiotics, cytostatics, and heavy metals are described. Also included are drugs which induce fibrosis and enhanced collagen accumulation.

Natural hormones and their synthetic analogues as well as drugs affecting hormonal secretion are known to affect collagen metabolism. They are described in Chap. 7. Some drugs able to produce the symptoms and signs of so-called drug-induced lupus-like syndrome have been found to affect connective tissue metabolism, including collagen. These processes are reviewed in Chap. 10.4.

## 24.2 Direct Inhibitors of Collagen Synthesis

Various stages of collagen biosynthesis have been used as targets of the pharmacological antifibrotic action. The most advanced studies focused on inhibition of the hydroxylation of proline and lysine residues, transportation of collagen outside the cell, and collagen maturation. Some efforts to inhibit conversion of procollagen into collagen have been carried out. The transcriptional and translational stages would seem to be very specific, but only a few studies suggest any opportunity here for pharmacological control. Glucocorticoids (see Chap. 7) are known to affect these stages. Retinoids (see below) or cytokines are believed to inhibit early steps of collagen synthesis. It is expected that in the future more specific antifibrotic therapy will be achieved using transcription and translation as its targets.

Certain substances tested for experimental control of fibrosis were found to exert their action on several stages of synthesis and/or simultaneously on the synthesis and breakdown of collagen. This effect can be additive, or the influence on one stage of collagen metabolism is reduced by the opposite action of the same substance on another stage.

### 24.2.1 Pharmacological Control of Hydroxylation of Proline and Lysine Residues

The importance of hydroxylation of some proline and lysine residues in the stabilization of collagen molecules as well as in the further processing of procollagen (e.g., incorporation of carbohydrates) is reviewed in earlier parts of this book. On the basis of these findings, hydroxylation is considered as a critical stage in the control of collagen synthesis. Decreased synthesis of collagen can be achieved by influencing the peptide substrate, cosubstrates, or cofactors of hydroxylation. The effect on the peptide substrate is based on the incorporation of proline or lysine analogues into the chains of procollagen that cannot be hydroxylated under normal conditions. This can be the result of chemical alterations which exclude hydroxylation or structural alterations which lead to a lack of susceptibility to the hydroxylases. The substances used for these purposes have to be able to penetrate the synthesizing cells and to be indistinguishable from proline and lysine by the metabolic systems involved in protein synthesis.

#### 24.2.1.1 Structural Analogues of Proline

The incorporation of proline and lysine analogues can dramatically alter the structure and metabolism of collagen. Their mechanism of antifibrotic activity includes blocking of triple-helix formation or destabilization of the helical conformation. Nonhelical polypeptides are more susceptible to intracellular degradation. Incorporation of lysine analogues also prevents glycosylation and

**Table 24.1.** Proline analogues that interfer with collagen synthesis

| |
|---|
| L-acetidine-2-carboxylic acid |
| 3,4-dehydroxy-L-proline |
| *cis*-4-fluoro-L-proline |
| *trans*-4-*cis*-hydroxy-L-proline |
| *trans*-4-fluoro-L-proline |
| thiazolinine-4-carboxylic acid |
| 4,4-difluoro-L-proline |
| 4,4-dimethyl-D,L-proline |
| *cis*-4-hydroxy-L-proline |
| *cis*-4-bromo-L-proline |

impairs cross-link formation (Uitto and Prockop 1974; Jimenez and Rosenbloom 1974; Tan et al. 1983; Eldridge et al. 1988).

Structural analogues of proline used as antifibrotic agents are listed in Table 24.1. Their ability to reduce the collagen content was investigated under various experimental conditions. A decrease in the breaking strength of peritendinous adhesions, diminished carbon-tetrachloride-induced fibrosis of the liver, and reduced collagen content in the estradiol-stimulated uterus were found in rats treated with D,L-3,4-dehydroxyproline (Bora et al. 1972; Lane et al. 1972; Salvador et al. 1976; Kerwar et al. 1976). *Trans*-4-*cis*-hydroxyproline diminished the mechanical properties of colonostomy wounds and prevented collagen accumulation in oxygen-damaged lungs (Daly et al. 1973; Riley et al. 1980). L-Azetidine-2-carboxylic acid was reported to be a successful inhibitor of experimental hepatic fibrosis (Rojkind 1973). On the other hand, a high toxicity of the proline analogues was seen associated with an inadequate influence on collagen synthesis when tested in animal models of fibrosis or when given to healthy animals. 3,4-Dehydroxyproline was found to be 15 times more toxic than L-proline (Madden et al. 1973). The toxicity included damage to hepatocytes and fibroblasts. Moreover, some of the analogues were reported to be ineffective when injected into animals.

Several investigators were not able to demonstrate any decrease in collagen content after treatment with proline analogues (Chvapil et al. 1974a; Fuller 1981). An increased synthesis of both collagen and noncollagenous proteins in skin wounds of rats treated with *N*-acetyl-*cis*-4-hydroxyproline was shown (Cohen and Diegelman 1978).

A low efficiency of proline analogues depends on the size of the free proline pool in the body. The incorporation of proline and its analogues is assumed to be a random process, and the exogenously added analogues are diluted by free proline. The amount of proline is only partially regulated by its intake in food, because it is synthesized endogenously by various metabolic pathways. Thus, elimination of proline from the diet has no influence upon the effectiveness of the antifibrotic activity of proline analogues (Madden et al. 1973). It is also possible that endogenous synthesis of proline bears some relation to collagen metabolism, and inhibition of collagen synthesis leads to an elevated production of proline. The increased protein catabolism which accompanies several pathological states may aggravate the dilution effect of proline and significantly reduce the inhibitory effectiveness of the structural proline analogues (Chvapil and Ryan 1973). The assumption that proline and its analogues are incorporated randomly may not be true. Some studies showed that *cis*-hydroxyproline was incorporated ten times more slowly than L-proline (Rosenbloom 1971), and 3,4-dehydroxyproline was incorporated about five times more slowly than the natural imino acid (Rosenbloom and Prockop 1970). The reduced rate may be caused by impaired transportation into the cells or slower direct incorporation into newly synthesized proteins. According to Lane et al. (1971), azetidine-2-carboxylic acid when administered in vivo was incorporated only as 2% of all proline into collagen.

The toxicity mechanisms of the proline analogues remain obscure. It can be hypothesized that the substances interfer with the incorporation of proline into proteins other than collagen including those which are very important for the biological activity of the cells and organs. An inhibitory effect on some enzymes has also been postulated (Chvapil 1982). It is of interest that this effect may be responsible for the protective activity of some structural analogues of proline in the carbon-tetrachloride-induced hepatic fibrosis. According to Chvapil (1982), it is possible that low doses of the analogues are insufficient to inhibit collagen synthesis directly but are able to affect the mixed oxidases of the liver. The latter inhibition decreases the harmful influence of carbon tetrachloride on the liver.

#### 24.2.1.2 Cosubstrates and Cofactors of Hydroxylation

Cosubtrates and cofactors of hydroxylation have been used as the target for the control of collagen synthesis. If was found that removal of the ferrous ions in in vitro systems was an effective measure to inhibit collagen synthesis. In cell or organ cultures $\alpha,\alpha'$-dipyridyl or *o*-phenantroline were successfully used (Chvapil 1974). Animal studies gave more controveral results. Reduced accumulation of collagen was found in animals with liver fibrosis treated with *o*-phenantroline (Chvapil et al. 1967a, b, 1968, 1974b; Chvapil and Hurych 1968; Chvapil and Brada 1971; Brada et al. 1972), skin granuloma (Whitson and Peacook 1969), and healing of colonostomy (Bora et al. 1972; Aronson and Rogerson 1972). The decrease in collagen deposition was not associated with reduced activity of prolyl hydroxylase when the enzyme was measured directly. Enhancement of the enzyme activity has been shown in some studies and in cell culture (Chvapil et al. 1974b). These results point out that chelating properties were not responsible for the diminished collagen content in tissues. Administration of the nonchelating analogue of *o*-phenantroline, *m*-phenantroline, also reduced liver fibrosis. The antifibrotic mechanism has not been elucidated. Interference with hepatic mixed function oxidases or inhibition of enzymes that are unrelated to collagen metabolism are hypothetical explanations of the observed results.

Some bivalent cations have been shown to inhibit prolyl 4-hydroxylase (and, to a lesser extent, other hydroxylases) with respect to ferrous ions. Zinc was reported to be the most effective. Administration of zinc led to a reduction of collagen content in the tissues of growing animals as well as to a decrease in collagen accumulation in carbon-tetrachloride-injured liver (Anttinen et al. 1984, 1985).

Structural analogues of 2-oxyglutarate have been used to inhibit the activity of prolyl hydroxylase. The most effective inhibitors are pyridine-2,5-dicarboxylate, pyridine-2,4-dicarboxylate, thiodiglycolate, 3,4-dihydroxybenzoate, and 2,3-dihydroxybenzoate. The last two are inhibitors with respect to ascorbic acid as well. Their structure as compared with 2-oxoglutarate is shown in Fig. 24.1.

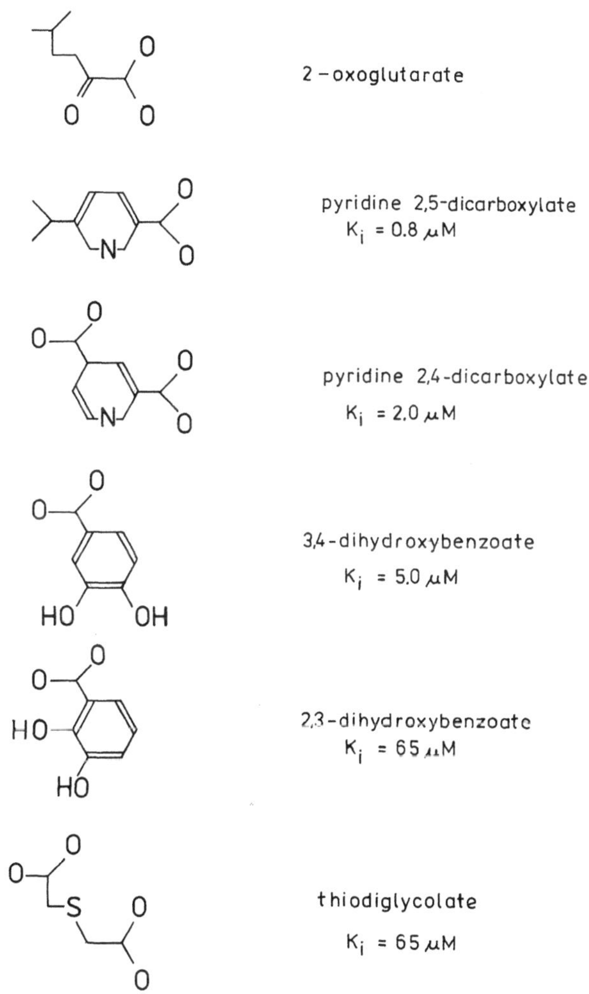


**Fig. 24.1.** Structure of 2-oxoglutarate and compounds inhibiting the prolyl 4-hydroxylase reaction competitively with respect to 2-oxoglutarate. (From Kivirikko and Savolainen 1988 with permission.)

The concept of the inhibitory mechanism has been elaborated on the basis of structural studies of the active center of prolyl 4-hydroxylase (see Chap. 3). Experiments in vivo on the role of inhibitor binding at the 2-oxoglutarate site of the enzyme have not been published except for a single report by Sasaki et al. (1987b). The described compounds have a low permeability and cannot penetrate the cell. It is possible that their structural characteristics will be a model for the synthesis of new inhibitors applicable in the pharmacological control of collagen biosynthesis. In the study of Sasaki et al. (1987b) a hydrophobic modification, an ethyl ester, 3,4-dihydroxybenzoic acid, was used. This compound did not affect the viability, proliferation, or plating efficiency of the cultured fibroblasts. A significant decrease in synthesis of types I and III collagens was found. The influence was specific, and the inhibitor had no effect on the production of noncollagenous proteins. Keloid fibroblasts were also shown to respond to the ethyl ester of 3,4-dihydroxybenzoic acid.

Coumaric acid (2-oxo-1,2*H*-pyran-5-carboxylic acid) was found to be a potent inhibitor of prolyl 4-hydroxylase and lysyl hydroxylase. A high concentration of 2-oxoglutarate can prevent the inactivation of the enzyme (Günzler et al. 1987).

Anthraquinones with at least two hydroxy groups placed ortho to each other are potent inhibitors of prolyl 4-hydroxylase. For example, 2,7,8-trihydroxyanthraquinone was found to inhibit the enzyme competitively with respect to 2-oxoglutarate but was a noncompetitive inhibitor with regard to ascorbate (Cunliffe and Franklin 1986; Franklin and Hitchen 1989).

The contribution of oxygen in the hydroxylation of proline and lysine provides another opportunity to inhibit the hydroxylation. A low oxygen or increased nitrogen content were used in in vitro experiments to reduce the rate of collagen synthesis. The effective decrease in oxygen partial pressure is far lower than that which leads to tissue damage, and obviously this method cannot be used in animal models of fibrosis.

The participation of oxygen in the form of the superoxide anion ($O_2^-$) has suggested the possible role of superoxide scavengers in the inhibition of hydroxylation. A classic superoxide scavenger, nitroblue tetrazolium, inhibits the hydroxylation of proline and lysine (Tuderman et al. 1977a). The other compound with the ability to remove superoxide anion, (+)catechin, has been reported to inhibit purified prolyl 4-hydroxylase. The inhibitory activity, $ID_{50}$, was found to be 100 $\mu M$. (+)Catechin inhibited the enzyme when applied to cultures of human skin fibroblasts, but there was no inhibition of hydroxylation when given to rats. This can be explained by an inability to penetrate the cells or rapid biotransformation once in the animal body (Lonati-Galligani et al. 1979). Other mechanisms of the effect of (+)catechin on connective tissue metabolism have also been reported (Niebes 1977).

#### 24.2.1.3 Fibrostatins

Fibrostatins are a recently discovered group of inhibitors of prolyl hydroxylase of microbial origin. In 1981 Okazaki et al. reported for the first time the isolation of an inhibitor, the anthraquinone glycoside P-1894B. The inhibitor

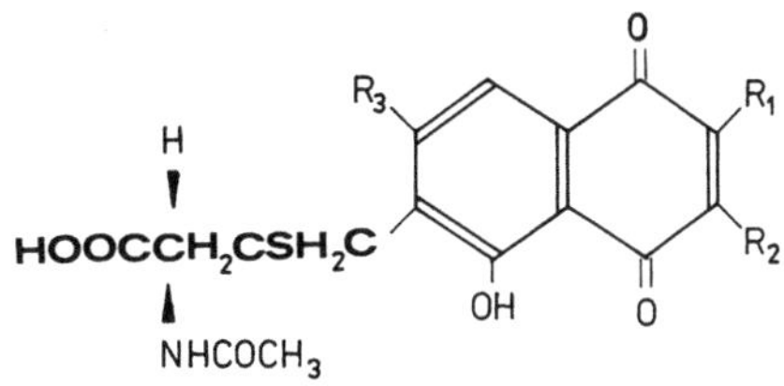


| | $R_1$ | $R_2$ | $R_3$ |
|---|---|---|---|
| Fibrostatin A | $-H$ | $-CH_3$ | $-OCH_3$ |
| Fibrostatin B | $-OCH_3$ | $-CH_3$ | $-OCH_3$ |
| Fibrostatin C | $-OCH_3$ | $-H$ | $-OCH_3$ |
| Fibrostatin D | $-OCH_3$ | $-CH_3$ | $-OH$ |
| Fibrostatin E | $-H$ | $-CH_2OH$ | $-OCH_3$ |
| Fibrostatin F | $-OCH_3$ | $-CH_2OH$ | $-OCH_3$ |

**Fig. 24.2.** Structure of fibrostatins. (From Ohta et al. 1987b with permission.)

known as vineomycin A1, an antitumor antibiotic, was obtained from *Streptomyces albogriseolus* subsp. no. 1894 (Ishimaru et al. 1982). It was found to reduce the total hydroxyproline content when applied topically to a cotton-pellet-induced granuloma and to skin lesions in rats. These results suggested that P-1894B could be used for topical treatment of hypertrophic scar tissue or keloid (Kubota et al. 1985). The clinical studies have not been reported. This discovery inspired investigations in *Streptomyces* as a source of inhibitors. Potent inhibitors of prolyl hydroxylase were found to be produced by *S. catenulae* subsp. *griseospora* no. 23924 and were named fibrostatins (Ishimaru et al. 1987). Six compounds were identified and their inhibitory activity determined. The inhibitory activity ($ID_{50}$ value) of fibrostatins A, B, C, D, E, and F against prolyl hydroxylase in vitro was 23, 39, 29, 180, 10, and 14 $\mu M$, respectively (Ohta et al. 1987b). The structure of fibrostatins is shown in Fig. 24.2.

Fibrostatin C was found to be a major component of the culture filtrate. It inhibited the activity of purified chick embryo prolyl hydroxylase by about 50% at a concentration of $2.9 \times 10^{-5}$ $M$. It was found that ferrous ions had little effect on the inhibition, which indicates that fibrostatin C does not chelate ferrous ions. An excess of ascorbate did not completely reverse the enzyme activity. This showed that fibrostatin C did not inhibit the hydroxylase acting as an antioxidant or an antagonist towards ascorbate. The possibility that fibrostatin C is a superoxide scavenger has also been excluded. The mechanism of the inhibition was of a mixed type with respect to $(Pro\text{-}Pro\text{-}Gly)_5$ and probably was related to the competition for the substrate binding center of the enzyme.

Studies in vivo of the antifibrotic activity of fibrostatin C were carried out in the uteri of immature rats treated with estradiol-17$\beta$. The inhibition of collagen synthesis was found when the substance was administered intraperitoneally (1 mg/kg daily) or orally (100 mg/kg daily) (Ishimaru et al. 1988). No further studies of antifibrotic activity of fibrostatins in different animal models were reported. The toxicity of these compounds has not been studied either.

### 24.2.2 Pharmacological Control of Collagen Secretion

Secretion of collagen to the extracellular space is regulated by the level of posttranslational and conformational changes in the molecule, including hydroxylation, glycosylation, and formation of the triple-helical structure. The process of transportation of collagen outside the cell is related to the activity of microtubules. Interference with microtubules impairs or inhibits collagen secretion.

The best explored drug which affects microtubules is colchicine. An alkaloid of the meadow saffron *Colchicum*, it has been known for centuries as a remedy for acute gout. It has antimitotic, antiinflammatory properties and increases urinary excretion of uric acid. The drug is very toxic. Its toxicity includes harmful effects on capillaries, bone marrow, liver, and kidneys.

Colchicine has several points of action on collagen metabolism. The most important is depolymerization of the microtubules of fibroblasts and probably other cells. In this way the drug inhibits collagen secretion. The presence of vacuoles containing collagen fibrils has been found in cells treated with colchicine (Mansour et al. 1988; Ishizeki et al. 1989). The other antifibrotic mechanism affects the synthesis of collagenase. This process is discussed in the next section. Some investigators reported a decreased synthesis of collagen in the liver tissue (Rojkind and Kershenobich 1975). This effect was nonselective as the synthesis of both collagenous and noncollagenous proteins was impaired. The possible explanation of its direct effect is the toxic influence on protein synthesis. Moreover, in granuloma tissue in vivo the enhanced synthesis of collagen and noncollagenous proteins after colchicine treatment was reported by Chvapil et al. (1980). Experimental studies showed that colchicine diminished collagen synthesis and deposition in the liver of rats treated simultaneously with carbon tetrachloride. The enhanced decrease in collagen content in the fibrotic liver was reported in rats with carbon-tetrachloride-induced hepatic fibrosis treated with colchicine 30 days after discontinuation of carbon tetrachloride injections (Rojkind and Kershenobich 1975).

Clinical studies have not confirmed the practical value of colchicine in the treatment of fibrotic processes. There was no improvement after colchicine treatment in patients with progressive sclerosis (scleroderma) or sarcoid arthritis (Harris et al. 1975; Alarcón-Segovia et al. 1974; Harris and Millis 1971; Guttadauria et al. 1977). Some improvement was noted in patients with cirrhosis (Alarcón-Segovia et al. 1974). In general, it is accepted that doses of the drug which may be effective in pharmacological inhibition of collagen secretion are too high to be tolerated in humans. Thus, the clinical application of colchicine as an antifibrotic drug cannot be recommended.

Several other substances are known or suggested to interfer with the microtubular structures of the cell. Local anesthetic agents have been reported to inhibit collagen secretion in fibroblast cultures (Eichhorn and Peterkofsky 1979). The effect of these drugs on microtubules and microfilaments has been shown in many studies. On the other hand, various mechanisms are possible to produce impaired synthesis and secretion of collagen. Local anesthetic agents are known to inhibit prolyl hydroxylase activity in granuloma tissue (Chvapil et al. 1979a). The drugs affect the cell membranes, produce displacement of calcium ions within the cells, and disturb the sodium conductance between the cell and its environment. The nonspecific decrease of protein synthesis produced by bupivacaine was found in skeletal muscle cells (Johnson and Jones 1978). There is no clinical indication for the use of local anesthetic agents as antifibrotic drugs.

A number of other substances affect collagen. Certain ones have been investigated in cell or tissue cultures in vitro and were shown to inhibit collagen secretion. It was found that cytochalasin B and vinblastine decreased collagen secretion. These substances are well documented agents producing depolymerization of the microtubules of the cell. Tunicamycin and monensin were reported to impair collagen production in their interference with protein secretion (Housley et al. 1980; Chvapil 1982). Other mechanisms, however, have been postulated, including the inhibition of glycosylation of collagen molecules.

### 24.2.3 Pharmacological Control of the Conversion of Procollagen to Collagen

As described earlier (Chap. 3), conversion of procollagen to collagen is associated with removal of the propeptides from the amino- and carboxyterminal ends of the molecule. The removal requires separate proteases. The amino acid sequence around the sites of cleavage has been identified. Several synthetic peptides with a sequence similar to that occurring around the sites have been tested. These peptides are able to inhibit conversion of procollagen to collagen but are active only at high concentrations, and their activity has been reported only in in vitro cell cultures. The inhibitory action of some polyamines on the cleavage of the carboxyterminal propeptide was described, but its mechanism remains unclear. It has been suggested that the inhibitors caused direct inhibition of the specific proteases.

At present, the inhibition of the conversion of procollagen to collagen has not been used for animal experiments. It is possible that further chemical modifications of these compounds will provide new effective measures for the control of fibrosis.

### 24.2.4 Pharmacological Control of Extracellular Processing of Collagen

Extracellular stages of the formation of collagenous structures, including cross-linking, form the main target of the pharmacological control of collagen maturation. Two classes of substances have been distinguished among the agents affecting cross-linking of collagen. The first includes lathyrogens, the compounds that produce the symptoms and signs of lathyrism. Lathyrogenic activity is associated with an irreversible inhibition of lysyl oxidase. The second class is represented by penicillamine. The action of penicillamine is more complex and includes direct interference with compounds involved in cross-link formation as well as the reversible inhibition of lysyl oxidase by the induction of copper deficiency. Other mechanisms have also been suggested. Copper deficiency leads to impaired collagen maturation, although it is useless as a practical method for the control of collagen metabolism.

#### 24.2.4.1 Lathyrism

The term "lathyrism" originated from the Latin name of the pea, *Lathyrus*. The harmful effects of *Lathyrus* species have been known since antiquity (Selye 1957). The extensive intake of pea flour and cereals, usually occurring in periods of famine, leads to so-called neurolathyrism or human lathyrism. The disorder is characterized by spastic paraplegia, increased pathologic reflexes, and other signs of nervous system involvement (Weaver 1967). Investigations on the experimental induction of lathyrism in animals led to the discovery of a new form of this disease, osteolathyrism. This was produced in animals by feeding with sweet pea and primarily affected the skin, bones, teeth, blood vessels, and tendons. Osteolathyrism has not been described in humans.

Chemical substances responsible for lathyrism have been identified. The most active osteolathyrogen is $\beta$-aminopropionitrile. Neurolathyrogens have been suggested to be its metabolic products. The hypothetic pathway of synthesis of $\beta$-cyano-L-alanine and $\alpha,\gamma$-diaminobutyric acid, the main neurolathyrogens, from osteolathyrogen was reported by Liener (1967). Several compounds other than $\beta$-aminopropionitrile exhibit the ability to induce osteolathyrism. The majority of them have been obtained via organic synthesis, and $\beta$-aminopropionitrile is almost the only naturally occurring osteolathyrogen. Compounds able to induce osteolathyrism are listed in Table 24.2.

New potent lathyrogens, 2-bromoethylamine and 2-chloroethylamine, have been described (Bailey and Light 1985). These substances are more potent than $\beta$-aminopropionitrile. Experimental studies on their toxicity have not been published.

Lathyrogens responsible for the induction of osteolathyrism form an irreversible complex with lysyl oxidase. The enzyme loses its activity. Thus, the formation of cross-links in collagen is impaired. Decreased maturation of collagen increases its solubility (Levene 1961; Tanzer and Gross 1964; Levene and Gross 1959; Fry et al. 1962; Mikkonen et al. 1960; Gross and Levene 1959; Wirtschafter and Bentley 1962; Aleo 1969; Pasquali-Ronchetti et al. 1981) and is associated with impaired mechanical properties of tissues, especially those with an extensive collagenous framework. Several studies showed a reduction in the tensile strength of the skin, aorta, and healing wounds, as well as dislocation of joints, deformities of the spinal column, a curled tail, and teeth de-

**Table 24.2.** Osteolathyrogens

| Chemical group | Compounds |
|---|---|
| Organic nitriles | β-Aminopropionitrile |
| | β-(γ-LGlutamyl)-aminopropionitrile |
| | Aminoacetonitrile |
| | Methyleneaminonitrile |
| | β,β'-Iminopropionitrile |
| | β-Hydrazinipropionitrile |
| Ureides | Semicarbazide |
| | Acetone semicarbazone |
| Hydrazides | Isonicotinic acid hydrazide |
| | Nicotinic acid hydrazide |
| | Cyanoacetic acid hydrazide |
| | Benzhydrazide |
| | γ-L-Glutamyl hydrazide |
| | Glycine hydrazide |
| | Carbohydrazide |
| | Thiocarbohydrazide |
| | Glutamic acid hydrazide |
| | p-Nitrobenzene acid hydrazide |
| Hydrazines | Hydrazine hydrate |
| | Dimethyl hydrazine |
| | Phenylhydrazine |
| Miscellaneous | Cyanamide |
| | 2-Cyanopropylamine |
| | Thiosemicarbazide |

Modified from Barrow et al. (1974) with permission

formities (Tanzer 1965). Levene and Gross (1959) found that the tensile strength of the skin was inversely related to the dose of the lathyrogen.

Lathyrogens applied in doses required for the inhibition of lysyl oxidase do not affect the rate of collagen synthesis (Aleo et al. 1974). Depressed synthesis was, however, reported in animals treated with higher doses of the agent. There are some suggestions that lathyrogens increase the rate of collagen degradation. This effect is probably secondary to the impaired maturation as collagen with a low level of cross-linking is more susceptible to the action of specific proteolytic enzymes. The interaction between defective collagen fibers and proteoglycans and structural glycoproteins is impaired. This mechanism is responsible for an increased urinary hydroxyproline excretion in animals with lathyrism. Hydroxyproline output has been found to correlate with the increase in soluble collagen content in tissues (Jasin and Ziff 1962; Orbison et al. 1965; Hartman 1966).

The osteolathyrogen, β-aminopropionitrile, has been used under various experimental conditions as an antifibrotic agent. The substance was effective in preventing contraction of the circular scar in the burned esophagus (Davis et al. 1972) and inpreserving the gliding function of an injured tendon (Peacock and Madden 1969).

In clinical trials lathyrogens were applied to patients with systemic scleroderma, peritendineal adhesions, and urethral strictures. The systemic administration of $\beta$-aminopropionitrile was associated with severe toxicity. A decrease in scar formation was also reported, although the toxicity and hypersensitive reactions common among these patients led to the discontinuation of treatment (Keiser and Sjoerdsma 1967; Peacock 1973; Peacock and Madden 1969).

As a result of systemic toxicity, lathyrogens have been used in the topical treatment of collagen overproduction (Chvapil 1988). The most frequent indications are peritendinous and perineural adhesions, joint stiffness, and burnt scar contractions. Animal experiments showed that the drugs penetrated the skin when applied topically and were effective. Only preliminary human studies have been reported, and further ones are needed to establish the clinical effectiveness and safety of topical treatment with lathyrogens.

#### 24.2.4.2 Penicillamine

D-Penicillamine is a metabolite of penicillin. The compound was initially introduced for the treatment of Wilson's disease due to its ability to chelate copper. This property is associated with the capacity to trap carbonyl compounds. It became apparent that penicillamine is an effective drug in the treatment of rheumatoid arthritis, although the mechanism of its action has not been elucidated.

The administration of penicillamine causes an accumulation of soluble collagen in various tissues (Nimni 1965, 1968; Nimni and Bavetta 1965; Harris and Sjoerdsma 1966b; Vogel 1975; Junker et al. 1981; Junker and Lorenzen 1983). An increase in collagen solubility is the result of impaired cross-link formation and depolymerization of incompletely cross-linked insoluble collagen.

The mechanisms of action of lathyrogens and penicillamine are different. Collagen extracted from tissues of animals treated with penicillamine contains higher than normal amounts of aldehydes and is able to form stable fibers in vitro. Penicillamine binds specifically to free aldehydes on collagen (Deshmukh and Nimni 1969). The collagen-penicillamine complex is in equillibrium with its constituents and can be completely dissociated by exhaustive dialysis. Nimni et al. (1972) found that low doses of penicillamine increased the aldehyde content in collagen. A maximal increase was reported with a dose of 200 mg/kg of body weight per day. Further increases in the dose led to a decrease in aldehyde content similar to that observed in animals treated with lathyrogens. These experiments showed that low doses of penicillamine blocked aldehydes, whereas at high doses the activity of lysyl oxidase declined. The mechanism of the blocking of aldehydes is shown in Fig. 24.3. The inhibitory effect on lysyl oxidase is probably caused by chelating of copper. Partial inhibition of the enzyme was found in vitro with $10^{-4}\,M$ of penicillamine, while $10^{-2}\,M$ caused complete inhibition.

In addition to preventing the cross-linking of newly synthesized collagen, penicillamine is able to enhance the solubility of collagen by labilizing Schiff-base-type cross-links (Nimni et al. 1967; Deshmukh and Nimni 1968). Other mechanisms have also been described. Penicillamine applied in high doses was

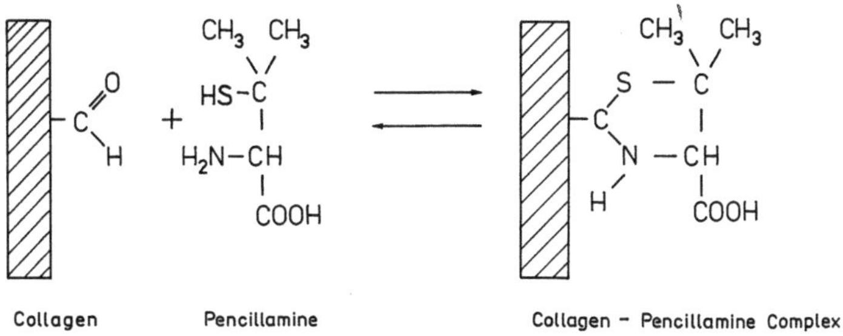


**Fig. 24.3.** A mechanism of the modes of action of penicillamine on collagen cross-linking. (From Nimni 1974 with permission.)

found in vitro to inhibit the hydroxylation of proline in procollagen (Chvapil et al. 1967b) and to diminish the glycosylation of collagen (Blumenkrantz and Asboe-Hansen 1973).

Penicillamine-induced alterations in collagen degradation remain obscure. Anderson (1969, 1970) noted an activating effect on collagenolytic enzymes. On the contrary, Francois et al. (1973) showed an inhibitory effect. There was no change in the urinary hydroxyproline output after penicillamine treatment.

Using $^{14}C$-labelled penicillamine, the binding of the drug to various rat tissues was investigated. It was found that a high level of radioactivity was observed in the skin, lungs, kidney, and liver, whereas in cartilage, bone, and muscle there was only a low level. Penicillamine was not metabolized in the body but was bound to albumin to a great extent (Planas-Bohne 1973). There was no correlation between the binding of penicillamine and collagen content. On the other hand, organs with a high turnover of collagen accumulate significantly more penicillamine that those with a slow turnover. Penicillamine was found to bind more rapidly to a soluble fraction of collagen than to an insoluble one (Grasedyck and Lindner 1975).

Penicillamine has been widely used in clinical practice either as an antirheumatic or antifibrotic agent. Its beneficial and adverse effects are related to its varied pharmacological activities, including dissociation of macroglobulins, stabilization of lysosomes, inactivation of prostaglandin synthetase, and some antitumor activity.

The antifibrotic activity of the drug has been tested under various experimental and clinical conditions. The increased solubility of cutaneous collagen was shown in patients with scleroderma treated with penicillamine (Harris and Sjoerdsma 1966b). A decrease in collagen content in the liver of patients with chronic hepatic disorders receiving penicillamine has been reported. It is, however, difficult to elucidate the mechanism under these conditions. The influence of penicillamine on immune functions may cause an indirect reduction of fibrosis (Wiontzek 1970; Wildhirt 1974; Sternlieb 1975; Resnick et al. 1975). Toxic effects of penicillamine that appeared after higher doses of the drug limit its clinical usage as an antifibrotic agent.

#### 24.2.4.3 Inhibitors of Lysyl Oxidase with Respect to Pyridoxal

Strong evidence exists that pyridoxal phosphate is a cofactor of lysyl oxidase. Pyridoxal is probably tightly bound to the enzyme protein (Fowler et al. 1970). Inhibition of lysyl oxidase was found in chicks on a pyridoxal-deficient diet (Murray et al. 1978). Several reagents which react with carboxyl groups (e. g., phenylhydrazine, hydroxylamine) inactivated the enzyme (Harris et al. 1974). Isonicotinyl hydrazine, known as isoniazid, an tuberculostatic agent, has been shown to antagonise pyridoxal. Isoniazid was found to inhibit lysyl oxidase (Arem and Misiorowski 1976) and increase collagen solubility in chick embryos (Carrington et al. 1984). This inhibition of collagen maturation can be reversed by pyridoxal. Further studies on the antifibrotic activity of isoniazid are needed.

### 24.2.5 Antifibroblast Serum

The concept of antifibroblast serum was created as a parallel to the idea of antilymphocyte serum. Attempts to control the fibroblast volume by specific antibodies directed against the fibroblast surface antigens failed because of the great cross-reactivity and low specificity of the serum, although a decrease in collagen content was observed in the granuloma of mice treated with the serum. The wide distribution of fibronectin was one of the reasons for the low specificity of this method (Steinbronn et al. 1974).

## 24.3 Pharmacological Control of Collagen Degradation

Modulation of collagen degradation can be used to obtain two contradictory aims. Increased breakdown of collagen is an effective measure to decrease net collagen deposition and can be used as an antifibrotic agent. On the other hand, some pathological conditions are associated with an elevated amount of degradation of collagen. Thus, pharmacological treatment under these conditions is directed against collagenase. The inhibitors of collagenase are a new group of recently investigated compounds.

### 24.3.1 Induction of Collagenolysis

Degradation of collagen is a complex sequence of events as described in Chap. 4. The rate of breakdown depends on the nature of the substrate as well as on the synthesis, secretion, and activation of procollagenase. The activity of the enzyme results from the equilibrium between activators and inhibitors. The role of other enzymes involved in the degradation of collagen remains unknown. Degradation takes place extracellularly, but large fragments of collagen can be phagocytized and digested within the cells. In light of the complex regulation of collagen degradation, it is very difficult to elucidate in vitro the mechanism of action of the agents which are suggested to increase collagen breakdown.

Colchicine was reported to stimulate de novo synthesis of collagenase in cultured synovial membrane and granuloma tissue (Harris and Krane 1971; Chvapil et al. 1980). It is thought to inhibit collagenase production by macrophages. The inhibition was related to the secretion of the enzyme and was not associated with blocking of the collagenase mediator, prostaglandin $E_2$ (Wahl and Winter 1984). Opposite results were shown by Gordon and Werb (1976). They described an enhanced secretion of collagenase by colchicine-treated macrophages. A similar influence of the drug was noted in cultured rheumatoid synovial cells (Harris and Krane 1971). The differences were caused by altered doses as well as by varying experimental protocols; colchicine inhibited collagenase production only when it was added at the onset of macrophage stimulation.

Stančikova et al. (1987) found an increased activity of cathepsins B and D in the liver of humans with cirrhosis treated with colchicine. The colchicine-induced inhibition of the release of lysosomal proteases from polymorphonuclear leukocytes has also been reported (Mikuliková and Trnavsky 1980).

There are contradictory reports on the effect of the chelating agent EDTA on collagenolysis. Administration of EDTA produces profound alterations in several ion concentrations. It was found that rats treated with EDTA excreted elevated amounts of hydroxyproline (Aronson and Rogerson 1972, cited in Chvapil 1982). It is suggested that EDTA produces sequestration of zinc ions and leads to labilization of lysosomes. The increased degradation of collagen has been postulated under conditions of treatment with EDTA on the basis of elevated urinary excretion of hydroxyproline. This is contradictory to studies on collagenase preparations. Mammalian collagenases are zinc enzymes and require calcium ions for their activity. EDTA as well as other chelating agents (e.g., 1,10-phenantrolene) are able to inhibit collagenase. The inhibition was found at concentrations in the millimolar range (Seltzer et al. 1977; Woolley et al. 1978a). Inhibitory studies were done in vitro. In vivo, it is very difficult to uncover the point of action of EDTA. Many stages of synthesis and degradation of collagen are modulated by the presence of various ions, and chelation of these ions produces multiple disturbances. It has to be mentioned that Tobin et al. (1974) was not able to detect any changes in the mechanical properties of skin wounds in rats treated chronically with EDTA despite the marked increase in the excretion of hydroxyproline. Inhibition of tumor collagenase by the chelating agent razoxane has been reported (Karakiulakis et al. 1989).

### 24.3.2 Inhibitors of Collagenase Secretion

Retinoids have been proposed as efficient inhibitors of collagenase production and as drugs for the treatment of rheumatoid arthritis. This suggestion is based on several experimental observations. A significant reduction of joint inflammation was reported in rats with adjuvant arthritis receiving orally 13-*cis*-retinoid (Brinckerhoff et al. 1983). The mechanism that reduced inflammation acid has been attributed to the diminished collagenase production by monocytes. A similar decrease was found in cultures of synovial cells treated with all-*trans*-retinoid acid, 13-*cis*-retinoid acid, and 4-hydroxyphenyl-retinamide (Brinckerhoff et al. 1980, 1982). On the other hand, in the rat models of arthritis induced by collagen, some retinoids augmented the inflammation (Trentham and Brinckerhoff 1982). Ohta et al. (1987a) investigated the influence of retinoids on collagenase production in human mononuclear cells. Incubation of mononuclear cells with all-*trans*-retinoid acid in the concentration range $10^{-7}-10^{-5}\,M$ resulted in a dose-dependent inhibition of collagenase production. All-*trans*-retinal was also noted to be a potent inhibitor. Less inhibition was produced by 13-*cis*-retinoid acid and trimethyl-methoxyphenyl retinoid acid ethyl ester (RO-10-9359). Inhibition was not caused by impaired total protein synthesis. The decrease in collagenase production was not associated with any visible signs of toxicity to the cells. Only preliminary clinical studies on retinoid therapy for rheumatoid arthritis have been reported (Harris 1984). Retinoids are thought to inhibit early stages of collagen synthesis, transcription, or translation (Forest et al. 1982).

Tunicamycin, an inhibitor of *N*-linked glycosylation, was found to inhibit collagenase production in cell cultures (Chu and Ladda 1987). There is no practical implication of this finding; collagenase is probably a glycoprotein and is synthesized via the dolichol phosphate pathway.

### 24.3.3 Inhibitors of Mammalian Collagenases

Several substances have been shown to inhibit mammalian collagenases. The mechanism is known only partially, and investigations of their structure-activity relationship have been carried out only in recent years.

Inhibitors of mammalian collagenase are either thiol compounds or have collagenlike structures. The former group includes cysteine, dithiothreitol, penicillamine, and naphthalene (Johnson et al. 1987). These compounds are nonspecific inhibitors, active at relatively low concentrations ($10-100\,\mu M$). Clark et al. (1985) described a thiol inhibitor, *N*-[[(5-chloro-2-benzothiazolyl)-thio]-phenyl-acetyl]-L-cysteine, known as WY-45,368. WY-45,368 is a potent inhibitor of human fibroblast collagenase, and its $I_{50}$ for collagenase against types I and II collagens is $10\,\mu M$. Human leukocyte elastase is also inhibited by WY-45,368.

The inhibitors with collagenlike structures follow the basis of the collagenase-cleaved sequence of collagen and the assumption that the zinc ion coordinates with the carbonyl oxygen of the cleaved amide bound. The incorporation of a zinc ligand at the cleavage site in the substrate is suggested to cause inhibition. Detailed studies on this class of inhibitors as well as computer models are reviewed by Johnson et al. (1987).

Actinonin, a pseudopeptide antibiotic isolated from actinomycete culture medium, and its structural analogues were found to inhibit synovial collagenase (Faucher et al. 1987; Lelièvre et al. 1989). Pharmacological studies of these groups of inhibitors have not been reported.

## 24.4 Drugs That Indirectly Affect Collagen Metabolism

A huge number of studies have been published on the influence of various drugs on collagen content and metabolism. The mechanism of their activity has not been documented in the majority of the reports. It is generally accepted that the drugs reviewed in this section do not directly affect the synthesis and degradation of collagen. Most exert their activity by the modulation of such phenomena as inflammation or proliferation of cells involved in the development of fibrosis. The inhibition of the inflammatory response and liberation of several humoral mediators leads to a decrease in collagen production. Similarly, cytostatics decrease cell proliferation and interfer with the activity of collagen-producing cells. The mechanisms of action of many of the drugs and toxins reviewed in this chapter are known only partially.

### 24.4.1 Nonsteroidal Antiinflammatory Drugs

Nonsteroidal antiinflammatory drugs are widely applied for the treatment of various rheumatic disorders. They are used as well as mild analgesic and antipyretic agents. Since the introduction of the first example, acetylsalicylic acid, a number of other compounds with different chemical structures have been described. The mechanism of their action is not fully understood, although there is interference with arachidonic acid metabolism. In particular, the inhibition of prostaglandin synthesis is believed to be the key phenomenon for the antiinflammatory and some immune effects (Goodwin and Ceuppens 1983).

The influence of these drugs on collagen metabolism has been investigated in various models of inflammation. The substances were reported to decrease total collagen content in carrageenan granuloma in rats. A significant reduction was found after flufenamic acid, clofenazone, phenylbutazone, and indomethacin administration (Trnavský and Trnavská 1971; Brettschneider et al. 1976a, b; Aalto and Kulonen 1972). Their effect on the solubility of collagen remains controversial. Some drugs were shown to produce a decrease in the soluble fractions (phenylbutazone, indomethacin, mefenamic acid), whereas flufenamic acid stimulated an increase (Brettschneider et al. 1976b). These results reflect an accumulated influence on the synthesis, maturation, and degradation of collagen. Several studies, however, suggest that the drugs increase maturation directly, although the mechanism of their action remains unclear. Trnavský et al. (1965) noted that sodium salicylate increased maturation in rats treated previously with β-aminopropionitrile. This effect was associated with an elevated urinary hydroxyproline output in healthy animals and an opposite effect in rats with lathyrism or adjuvant arthritis. Sodium salicylate was also reported to normalize the impaired maturation and enhanced degradation of newly formed collagen in rats with adjuvant arthritis (Trnavská and Trnavský 1974). Phenylbutazone under the same conditions was found to accelerate the maturation, but it had no effect on degradation. Acetylsalicylic acid was found to decrease the urinary excretion of hydroxyproline when applied to healthy human beings (Liakakos et al. 1971, 1973). Similar results were found in patients with rheumatoid arthritis (Dormidontov et al. 1979).

Interestingly, naproxen was found to enhance collagen synthesis and to depress the biosynthesis of noncollagenous proteins as well as glycosaminoglycans (Suzuki et al. 1974, 1976). The mechanism of this influence is obscure. It is possible that indirect effects mediated through prostaglandin production take part.

An increase in the type III collagen content in the skin of rats treated with antiinflammatory drugs was reported by Pałka and Galewska (1985). A decreased ratio of type I/type III collagens in the liver of rats receiving acetylsalicylic acid or naproxen was shown by Kucharz and Hawryluk (1987). Collagen alterations in the liver of animals treated with high doses of the drugs were caused by two opposed mechanisms, liver damage associated with early stages of fibrosis and inhibition of collagen synthesis by salicylic acid derivatives.

Human leukocyte collagenase was shown to be inhibited in vitro by acetylsalicylic acid, phenylbutazone, and indomethacin. Flufenamic acid was a less potent inhibitor (Wojtecka-Łukasik and Dancewicz 1974). Collagenolytic cathepsin was inhibited in granulation tissue by phenylbutazone and indomethacin. Flufenamic acid had a very weak inhibitory effect (Stančikova et al. 1977). The inhibition of collagenolytic cathepsin in the liver was described by Kucharz et al. (1982b).

### 24.4.2 Tissue Extracts

Extracts from healthy cartilage and bone marrow of calves are used in the therapy of osteoarthrosis. These biostimulants are known under the trade name Rumalon (Robapharm, Basel, Switzerland). There are a few reports on their effect on collagen metabolism. Adam et al. (1978) showed that they significantly enhanced collagen formation in the articular and sternal cartilage of chick embryos. A similar increase was reported in the cornea and sclera as well as in sponge-induced granuloma, although the increase in cartilage was more pronounced. An increase in synthesis of glycosaminoglycans was also observed in the cartilage slices treated with the cartilage-bone marrow extract (Trnavský and Vykydal 1976). Other tissue extracts, i. e., heparinoid composed of proteoglycans and aprotinin, a polypeptide protease inhibitor, were found to inhibit collagenolytic cathepsin activity (Stančikova et al. 1977). Heparinoid also stimulated collagen synthesis in cartilage (Deyl and Adam 1982).

### 24.4.3 Heavy Metals

There is a long history of the usage of heavy metals as drugs. Most of them have been replaced by more specific and less toxic drugs. An exception is the organic compounds of gold that have been used in the treatment of rheumatoid arthritis.

According to Adam et al. (1970), there are two ways in which heavy metals can interact with collagen in vivo. The metals can bind directly to the collagenous structures. Some can be bound in this way and increase polymerization of collagen fibers. The other way is to interfer with synthesis and breakdown through the replacing ions needed for the proper activity of certain enzymes. The indirect effect of heavy metals on collagen metabolism is associated with their toxicity and cellular changes. Some of these changes are followed by scar formation or replacement of the parenchymal cells with fibrous tissue.

Direct binding of metals to native collagen fibers arises after the administration of the metal in the form of electropositive complexes or electronegative anions. The binding can be visualized under electron microscopy.

Gold was found to increase collagen polymerization and the shrinking temperature of collagen preparations. In vitro, a short treatment of collagen samples did not influence the stabilization of collagen structures. When the exposure was long-lasting, the gold complex was destroyed. Thus, the metal atom reacted with the protein molecule. These processes led to cross-link formation, and changes in cross-striation were observed on electron micrographs. Similar phenomena were suggested to occur in vivo.

Goldberg et al. (1980) reported that gold sodium thiomalate decreased the percentage of type III collagen produced by rheumatoid synovial cells in culture. The decrease in proliferation of these cells was also described (Goldberg et al. 1981). In cultures of human synovial cells, gold (10 μM increased the amount of collagen produced per cell, and the percentage of collagen to total protein was elevated by 50%. This effect is probably a unique property of gold, as other heavy metals inhibited collagen synthesis under the same conditions (Goldberg et al. 1983).

There are some controversies in the studies on the effect of gold compounds on collagenase activity. Gold thiomalate and gold thioglucose were shown to be potent activators of latent human leukocyte collagenase (Lindy et al. 1986, 1988). Inhibition of the human neutrophil collagenase by several gold compounds, including the drugs aurothioglucose, sodium gold thiomalate, sodium gold thiosulfate, and auranofin (l-thio-$\beta$-D-glucopyranose-2,3,4,6-tetraacetato-

*S*-triethylphosphine gold) was reported by Mallya and Van Wart (1987). The mechanism has been described. The gold compounds inhibited both active and latent collagenase previously activated reversibly by *p*-chloromercuribenzoate. It was found that the human neutropphil collagenase possesses a number of transitional metal ions. There is at least one binding site whose occupancy by gold (I), cadmium (II), mercury (II), or copper (II) causes noncompetitive inhibition. When zinc (II) is bound to this site, the enzyme activity is retained. The others are active sites that are occupied by zinc (II) and activation sites whose occupancy by enzyme activators leads to the activation of the latent enzyme (Mallya and Van Wart 1989). Gold from chrysotherapeutic agents penetrates to the synovial fluid and inhibits synovial collagenase (Wojtecka-Łukasik et al. 1986). The inhibition can be an important mechanism for chrysotherapy. The earlier described controversy, an activation of collagenase by some gold compounds, may be related to the effect of the organic compounds instead of to gold as a metal. It is possible that thiol groups of the gold thiomalate influence the latent enzyme.

Cadmium produces a variety of pathological effects including alterations in the connective tissue. Experimental studies showed that cadmium treatment of rats produced an increase in urinary excretion of hydroxyproline and hydroxylysine (Nagai et al. 1982; Kucharz 1988a). An elevated ratio of glucosyl-galactosyl-hydroxylysine to galactosyl-hydroxylysine was found in the urine of cadmium-treated rats (Nagai et al. 1982). Similar results were reported in humans with so-called ouch-ouch disease, occurring in Japan, and were probably caused by chronic cadmium poisoning (Iguchi and Sano 1974). A disturbed structure of bones after intoxication with cadmium was caused by impaired collagen metabolism (Yoshiki et al. 1978, 1985). Subchronic exposure of rats to cadmium produced a decrease in the total collagen content in the skin, heart muscle, liver, and kidneys. This was produced by the loss of insoluble collagen. In the lungs, a slight increase in total collagen content without significant changes in its solubility was shown. The tissue alterations were accompanied by elevated serum levels of hydroxyproline and hydroxylysine (Kucharz 1988a). It was also found that fungal collagenase activity was affected by cadmium (Rosenzweig and Pramer 1986).

Vanadium has been reported to decrease the total collagen content in the lungs of rats. The reduction was associated with a low level of soluble collagen and an increase in the insoluble fraction (Kowalska 1988, 1989).

Toxic effects of lead were associated with a rise in the total and insoluble collagen content in the liver (Kucharz 1986b). It is possible that these changes were caused by hepatocyte damage and subsequent fibrosis (Kucharz and Stawiarska-Pięta 1986). Similar findings were described after intoxication with mercury (Olczyk et al. 1990). Lead, mercury, cadmium, and silver were shown to inhibit collagen synthesis in cultured synovial cells (Goldberg et al. 1983). An enhanced collagen content was shown in rats treated with mercury (Rana and Prakash 1986).

Molybdenum, administered as ammonium molybdate, was found to affect collagen content and metabolism when given chronically to rats. An increase in collagen content in the liver, kidneys, and testes was shown. There were no changes in collagen content in the striated muscle. The contraction-relaxation behavior of rat tail tendon collagen suggested that molybdenum impaired the cross-link formation of collagen. High amounts of molybdenum were found to be bound to collagen. The presence of the metal was shown by electron microscopy, and a distinct cross-striation pattern was observed in collagen without additional staining. The relative positions of individual bands along the unit were similar to those observed with phosphotungstic acid staining. Binding of molybdenum to collagen was also confirmed with physicochemical methods. Bíbr et al. (1977) hypothesized that a direct interaction between collagen and molybdate ions occurred at a cross-linking site of the collagen molecule. Binding of the metal to $\delta$-aminoadipic acid was suggested as the mechanism responsible for the defective maturation of collagen. Tungsten was found to induce very similar alterations in collagen metabolism like those after molybdenum treatment (Bíbr et al. 1987).

Samar, the metal from the lanthanide series, was found to enhance the rate of spontaneous aggregation, i.e., polymerization, of collagen in vitro. Samar ions accelerated both the nucleation and the growth phases of polymerization. Collagen fibrils formed in the presence of these ions were thinner and more thermoresistant than control fibrils (Evans and Drouven 1983). The effect of lanthanides on collagen metabolism in vivo is unknown.

### 24.4.4 Fluoride

Fluoride is an element necessary for normal metabolism in animals and humans. It takes part in the development of the crystalline structure of dental enamel and calcification of bones. A small dose is applied in the general prophylaxis of caries and has been found to be highly effective. On the other hand, treatment of animals and humans with higher than optimal doses leads to severe toxic reactions. Collagen metabolism has been mostly investigated in animals receiving various high doses.

The effect of fluoride on the fetal and postnatal development of rats was studied by Drożdż et al. (1980, 1981b, 1984a). Female rats were exposed to hydrofluoride 7 weeks before pregnancy, during pregnancy and lactation, as well as with their offspring up to 6 months after birth. Increased serum and urinary levels of hydroxyproline and hydroxylysine were found in the offspring. A decrease in total collagen content and impaired maturation of collagen in the lungs and skin of the offspring were reported. The cumulation after oral and inhalatory exposure to fluoride compounds on the influence of collagen metabolism was also reported (Kucharz et al. 1978). The results obtained in growing rats are in agreement with the experiments on the fluoride treatment of adult animals (Janecki et al. 1982). It was shown that low doses of fluoride applied to pigs or monkeys did not alter the collagen content in the skin and bone (Reddy and Srikantia 1971; Ammitzbøll et al. 1988b). High doses of fluoride reduced the collagen content in the tissues of rabbits and rats (Sharma 1982; Helgeland 1977a, b).

Changes in collagen were accompanied by alterations in glycosaminoglycans and glycoproteins of the connective tissue (Grucka-Mamczar et al. 1982; Kucharz et al. 1986a). Growing animals were found to be more sensitive to the toxic effects of fluoride than adult ones.

Fluoride effectivly increases bone density in osteoporotic patients. Bone morphometric studies indicated that the osteogenic action of fluoride is mediated through increased osteoblast proliferation (Briancon and Meunier 1981; Harrison et al. 1981). Long-term treatment of osteoporotic patients with fluoride is associated with a significant increase in the nondialyzable fraction of urinary hydroxyproline. Total urinary hydroxyproline levels remain unchanged. The increase persisted for at least 6 months after the cessation of fluoride therapy (Manzke et al. 1977).

### 24.4.5 Cytostatic Drugs

Cytostatic drugs are used in the therapy of neoplastic disorders and certain diseases associated with autoimmune phenomena. Changes in collagen content in experimental granuloma in rats and in the skin of intact animals treated with cyclophosphamide were reported by Hansen and Lorenzen (1975). These changes were associated with the inhibition of granuloma formation. Any alterations in collagen cross-linking were not detected. The effect of cyclophosphamide or azathioprine on collagen content and solubility in the tissues of guinea pig was investigated by Kucharz and Drożdż (1978b). Azathioprine caused a decrease in the total collagen content in the skin, liver, and bones. Cyclophosphamide produced a decrease in the collagen content of the skin and liver, while in bones an increase was shown. On the basis of these experiments, it was suggested that azathioprine acts mostly on collagen biosynthesis whereas cyclophosphamide affects mainly collagen degradation (Drożdż and Kucharz 1977). High doses of cyclophosphamide when applied to rats suppressed collagen synthesis and solubility in bone and skin wounds (Wie and Beck 1981a). Contrary to the earlier studies, it was reported that collagen maturation was impaired at low dose levels of cyclophosphamide whereas high doses inhibited collagen synthesis (Wie and Beck 1981a). The recovery after high-dose cyclophosphamide administration was related to the tissue turnover of collagen. A depression was found in bones 9 days after cessation of the treatment but not in granuloma tissue, where recovery was very quick (Wie and Beck 1981b).

Treatment with cytostatic drugs was found to decrease the hydroxyproline level in the serum and urine in guinea pigs (Kucharz and Drożdż 1978b). In patients with chronic glomerulonephritis, an increased hydroxyproline excretion after azathioprine treatment was reported (Kidawa 1978).

In general, the effect of cytostatic drugs on collagen metabolism is mainly related to the inhibition of cell proliferation. It is possible that collagen metabolism is directly affected, but no clear evidence has been given.

An interesting exception among the cytostatic drugs are bleomycins. These are a group of chemotherapeutic agents isolated from cultures of *Streptomyces verticillus*, and bleomycin sulfate is widely used as an antineoplastic drug. The major toxic effect of bleomycin is pulmonary and cutaneous fibrosis (sclerodermalike syndrome) (Cohen et al. 1972; Blum et al. 1973; Keifer 1973; Bork and Korting 1983; Finch et al. 1984). Both fibrotic states are characterized by an increase in collagen accumulation. The mechanism is not fully understood. It was found that treatment with bleomycin led to increased synthesis of procollagen in rat granuloma fibroblasts and human skin and lung fibroblasts (Otouka et al. 1976; Clark et al. 1980). This stimulation has been shown to be specific as compared with noncollagenous protein synthesis. Furthermore, procollagen synthesis directed by polysomes isolated from bleomycin-treated human embryonic lung fibroblasts was elevated as compared with the production of other proteins (Sterling et al. 1982). An augmentation in prolyl hydroxylase activity was also reported under the same conditions (Sterling et al. 1982). Bleomycin did not increase the total cellular procollagen content but caused a redistribution of type I procollagen mRNAs. Both the nuclear and cytoplasmic procollagens were reduced, and polysomal procollagen was significantly increased. These changes were reversed by dexamethasone, and glucocorticoids have been suggested to be useful in the treatment of the toxic effects of bleomycin (Sterling et al. 1983b).

### 24.4.6 Immunomodulants

Levamisole is one of the drugs found to stimulate the immune response. It inhibits collagen synthesis in skin samples incubated in vitro. This effect is probably nonspecific and related to reduced cell proliferation (Trnavská et al. 1978). An influence of other immunomodulatory drugs upon collagen metabolism has not been described.

### 24.4.7 Antibiotics

Single reports have been published on the effect of antibiotics on collagen metabolism. Antibiotics of the tetracycline group impair the growth of bones, but the nature of this impairment remains unknown (Gudmundson 1971a, b). The effect of tetracycline is not limited to bones. It was found that the drug reduced the mechanical properties of skin. (Engesaeter and Skar 1978). This finding focused attention on collagen. Engesaeter et al. (1980a) treated rats with oxytetracycline for 14 days. At the end of this period the soluble fraction of collagen was elevated in the skin and bones. Three weeks later, no significant difference in collagen solubility between oxytetracycline-treated and control rats was found. The amount of collagen in the bones and skin was not affected by tetracycline treatment. Administration of the drug to animals did not change collagen synthesis. On the other hand, a decreased synthesis of collagen was found in tissue cultures (Halme et al. 1969; Uitto 1975). Earlier, Chvapil et al. (1967a) noted an inhibition of proline hydroxylation by tetracycline. Depressed collagen synthesis may be caused by the nonspecific inhibition of protein synthesis by high doses of tetracycline (Vazquez 1974) or its chelating properties. In animal models, the level of the drug in tissues was significantly lower than that responsible for protein inhibition in mammalian cells.

Oxytetracycline was found to decrease the collagen content in healing surgical wounds of the stomach in rats. The decrease was associated with a lowered mechanical strength of the wounds. In rats receiving tetracycline, increased serum levels of hydroxyproline were reported (Rudnicki and Wojtyczka 1984). Decreased mineralization of bones was noted in young rats receiving oxytetracycline in another study (Engesaeter et al. 1980b). Inhibition of collagenase from polymorphonuclear leukocytes was described by Golub et al. (1983, 1984a, b). The possible role of this inhibition in the therapeutic properties of the drug remains unknown.

Administration of cloxacillin, fusidic acid, and lincomycin was shown to have no effect on collagen solubility in the skin of rats (Engesaeter et al. 1980c). Chloramphenicol was shown to reduce collagen synthesis; both nonspecific protein synthesis and hydroxylation of proline residues were impaired (Konno and Tetsuka 1964). More detailed studies on the effect of antibiotics on the metabolism of collagen have not been reported. It is important to note that tetracyclines impair collagen maturation and that these drugs can restrict the healing of wounds.

# 25 Indices of Collagen Metabolism in Clinical Practice

Collagen abnormalities are shown to accompany pathophysiological changes of almost all organs of the body. The complex role of collagen in the structural integrity and function of the body is associated with collagen involvement in the etiology and progression of a number of disorders. Constant efforts to elaborate credible and applicable methods of the determination and characterization of collagen abnormalities have been carried out since early discoveries in connective tissue metabolism. Several procedures have been reported, but only a few have been accepted for standard usage. The indices are classified for practical purposes into the following groups: (1) those measured in body fluids; (2) those determined in tissue samples; and (3) those measured in in vitro cultured cells or tissues. Serum and urine are the biological materials most commonly used in clinical practice. The serum and urine markers of collagen metabolism are listed in Table 25.1. The urinary output of hydroxyproline has been employed as a collagen marker since the early 1960s. There is a long list of studies which indicate a number of factors affecting the urinary excretion of hydroxyproline (for review, see Prockop and Sjoerdsma 1961; Prockop and Kivirikko 1967; Weiss and Klein 1969; Kivirikko 1970; Drożdż and Kucharz 1972), although the practical value of this index is relatively low. It is used in the estimation of bone metabolism and as an additional index in the evaluation of some disorders (e.g., thyroid dysfunctions in children). It is difficult to conclude anything about collagen metabolism from the determination of hydroxyproline in urine. Determination of free and peptide-bound hydroxyproline levels in urine is without value for clinical practice. Other markers are rarely recommended (Kucharz and Drożdż 1977b). The possible role of hydroxylysine glycosides in the determination of the tissue location of degraded collagen has not been established (Askenasi 1974, 1975). Serum levels of collagen-degrading products have not been found to be useful for diagnosis. The determination of the terminal peptides of procollagen is applied to measure collagen synthesis in patients with various disorders. This marker correlates with synthesis, but a lack of tissue specificity limits its practical value. Estimation of the aminoterminal propeptide of type III collagen can be useful in monitoring certain disorders associated with altered collagen synthesis (Hahn 1984). It is important to note that the aminoterminal propeptides are cleared from the circulation by endocytosis in the liver, and hepatic function influences their serum level (Smedsrød 1988).

The evaluation of enzymes of collagen synthesis requires the application of difficult procedures and is not considered a routine laboratory test. The avail-

**Table 25.1.** Serum and urine indices of collagen metabolism

*Serum*
Metabolites of collagen
- Total, free, and protein-bound hydroxyproline
- Total, free, and protein-bound hydroxylysine
- Total and free proline
- Collagenlike protein
- Aminoterminal propeptide of type III procollagen
- 7S-antigen, a fragment of type IV collagen

Enzymes of collagen synthesis
- Activity of prolyl 4-hydroxylase
- Immunoreactive protein of prolyl 4-hydroxylase
- Activity of lysyl hydroxylase
- Activity of glycosylases of procollagen
- Activity of lysyl oxidase

Enzymes of collagen degradation and their inhibitors
- Activity of collagenase
- Activity of collagen peptidase
- Total serum inhibitory activity against collagenolysis
- $\beta_1$-Anticollagenase
- $\alpha_2$-Macroglobulin

*Urine*
- Total, free, and peptide-bound hydroxyproline
- Total, free, and peptide-bound hydroxylysine
- Galactosylhydroxylysine, glucosylgalactosylhydroxylysine, and their mutual ratio

**Table 25.2.** Indices of collagen metabolism measured in the tissue samples

Total hydroxyproline content
Collagen type pattern determined with biochemical or immunohistochemical methods
Collagen solubility
Collagen cross-links
Activity of prolyl 4-hydroxylase
Activity of lysyl hydroxylase
Activity of procollagen glucosyltransferases
Activity of lysyl oxidase
Activity of collagenase
Latent/active collagenase ratio
Activity of collagen peptidase
Activity of collagenolytic cathepsin

able reports indicate that the enzyme activity is elevated in patients with fibrotic disorders. These measurements do not have a greater value than other markers (e.g., hydroxyproline, aminoterminal propeptide of type III procollagen) (Hahn and Martini 1980b).

The activity of serum enzymes as related to collagen degradation and their inhibitors has been shown to alter in patients with abnormalities in collagen metabolism. Liver function is suggested to affect levels of the enzymes and the inhibitors. The determination of collagen peptidase has been suggested to be useful in the diagnosis of liver disorders (Kucharz et al. 1981).

It is significant that almost all serum and urine markers of collagen metabolism have been found to be related to the age of the individual. Thus, it is difficult to elaborate creditable normal values for the healthy population. Several reports showed the relationship of markers to age (Tuderman and Kivirikko 1977; Minisola et al. 1985). Biological fluids other than serum or urine (e.g., synovial fluid) have been rarely used in the determination of collagen markers.

Tissue samples are considerably less available for diagnostic purposes than serum or urine. The measured indices here are listed in Table 25.2. Biochemical estimations of the collagen metabolism in tissues have no greater value than the ultrastructural characteristics in the clinical diagnosis.

In vitro culturing of the cells (usually fibroblasts) is used for the determination of inborn defects in collagen metabolism. This procedure is not widely used in clinical practice.

The major limitation in the clinical application of the markers of collagen metabolism is the lack of organ specificity. Another limitation is the number of factors that affect the level of the indices. They include biological age, nutrition, physical exercise, medication, and several pathological conditions (e.g., Wheat et al. 1989). There is no disorder specificity as the markers reflect collagen metabolism and are not related to any particular disease. In conclusion, the biochemical markers of collagen metabolism are applicable in monitoring fibrosis or enhanced collagen degradation, but their diagnostic value is relatively low. The only exception involves the hereditary disorders of collagen. Diagnosis of these diseases requires profound insight into collagen metabolism at the cellular level.

# References


Aalto M, Kulonen E (1972) Effects of serotonin, indomethacin and other antirheumatic drugs on the synthesis of collagen and other proteins in granulation tissue slices. Biochem Pharmacol 21:2835–2840

Aalto M, Potila M, Kulonen E (1976) The effect of silica-treated macrophages on the synthesis of collagen and other proteins in vitro. Exp Cell Res 97:193–202

Abbott J, Schiltz J, Dienstman S, Holtzer H (1974) The phenotypic complexity of myogenic clones. Proc Natl Acad Sci USA 71:1506–1510

Abe S, Shinmei M, Nagai Y (1973) Synovial collagenase and joint diseases: the significancy of latent collagenase with special reference to rheumatoid arthritis. J Biochem 73:1007–1011

Abergel RP, Uitto J (1987) Fibrotic skin diseases. Keloids as a model to study the mechanisms of collagen deposition in tissues. In: Uitto J, Perejda AJ (eds) Connective tissue disease. Molecular pathology of the extracellular matrix. Dekker, New York, pp 345–366

Abergel RP, Pizzuro D, Meeker CA, Lask G, Matsuoka LY, Minor RR, Chu ML, Uitto J (1985) Biochemical composition of the connective tissue in keloids and analysis of collagen metabolism in keloid fibroblast cultures. J Invest Dermatol 84:384–390

Abrahams C, Janicki JS, Weber KT (1987) Myocardial hypertrophy in Macaca fascicularis. Structural remodeling of the collagen matrix. Lab Invest 56:676–683

Abramson M, Schilling RW, Huang CC, Salome RG (1975) Collagenase activity in epidermoid carcinoma of the oral cavity and larynx. Ann Otol Rhinol Laryngol 84:158–162

Abrass CK, Peterson CV, Raugi GJ (1988) Phenotypic expression of collagen types in mesangial matrix of diabetic and nondiabetic rats. Diabetes 37:1695–1702

Achard C (1902) Arachnodactylie. Bull Mém Soc Méd Hôp Paris 19:834–843

Ackermann W, Pott G, Voss B, Müller KM, Gerlach U (1981) Serum concentration of procollagen-III-peptide in comparison with the serum activity of N-acetyl-$\beta$-glucosaminidase for diagnosis of the activity of liver fibrosis in patients with chronic active liver diseases. Clin Chim Acta 112:365–369

Adam M, Deyl Z (1983) Altered expression of collagen phenotype in osteoarthrosis. Clin Chim Acta 133:25–32

Adam M, Deyl Z (1984) Degenerated annulus fibrosus of the intervertebral disc contains collagen type III. Ann Rheum Dis 43:258–263

Adam M, Deyl Z, Rosmus J (1970) Interaction of collagen with metals in vivo. Academia, Praha

Adam M, Brettschneider I, Musilová J, Praus R (1978) Effect of cartilage bone-marrow extract on articular collagen formation. Pharmacology 16:49–53

Adam M, Novotná J, Deyl Z (1984) Changes in collagen metabolism – another look at osteoarthrosis. Acta Biol Hung 35:181–187

Adams E (1978) Invertebrate collagens. Marked differences from vertebrate collagens appear in only a few invertebrate groups. Science 202:591–598

Adams P, Eyre DR, Muir H (1977) Biochemical aspects of development and ageing of human lumbar intervertebral discs. Rheumatol Rehabil 16:22–29

Adamson ED (1982) The effect of collagen on cell division, cellular differentiation and embryonic development. In: Weiss JB, Jayson MIV (eds) Collagen in health and disease. Churchill Livingstone, Edinburgh, pp 218–243

Adelmann BC, Kirrane JA, Glynn LE (1972) The structural basis of cell-mediated immunological reactions of collagen. Characteristics of cutaneous delayed hypersensitivity reactions in specifically sensitized guinea pigs. Immunology 23:723–738

Aer J (1968) Effect of thyrocalcitonin on urinary hydroxyproline and calcium in rat. Endocrinology 83:379–380

Aer J, Halme J, Kivirikko KI, Laitinen O (1968) Action of growth hormone on the metabolism of collagen in the rat. Biochem Pharmacol 17:1173–1180

Aggeler J, Engvall E, Werb Z (1981) An irreversible tissue inhibitor of collagenase in human amniotic fluid: characterization and separation from fibronectin. Biochem Biophys Res Commun 110:1195–1201

Agrawal K, Fox H (1972) Subepithelial endometrial collagen. Am J Obstet Gynecol 114:172–175

Ainamo J (1970) Morphogenetic and functional characteristics of coronal cementum in bovine molars. Scand J Dent Res 78:378–392

Åkerbloom HK, Martin JM, Garay GL, Moscarello M (1972) Experimental hypersomatotropism. II. Metabolic effects in rats bearing the MtT-WI5 tumor. Horm Metab Res 4:15–21

Al-Adnani MS, Kirrane JA, McGee JOD (1975) Inappropriate production of collagen and prolyl hydroxylase by human breast cancer cells in vivo. Br J Cancer 31:653–660

Alarcón-Segovia D (1969) Drug-induced lupus syndromes. Mayo Clin Proc 44:664–681

Alarcón-Segovia D (1975) Drug-induced systemic lupus erythematosus and related syndromes. Clin Rheum Dis 1:573–582

Alarcón-Segovia D, Kraus A (1984) Drug-related lupus syndromes. In: Proceedings of the 2nd world conference on clinical pharmacology and therapeutics. American Society for Pharmacology and Experimental Therapeutics, Bethesda, pp 187–206

Alarcón-Segovia D, Wakim KG, Worthington JW, Ward LE (1967) Clinical and experimental studies on the hydralazine syndrome and its relationship to systemic lupus erythematosus. Medicine 46:1–33

Alarcón-Segovia D, Ibánez G, Kershenobich D, Rojkind M (1974) Treatment of scleroderma. Lancet 1:1054–1055

Albrechtsen R, Nielsen M, Wewer U, Engvall E, Ruoslahti E (1981) Basement membrane changes in breast cancer detected by immunochemical staining for laminin. Cancer Res 41:5076–5081

Aleo JJ (1969) Collagen synthesis in cultured cells: the influence of beta-aminopropionitrile. Proc Soc Exp Biol Med 30:451–454

Aleo JJ, Novack R, Levy E (1974) Collagen synthesis by lathyrogen-treated 3T6 fibroblasts. Connect Tissue Res 2:91–93

Alitalo K, Vaheri A (1982) Pericellular matrix in malignant transformation. Adv Cancer Res 37:111–158

Alitalo K, Bornstein P, Vaheri A, Sage H (1983) Biosynthesis of an unusual collagen type by human astrocytoma cells in vitro. J Biol Chem 258:2653–2661

Allmann H, Fietzek P, Glanville RW, Kühn K (1979) The covalent structure of calf skin type III collagen. VI. The amino acid sequence of the carboxyterminal cyanogen bromide peptide $\alpha$1(III) CB9B (position 928–1028). Z Physiol Chem 360:861–868

Alvarez OM (1987) Pharmacological and environmental modulation of wound healing. In: Uitto J, Perejda AJ (eds) Connective tissue disease. Molecular pathology of the extracellular matrix. Dekker, New York, pp 367–384

Alvarez-Morujo AJ, Alvarez-Morujo A (1978) The collagen of the gingiva and of its blood vessels. Acta Anat 101:66–75

Al-Zaid NS (1988) Muscle collagen content in diabetic rats. Acta Physiol Hung 71:41–43

Al-Zaid NS, Bou-Resli MN, Ibrahim MEA (1980a) Study of the connective tissue of human fetal membranes. J Reprod Fertil 59:383–386

Al-Zaid NS, Bou-Resli MN, Goldspink G (1980b) Bursting pressure of fetal membranes and their relation to premature rupture of membranes. Br J Obstet Gynaecol 87:227–229

Al-Zaid NS, Gumaa KA, Bou-Resli MN, Ibrahim MEA (1988) Premature rupture of fetal membranes changes in collagen type. Acta Obstet Gynecol Scand 67:291–295

Al-Zuhair AGH, Al-Adnani MS, Al-Bader AA, Abdulla MA (1986) Distribution and ultra-structural characterization of the stroma in scirrhous "infiltrating" carcinoma of the human breast. J Submicrosc Cytol 18:409–416

Amenta PS, Gay S, Vaheri A, Martinez-Hernandez A (1986) The extracellular matrix is an integrated unit: ultrastructural localization of collagen types I, III, IV, V, VI, fibronectin, and laminin in human term placenta. Coll Relat Res 6:125–152

Amenta PS, Gil J, Martinez-Hernandez A (1988) Connective tissue of rat lung. II. Ultrastructural localization of collagen types III, IV, and VI. J Histochem Cytochem 36:1167–1173
Amma MKP, Singh J, Sareen P, Sareen K (1978) Changes in rat plasma hydroxyproline during estrus and estrogen treatment. Indian J Exp Biol 16:806–808
Ammitzbøll T, Serup J (1984) Collagen metabolites in urine in localized scleroderma (morphoea). Acta Derm Venerol (Stockh) 64:534–569
Ammitzbøll T, Pedersen SS, Espers F, Schiøler H (1988a) Excretion of urinary collagen metabolites correlates to severity of pulmonary disease in cystic fibrosis. Acta Paediatr Scand 77:842–846
Ammitzbøll T, Richards A, Kragstrup J (1988b) Collagen and glycosaminoglycans in fluoride-exposed pigs. Pharmacol Toxicol 62:239–240
Ampe J, Dequeker J, Gevers G (1986) Regional variation of bone matrix components in osteoarthrotic and normal femoral heads. Clin Rheumatol 5:225–230
Anastassiades T, Anastassiades PA, Denstedt OF (1972) Changes in the connective tissue of the atherosclerotic intima and media of the aorta. Biochim Biophys Acta 261:418–427
Anderson AJ (1969) Effects of lysosomal collagenolytic enzymes, antiinflammatory drugs, and other substances on some properties of insoluble collagen. Biochem J 113:457–463
Anderson AJ (1970) Lysosomal enzyme activity in rats with adjuvant-induced arthritis. Ann Rheum Dis 29:307–313
Anderson JC, Labedz RI, Kewley MA (1977) The effect of bovine tendon glycoprotein on the formation of fibrils from collagen solutions. Biochem J 168:345–351
Andreassen TT, Oxlund H (1985) Thermal stability of collagen in relation to in vitro non-enzymatic glycosylation and browning in vitro. Diabetologia 28:687–691
Andreassen TT, Oxlund H (1987) Changes in collagen and elastin of the rat aorta induced by experimental diabetes and food restriction. Acta Endocrinol (Copenh) 115:338–344
Andreassen TT, Seyer-Hansen K, Oxlund H (1981) Biomechanical changes in connective tissues induced by experimental diabetes. Acta Endocrinol (Copenh) 98:432–436
Andrén L (1960) Instability of the pubic symphysis and congenital dislocation of the hip in newborns. Acta Radiol 54:123–128
Andriopoulos NA, Mestecky J, Miller EJ, Bradley EL (1976a) Antibodies to native and denatured collagens in sera of patients with rheumatoid arthritis. Arthritis Rheum 19:613–617
Andriopoulos NA, Mestecky J, Wright GP, Miller EJ (1976b) Characterization of antibodies to the native collagens and to their component $\alpha$ chains in the sera and joint fluids of patients with rheumatoid arthritis. Immunochemistry 13:709–712
Andriopoulos NA, Mestecky J, Miller EJ, Bennett JC (1976c) Antibodies to human native and denatured collagens in synovial fluids of patients with rheumatoid arthritis. Clin Immunol Immunopathol 6:209–212
Ansay M, Gillet A, Hanset R (1968) La dermatosparaxie hereditaire des bovides: biochimie descriptive de la peau. Ann Med Vet 112:449–451
Anttinen H (1977) Collagen glucosyltransferase activity in human serum. Clin Chim Acta 77:323–330
Anttinen H, Ryhänen L, Puistola U, Arranto A, Oikarinen A (1984) Decrease in liver collagen accumulation in carbon tetrachloride-injured and normal growing rats upon administration of zinc. Gastroenterology 86:532–539
Anttinen H, Oikarinen A, Puistola U, Pääkkö P, Ryhänen L (1985) Prevention by zinc of rat lung collagen accumulation in carbon tetrachloride injury. Am Rev Respir Dis 132:536–540
Araszkiewicz Z, Styczewska H (1972) Wydalanie hydroksyproliny i wapnia w moczu chorych z nadczynnością tarczycy (Urinary excretion of hydroxyproline and calcium in patients with hyperthyroidism). Pol Arch Med Wewn 48:557–561 (in Polish)
Arem AJ, Misiorowski R (1976) Lathyritic activity of isoniazid. J Med Westbury NY 7:239–248
Arheden K, Mandahl N, Heim S, Mitelman F (1989) In situ hydridization localizes the human type II alpha 1 collagen gene (COL2A1) to 12q13. Hereditas 110:165–167
Armendáriz-Borunda J, Greenwel P, Rojkind M (1989) Kupffer cells from $CCl_4$-treated rat livers induce skin fibroblast and liver fat-storing cell proliferation in culture. Matrix 9:150–158
Arneson MA, Hammerschmidt DE, Furcht LT, King RA (1980) A new Ehlers-Danlos syndrome: fibronectin corrects defective platelet function. JAMA 244:44–147

Aronson AL, Rogerson KM (1972) Effect of calcium and chromium chelates of ethylenediaminetetraacetate on intestinal permeability and collagen metabolism in the rat. Toxicol Appl Pharmacol 21:440–453
Arrago JP, Poirier O, Najean Y (1984) Evolution des polyglobulies primitives vers la myelofibrosis. Surveillance par le dosage de l'aminopropeptide du' procollagen III. Presse Med 13:2429–2432
Asboe-Hansen G (ed) (1954) Connective tissue in health and disease. Munksgaard, Copenhagen
Asboe-Hansen G (ed) (1966) Hormones and connective tissue. Munksgaard, Copenhagen
Asbury WT (1940) The molecular structure of the fibers of the collagen group. J Int Soc Leather Trades Chem 24:69–92
Askenasi R (1974) Urinary hydroxylysine and hydroxylysyl glycoside excretions in normal and pathological states. J Lab Clin Med 83:673–679
Askenasi R (1975) Urinary excretion of hydroxylysyl glycosides as an index of collagen metabolism in disease. In: Peeters H (ed) Protides of the biological fluids, vol 22. Pergamon, Oxford, pp 121–125
Askenasi R, Demeester-Mirkine N (1975) Urinary excretion of hydroxylysyl glycosides and thyroid function. J Clin Endocrinol 40:342–344
Aspden RM, Yarker YE, Hukins DWL (1985) Collagen orientations in the meniscus of the knee joint. J Anat 140:371–380
Aula P, Autio S, Raivio KO, Rapola J (1982) Aspartylglycosaminuria. In: Durand P, O'Brien JS (eds) Genetic errors of glycoprotein metabolism. Edi-Ermes, Milano, pp 123–152
Aumailley M, Bricaud H (1981) Collagen synthesis in organ culture of normal and atherosclerotic aortas. Arteriosclerosis 39:1–9
Aumailley M, von der Mark H, Timpl R (1985) Size and domain structure of collagen VI produced by cultured fibroblasts. FEBS Lett 182:499–502
Aumailley M, Pöschl E, Martin GR, Yamada Y, Müller PK (1988) Low production of procollagen III by skin fibroblasts from patients with Ehlers-Danlos syndrome type IV is not caused by decreased levels of procollagen III mRNA. Eur J Clin Invest 18:207–212
Autio S (1972) Aspartylglycosaminuria. Analysis of thirty-four patients. J Ment Defic Res Monogr Ser 1:1–93
Ayad S, Weiss JB (1984) A new look at vitreous-humour collagen. Biochem J 218:835–840
Ayad S, Abedin MZ, Grundy SM, Weiss JB (1981) Isolation and characterisation of an unusual collagen from hyaline cartilage and intervertebral disc. FEBS Lett 123:195–199
Ayad S, Abedin MZ, Weiss JB, Grundy SM (1982) Characterisation of another short-chain disulphide-bonded collagen from cartilage, vitreous and intervertebral disc. FEBS Lett 139:300–304
Ayad S, Evans H, Weiss JB, Holt L (1984) Type VI collagen but not type V collagen is present in cartilage. Coll Rel Res 4:165–168
Babbs C, Hunt LP, Haboubi NY, Smith A, Rowan BP, Warnes TW (1988) Type III procollagen peptide: a marker of disease activity and prognosis in primary biliary cirrhosis. Lancet 1:1022–1024
Babel W, Glanville RW (1984) Structure of human-basement membrane (type IV) collagen. Complete amino-acid sequence of a 914-residue-long pepsin fragment from $\alpha$1(IV) chain. Eur J Biochem 143:543–556
Baccetti B (1985) Collagen and animal phylogeny. In: Bairati A, Garrone R (eds) Biology of invertebrate and lower vertebrate collagens. Plenum, New York, pp 29–47
Bachinger HP, LeRoy EC (1986) Connective tissue in scleroderma (systemic sclerosis). In: Kühn K, Krieg T (eds) Connective tissue: biological and clinical aspects. Karger, Basel, pp 430–450
Baer E, Gathercole LJ, Keller A (1974) Structure hierarchies in tendon collagen: an interim summary. In: Atkins EDT, Keller A (eds) Structure of fibrous biopolymers. Butterworths, London, pp 189–195
Baer E, Cassidy JJ, Hiltner A (1988) Hierarchical structure of collagen and its relationship to the physical properties of tendon. In: Nimni ME (ed) Collagen, vol II. CRC Press, Boca Raton, pp 177–199
Baer RW, Taussig HB, Oppenheimer EH (1943) Congenital aneurysmal dilatation of the aorta associated with arachnodactyly. Bull Johns Hopkins Hosp 72:309–311

Baich A, Chen PC (1979) Cirrhosis and the synthesis of proline and glycine in rat liver. Physiol Chem Phys 11:377–383
Bailey AJ (1968a) Effect of ionizing radiation on connective tissue components. Int Rev Connect Tissue Res 4:233–281
Bailey AJ (1968b) The nature of collagen. In: Florkin M, Stotz EH (eds) Comprehensive biochemistry, vol 26B. Elsevier, Amsterdam, pp 297–423
Bailey AJ (1985) The collagen of the Echinodermata. In: Bairati A, Garrone R (eds) Biology of invertebrate and lower vertebrate collagens. Plenum, New York, pp 369–388
Bailey AJ, Etherington DJ (1980) Metabolism of collagen and elastin. In: Florkin M, Stotz E (eds) Comprehensive biochemistry, vol 19B. Elsevier, Amsterdam, pp 299–460
Bailey AJ, Light ND (1985) Intermolecular cross-linking in fibrotic collagen. Ciba Found Symp 114:80–96
Bailey AJ, Sims TJ (1976) Chemistry of the collagen crosslinks. Nature of the crosslinks. Biochem J 153:211–215
Bailey AJ, Tromans WJ (1964) Effects of ionizing radiation on the ultrastructure of collagen fibrils. Radiat Res 23:145–155
Bailey AJ, Bendall JR, Rhodes DN (1962) The effect of irradiation on the shrinkage temperature of collagen. Int J Appl Radiat Isot 13:131–136
Bailey AJ, Bazin S, Delaunay A (1973) Changes in the nature of the collagen during development and resorption of granulation tissue. Biochim Biophys Acta 328:383–390
Bailey AJ, Bazin S, Sims TJ, LeLous M, Nicoletis C, Delaunay A (1975) Characterization of the collagen in human hypertrophic and normal scars. Biochim Biophys Acta 405:412–421
Bailey AJ, Sims TJ, Gabbiani G, Bazin S, LeLous M (1977) Collagen of Dupuytren's disease. Clin Sci Mol Med 53:499–502
Bailey AJ, Gathercole LJ, Dlugosz J, Keller A, Voyle CA (1982) Proposed resolution of the paradox of extensive crosslinking and low tensile strength of cuvierian tubule collagen of the sea cucumber Holothuria forskali. Int J Biol Macrolol 4:329–335
Bairati A (1972) Collagen: an analysis of phylogenetic aspects. Boll Zool 39:205–212
Bairati A, Garrone R (eds) (1985) Biology of invertebrate and lower vertebrate collagens. Plenum, New York
Bairati A, Amante L, De Petris S, Pernis S (1964) Studies on the ultrastructure of the lymph nodes. I. The reticular network. Z Zellforsch 63:644–649
Balazs EA (ed) (1970) Chemistry and molecular biology of the intercellular matrix. Academic, New York
Ballard JB, Butler WT (1974) Proteins of the periodontium. Biochemical studies on the collagen and noncollagenous proteins of human gingivae. J Oral Pathol 3:176–184
Balleisen LS, Gay S, Marx R, Kühn K (1975) Comparative investigation on the influence of human and bovine collagen type I, II and III on the aggregation of human platelets. Klin Wochenschr 53:903–905
Banes AJ, Yamauchi M, Mechanic GJ (1983) Nonmineralized and mineralized compartments of bone. The role of pyridinoline in nonmineralized collagens. Biochem Biophys Res Commun 113:975–981
Banfield WG (1958) Effect of growth hormone on acetic acid-extractable collagen of hamster skin. Proc Soc Exp Biol Med 97:309–312
Banga I (1966) Structure and function of elastin and collagen. Akademiai Kiado, Budapest
Bańkowski E (1972) Biochemiczne i kliniczne aspekty metabolizmu kolagenu (Biochemical and clinical aspects of collagen metabolism). Postępy Hig Med Dośw 26:207–223 (in Polish)
Bańkowski E, Mitchell WM (1973) Human procollagen I. An anionic tropocollagen precursor from skin fibroblasts in culture. Biophys Chem 1:73–86
Bańkowski E, Pawlicka E (1989) An increase of collagen biosynthesis in livers of rats submitted to chronic intoxication with ethanol. Mol Cell Biochem 85:43–48
Bańkowski E, Rzeczycki W, Nowak HF, Jodczyk KJ (1977) Badania nad biosyntezą kolagenu przez fibroblasty transformowane wirusem SV-40 (Studies on collagen biosynthesis in fibroblasts transformed with the SV-40 virus). In: Nowak F (ed) Onkologia kliniczna (Clinical oncology). Akademia Medyczna, Białystok, pp 51–54 (in Polish)

Bańkowski E, Nowak HF, Jodczyk KJ (1980) Stimulatory effect of platelet homogenate on collagen biosynthesis in L-929 cells. Exp Pathol 18:462–467

Bańkowski E, Pawlicka E, Jodczyk KJ (1986) Prolyl hydroxylase activity as an index of liver damage induced by ethanol. Cell Biochem Funct 5:211–215

Barnard K, Light ND, Sims TJ, Bailey AJ (1987) Chemistry of collagen cross-links. Origin and partial characterization of a putative mature cross-link of collagen. Biochem J 244:303–309

Barnes MJ (1969) Ascorbic acid and the biosynthesis of collagen and elastin. Bibl Nutr Dieta 13:86–98

Barnes MJ (1982) The collagen-platelet interaction. In: Weiss JB, Jayson MIV (eds) Collagen in health and disease. Churchill Livingstone, Edinburgh, pp 179–197

Barnes MJ (1985a) Collagens in atherosclerosis. Coll Relat Res 5:65–97

Barnes MJ (1985b) Function of ascorbic acid in collagen metabolism. Ann NY Acad Sci 460:264–277

Barnes MJ (1988) Collagen od normal and diseased blood vessel wall. In: Nimni ME (ed) Collagen, vol I. CRC Press, Boca Raton, pp 275–290

Barnes MJ, Kodicek E (1972) Biological hydroxylations and ascorbic acid with special regard to collagen metabolism. Vitam Horm 30:1–43

Barnes MJ, MacIntyre DE (1979) Collagen-induced platelet aggregation. The activity of basement membrane collagens relative to other collagen types. Front Matrix Biol 7:246–252

Barnes MJ, Constable BJ, Kodicek E (1969) Excretion of hydroxyproline and other amino acids in scorbutic guinea pigs. Biochim Biophys Acta 184:358–365

Barnes MJ, Constable BJ, Morton LF, Kodicek E (1970) Studies in vivo on the biosynthesis of collagen and elastin in ascorbic acid-deficient guinea pigs. Biochem J 119:575–585

Barnes MJ, Constable BJ, Morton LF, Kodicek E (1973a) The influence of dietary calcium deficiency and parathyroidectomy on bone collagen structure. Biochim Biophys Acta 328:373–388

Barnes MJ, Constable BJ, Morton LF, Kodicek E (1973b) Bone collagen metabolism in vitamin D-deficiency. Biochem J 132:113–115

Barnes MJ, Bailey AJ, Gordon JL, MacIntyre DE (1980) Platelet aggregation by basement membrane-associated collagens. Thromb Res 18:375–388

Baron R, Neff L, Louvard D, Courtoy PJ (1985) Cell-mediated extracellular acidification and bone resorption: evidence for a low pH in resorbing lacunae and localization of a 100 kD lysosomal membrane protein at the osteoclast ruffled border. J Cell Biol 101:2210–2222

Barrow MV, Simpson CF, Miller EJ (1974) Lathyrism: a review. Quart Rev Biol 49:101–128

Barrows GH, Schrodt GR, Greenberg RA, Tamburro CH (1980) Changes in stainable collagen in the ageing normal human liver. Gastroenterology 79:1099

Barsh GS, Byers PH (1981) Reduced secretion of structurally abnormal type I procollagen in a form of osteogenesis imperfecta. Proc Natl Acad Sci USA 78:5142–5146

Barsh GS, Roush CL, Bonadio J, Byers PH, Gelinas RE (1985) Intron mediated recombination may cause a deletion in an alpha type I collagen chain in a lethal form of osteogenesis imperfecta. Proc Natl Acad Sci USA 82:2870–2874

Barsky SH, Grotendorst GR, Liotta RA (1982) Increased content of type V collagen in desmoplasia of human breast carcinoma. Am J Pathol 108:276–283

Barsky SH, Siegal GP, Jannotta F, Liotta LA (1983) Loss of basement membrane components by invasive tumors but not by their benign counterparts. Lab Invest 49:140–147

Bartholomew JS, Anderson JC (1983) Investigation of relationship between collagens, elastin and proteoglycans in bovine thoracic aorta by immunofluorescence techniques. Histochem J 15:1177–1190

Bartlet CP, Heale G, Levene CI (1985) Some factors regulating collagen polymorphism in cultured porcine and bovine aortic endothelium. Arteriosclerosis 54:301–309

Bartlett MK, Jones CM, Ryan AE (1942) Vitamin C and wound healing: ascorbic acid content and tensile strength of healing wounds in human beings. N Engl J Med 226:474–481

Bartlett MW, Egelstaff PA, Holden TM, Stinson RH, Sweeny PR (1973) Structural changes in tendon collagen resulting from muscular dystrophy. Biochim Biophys Acta 328:213–220

Bartos F, Ledvina M (1979) Collagen, elastin and desmosines in three layers of bovine aortas of different ages. Exp Gerontol 14:21–26

Bartošová D, Chvapil M, Korecký B, Poupa O, Rakušan K, Turek Z, Vízek M (1969) The growth of the muscular and collagenous parts of the rat heart in various forms of cardiomegaly. J Physiol (Lond) 200:285–295

Barzansky BL, Lenhoff HM, Bode H (1975) Hydra mesogloea: similarity of its amino acid and neutral sugar composition to that of vertebrate basal lamina. Comp Biochem Physiol 50B:419–424

Bashey RI, Jimenez SA (1977a) Collagen biosynthesis in human heart values. Fed Proc 36:1068A

Bashey RI, Jimenez SA (1977b) Increased sensitivity of scleroderma fibroblasts in culture to stimulation of protein and collagen synthesis by serum. Biochem Biophys Res Commun 76:1214–1222

Bashey RI, Jimenez SA (1988) Collagens in heart valves. In: Nimni ME (ed) Collagen, vol I. CRC Press, Boca Raton, pp 257–274

Bashey RI, Torii S, Angrist A (1967) Age-related collagen and elastin content of human heart valves. J Gerontol 22:203–208

Bashey RI, Bashey HM, Jimenez SA (1978) Characterization of pepsin-solubilized bovine heart valve collagen. Biochem J 173:885–894

Bashey RI, Jimenez SA, Balazs T (1980) The effect of hydralazine on collagen biosynthesis and secretion in isolated embryonic chick tendon cells. In: Holmstedt B, Lauwerys R, Mercier M, Robertford M (eds) Mechanisms of toxicity and hazard evaluation. Elsevier, Amsterdam, pp 647–650

Bashey RI, Cox R, McCann J, Jimenez SA (1989) Changes in collagen biosynthesis, types, and mechanics of aorta in hypertensive rats. J Lab Clin Med 113:604–611

Bassuk JA, Berg RA (1989) Protein disulphide isomerase, a multifunctional endoplasmic reticulum protein. Matrix 9:244–258

Bateman JF, Chan D, Mascara T, Rogers JG, Cole WG (1986) Collagen defects in lethal perinatal osteogenesis imperfecta. Biochem J 240:699–708

Bateman JF, Mascara T, Chan D, Cole WG (1987a) A structural mutation of the collagen $\alpha$1(I)CB7 peptide in lethal perinatal osteogenesis imperfecta. J Biol Chem 262:4445–4451

Bateman JF, Chan D, Walker ID, Rogers JG, Cole WG (1987b) Lethal perinatal osteogenesis imperfecta due to the substitution of arginine for glycine at residue 391 of the $\alpha$1(I) chain of type I collagen. J Biol Chem 262:7021–7027

Bates CJ (1977) Proline and hydroxyproline excretion and vitamin C status in elderly human subjects. Clin Sci Mol Med 52:535–543

Bates CJ (1979) Vitamin C deficiency in guinea pigs: variable sensitivity of collagen at different sites. Int J Vitam Nutr Res 49:77–86

Bates CJ, Prynne CJ, Levene CI (1972) The synthesis of unhydroxylated collagen by 3T6 mouse fibroblasts in culture. Biochim Biophys Acta 263:397–405

Bauer EA, Eisen AZ (1978) Recessive dystrophic epidermolysis bullosa. Evidence for increased collagenase as a genetic characteristic in cell culture. J Exp Med 148:1378–1387

Bauer EA, Uitto J (1979) Collagen in cutaneous diseases. Int J Dermatol 18:251–270

Bauer EA, Uitto J (1982) Skin. In: Weiss JB, Jayson MIV (eds) Collagen in health and disease. Churchill Livingstone, Edinburgh, pp 472–487

Bauer EA, Gedde-Dahl T Jr, Eisen AZ (1977a) The role of human skin collagenase in epidermolysis bullosa. J Invest Dermatol 68:119–124

Bauer EZ, Gordon JM, Reddick ME, Eisen AZ (1977b) Quantitation and immunocytochemical localization of human skin collagenase in basal cell carcinoma. J Invest Dermatol 69:363–367

Baum BJ, Moss J, Breul SD, Berg RA, Crystal RG (1980) Effect of cyclic AMP on the intracellular degradation of newly synthesized collagen. J Biol Chem 255:2843–2847

Baumgartner HR (1977) Platelet interaction with collagen fibrils in flowing blood. I. Reaction of human platelets with $\alpha$ chymotrypsin-digested subendothelium. Thromb Haemost 37:1–16

Bavetta LA, Bekhor I, Shah R, O'Day P, Nimni ME (1962) Metabolic and anti-inflammatory properties of 6-methylprednisolone alone and in combination with anabolic hormones. Endocrinology 71:221–226

Baxter JD, Forsham PH (1972) Tissue effects of glucocorticoids. Am J Med 53:573–589

Bazin S, Briquelet N, Allain JC, Delaunay A (1975) Effet exercé par du tétrachlorure de carbone sur la qualité du collagène hépatique. CR Acad Sci Paris 280:517–520

Bazin S, Le Louis M, Delaunay A (1976) Collagen in granulation tissues. Agents Actions 6:272–276
Beachey EH, Chiang TM, Kang AH (1979) Collagen-platelet interaction. Int Rev Connect Tissue Res 8:1–21
Bear RS (1942) Long x-ray diffraction spacing of collagen. J Am Chem Soc 64:727
Beard HK, Stevens RL (1980) Biochemical changes in the intervertebral disc. In: Jayson MIV (ed) The lumbar spine and back pain. Pitman Medical, London, pp 407–436
Beard HK, Faulk WP, Conochie LB, Glynn LE (1977) Some immunological aspects of collagen. Prog Allergy 22:45–106
Beard HK, Ueda M, Faulk WP, Glynn LE (1978) Cell-mediated and humoral immunity to chick type II collagen and its cyanogen bromide peptides in guinea pigs. Immunology 34:32–335
Behera HN, Patnaik BK (1979) In vivo and in vitro effects of alloxan on collagen characteristics of bone, skin and tendon of Swiss mice. Gerontology 25:255–260
Behera HN, Patnaik BK (1981a) Effect of alloxan diabetes on the characteristics of collagen of cotton-pellet granuloma and dorsal skin of Swiss mice. Gerontology 27:209–215
Behera HN, Patnaik BK (1981b) Recovery from alloxan diabetes as revealed by collagen characteristics of bone, skin and tendon of Swiss mice. Gerontology 27:32–36
Behera HN, Patnaik BK (1982) Increased stability of collagen following alloxan diabetes in Swiss mice. Gerontology 28:163–167
Beighton P, Uersfeld GA (1985) On the paradoxically high relative prevalence of osteogenesis imperfecta type III in the black population of South Africa. Clin Genet 27:398–401
Beil W, Timpl R, Furthmayr H (1973) Conformation dependence of antigenic determinants on the collagen molecule. Immunology 24:13–19
Beisswenger PJ, Spiro RG (1973) Studies on the human glomerular basement membrane. Composition, nature of the carbohydrate units and chemical changes in diabetes mellitus. Diabetes 22:180–193
Beldekas JC, Smith B, Gerstenfeld LC, Sonenshein GE, Franzblau C (1981) Effects of 17$\beta$-estradiol on the biosynthesis of collagen in cultured bovine aortic smooth muscle cells. Biochemistry 20:2162–2167
Bellamy G, Bornstein P (1971) Evidence for procollagen, a biosynthetic precursor of collagen. Proc Natl Acad Sci USA 68:1138–1142
Belton JC, Crise N, McLaughlin RF, Tueller EE (1977) Ultrastructural alterations in collagen associated with microscopic foci of human emphysema. Hum Pathol 8:669–677
Ben-Bassat M, Cohen L, Rosenfeld J (1971) The glomerular basement membrane in the nail-patella syndrome. Arch Pathol 92:350–355
Bender E, Silver FH, Hayashi K (1983) Model conformations of the carboxyl telopeptides in vivo based on type I collagen fibral banding patterns. Coll Relat Res 3:407–418
Bennett WM, Murgrave JE, Campbell RA, Elliot D, Cox R, Brooks RE, Lovrien EW, Beals RK, Porter GA (1973) The nephropathy of the nail-patella syndrome. Clinicopathological analysis of 11 kindred. Am J Med 54:304–319
Bennick A, Hunt AM (1967) Collagenase activity in oral tissue. Arch Oral Biol 12:1–9
Benoit FL, Theil GB, Watten RH (1963) Hydroxyproline excretion in endocrine disease. Metabolism 12:1072–1082
Benson PF (1965) Hydroxyproline excretion in scoliosis. In: Zorab PA (ed) Proceedings of a symposium on scoliosis. Institute of Diseases of the Chest, London, pp 47–49
Benson PF (1972) Hydroxyproline excretion in idiopathic, congenital and paralytic scoliosis. Arch Dis Child 47:476
Benson SC, LuVulle PA (1981) Inhibition of lysyl oxidase and prolyl hydroxylase activity in glucocorticoid treated rats. Biochem Biophys Res Commun 99:557–562
Bentley SA (1982) Collagen synthesis by bone marrow stromal cells: a quantitative study. Br J Haematol 50:491–497
Bentley SA, Foidart JM (1980) Some properties of marrow derived adherent cells in tissue culture. Blood 56:1006–1012
Bentley SA, Alabaster O, Foidart JM (1981) Collagen heterogeneity in normal human bone marrow. Br J Haematol 48:287–291

Bentsen KD, Hørslev-Petersen K, Junker P, Juhl E, Lorenzen I (1987) Serum aminoterminal procollagen type III peptide in acute viral hepatitis. A long-term follow-up study. Liver 7:96–105

Bentz H, Morris NP, Murray LW, Sakai LY, Hollister DW, Burgeson RE (1983) Isolation and partial characterization of a new human collagen with an extended triple-helical structure domain. Proc Natl Acad Sci USA 80:3168–3172

Benya PD (1980) EC collagen: biosynthesis by corneal endothelial cells and separation from type IV without pepsin treatment or denaturation. Renal Physiol 3:30–35

Benya PD, Brown PD (1986) Modulation of the chondrocyte phenotype in vitro. In: Kuettner KE, Schleyerbach R, Hascall VC (eds) Articular cartilage biochemistry. Raven, New York, pp 219–234

Benya PD, Padilla SR, Nimni ME (1978) Independent regulation of collagen types by chondrocytes during the loss of differentiated function in culture. Cell 15:1313–1321

Ben-Zvi A, Rodrigues MM, Krachmer JH, Fujikawa LS (1986) Immunohistochemical characterization of extracellular matrix in the developing human cornea. Curr Eye Res 5:105–117

Berenson GS, Geer JC (1963) Heart disease in the Hurler and Mafran syndrome. Arch Intern Med 111:58–69

Berenson GS, Radhakrishnamurthy B, Dalfers ER Jr, Ruiz H, Srinivasan SR, Plavidal F, Brickman F (1972) Connective tissue macromolecular changes in rats with experimentally induced diabetes and hyperinsulinism. Diabetes 21:733–743

Berg RA (1986) Intracellular turnover of collagen. In: Mecham R (ed) Regulation of matrix accumulation. Academic, New York, pp 29–52

Berg RA, Prockop DJ (1973a) The thermal transition of a non-hydroxylated form of collagen. Evidence for a role for hydroxyproline in stabilizing the triple-helix of collagen. Biochem Biophys Res Commun 52:115–120

Berg RA, Prockop DJ (1973b) Affinity column purification of procollagen proline hydroxylase from chick embryos and further characterization of the enzyme. J Biol Chem 248:1175–1182

Berg RA, Schwartz ML, Crystal RG (1980) Regulation of the production of secretory proteins: intracellular degradation of newly synthesized "defective" collagen. Proc Natl Acad Sci USA 77:4746–4750

Berg RA, Schwartz ML, Rome LH, Crystal RG (1984) Lysosomal function in the degradation of defective collagen in culture lung fibroblasts. Biochemistry 23:2134–2148

Berman MB (1975) Collagenase inhibitors: rationale for their use in treating corneal ulceration. Int Ophthalmol Clin 15:49–66

Berman MB (1976) The role of $\alpha$-macroglobulins in corneal ulceration. In: Jamieson GA, Greenwalt TJ (eds) Trace components of plasma: isolation and clinical significance. Liss, New York, pp 225–253

Berman MB (1980) Collagenase and corneal ulceration. In: Woolley DE, Evanson JM (eds) Collagenase in normal and pathological connective tissues. Wiley, Chichester, pp 141–174

Berman MB, Dohlman C (1975) Collagenase inhibitors. Arch Ophthalmol (Paris) 35:95–108

Berman JJ, Foidart JM (1980) Synthesis of collagen by rat liver epithelial cultures. Ann NY Acad Sci 348:153–164

Berman MB, Manabe R (1973) Corneal collagenases: evidence for zinc metalloenzymes. Ann Ophthalmol 5:1193–1209

Berman MB, Dohlman CH, Gnädinger M, Davison PF (1971) Characterization of collagenolytic activity in the ulcerating cornea. Exp Eye Res 11:255–257

Berman MB, Barber JC, Talamo RC, Langley CE (1973a) Corneal ulceration and the serum antiproteases. I. $\alpha_1$-Antitrypsin. Invest Ophthalmol 12:759–779

Berman MB, Kerza-Kwiatecki AP, Davison PF (1973b) Characterization of human corneal collagenase. Exp Eye Res 15:367–373

Berman MB, Gordon J, Garcia LA, Gage J (1975) Corneal ulceration and the serum antiproteases. II. Complexes of corneal collagenases and $\alpha$-macroglobulins. Exp Eye Res 20:231–244

Bernacka K, Kuryliszyn-Moskal A, Sierakowski S (1988) The levels of alpha-1-antitrypsin and alpha-1-antichymotrypsin in the sera of patients with gastrointestinal cancers during diagnosis. Cancer 62:1188–1193

Bernstein J (1987) The glomerular basement membrane abnormality in Alport's syndrome. Am J Kidney Dis 10:222–229

Bertolotto A, Magrassi ML (1984) Cellulose acetate electrophoresis of glycosaminoglycans in the central nervous system. Electrophoresis 5:97–101
Beutner EH, Trifshauser C, Hazen SP (1966) Collagenase activity of gingival tissue from patients with periodontal disease. Proc Soc Exp Biol Med 121:1082–1088
Bhatnagar RS, Rapaka SR (1976) Synthetic polypeptide models of collagen: synthesis and applications. In: Ramachandran GN, Reddi AH (eds) Biochemistry of collagen. Plenum, New York, pp 479–523
Bhatnagar RS, Rapaka SR, Liu TZ, Wolfe SM (1972) Hydralazine-induced disturbances in collagen synthesis. Biochim Biophys Acta 271:125–132
Bíbr B, Deyl Z, Lener J, Adam M (1977) Investigation on the reaction of molybdenum with collagen in vivo. Int J Pept Protein Res 10:190–196
Bíbr B, Deyl Z, Lener J, Kučera J, Simkova M (1987) The mechanism of action of molybdenum and tungsten upon collagen structures in vivo. Physiol Bohemoslov 36:417–424
Bieglmayr C, Hofer G (1989) Radioimmunoassay for immunoreactive non-collagenous domain of type IV collagen in serum: normal pregnancy and preeclampsia. J Clin Chem Clin Biochem 27:163–167
Biempica LA, Morecki R, Wu CH, Giambione MA, Rojkind M (1977) Isolation and immunohistochemical localization of a component of basement membrane collagen in human liver. Gastroenterology 73:1213
Biempica L, Morecki R, Wu CH, Giambone MA, Rojkind M (1980) Immunocytochemical localization of type B collagen. A component of basement membrane in human liver. Am J Pathol 98:591–602
Bienkowski RS (1984) Collagen degradation in human lung fibroblasts: extent of degradation, role of lysosomal proteases, and evaluation of an alternate hypothesis. J Cell Physiol 121:152–158
Bienkowski RS, Baum BJ, Crystal RG (1978) Fibroblasts degrade newly synthesized collagen within the cell before secretion. Nature 276:413–416
Bihari-Varga M (1986) Collagen, aging, and atherosclerosis. In: Gotto AM, Paoletti R (eds) Atherosclerosis reviews, vol 14. Raven, New York, pp 171–181
Bihari-Varga M, Simon J, Feher J, Gero S (1969) Thermal investigations on structural glycosaminoglycans and proteins. II. The influence of atherosclerosis on the thermal decomposition of aortic intima. Acta Biochim Biophys 4:279–285
Billingham MEJ, Davies GE (1979) Experimental models of arthritis in animals as screening tests for drugs to treat arthritis in man. In: Vane JR, Ferreira SH (eds) Anti-inflammatory drugs. Springer, Berlin Heidelberg New York, pp 108–144 (Handbook of experimental pharmacology, vol 50, part 2)
Birbeck MSC, Wheatley DN (1965) An electron microscopic study of the invasion of ascites tumor cells into the abdominal wall. Cancer Res 25:490–497
Birk DE, Lande MA (1981) Corneal and scleral collagen fiber formation in vitro. Biochim Biophys Acta 670:362–369
Birkedal-Hansen H (1982) Teeth. In: Weiss JB, Jayson MIV (eds) Collagen in health and disease. Churchill Livingstone, Edinburgh, pp 488–497
Birkedal-Hansen H, Butler WT, Taylor RE (1977) Proteins of the periodontium. Calcif Tissue Res 23:39–45
Birkedal-Hansen H, Taylor RE, Bhown AS, Katz J, Lin HY, Wells R (1985) Cleavage of bovine skin type III collagen by proteolytic enzymes. Relative resistance of the fibrillar form. J Biol Chem 260:16411–16417
Biswas C (1982) Host-tumor cell interactions and collagenase activity. In: Liotta LA, Hart IR (eds) Tumor invasion and metastasis. Nijhoff, The Hague, pp 405–425
Biswas C (1984) Collagenase stimulation in cocultures of human fibroblasts and human tumor cells. Cancer Lett 24:201–207
Biswas C (1985) Matrix influence on the tumor cell stimulation of fibroblast collagenase production. J Cell Biochem 28:39–45
Black MM, Bottoms E, Shuster S (1970) Skin collagen content and thickness in systemic sclerosis. Br J Dermatol 83:552–555
Bloodwork JMB Jr (1980) Pathology of the diabetic kidney. In: Friedman EA, L'Esperance FA Jr (eds) Diabetic renal-retinal syndrome. Grune and Stratton, New York, pp 159–174

Blum RH, Carter SK, Agre K (1973) A clinical review of bleomycin – a new antineoplastic agent. Cancer 31:903–914

Blumgart HL, Gilligan DR, Schlesinger MJ (1940) The degree of myocardial fibrosis in normal and pathological hearts as estimated chemically by the collagen content. Trans Assoc Am Physicians 55:313–315

Blumenkrantz N, Asboe-Hansen G (1973) Biosynthesis of collagen $^{14}C$-hydroxyproline and $^{14}C$-hydroxylysine during maturation of chick embryos. Mech Ageing Dev 1:445–450

Blumenkrantz N, Asboe-Hansen G (1974a) Connective tissue effects of drugs causing lupoid syndrome. Acta Derm Venerol (Stockh) 54:35–38

Blumenkrantz N, Asboe-Hansen G (1974b) Effect of diphenylhydantoin on connective tissue. Acta Neurol Scand 50:302–306

Blumenkrantz N, Asboe-Hansen G (1976) Cortisol effect on collagen biosynthesis in embryonic explants and in vitro hydroxylation of protocollagen. Acta Endocrinol (Copenh) 83:665–672

Boerger F (1914) Über zwei Fälle von Arachnodaktylie. Z Kinderheilkd 12:161–184

Bogue RH (1922) The chemistry and technology of gelatin and glue. McGraw-Hill, New York, pp 1–6

Boguta G, Dancewicz AM (1983) Radiolytic and enzymatic dimerization of tyrosyl residues in insulin, ribonuclease, papain and collagen. Int J Radiat Biol 43:249–265

Bohan A, Peters JB (1975) Polymyositis and dermatomyositis. N Engl J Med 292:403–407

Böhmer T (1971) The biosynthesis of collagen. Acta Rheum Scand 17:209–222

Bolarin DM, Savolainen ER, Kivirikko KI (1982) Enzymes of collagen synthesis and type III procollagen amino-propeptide in serum from Nigerians with hepatocellular carcinoma and other malignant diseases. Int J Cancer 29:401–405

Bolarin DM, Savolainen ER, Kivirikko KI (1984) Three serum markers of collagen biosynthesis in Nigerians with cirrhosis and various infectious diseases. Eur J Clin Invest 14:90–95

Bolarin DM, Barker K, Fuller GC (1987) Enzyme markers of collagen synthesis in carbon tetrachloride-induced fibrosis and during colchicine modification of $CCl_4$-induced injury. Exp Mol Pathol 46:145–152

Bolokin D, Tate RL, Luber JH, Kohn LD, Winard RJ (1975) Experimental exophthalmos: binding of thyrotropin and an exophthalmogenic factor dervived from thyrotropin to retro-orbital tissue plasma membranes. J Biol Chem 250:6516–6521

Boman TE, Balian G (1989) Extracellular matrix in bone marrow cell cultures. Synthesis of a 45K non-collagenous protein and a 17K protein sensitive to bacterial collagenase. Matrix 9:99–108

Bonadonna G, Sonenberg M, Merlino MJ (1965) Total urinary hydroxyproline excretion in diabetics before and after hypophysectomy and after growth hormone in adults hypopituitarism. Metabolism 14:832–835

Bond MD, Van Wart HE (1984a) Purification and separation of individual collagenases of Clostridium histolyticum using red dye ligand chromatography. Biochemistry 23:3077–3085

Bond MD, Van Wart HE (1984b) Relationship between the individual collagenases of Clostridium histolyticum: evidence for evolution by gene duplication. Biochemistry 23:3092–3099

Bond MD, Van Wart HE (1984c) Characterization of the individual collagenases from Clostridium histolyticum. Biochemistry 23:3085–3091

Bonnet J, Aumailley M, Gardères PE, Benchimol D, Morcau C, Brottier L, Mézard GS, Crockett R, Larrue J, Bricaud H (1987) Exploration du métabolisme du collagène chez le coronarien. Arch Mal Coeur 80:1577–1584

Bonnet J, Gardères PE, Aumailley M, Moreau C, Gouverneur G, Benchimol D, Crockett R, Larrue J, Bricaud H (1988) Serum type III procollagen peptide levels in coronary artery diseases (a marker of atherosclerosis). Eur J Clin Invest 18:18–21

Bonnin CM, Sparrow MP, Taylor RR (1981) Collagen synthesis and content in right ventricular hypertrophy in the dog. Am J Physiol 241:H708–H713

Bora FW Jr, Lane JM, Prockop DJ (1972) Inhibitors of collagen biosynthesis as a means of controlling scar formation in tendon injury. J Bone Joint Surg 54A:1501–1508

Boransky R (1950) Guide to the literature on collagen. Eastern Regional Research Laboratory, Philadelphia

Borel JP, Vieillard A, Randoux A (1976) Effect of some purified plasma proteins on collagen biosynthesis in vitro. Agents Actions 6:207–210

Borel JP, Maquart FX, Randoux A (1984) Contrôle hormonal du métabolisme du collagène. Pathol Biol 32:795–812
Borg TK, Caulfield JB (1979) Collagen in the heart. Tex Rep Biol Med 39:321–333
Borg TK, Caulfield JB (1981) The collagen matrix of the heart. Fed Proc 40:2037–2041
Bork K, Korting GW (1983) Symptomatische Sklerodermie durch Bleomyzin. Hautarzt 34:10–12
Bornstein P (1974) The biosynthesis of collagen. Ann Rev Biochem 43:567–603
Bornstein P, Byers PH (1980) Disorders of collagen metabolism. In: Bondy P, Rosenberg LE (eds) Metabolic control of disease, 7th edn. Saunders, Philadelphia, pp 1089–1153
Bornstein P, Piez KA (1966) The nature of the intramolecular cross-link in collagen. Biochemistry 5:3460–3473
Bornstein P, Sage H (1980) Structurally distinct collagen types. Ann Rev Biochem 49:957–1003
Bornstein P, Traub W (1979) The chemistry and biology of collagen. In: Neurath H, Hill RL (eds) The proteins, vol 4, 3rd edn. Academic, London, pp 441–632
Bornstein P, McKay J, Liska DAJ, Apone S, Devarayalu S (1988a) Interactions between the promoter and first intron are involved in transcriptional control of $\alpha$ 1(I) collagen gene expression. Mol Cell Biol 8:4851–4857
Bornstein P, McKay J, Devarayalu S, Cook SC (1988b) A highly conserved, 5'-untranslated, inverted repeat sequence is ineffective in translational control of the $\alpha$ 1(I) collagen gene. Nucleic Acids Res 16:9721–9736
Botstein GR, Sherer GK, LeRoy EC (1982) Fibroblast selection in scleroderma. An alternative model of fibrosis. Arthritis Rheum 25:189–195
Boucek RJ, Noble NL, Gunya-Smith Z, Butler WT (1981) The Marfan syndrome: a deficiency of chemically stable crosslinks. N Engl J Med 305:988–991
Bounameaux V (1959) L'accolement des plaquettes fibres sousendothéliales. C R Séan Soc Biol (Paris) 153:865–867
Bou-Resli MN, Al-Zaid NS, Ibrahim MEA (1981) Full-term and prematurely ruptured fetal membranes. An ultrastructural study. Cell Tissue Res 220:263–268
Bowes JH, Moss JA (1962) The effect of gamma radiation on collagen. Radiat Res 16:211–223
Boyd CD, Weliky K, Toth-Fejel S, Deak SB, Christiano AM, Mackenzie JW, Sandell LJ, Tryggvason K, Magenis E (1986) The single copy gene coding for human $\alpha$ 1(V) procollagen is located at the terminal end of the long arm of chromosome 13. Hum Genet 74:121–125
Braams R (1961) The effect of electron radiation on the tensile strength of tendon, part I. Int J Radiat Biol 4:27–31
Braams R (1963) The effect of electron radiation on the tensile strength of tendon, part II. Int J Radiat Biol 7:29–39
Brada Z, Bulba S, Chen MC (1972) Influence on some nitrogen derivatives of phenanthrene on chronic and acute effect of ethionine. Fed Proc 31:842
Bradley KH, McConnell SD, Crystal RG (1974) Lung collagen composition and synthesis: characterization and changes with age. J Biol Chem 249:2674–2683
Bradley KH, McConnell-Breul S, Crystal RG (1975) Collagen in the human lung: quantitation of rates of synthesis and partial characterization of composition. J Clin Invest 55:543–550
Bradway W (1929) The morphogenesis of the thyroid follicles of the chick. Anat Rec 42:157–167
Brady AH (1975) Collagenase in scleroderma. J Clin Invest 56:1175–1180
Branford White CJ, Hudson M (1977) Characterization of a chondroitin sulphate proteoglycan from bovine brain. J Neurochem 28:581–588
Brass LF, Bensusan HB (1974) The role of collagen quarternary structure in the platelet collagen interaction. J Clin Invest 54:1480–1487
Braun-Falco O (1957) Über das Verhalten der interfibrillaren Grundsubstanz bei Sklerodermie. Dermatol Monatsschr 136:1085–1090
Brenner DA, O'Hara M, Angel P, Chojkier M, Karin M (1989a) Prolonged activation of jun and collagenase genes by tumor necrosis factor-$\alpha$. Nature 337:661–663
Brenner DA, Rippe RA, Veloz L (1989b) Analysis of the collagen $\alpha$ 1(I) promoter. Nucleic Acids Res 17:6055–6064
Bretlau P, Balle V, Causse JB, Hørslev-Petersen K, Sørensen CH, Sølvsteen M (1987) Is otosclerosis an autoimmune disease? Immunobiology, histophysiology and tumor immunology in

otolaryngology. Proceedings of the 2nd international academic conference, Utrecht, Aug 26–29, 1986. Kugler, Amsterdam, pp 201–206
Brettschneider I, Praus R, Musilová J, Adam M (1976a) Effect of some antirheumatics on connective tissue components. Arzneimittelforschung 26:846–848
Brettschneider I, Musilová J, Olsovska Z, Adam M, Praus R (1976b) The effect of diftalone on collagen and glycosaminoglycans metabolism. Pharmacology 14:414–421
Breul SD, Bradley KH, Hance AJ, Schafer MP, Berg RA, Crystal RG (1980) Control of collagen synthesis by human diploid lung fibroblasts. J Biol Chem 255:5250–5260
Briancon D, Meunier PJ (1981) Treatment of osteoporosis with fluoride, calcium, and vitamin D. Orthop Clin North Am 12:629–648
Brinckerhoff C, Harris ED Jr (1981) Modulation by retinoic acid, and corticosteroids of collagenase production by rabbit synovial fibroblasts treated with phorbol myristate acetate or poly(ethylene glycol). Biochim Biophys Acta 677:424–432
Brinckerhoff CE, McMillan RM, Fahey JV, Harris ED Jr (1979) Collagenase production by synovial fibroblasts treated with phorbol myristate acetate. Arthritis Rheum 22:1109–1116
Brinckerhoff CE, McMillan RM, Dayer JM, Harris ED Jr (1980) Inhibition by retinoic acid of collagenase production in rheumatoid synovial cells. N Engl J Med 303:432–436
Brinckerhoff CE, Nagase H, Nagel JE, Harris ED Jr (1982) Effects of all-trans-retinoic acid (retinoic acid) and 4-hydroxyphenylretinamide on synovial cells and articular cartilage. J Am Acad Dermatol 6:591–602
Brinckerhoff CE, Coffey JW, Sullivan AC (1983) Inflammation and collagenase production in rats with adjuvant arthritis reduced with 13-cis-retinoic acid. Science 221:756–758
Brinckerhoff CE, Mitchell TI, Karmilowicz MJ, Kluve-Beckerman B, Benson MD (1989) Autocrine induction of collagenase by serum amyloid $\alpha$-like and $\beta_2$-microglobulin-like proteins. Science 243:655–657
Brock DG, Neubauer HP, Strecker H (1985) Type IV collagen antigens in serum of diabetic rats: a marker for basement membrane collagen biosynthesis. Diabetologia 28:928–932
Broek DL, Madri J, Eikenberry EF, Brodsky B (1985) Characterization of the tissue form of type V collagen from chick bone. J Biol Chem 260:555–562
Brooks C, Simons ER (1978) Effect of age on collagen fibril formation. Gerontology 24:161–168
Brown CH (1975) Structural materials in animals. Pirman, London
Brown R (1982) Placenta. In: Weiss JB, Jayson MIV (eds) Collagen in health and disease. Churchill Livingstone, Edinburgh, pp 456–465
Brown MD (1983) Intradiscal therapy. Chymopapain or collagenase. Year Book Medical Publishers, Chicago
Brown SI, Weller CA (1970) Cell origin of collagenase in normal and wounded corneas. Arch Ophthalmol 83:74–77
Brown SI, Weller CA, Wassermann HE (1969) Collagenolytic activity of alkali-burned corneas. Arch Ophthalmol 81:370–373
Brown KS, Cranley RE, Greene R, Kleinman HK, Pennypacker JP (1981) Disproportionate micromelia: an incomplete dominant mouse dwarfism with abnormal cartilage matrix. J Embryol Exp Morphol 62:165–182
Browning MC, Weidner N, Lorentz WB Jr (1988) Renal histopathology of the nail-patella syndrome in a two-year-old boy. Clin Nephrol 29:210–213
Bruckner P, Vaughan L, Winterhalter KH (1985) Type IX collagen from sternal cartilage of chicken embryo contains covalently bound glycosaminoglycans. Proc Natl Acad Sci USA 82:2608–2612
Bruns RR (1969) A symmetrical extracellular fibril. J Cell Biol 42:418–430
Buccino RA, Harris E, Spann JF Jr, Sonnenblick EH (1969) Response of myocardial connective tissue to development of experimental hypertrophy. Am J Physiol 216:425–428
Buckingham RB, Prince RK, Rodnan GP, Taylor F (1978) Increased collagen accumulation in dermal fibroblast cultures from patients with progressive systemic sclerosis (scleroderma). J Lab Clin Med 92:5–21
Buckingham BA, Uitto J, Sandberg C, Keens T, Roe T, Costin G, Kaufman F, Bernstein B, Landing B, Castellano A (1984) Scleroderma-like changes in insulin-dependent diabetes mellitus: clinical and biochemical study. Diabetes Care 7:163–169

Bullough PG, Munuera L, Murpha J, Weinstein AM (1970) The strength of the menisci of the knee as it relates to fine structure. J Bone Joint Surg 52B:564–570

Bunge MB, Williams AK, Wood PM, Uitto J, Jeffrey JJ (1980) Comparison of nerve cell and nerve cell plus Schwann cell cultures, with particular emphasis on basal lamina and collagen formation. J Cell Biol 84:184–202

Bunge RP, Bunge MB (1978) Evidence that contact with connective tissue matrix is required for normal interaction between Schwann cell and nerve fibers. J Cell Biol 78:943–950

Bunge RP, Bunge MB, Williams AK, Wartels LK (1982) Does the dystrophic mouse nerve lesion result from an extracellular matrix abnormality? In: Schotland DL (ed) Disorders of the motor unit. Wiley, New York

Burgeson R (1982) Articular cartilage, intervertebral disc, synovia. In: Weiss JB, Jayson MIV (eds) Collagen in health and disease. Churchill Livingstone, Edinburgh, pp 335–361

Burgeson RE (1987a) The collagens of skin. Curr Probl Dermatol 17:61–75

Burgeson RE (1987b) Type VII collagen. In: Mayne R, Burgeson RE (eds) Structure and function of collagen types. Academic, Orlando, pp 145–172

Burgeson RE, Hollister DW (1979) Collagen heterogeneity in human cartilage: identification of several new collagen chains. Biochem Biophys Res Commun 87:1124–1131

Burgeson RE, El Adli FA, Kaitila II, Hollister DW (1976) Fetal membrane collagens: identification of two new collagen $\alpha$ chains. Proc Natl Acad Sci USA 73:2579–2583

Burke JM, Ross R (1977) Collagen synthesis by monkey arterial smooth muscle cells during proliferation and quiescence in culture. Exp Cell Res 107:387

Burleigh PMC, Poole AR (eds) (1974) Dynamics of connective tissue macromolecules. North-Holland, Amsterdam, pp 1–434

Burleigh MC, Werb Z, Reynolds JJ (1977) Evidence that species specificity and rate of collagen degradation are properties of collagen, not collagenase. Biochim Biophys Acta 494:198–208

Burnell JM, Baylkink DJ, Chestnut CH III, Mathews MW, Teubner EJ (1982) Bone matrix and mineral abnormalities in postmenopausal osteoporosis. Metabolism 31:1113–1120

Burri PH, Dbaly J, Weibel ER (1974) The postnatal growth of the rat lung. I. Morphology. Anat Rec 178:711–730

Bushell GR, Ghosh P, Taylor TKF (1978) Collagen defect in idiopathic scoliosis. Lancet 2:94–95

Butkowski RJ, Langeveld JPM, Wieslander J, Hamilton J, Hudson BG (1987) Localization of the Goodpasture epitope to a novel chain of basement membrane collagen. J Biol Chem 262:7874–7877

Butler WT (1984) Matrix macromolecules of bone and dentin. Coll Relat Res 4:297–307

Butler WT, Fujioka T (1972) Fine structure of the annulus fibrosus of the intervertebral disc of the cat. Anat Anz 132:454–464

Butler WT, Birkedal-Hansen H, Beegle WF, Taylor RE, Chung E (1975) Proteins of the periodontium. Identification of collagens with the $\alpha 1(I)_2\alpha 2$ and $\alpha 1(III)_3$ structures in bovine periodontal ligament. J Biol Chem 250:8907–8912

Byers PH (1983) Inherited disorders of collagen biosynthesis: Ehlers-Danlos syndrome, the Marfan syndrome and osteogenesis imperfecta. In: Spittell P (ed) Clinical medicine, vol 12. Harper and Row, Philadelphia, pp 1–41

Byers PH, Bonadio JF (1985) The molecular basis of clinical heterogeneity in osteogenesis imperfecta: mutation in type I collagen genes have different effects on collagen processing. In: Lloyd J, Scriver CR (eds) Metabolic and genetic diseases in pediatrics. Butterworths, London, pp 56–90

Byers PH, Narayanan AS, Bornstein P, Hall JG (1976) An X-linked form of cutis laxa due to deficiency of lysyl oxidase. Birth Defects 12:293–298

Byers PH, Holbrook KA, Hall JG, Bornstein P, Chandler JW (1978) A new variety of spondyloepiphyseal dysplasia characterized by punctate corneal dystrophy and abnormal dermal collagen fibrils. Hum Genet 40:157–169

Byers PH, Holbrook KA, McGillivray B, MacLeod PM, Lowry RB (1979) Clinical and ultrastructural heterogeneity of type IV Ehlers-Danlos syndrome. Hum Genet 47:141–150

Byers PH, Siegel RC, Holbrook KA, Narayanan AS, Bornstein P, Hall JG (1980) X-linked cutis laxa: defective collagen crosslink formation due to decreased lysyl oxidase activity. N Engl J Med 303:61–65

Byers PH, Holbrook KA, Barsh GS (1981a) Type IV Ehlers-Danlos syndrome. In: Akenson W, Bornstein P, Glimcher MJ (eds) Proceedings of the workshop on heritable disorders of connective tissue. Mosby, St Louis

Byers PH, Siegel RC, Petersen KE, Rowe DW, Holbrook KA, Smith LT, Chang YH, Fu JCC (1981b) Marfan syndrome: abnormal $\alpha 2$ chain type I collagen. Proc Natl Acad Sci USA 78:7455–7459

Byers PH, Holbrook KA, Barsh GS, Smith LT, Bornstein P (1981c) Altered secretion of type III procollagen in a form of type IV Ehlers-Danlos syndrome: biochemical studies in cultured fibroblasts. Lab Invest 44:336–341

Byers PH, Holbrook KA, Barsh GS (1983) Ehlers-Danlos syndrome. In: Emery AEH, Rimoin DL (eds) Principles and practice of medical genetics, vol 2. Churchill Livingstone, Edinburgh, pp 836–850

Čabak V, Dickerson JWT, Widdowson EM (1963) Response of young rats to deprivation of protein or of calories. Br J Nutr 17:601–616

Cajander S, Bjersing L (1976) Further studies of the surface epithelium covering preovulatory rabbit follicles with special reference to lysosomal alteration. Cell Tissue Res 169:129–141

Callis AH, Sohnle PG, Mandel GS, Weissner J, Mandel NS (1985) Kinetics of inflammatory and fibrotic pulmonary changes in a murine model of silicosis. J Lab Clin Med 105:547–553

Campbell PM, LeRoy EC (1975) Pathogenesis of systemic sclerosis: vascular hypothesis. Arthritis Rheum 18:351–369

Cannon DJ, Cintron C (1975) Collagen cross-linking in corneal scar formation. Biochim Biophys Acta 412:18–25

Cannon DJ, Foster CS (1978) Collagen cross-linking in keratoconus. Invest Ophthalmol Vis Sci 17:63–65

Cannon RO III, Butany JW, McManus BM, Speir E, Kravitz AB, Bolli R, Ferrans VJ (1983) Early degradation of collagen after acute myocardial infarction in the rat. Am J Cardiol 52:390–395

Caplan RM (1967) Visceral involvement in lipoid proteinosis. Arch Dermatol 95:149–155

Capodici C, Berg RA (1989) Cathepsin G degrades denatured collagen. Inflammation 13:137–145

Capodici C, Muthukumaran G, Amoruso MA, Berg RA (1989) Activation of neutrophil collagenase by cathepsin G. Inflammation 13:245–258

Cappelli V, Forni R, Poggesi C, Reggiani C, Ricciardi L (1984) Age-dependent variations of diastolic stiffness and collagen content in rat ventricular myocardium. Arch Int Physiol Biochim 92:93–106

Carey DJ, Eldridge CF, Cornbrooks CJ, Timpl R, Bunge RP (1983) Biosynthesis of type IV collagen by cultured rat Schwann cells. J Cell Biol 98:473–479

Carmichael DJ, Dodd CM, Nawrot CF (1974) Studies on matrix proteins of normal and lathyritic rat bone and dentine. Calcif Tissue Res 14:177–182

Carmichael DJ, Dodd CM, Veis A (1977) The solubilization of bone and dentin collagens by pepsin. Effect of cross-linkages and non-collagen components. Biochim Biophys Acta 491:177–183

Carr I, Levy M, Orr K, Bruni J (1985) Lymph node metastasis and cell movement: ultrastructural studies on the rat 13762 mammary carcinoma and Walker carcinoma. Clin Exp Metastasis 3:125–139

Carrington MJ, Bird TA, Levene CI (1984) The inhibition of lysyl oxidase in vivo by isoniazid and its reversal by pyridoxal. Effect on collagen cross-linking in the chick embryo. Biochem J 221:837–843

Cârsteanu M, Vlădescu C (1982) Colagenul biochimie şi fiziologie. Editura Academiei Rep Soc România, Bucuresti

Carter WG (1982a) The cooperative role of the transformation-sensitive glycoproteins, GP 140 and fibronectin in cell attachment and spreading. J Biol Chem 257:3249–3257

Carter WG (1982b) Transformation-dependent alterations in glycoproteins of the extracellular matrix fibroblasts. Characterization of GP250 and the collagen-like GP140. J Biol Chem 257:13805–13815

Carter C, Wilkinson J (1964) Persistent joint laxity and congenital dislocation of the hip. J Bone Joint Surg 46B:40–45

Carter EA, McCarron MJ, Alpert E, Isselbacher KJ (1980) Lysyl oxidase and collagenase activities in hepatic fibrosis. Gastroenterology 78:1303
Carvier JP, Blout ER (1967) Polypeptide models for collagen. In: Ramachandran GN (ed) Treatise on collagen, vol I. Academic, London, pp 441–523
Caspari PG, Gibson K, Harris P (1975) Collagen and the myocardium. A study of their normal development and relationship in the rabbit. Cardiovasc Res 9:187–189
Caspari PG, Newcomb M, Gibson K, Harris P (1977) Collagen in the normal and hypertrophied human ventricle. Cardiovasc Res 11:554–558
Cassel J (1959) Effects of γ-radiation on collagen. J Am Leather Chem Assoc 54:432–451
Casuccio C (1962) An introduction to the study of osteoporosis. Biochemical and biophysical research in bone ageing. Proc R Soc Med 55:663–668
Castagnola P, Dozin B, Moro G, Cancedda R (1988) Changes in the expression of collagen genes show two stages in chondrocyte differentiation in vitro. J Cell Biol 106:461–467
Castejón HV (1970) Histochemical demonstration of acid glycosaminoglycans in the nerve cell cytoplasm of mouse central nervous system. Acta Histochem 35:161–172
Caulfield JB, Borg TK (1979) The collagen network of the heart. Lab Invest 40:364–372
Caulfield JB, Tao SB, Nichtigal M (1984) Ventricular collagen matrix and alterations. Adv Myocardial 5:257–269
Caulfield JP, Hein A, Dynesius-Trentham R, Trentham DE (1982) Morphological demonstration of two stages in the development of type II collagen-induced arthritis. Lab Invest 46:321–343
Cavalleri A, Gobba F, Bacchella L, Luberto F, Ziccardi A (1988) Serum type III procollagen peptide in asbestos workers: an early indicator of pulmonary fibrosis. Br J Ind Med 45:818–823
Cawston TE, Murphy G, Mercer E, Galloway WA, Hazleman BL, Reynolds JJ (1983) The interaction of purified rabbit bone collagenase with purified rabbit bone metalloproteinase inhibitor. Biochem J 211:313–318
Centrella M, Canalis E, McCarthy TL, Stewart AF, Orloff JJ, Insogna KL (1989a) Parathyroid hormone-related protein modulates the effect of transforming growth factor-β on deoxyribonucleic acid and collagen synthesis in fetal rat bone cells. Endocrinology 125:199–208
Centrella M, McCarthy TL, Canalis E (1989b) Platelet-derived growth factor enhances deoxyribonucleic acid and collagen synthesis in osteoblast-enriched cultures from fetal rat parietal bone. Endocrinology 125:13–19
Chamson A, Berbis P, Fabre JF, Privat Y, Frey J (1987) Collagen biosynthesis and isomorphism in a case of Ehlers-Danlos syndrome type VI. Arch Dermatol Res 279:303–307
Charrière G, Hartmann DJ, Vignon E, Ronzière MC, Herbage D, Ville G (1988) Antibodies to types I, II, IX, and XI collagen in the serum of patients with rheumatic diseases. Arthritis Rheum 31:325–332
Charron D, Roberts L, Couty MC, Binet JL (1979) Biochemical and histological analysis of bone marrow collagen in myelofibrosis. Br J Haematol 41:151–161
Cheah KSE (1985) Collagen genes and inherited connective tissue disease. Biochem J 229:287–303
Cheah KSE, Grant ME (1982) Procollagen genes and messenger RNAs. In: Weiss JB, Jayson MIV (eds) Collagen in health and disease. Churchill Livingstone, Edinburgh, pp 73–100
Chen KH, Paz MA, Gallop PM (1977) Collagen prolyl hydroxylation in WI-38 fibroblasts: action of hydralazine. In Vitro 13:49–54
Chen TS, Boesch CL, Leevy CM (1983) Hepatic proline after bile duct ligation in rats. Experientia 39:585–586
Chen-Kiang S, Cardinale GJ, Udenfriend S (1977) Homology between a prolyl hydroxylase subunit and a tissue protein that crossreacts immunologically with the enzyme. Proc Natl Acad Sci USA 74:4420–4424
Cheung HS (1987) Distribution of type I, II, III and V in the pepsin solubilized collagens in bovine menisci. Connect Tissue Res 16:343–356
Cheung HS, Halverson PB, McCarty DJ (1981) Release of collagenase, neutral protease, and prostaglandins from cultured mammalian synovial cells by hydroxyapatite and calcium pyrophosphate dihydrate crystals. Arthritis Rheum 24:1338–1344
Chiang TM, Kang AH (1982) Isolation and purification of a collagen α1(I) receptor from human platelet membrane. J Biol Chem 257:7581–7586

Chiang TM, Mainardi CL, Seyer JM, Kang AH (1980) Collagen-platelet interaction. Type V (AB) collagen induces platelet aggregation. J Lab Clin Med 95:99–107

Chiang TM, Kang AH, Dale JB, Beachey EH (1984) Immunochemical studies of the purified human platelet receptor for the $\alpha 1$(I) chain of chick skin collagen. J Immunol 133:872–876

Chiang TM, Jin A, Hasty KA, Kang AH (1989) Collagen-platelet interaction: inhibition by a monoclonal antibody which binds a 90000 dalton platelet glycoprotein. Thromb Res 53:129–143

Child A, Handler C, Light N, Dorrance D (1986) Increased prevalence of mitral valve prolapse associated with an elevated skin type III/III + I collagen ratio in joint hypermobility syndrome. Agents Actions [Suppl] 18:125–129

Chiumello G, del Guercio MJ (1965) Modificazioni della idrossiprolinuria nel trattamento dell'ipotiroidismo e del nanismo ipopituitarico. Minerva Pediatr 17:1069–1073

Chmiel J, Suszka B, Prętczak J, Ignatowicz E, Błażejewska B (1980) Glikozaminoglikany w moczu dzieci ze skoliozą idiopatyczną (Urinary glycosaminoglycans in children with idiopathic scoliosis). Chir Narządów Ruchu Ortop Pol 45:37–41 (in Polish)

Chojkier M, Brenner DA (1988) Therapeutical strategies for hepatic fibrosis. Hepatology 8:176–182

Christner P (1980) Collagenase in the human periodontal ligament. J Pedriodontol 51:455–461

Christner P, Robinson P, Clark CC (1977) A preliminary characterization of human cementum collagen. Calcif Tissue Res 23:147–153

Chu ML, Williams CJ, Pepe G, Hirsch JL, Prockop DJ, Ramirez F (1983) Internal deletion in a collagen gene in a perinatal lethal form of osteogenesis imperfecta. Nature 304:78–80

Chu ML, Garguilo V, Williams CJ, Ramirez F (1985) Multiexon deletion in an osteogenesis imperfecta variant with increased type III collagen mRNA. J Biol Chem 260:691–694

Chu ML, Conway D, Pan TC, Baldwin C, Mann K, Deutzmann R, Timpl R (1988) Amino acid sequence of the triple-helical domain of human collagen type VI. J Biol Chem 263:18601–18606

Chua CC, Ladda RL (1987) Effect of tunicamycin on the biosynthesis of human fibroblast collagenase. Coll Relat Res 7:285–293

Chua CC, Barritault D, Geiman DE, Ladda RL (1987) Induction and suppression of type I collagenase in cultured human cells. Coll Relat Res 7:277–284

Chung E, Rhodes RK, Miller EJ (1976) Isolation of three collagenous components of probable basement membrane origin from several tissues. Biochem Biophys Res Commun 71:1167–1174

Church RL (1980) Procollagen and collagen produced by normal bovine corneal stroma fibroblasts in cell culture. Invest Ophthalmol Vis Sci 19:192–202

Chvapil M (1960) Biochemický vývoj vaziva (Biochemistry of connective tissue). Státni zdravotnické nakl, Praha (in Czech)

Chvapil M (1961) Fysiologie vaziva (Physiology of connective tissue). Cesk Fysiol 10:135–154 (in Czech)

Chvapil M (1967) The physiology of connective tissue. Butterworth, London

Chvapil M (1974) Pharmacology of fibrosis: definitions, limits and perspectives. Life Sci 16:1345–1362

Chvapil M (1982) Experimental modifications of collagen synthesis and degradation and their therapeutic application. In: Weiss JB, Jayson MIV (eds) Collagen in health and disease. Churchill Livingstone, Edinburgh, pp 206–217

Chvapil M (1988) Method of treatment of fibrotic lesions by topical administration of lathyrogenic drugs. In: Nimni ME (ed) Collagen, vol II. CRC Press, Boca Raton, pp 161–176

Chvapil M, Hurych J (1968) Control of collagen biosynthesis. Int Rev Connect Tissue Res 4:67–196

Chvapil M, Brada Z (1971) Protective effect of 1,10-phenanthroline on two models of liver fibrosis. Fed Proc 30:230 (abs no 209)

Chvapil M, Koopman CF (1984) Scar formation: physiology and pathological states. Otolaryngol Clin North Am 17:265–272

Chvapil M, Ryan JN (1973) The pool of free proline in acute and chronic liver injury and its effect on the synthesis of collagen and globular proteins. Agents Actions 3:38–44

Chvapil M, Hurych J, Ehlichová E, Čmuchalova B (1967a) Effects of various chelating agents, quinones, diazoheterocyclic compounds and other substances on proline hydroxylation and synthesis of collagenous and non-collagenous proteins. Biochim Biophys Acta 140:339–348

Chvapil M, Hurych J, Ehrlichová E, Tichy M (1967b) Mechanism of the action of chelating agents on proline hydroxylation and its incorporation into collagenous and non-collagenous proteins. Eur J Biochem 2:229–235

Chvapil M, Hurych J, Ehrlichová E (1968) Effect of long term in vivo application of phenenthroline, penicillamine and further chelating agents on the synthesis of collagenous and non-collagenous proteins in fibrotic liver and wound granulation tissue. Z Physiol Chem 349:218–222

Chvapil M, Madden JW, Carlson ED (1974a) Effect of cis-hydroxy-proline on collagen and other proteins in skin wounds, granuloma tissue, and liver of mice and rats. Exp Mol Pathol 20:363–373

Chvapil M, McCarthy D, Madden JW, Peacock EE Jr (1974b) Effect of 1,10-phenanthroline and desferrioxamine in vivo on prolyl hydrxylase and hydroxylation of collagen in various tissues of rats. Biochem Pharmacol 23:2165–2173

Chvapil M, Hameroff SR, O'Dea K, Peacock EE Jr (1979a) Local anesthetics and wound healing. J Surg Res 27:367–371

Chvapil M, Eskelson CD, Stiffel V, Owen JA (1979b) Early changes in the chemical composition of the rat lung after silica administration. Arch Environ Health 34:402–406

Chvapil M, Peacock EE Jr, Carlson EC, Blau S, Steinbronn K, Morton D (1980) Colchicine and wound healing. J Surg Res 28:49–56

Chyun YS, Raisz LG (1984) Stimulation of bone formation by prostaglandin $E_2$. Prostaglandins 27:96–103

Cidadão AJ (1989) Interactions between fibronectin, glycosaminoglycans and native collagen fibrils-: an EM study in artificial three-dimensional extracellular matrices. Eur J Cell Biol 48:303–312

Cidadão AJ, Thorsteinsdóttir S, David-Ferreira JF (1988) Reevaluation of fibronectin-collagen interactions in tissues: an immunocytochemical and immunochemical study. J Histochem Cytochem 36:639–648

Cintron C (1974) Hydroxylysine glycosides in the collagen of normal and scarred rabbit corneas. Biochem Biophys Res Commun 60:288–294

Cintron C, Hong BS (1988) Heterogeneity of collagens in rabbit cornea: type VI collagen. Invest Ophthalmol Vis Sci 29:760–766

Cintron C, Hassinger LC, Kublin CL, Cannon DJ (1978) Biochemical and ultrastructural changes in collagen during corneal would healing. J Ultrastruct Res 65:13–22

Cintron C, Hong BS, Kublin CL (1981) Quantitative analysis of collagen from normal developing corneas and corneal scars. Curr Eye Res 1:1–8

Cintron C, Hong BS, Covington HI, Macarak EJ (1988) Heterogeneity of collagens in rabbit cornea: type III collagen. Invest Ophthalmol Vis Sci 29:767–775

Clague RB, Shaw MJ, Holt PJL (1980) Incidence of serum antibodies to native type I and type II collagens in patients with inflammatory arthritis. Ann Rheum Dis 39:201–206

Clague RB, Shaw MJ, Holt PJL (1981) Incidence and correlation between serum IgG and IgM antibodies to native type II collagen in patients with chronic inflammatory arthritis. Ann Rheum Dis 40:6–10

Clague RB, Firth SA, Holt PJL, Skingle J, Greenbury CL, Webley M (1983) Serum antibodies to type II collagen in rheumatoid arthritis: comparison of 6 immunological methods and clinical features. Ann Rheum Dis 42:537–544

Claire M, Jacotot B, Robert L (1976) Characterization of lipids associated with macromolecules of the intercellular matrix of human aorta. Connect Tissue Res 4:61–71

Clark CC, Richards CF (1985) Isolation and partial characterization of precursors to minor cartilage collagens. Coll Relat Res 5:205–223

Clark DE, Wei P, Grant NH (1985) A novel inhibitor of mammalian collagenase. Life Sci 37:575–578

Clark JG, Starcher BC, Uitto J (1980) Bleomycin-induced synthesis of type I procollagen by human lung and skin fibroblasts in culture. Biochim Biophys Acta 631:359–370

Clark JG, Dedon TF, Wayner EA, Carter WG (1989) Effects of interferon-$\gamma$ on expression of cell surface receptors for collagen and deposition of newly synthesized collagen by cultured human lung fibroblasts. J Clin Invest 83:1505–1511
Clark RAF (1988) Overview and general considerations of wound repair. In: Clark RAF, Henson PM (eds) The molecular and cellular biology of wound repair. Plenum, New York, pp 3–33
Clark RAF, Henson PM (eds) (1988) The molecular and cellular biology of wound repair. Plenum, New York
Clark RAF, Lanigan BS, Dellapelle P, Manseau E, Dvorak HF, Colvin RB (1982) Fibronectin and fibrin provide a provisional matrix for epidermal cell migration during wound reepithelialization. J Invest Dermatol 79:264–269
Clausen B (1962) Influence of age on connective tissue. Lab Invest 11:229–234
Clement B, Guguen-Guillouzo C, Grimaud JA, Rissel M, Guillouzo A (1988a) Effect of hydrocortisone on deposition of types I and IV collagen in primary culture of rat hepatocytes. Cell Mol Biol 34:449–460
Clement B, Laurent M, Guguen-Guillouzo C, Lebean G, Guillouzo A (1988b) Types I and IV procollagen gene expression in cultured rat hepatocytes. Coll Relat Res 8:349–359
Clemmons DR, Van Wyk JJ (1981) Somatomedin-C and platelet-derived growth factor stimulate human fibroblast replication. J Cell Physiol 106:361–367
Clore JN, Cohen IK, Diegelmann RF (1979) Quantitative assay of type I and III collagen synthesized by keloid biopsies and fibroblasts. Biochim Biophys Acta 586:384–390
Cohen G, Fehr K, Wagenhäuser FJ (1983a) Leukocyte elastase and free collagenase activity in synovial effusions: relation to number of polymorphonuclear leukocytes. Rheumatol Int 3:89–95
Cohen I, Mosher M, O'Keefe E (1972) Cutaneous toxicity of bleomycin therapy. Arch Dermatol 107:553–555
Cohen IK, Diegelmann RF (1978) Effect of N-acetyl-cis-4-hydroxyproline on collagen synthesis. Exp Mol Pathol 28:58–64
Cohen S, Johnson AR, Hurd E (1983b) Cytotoxicity of sera from patients with scleroderma. Arthritis Rheum 26:170–178
Cohn DH, Apone S, Eyre DR, Starman BJ, Andreassen P, Charbonneau H, Nicholls AC, Pope FM, Byers PH (1988) Substitution of cysteine for glycine within the carboxyl-terminal telopeptide of the $\alpha 1$ chain of type I collagen produces mild osteogenesis imperfecta. J Biol Chem 263:14605–14607
Cole WG (1988) Osteogenesis imperfecta. Baillieres Clin Endocrinol Metab 2:243–265
Cole BC, Washburn LR, Samuelson CO Jr, Ward JR (1982) Experimental models of rheumatoid arthritis, systemic lupus erythematosus and scleroderma. In: Panayi GS (ed) Scientific basis of rheumatology. Churchill Livingstone, Edinburgh, pp 22–52
Cole WG, Chan D, Hickey AJ, Wilcken DEL (1984) Collagen composition of normal and myxomatous human mitral heart valves. Biochem J 219:451–460
Collier LL, Leathers CW, Counts DF (1980) A clinical description of dermatosparaxis in a Himalayan cat. Feline Pract 10:25–36
Collins D, Lindberg K, McLees B, Pinnell S (1977) The collagen of heart valve. Biochim Biophys Acta 495:129–139
Collins JF, Jones MA (1978) Connective tissue proteins of the baboon lung: concentration, content and synthesis of collagen in the normal lung. Connect Tissue Res 5:211–215
Comens P (1961) Hydralazine lupus syndrome. In: Mills LC, Moyer JH (eds) Inflammation and diseases of connective tissue. Saunders, Philadelphia, pp 190–200
Conen PF (1971) Histopathological study of tissues in scoliotic patients. J Bone Joint Surg 53A:199
Conrad GW, Dessau W, von der Mark K (1980) Synthesis of type III collagen by fibroblasts from the embryonic chick cornea. J Cell Biol 84:501–512
Cornbrooks CJ, Mitheu F, Cochran JM, Bunge RP (1983) Factors affecting Schwann cell basal lamina formation cultures of dorsal root ganglia from mice with muscular dystrophy. Dev Brain Res 6:57–67
Counte N, Scandellari C, Guido R (1970) L'escrezione urinaria di idrossiprolina in corso di ipertirodismo. Omnia Med Ther 48:497–506

Counts DF, Knighten P, Hegreberg GA (1977) Biochemical changes in the skin of mink with Ehlers-Danlos syndrome: increased collagen biosynthesis in the dermis of affected mink. J Invest Dermatol 69:521–526

Counts DF, Rojas FJ, Cutroneo KR (1979) Glucocorticoids decrease prolyl hydroxylase activity without the cellular accumulation of underhydroxylated collagen. Mol Pharmacol 15:99–107

Counts DF, Byers PH, Holbrook KA, Hegreberg GA (1980) Dermatosparaxis in a Himalayan cat. I. Biochemical studies of dermal collagen. J Invest Dermatol 74:96–99

Cournil I, Leblond CP, Pomponio J, Hand AR, Sederlof L, Martin GR (1979) Immunohistochemical localization of procollagens. J Histochem Cytochem 27:1059–1069

Coventry MB (1969) Anatomy of the intervertebral disk. Clin Orthop 67:9–15

Cremer MA, Pitcock JA, Stuart JM, Kang AH, Townes AS (1981) Auricular chondritis in rats: an experimental model of relapsing polychondritis induced with type II collagen. J Exp Med 154:535–540

Cretius K (1959) Der Kollagengehalt menschlicher Uterusmuskulatur. Bibl Gynaecol 20:68–89

Cretius K (1965) Zur molekularen Struktur des Bindegewebes im menschlichen Uterus. Arch Gynakol 202:43–46

Cullen BM, Harkness RD (1960) The effect of hormones on the physical properties and collagen content of the rat's uterine cervix. J Physiol (Lond) 152:419–436

Cunliffe CJ, Franklin TJ (1986) Inhibition of prolyl 4-hydroxylase by hydroanthraquinones. Biochem J 239:311–315

Cupo LN, Pyeritz RE, Olson JL, McPhee SJ, Hutchins GM, McKusick VA (1981) Ehlers-Danlos syndrome with abnormal collagen fibrils, sinus of valsalva aneurysm, myocardial infarction, panacinar emphysema and cerebral heterotopias. Am J Med 71:1051–1058

Cuppage FE, Neagoy DR, Tate A (1967) Repair of the nephron following temporary occlusion of the renal pedicle. Lab Invest 17:660–674

Cuschieri A, Felgate RA (1972) Urinary hydroxyproline excretion in carcinoma of the breast. Br J Exp Pathol 53:237–241

Cush JJ, Goldings EA (1985) Drug-induced lupus: clinical spectrum and pathogenesis. Am J Med Sci 290:36–45

Cutroneo KR, Counts DF (1975) Anti-inflammatory steroids and collagen metabolism: glucocorticoid-mediated alterations of prolyl hydroxylase activity and collagen synthesis. Mol Pharmacol 11:632–639

Cutroneo KR, Costello D, Fuller GC (1971) Alteration of proline hydroxylase activity by glucocorticoids. Biochem Pharmacol 20:2797–2804

Cutroneo KR, Stassen FLH, Cardinale GJ (1975) Anti-inflammatory steroids and collagen metabolism: glucocorticoid-mediated decrease of prolyl hydroxylase. Mol Pharmacol 11:44–51

Cutroneo KR, Rokowski R, Counts DF (1981) Glucocorticoids and collagen synthesis: comparison of in vivo and cell culture studies. Coll Relat Res 1:557–568

Cutroneo KR, Sterling KM Jr, Shull S (1986) Steroid hormone regulation of extracellular matrix proteins. In: Mecham R (ed) Regulation of matrix accumulation. Academic, New York, pp 119–175

Dabbous MK, Seif M, Brinkley SB, Butler T, Braswell CW (1979) Gingival matrix proteins – the nature of insoluble bovine gingival collagen. J Periodont Res 14:204–209

Dąbrowski R (1987) Alteration in thermal stability of collagen and collagen-chondroitin-6-sulphate complex induced by histamine. Acta Physiol Pol 38:439–444

Dąbrowski R, Maśliński C (1970) The effect of histamine on collagen formation and collagen polymerization in the skin wound healing of guinea pigs. Life Sci 9:189–202

Dąbrowski R (1978) Udział histaminy zewnątrz i wewnątrzpochodnej w rozwoju tkanki łącznej (Contribution of exo- and endogenic histamine in development of the connective tissue). Reumatologia 16:425–428 (in Polish)

Dąbrowski R, Szczepanowska A (1984) Alterations in histamine and collagen induced in chick embryos aminoguanidine and 48/80. Agents Actions 14:458–460

Dąbrowski R, Maśliński C, Górski P (1975) The effects of histamine liberators and exogenous histamine on wound healing in rat. Agents Actions 5:311–314

Dahmen G (1961) Feingewebliche und submikroskopische Befunde beim Morbus Dupuytren. Z Orthop 104:247–254

Dairiki-Shortliffe LM, Freiha FS, Kessler R, Stamey TA, Constantinou CE (1989) Treatment of urinary incontinence by the periurethral implantation of glutaraldehyde cross-linked collagen. J Urol 141:538–541
Dalgleish R, Williams G, Hawkins JR (1986) Length polymorphism in the pro $\alpha 2$(I) collagen gene: an alternative explanation in a case of Marfan syndrome. Hum Genet 73:91–92
Dalgleish R, Hawkins JR, Keston M (1987) Exclusion of the $\alpha 2$(I) and $\alpha 1$(III) collagen genes as the mutant loci in a Marfan syndrome family. J Med Genet 24:148–151
Daly JM, Steiger E, Prockop DJ, Dudrick SJ (1973) Inhibition of collagen synthesis by the proline analogue cis-4-hydroxyproline. J Surg Res 14:551–555
Damle SP, Minor RR, Gregory JD (1985) Proteoglycans in dog skin in a genetic dominant trait that shows defects in collagen fibrillogenesis. In: Davidson N, Williams P, De Ferrante N (eds) Glycoconjugates, vol 1. Prager, New York, pp 301–312
Dandona P, El Kabib DJ (1970) The effect of the injections of exophthalmic sera on the Harderian gland of the guinea-pig. Clin Sci 38:2P
Danforth DN, Buckingham JC (1964) Connective tissue mechanisms and their relation to pregnancy. Obstet Gynecol Surv 19:715–732
Danlos M (1908) Un cas de cutis laxa avec tumeurs par contusion chronique des coudes et des Mace de Lepinay. Bull Soc Fr Dermatol 19:70–72
Daughaday WH, Kozak Mariz I (1962) Conversion of proline-U-$C^{14}$ to labeled hydroxyproline by rat cartilage in vitro: effects of hypophysectomy, growth hormone, and cortisol. J Lab Clin Med 59:741–752
Davídková E, Švadlenka I, Deyl Z (1975) Interactions of malonaldehyde with collagen. IV. Localization of malonaldehyde binding sites in collagen molecule. Z Lebensm Unters Forsch 158:279–282
Davis BH (1987) Hepatic fibrogenesis: new developments? Am J Gastroenterol 82:544–546
Davis BP, Jeffrey JJ, Eisen AZ, Derby A (1975) The induction of collagenase by thyroxine in resorbing tadpole tailfin in vitro. Dev Biol 44:217–222
Davis JM, Boswell BA, Bächinger HP (1989) Thermal stability and folding of type IV procollagen and effect of peptidyl-prolyl cis-trans-isomerase on the folding of the triple helix. J Biol Chem 264:8956–8962
Davis WM, Madden JW, Peacock EE Jr (1972) A new approach to the control of esophageal stenosis. Ann Surg 176:469–476
Davison PF (1982) Tendon. In: Weiss JB, Jayson MIV (eds) Collagen in health and disease. Churchill Livingstone, Edinburgh, pp 498–505
Davison PF (1989) The contribution of labile crosslinks to the tensile behavior of tendons. Connect Tissue Res 18:293–305
Davison PF, Berman MB (1973) Corneal collagenase: specific cleavage of types $(\alpha 1)_2\alpha 2$ and $(\alpha 1)_3$ collagens. Connect Tissue Res 2:57–64
Davison PF, Patel A (1975) Age-related changes in aldehyde location on rat tail tendon collagen. Biochem Biophys Res Commun 65:983–989
Dayer JM, Demczuk S (1984) Cytokines and other mediators in rheumatoid arthritis. Springer Semin Immunopathol 7:387–413
Dayer JM, de Rochemonteix B, Burrus B, Demczuk S, Dinarello CA (1986) Human recombinant interleukin 1 stimulates collagenase and prostaglandin $E_2$ production by human synovial cells. J Clin Invest 77:645–648
DeGroot LJ, Refetoff S, Bernal J, Rue PA, Coleoni AH (1978) Nuclear receptors for thyroid hormone. J Endocrinol Invest 1:79–88
Delbrück A, Schröder H (1983) Metabolism and proliferation of cultured fibroblasts from specimens of human palmar fascia and Dupuytren's contracture. The pathobiochemistry of connective tissue proliferation. J Clin Chem Biochem 21:11–17
DeLustro F, Carlson RP, Datko LJ, DeLustro B, Lewis AJ (1984) The absence of antibodies to type II collagen in established adjuvant arthritis in rats. Agents Actions 14:673–679
DeMichelle SJ, Brown RG (1984) Connective tissue metabolism in muscular dystrophy. Levels of collagen and mucopolysaccharides in embryonic chicken with genetic muscular dystrophy. Comp Biochem Biophys 79B:203–209

Denholm LJ, Cole WW (1983) Heritable bone fragility, joint laxity and dysplastic dentin in Friesian calves: a bovine syndrome of osteogenesis imperfecta. Aust Vet J 60:9–17
Denning S, Pinnell SR (1979) Heart valve collagen – identification of two new species. Clin Res 27:162A
De Paepe A, Nicholls A, Narcisi P, De Keyser F, Quatacker J, Van Staey M, Matton M, Pope FM (1987) Ehlers-Danlos syndrome type I: a clinical and ultrastructural study of a family with reduced amounts of collagen type III. Br J Dermatol 117:89–97
DePalma AF, Rothman RH (1970) The intervertebral disc. Saunders, Philadelphia
Dequeker J, Merlevede W (1971) Collagen content and collagen extractability pattern of adult human trabecular bone according to age, sex and amount of bone mass. Biochim Biophys Acta 244:410–420
Dequeker J, Remans J, Franssen R, Waes J (1971) Ageing patterns of trabecular and cortical bone and their relationship. Calcif Tissue Res 7:23–30
De Rycker C, Vandalem JL, Hennen G (1984) Effects of 3,5,3′-triiodothyronine on collagen synthesis by cultured human skin fibroblasts. FEBS Lett 174:34–37
Deshmukh K, Ninmi ME (1968) Soluble collagen with high aldehyde content extracted from insoluble collagen with mercaptoethylamine. Biochim Biophys Acta 154:258–260
Deshmukh K, Nimni ME (1969) A defect in the intramolecular and intermolecular cross-linking of collagen caused by penicillamine. II. Functional groups involved in the interaction process. J Biol Chem 244:1787–1795
De Simone DP, Parson DB, Johnson KE, Jacobs RP (1983) Type II collagen induced arthritis, a morphologic and biochemical study of articular cartilage. Arthritis Rheum 26:1245–1251
Desjardins M, Bendayan M (1989) Heterogenous distribution of type IV collagen, entactin, heparan sulfate proteoglycan, and laminin among renal basement membranes as demonstrated by quantitative immunocytochemistry. J Histochem Cytochem 37:885–897
Deskmukh-Phadke K, Lawrence M, Nanda S (1978) Synthesis of collagenase and neutral proteases by articular chondrocytes: stimulation by a macrophage-derived factor. Biochem Biophys Res Commun 85:490–496
de Vries WN, de Wet WJ (1986) The molecular defect in an autosomal dominant form of osteogenesis imperfecta. J Biol Chem 261:9056–9064
de Wet WJ, Pihlajaniemi T, Myers J, Kelly TE, Prockop DJ (1983) Synthesis of a shortened pro $\alpha$2(I) chain and decreased synthesis of pro $\alpha$2(I) chains in a proband with osteogenesis imperfecta. J Biol Chem 258:7721–7728
de Wet WJ, Sippola M, Tromp G, Prockop DJ, Chu ML, Ramirez F (1986) Use of R-loop mapping for the assessment of human collagen mutations. J Biol Chem 261:3857–3862
De Witt MT, Handley CJ, Oakes BW, Lowther DA (1984) In vitro response of chondrocytes to mechanical loading. The effect of short term mechanical tension. Connect Tissue Res 12:97–109
Dexter TM, Allen TD, Lajtha LG (1977) Conditions controling the proliferation of hemopoietic stem cells in vitro. J Cell Physiol 91:335–344
Deyl Z, Adam M (1977) Quantitative changes in insoluble collagen during ontogeny in rodents (collagen type I and type III). Mech Ageing Dev 6:25–33
Deyl Z, Adam M (eds) (1981) Connective tissue research. Chemistry, biology and physiology. Liss, New York
Deyl Z, Adam M (1982) Collagen in ageing and disease. Academia, Praha
Deyl Z, Juřicová M, Rosmus J, Adam M (1971) Aging of the connective tissue: collagen cross linking in animals of different species and equal age. Exp Gerontol 6:227–233
Deyl Z, Jelínek J, Rosmus J, Adam M (1972a) Changes in the collagenous stroma in rat kidney in relation to age. Exp Gerontol 7:353–358
Deyl Z, Rosmus J, Adam M (1972b) Hormonal control of collagen ageing. Exp Gerontol 7:38–43
Deyl Z, Macek K, Adam M (1979) Collagen $\alpha$ and $\beta$ chains constitute two separate molecular species. Biochem Biophys Res Commun 89:627–634
Dickson IR, Happey F, Pearson CH, Naylor A, Turner RL (1967) Variations in the protein of human intervertebral disc with age. Nature 215:52–53

Dieppe PA, Cawston T, Mercer E, Campion GV, Hornby J, Hutton CW, Doherty M, Watt I, Woolf AD, Hazleman B (1988) Synovial fluid collagenase in patients with destructive arthritis of the shoulder joint. Arthritis Rheum 31:882–890

Di Ferrante N, Leachman RD, Angelini P, Donnelly PV, Francis G, Almazan A, Segni G (1975) Lysyl oxidase deficiency in Ehlers-Danlos syndrome type V. Connect Tissue Res 3:49–53

Dinarello CA (1984) Interleukin 1. Rev Inf Dis 6:51–95

Dodd CM, Carmichael DJ (1979) The collagenous matrix of bovine predentine. Biochim Biophys Acta 577:117–123

Dogliotti GC (1931) La struttura del miocardio dell'uomo nei vari individui e nelle varie eta. Z Anat Entwicklung 96:680–722

Doillon CJ, Dunn MG, Silver FH (1988) Relationship between mechanical properties and collagen structure of closed and open wounds. J Biomech Eng 110:352–356

Dölz R, Engel J, Kühn K (1988) Folding of collagen IV. Eur J Biochem 178:357–366

Dombi GW (1986) Collagen fibril formation in the presence of dexamethasone disodium phosphate. Connect Tissue Res 15:257–268

Donaldson CL, Hulley SB, Vogel JM, Hattner RS, Bayers JH, McMillian DE (1970) Effect of prolonged bed rest on bone mineral. Metabolism 19:1071–1084

Dorfman A, Ho PL (1970) Synthesis of acid mucopolysaccharides by glial tumor cells in tissue culture. Proc Natl Acad Sci USA 66:495–499

Dorfmann H, Kahn MF, de Sèze S (1972a) Les lupus iatrogènes: état actuel de la question. I. Étude clinique et principaux produits incrimés. Nouv Presse Med 1:2907–2912

Dorfmann H, Kahn MF, de Sèze S (1972b) Les lupus iatrogènes: état actuel de la question. II. Physiopathologie des lupus induits. Nouv Presse Med 1:2967–2970

Dormidontov Ye N, Baranova E Ya, Korshunov NI (1979) Vliyaniye protivorevmaticheskoy terapii na obmen kollagena u bolnikh revmatoidnim artritom (Influence of anti-rheumatic treatment on collagen metabolism in patients with rheumatoid arthritis). Ter Arkh 51(7):44–47 (in Russian)

Dresden MH, Heilman SA, Schmidt JD (1972) Collagenolytic enzymes in human neoplasms. Cancer Res 32:993–996

Dreux C, Fiet J, Dakkak R (1971) L'hydroxyproline libre et peptidique issue des collagènes. Mèthodes de séparation et de dosage: application en biologie clinique. Ann Derm Syph 98:529–533

Drickamer K, Dordal MS, Reynolds L (1986) Mannose-binding proteins isolated from rat liver contain carbohydrate-binding domains linked to collagenous tails complete primary structures and homology with pulmonary surfactant. J Biol Chem 261:6878–6887

Dróżdż M, Kucharz EJ (1972) Wydalanie hydroksyproliny w moczu wykładnikiem metabolizmu kolagenu ustrojowego (Urinary excretion of hydroxyproline as an index of systemic collagen metabolism). Pol Tyg Lek 27:1655–1657 (in Polish)

Dróżdż M, Kucharz EJ (1976) Effect of chronic exposure to nitrogen dioxide on collagen fractions content in skin of guinea pigs. Acta Biol Med Germ 35:1023–1025

Dróżdż, Kucharz EJ (1977) The effect of cytostatic drugs on collagen metabolism in guinea pigs. Arch Immunol Ther Exp 25:773–778

Dróżdż M, Kucharz EJ (1978) Serum hydroxyproline and hydroxylysine levels in vitral hepatitis. Med Interne 16:79–82

Dróżdż M, Kucharz EJ (1979) Collagen content in the injured heart muscle of rats with collagen-like syndrome. Rev Roum Biochim 16:267–271

Dróżdż M, Kucharz EJ, Samek E (1975) Dziedziczne zaburzenia metabolizmu kolagenu (Hereditary disorders of collagen metabolism). Przegl Lek 32:283–289 (in Polish)

Dróżdż M, Luciak M, Kucharz EJ, Ludyga K, Olczyk K (1976) Badania nad metabolizmem tkanki łącznej świnek morskich przewlekle eksponowanych na działanie odlotowych gazów przemysłowych (Studies on metabolism of the connective tissue in guinea pigs after a long-term exposure to industrial exhaust gases). Med Pr 27:173–186 (in Polish)

Dróżdż M, Kucharz EJ, Szyja J (1977) Effect of chronic exposure to nitrogen dioxide on collagen content in lung and skin of guinea pigs. Environ Res 13:369–377

Dróżdż M, Kucharz EJ, Grucka-Mamczar E (1979a) Influence of thyroid hormones on collagen content in tissues of guinea pigs. Endokrinologie 73:105–111

Dróżdż M, Kucharz EJ, Olczyk K (1979b) Effect of catecholamines on the activity of lysosomal enzymes in the blood serum of rats chronically treated with hydrazinophthalazines. Folia Biol (Kraków) 27:9–16

Dróżdż M, Kucharz EJ, Olczyk K (1979c) Age-related changes in the content of fibrous proteins in the rat tissues. Acta Physiol Pol 30:359–363

Dróżdż M, Kucharz EJ, Grucka-Mamczar E (1980) Effect of sodium fluoride on collagen content in skin and lungs of growing rats. Acta Biol Med Germ 39:287–293

Dróżdż M, Kucharz EJ, Głowacki A (1981a) Effect of isoprenaline-induced myocardiopathy on glycosaminoglycans content in the heart muscle, aortic wall, blood plasma and urine or normal and hydrazinophthalazine-treated rats. Rev Roum Biochim 18:262–266

Dróżdż M, Kucharz EJ, Grucka-Mamczar E (1981b) Studies on the influence of fluoride compounds upon connective tissue metabolism in growing rats. I. Effect of hydrofluoride on collagen metabolism. Eur Toxicol Res 3:237–241

Dróżdż M, Kucharz EJ, Mamczar A (1982) Collagen content in tissues of irradiated rats. Exp Pathol 22:125–127

Dróżdż M, Piwowarczyk B, Olczyk K, Wieczorek M, Wróblewska-Adamek I (1983) Leukocyte collagenolytic activity in normal and hydrazinophthalazine-treated rats during isoprenaline-induced cardiomyopathy. Exp Pathol 23:253–258

Dróżdż M, Kucharz EJ, Stawiarska B (1984a) Studies on the influence of fluoride compounds upon connective tissue metabolism in growing rats. II. Effect of oral administration of sodium fluoride with and without simultaneous exposure to hydroxyfluoride on collagen metabolism. J Toxicol Med 4:151–157

Dróżdż M, Kucharz EJ, Głowacki A, Olczyk K (1984b) Effect of antiinflammatory steroids upon glycosaminoglycan metabolism in rats with carbon tetrachloride-induced hepatic fibrosis. Rev Roum Biochim 21:175–176

Dróżdż M, Petelenz T, Łępkowski A, Kucharz EJ, Słomińska-Petelenz T, Drążkiewicz U, Krombholc-Pawełek D, Żuk-Popiołek I (1989) Collagen content in the heart of patients operated on for valvular heart disease caused mainly by infective endocarditis. Mater Med Pol 29:111–115

Duance VC, Restall DJ, Beard H, Bourne FJ, Bailey AJ (1977) The location of three collagen types in skeletal muscle. LEBS Lett 79:248–252

Duance VC, Stephens HR, Dunn M, Bailey AJ, Dubowitz V (1980a) A role for collagen in the pathogenesis of muscular dystrophy? Nature 284:470–472

Duance VC, Black CM, Dubowitz V, Hughes GRV, Bailey AJ (1980b) Polymyositis – an immunofluorescence study on the distribution of collagen types. Muscle Nerve 3:487–490

Duance VC, Wotton SF, Bailey AJ (1985) Isolation and characterization of mammalian parent type M (IX) collagen. Ann NY Acad Sci 460:422–425

Dublet B, van der Rest M (1987) Type XII collagen is expressed in embryonic chick tendons. Isolation of pepsin-derived fragments. J Biol Chem 262:17724–17727

Dublet B, Dixon E, de Miguel E, van der Rest M (1988) Bovine type XII collagen: amino acid sequence of a 10 kDa pepsin fragment from periodontal ligament reveals a high degree of homology with the chicken $\alpha$ 1(XII) sequence. FEBS Lett 233:177–180

Dublet B, Oh S, Sugrue SP, Gordon MK, Gerecke DR, Olsen BR, Van der Rest M (1989) The structure of avian type XII collagen. $\alpha$ 1(XII) chains contain 190-kDa non-triple helical amino-terminal domains and form homotrimeric molecules. J Biol Chem 264:13150–13156

Dull TA, Henneman PH (1963) Urinary hydroxyproline as an index of collagen turnover in bone. N Engl J Med 268:132–134

Duncan MR, Berman B (1989a) Differential regulation of collagen, glycosaminoglycan, fibronectin, and collagenase activity production in cultured human adult dormal dermal fibroblasts by interleukin 1 alpha and beta and tumor necrosis factor-alpha and beta. J Invest Dermatol 92:699–706

Duncan MR, Berman B (1989b) Differential regulation of glycosaminoglycan, fibronectin, and collagenase production in cultured human dermal fibroblasts by interferon-alpha, -beta, and -gamma. Arch Dermatol Res 281:11–18

Duncan GW, Yen EHK, Pritchard ET, Suga DM (1984) Collagen and prostaglandin synthesis in force-stressed periodontol ligament in vitro. J Dent Res 63:665–669

Dunn MA, Kamel R (1981) Hepatic schistosomiasis. Hepatology 1:653–661

Dvorak HF, Kaplan AP, Clark RAF (1988) Potential functions of the clotting system in wound repair. In: Clark RAF, Henson PM (eds) The molecular and cellular biology of wound repair. Plenum, New York, pp 57–86

Dyer RF, Sodek J, Heersche JNM (1980) The effect of 17$\beta$-estradiol on collagen and noncollagenous protein synthesis in the uterus and some periodontal tissues. Endocrinology 107:1014–1021

Eachempati V, Ostapowicz F, Lazzara JV (1975) Urinary hydroxyproline in pregnancy: an index of placental function. Obstet Gynecol 15:12–17

Eady RAJ, Tidman MJ, Heagerty AHM, Kennedy AR (1987) Approaches to the study of epidermolysis bullosa. Curr Probl Dermatol 17:127–141

Eastoe JE (1968) Chemical aspects of the matrix concept in calcified tissue organization. Calcif Tiss Res 2:1–6

Eeckhout Y, Vaes G (1977) Further studies on the activation of procollagenase, the latent precursor of bone collagenase. Effects of lysosomal cathepsin B, plasmin and kallikrein, and spontaneous activation. Biochem J 166:21–31

Ehlers E (1901) Cutix laxa. Neigung zu Hemorrhagien in der Haut. Lockerung mehrerer Artikulationen. Dermatol Z 8:173–174

Ehrenberg R, Winnecken HG, Bielerichen H (1954) Der Alternsgang des Bindegewebes in menschlichen Organen (Herz und Leber). Z Naturforsch 9:492–495

Ehrhart LA, Holderbaum D (1977) Stimulation of aortic protein synthesis in experimental rabbit atherosclerosis. Atherosclerosis 27:477–485

Ehrhart LA, Holderbaum D (1980) Aortic collagen, elastin and nonfibrous protein synthesis in rabbits fed cholesterol and peanut oil. Atherosclerosis 37:423–432

Ehrlich GE (1984a) Diffuse collagen disease. JAMA 251:1595–1596

Ehrlich HP (1984b) The role of connective tissue matrix in hypertrophic scar contracture. In: Hunt TK, Heppenstall RB, Pines E, Rovee D (eds) Soft and hard tissue repair. Praeger, New York, pp 533–553

Ehrlich PH, Tarver H, Hunt TK (1973) Effects of vitamin A and glucocorticoids upon inflammation and collagen synthesis. Ann Surg 177:222–227

Eichhorn JH, Peterkofsky B (1979) Local anesthetic-induced inhibition of collagen secretion in cultured cells under conditions where microtubules are not depolymerized by these agents. J Cell Biol 81:26–42

Eisele CW, Eichelberger L (1945) Water, electrolyte and nitrogen content of human skin. Proc Soc Exp Biol NY 58:97–100

Eisen AZ (1969) Human skin collagenase: relationship to the pathogenesis of epidermolysis bullosa dystrophica. J Invest Dermatol 52:449–453

Eisen AZ, Bauer EA, Jeffrey JJ (1970) Animal and human collagenases. J Invest Dermatol 55:359–373

Eisner M, Kondraciuk A (1967) Przydatość kliniczna oznaczania poziomu proliny i hydroksyproliny w surowicy krwi i moczu w chorobach tarczycy (Clinical application of determination of serum level of proline and hydroxyproline in diseases of the thyroid gland). Pol Arch Med Wewn 38:575–581 (in Polish)

Ekman G, Uldbjerg N, Malmström A, Ulmsten U (1983) Increased postpartum collagenolytic activity in cervical connective tissue from women treated with prostaglandin $E_2$. Gynecol Obstet Invest 16:292–298

Eldridge CF, Bunge RP, Bunge MB (1988) Effects of cis-4-hydroxy-L-proline, an inhibitor of Schwann cell differentiation, on secretion of collagenous and noncollagenous proteins by Schwann cells. Exp Cell Res 174:491–501

Elliott DH (1965) Structure and function of mammalian tendon. Biol Rev 40:392–421

El Meneza S, Olds GR, Kresina TF, Mahmoud AAF (1989) Dynamics of hepatic connective tissue matrix constituents during murine Schistosoma mansoni infection. Hepatology 9:50–56

Elsas LJ, Miller RL, Pinnell SR (1978) Inherited human collagen lysyl hydroxylase deficiency: ascorbic acid response. J Pediatr 92:378–384

Elsdale T, Bard J (1972) Collagen substrata for studies on cell behavior. J Cell Biol 54:626–637

Elwy AM (1967) Pathological aspects of bilharziasis in Egypt. In: Mostofi KF (ed) Bilharziasis. Springer, Berlin Heidelberg New York, pp 39–44

Emanuel BS, Cannizzaro LA, Seyer JM, Myers JC (1985) Human $\alpha$1(III) and $\alpha$2(V) procollagen genes are located on the long arm of chromosome 2. Proc Natl Acad Sci USA 82:3385–3389

Emanuel BS, Sellinger BT, Gudas LJ, Myers JC (1986) Localization of the human procollagen $\alpha$1(IV) gene to chromosome 13q34 by in situ hybridization. Am J Hum Genet 38:38–44

Emmrich R (1970) Die Hydroxyprolinausscheidung im Harn bei chronischen Krankheiten des Bindegewebes. Z Gesamte Inn Med 25:1062–1065

Emmrich R, Häntzschel HJ, Hantzschel H (1967) Die klinische Bedeutung der Hydroxyprolinausscheidung im Harn. Z Gesamte Inn Med 22:193–199

Emonard H, Grimaud JA (1989) Active and latent collagenase activity during reversal of hepatic fibrosis in murine schistosomiasis. Hepatology 10:77–83

Engesaeter LB, Skar AG (1978) Effects of oxytetracycline on the mechanical properties of bone and skin in young rats. Acta Orthop Scand 49:529–534

Engesaeter LB, Underdal T, Løvstad R (1980a) Effect of oxytetracycline on solubility and synthesis of collagen in young rats. Acta Orthop Scand 51:43–48

Engesaeter LB, Underdal T, Langeland N (1980b) Effects of oxytetracycline on mineralization of bone in young rats. Acta Orthop Scand 51:459–465

Engesaeter LB, Underdal T, Lyngaas KHN (1989c) Effects of cloxacillin, doxycycline, fusidic acid and lincomycin on mineralization and solubility of collagen in young rats. Acta Orthop Scand 51:467–470

Englert ME, Landes MJ, Oronsky AL, Kerwar SS (1984) Suppression of type II collagen-induced arthritis by the intravenous administration of type II collagen or its constituent peptide $\alpha_1$(II) $CB_{10}$. Cell Immunol 87:357–365

Engvall E, Rouslahti E, Miller EJ (1978) Affinity of fibronectin to collagens of different genetic types and to fibrinogen. J Exp Med 147:1584–1595

Ernst M, Schmid C, Froesch ER (1988) Enhanced osteoblast proliferation and collagen gene expression by estradiol. Proc Natl Acad Sci USA 85:2307–2310

Ertel J, Isseroff H (1974) Proline in fascioliasis. I. Comparative activities of ornitine-$\delta$-transaminase and proline oxidase in Fasciola and mammalian livers. J Parasitol 60:574–577

Ertel J, Isseroff H (1976) Proline in fascioliasis. II. Characteristics of a partially purified ornithine-$\delta$-transaminase from Fasciola. Rice Univ Stud 62:97–109

Espey LL (1967) Tenacity of porcine Graafian follicle as it approaches ovulation. Am J Physiol 212:1397–1401

Espey LL (1980) Ovulation as an inflammatory reaction – a hypothesis. Biol Reprod 22:73–106

Espey LL, Coons PJ (1976) Factors which influence ovulatory degradation of rabbit ovarian follicles. Biol Reprod 14:233–245

Espey LL, Rondell P (1967) Estimation of mammalian collagenolytic activity with a synthetic substrate. J Appl Physiol 23:757–761

Etherington DJ (1972) The nature of the collagenolytic cathepsin of rat liver and its distribution in other rat tissues. Biochem J 127:685–692

Etherington DJ (1973) Collagenolytic-cathepsin and acid-proteinase activities in the rat uterus during post partum involution. Eur J Biochem 32:126–128

Etherington DJ (1974) The purification of bovine cathepsin B1 and its mode of action on bovine collagens. Biochem J 137:547–557

Etherington DJ (1977) The dissolution of insoluble bovine collagens by cathepsin B1, collagenolytic cathepsin and pepsin. The influence of collagen type, age and chemical purity on susceptibility. Connect Tissue Res 5:135–145

Etherington DJ, Birkedahl-Hansen H (1987) The influence of dissolved calcium salts on the degradation of hard-tissue collagens by lysosomal cathepsins. Coll Relat Res 7:185–199

Etherington DJ, Pugh D, Silver IA (1981) Collagen degradation in an experimental inflammatory lesion: studies on the role of the macrophages. Acta Biol Med Germ 40:1625–1636

Etkin W (1968) Tadpole tailfin resorption. In: Etkin W, Gilbert W (eds) Metamorphosis. Appleton-Century-Crofts, New York, pp 313–348

Evans CH, Drouven BJ (1983) The promotion of collagen polymerization by lanthanide and calcium ions. Biochem J 213:751–758

Evanson JM, Jeffrey JJ, Krane SM (1967) Human collagenase: identification and characterization of an enzyme from rheumatoid synovium in culture. Science 158:499–502

Exer B, Krupp P, Menassé R, Riesterer L (1976) Influence of adjuvant arthritis on connective tissue metabolism. Agents Actions 6:651–656
Eyre DR (1979) Biochemistry of the intervertebral disc. Int Rev Connect Tissue Res 8:227–291
Eyre DR (1980) Collagen: molecular diversity in the body's protein scaffold. Science 207:1315–1322
Eyre DR (1987) Collagen cross-linking amino acids. In: Methods in enzymology, vol 144. Academic, London, pp 115–139
Eyre DR, Glimcher MJ (1972) Reducible crosslinks in hydroxylysine-deficient collagens of a heritable disorder of connective tissue. Proc Natl Acad Sci USA 69:2594–2598
Eyre DR, Muir H (1975) The distribution of different molecular species of collagen in fibrous, elastic and hyaline cartilage of the pig. Biochem J 151:592–602
Eyre DR, Muir H (1977) Quantitative analysis of types I and II collagens in human intervertebral discs at various ages. Biochim Biophys Acta 492:29–42
Eyre DR, Oguchi H (1980) The hydroxypyridinium crosslinks of skeletal collagens: their measurement, properties and a proposed pathway of formation: their measurement, properties and a proposed pathway of formation. Biochem Biophys Res Commun 92:403–410
Eyre DR, Wu JJ (1983) Collagen of fibrocartilage: a distinctive molecular phenotype in bovine meniscus. FEBS Lett 158:265–270
Eyre DR, Wu JJ (1987) Type XI or $1\alpha 2\alpha 3\alpha$ collagen. In: Mayne R, Burgeson RE (eds) Structure and function of collagen types. Academic, Orlando, pp 261–281
Eyre DR, Koob TJ, Van Ness KP (1984a) Quantitation of hydroxy-pyridinium crosslinks in collagen by high-performance liquid chromatography. Anal Biochem 137:380–388
Eyre DR, Paz MA, Gallop PM (1984b) Cross-linking in collagen and elastin. Annu Rev Biochem 53:717–748
Eyre DR, Upton MP, Shapiro FD, Wilkinson RH, Vawter GF (1986) Nonexpression of cartilage type II collagen in a case of Langer-Saldino achondrogenesis. Am J Hum Genet 39:52–67
Eyre DR, Apon S, Wu JJ, Ericsson LH, Walsh KA (1987a) Collagen type IX: evidence for covalent linkages to type II collagen in cartilage. FEBS Lett 220:337–341
Eyre DR, Wu JJ, Apone S (1987b) A growing family of collagens in articular cartilage: identification of 5 genetically distinct types. J Rheumatol 14:25–27
Fabier JL, Viau M, Constans J, Clerc A (1976) Analyse chromatographique des amides amines urinaires au cours de la maladie de Paget. Rev Rhum 43:249–253
Faglia G, Norbiato G, Tirinnanzi ML (1966) L'idrossiprolina totale urinaria in stati di nanismo di varia origine. Atti Accad Med Lomb 21:501–507
Falk CT, Schwartz RC, Ramirez F, Tsipouras P (1986) Use of molecular haplotypes specific for the human pro$\alpha$2(I) collagen gene in linkage analysis of the mild autosomal dominant forms of osteogenesis imperfecta. Am J Hum Genet 38:269–279
Farquharson C, Robins SP (1989) Immunolocalization of collagen types I and III in the arterial wall of the rat. Histochem J 21:172–178
Fassbender HG (1983) Histomorphological basis of articular cartilage destruction in rheumatoid arthritis. Coll Relat Res 3:141–155
Faucher D, Lelièvre Y, Cartwright T (1987) An inhibitor of mammalian collagenase active at micromolar concentrations from an actinomycete culture broth. J Antibiot 40:1757–1762
Fauvel F, Legrand YJ, Bentz H, Fietzek PP, Kühn K, Caen P (1978) Platelet-collagen interaction: adhesion of human blood platelets to purified (CB4) peptide from type III collagen. Thromb Res 12:841–845
Feingold JE, Bois AE, Chonipret A, Broyer M, Gubler MC, Grunfeld JP (1985) Genetic heterogeneity of Alport syndrome. Kidney Int 27:672–677
Feit H, Kawai M, Mostafapour AS (1989) The role of collagen cross-linking in the increased stiffness of avian dystrophic muscle. Muscle Nerve 12:486–492
Felländer M, Gladnikoff H, Jacobson E (1970) Instability of the hip in the newborn. Acta Orthop Scand [Suppl]130:36–54
Fels IG (1958) The determination of hydroxyproline in liver. Clin Chem 4:62–67
Fernandez-Madrid F (1970) Collagen biosynthesis. Clin Orthop 68:163–181
Fessler JH, Fessler LI (1978) Biosynthesis of procollagen. Annu Rev Biochem 47:129–162

Fessler JH, Fessler LI (1987) Type V collagen. In: Mayne R, Burgeson RE (eds) Structure and function of collagen types. Academic, Orlando, pp 81–103
Fessler LI, Morris NP, Fessler JH (1975) Procollagen: biological scission of amino and carboxyl extension peptides. Proc Natl Acad Sci USA 72:4905–4909
Fessler JH, Shigaki N, Fessler LI (1985) Biosynthesis and properties of procollagens V. Ann NY Acad Sci 460:181–186
Fietzek PP, Kühn K (1976) The primary structure of collagen. Int Rev Connect Tissue Res 7:1–60
Fietzek PP, Kell I, Kühn K (1972) The covalent structure of collagen. Amino acid sequence of the N-terminal region of $\alpha$2-CB4 from calf and rat skin collagen. FEBS Lett 26:66–68
Finch WR, Buckingham RB, Rodnan GP, Prince RK, Winkelstein A (1984) Scleroderma induced by bleomycin. In: Black CH, Muers AR (eds) Systemic sclerosis (scleroderma). Gower Medical, New York, pp 114–121
Finerman GAM, Downing S, Rosenberg LE (1967) Amino acid transport in bone. II. Regulation of collagen synthesis by perturbation of proline transport. Biochim Biophys Acta 135:1008–1015
Fischer GM (1976) Effects of spontaneous hypertension and age on arterial connective tissue in the rat. Exp Gerontol 11:209–215
Fischer GM, Llaurado JG (1967) Connective tissue composition of canine arteries: effects of renal hypertension. Arch Pathol 84:95–101
Fischer GM, Swain ML (1978) In vivo effects of sex hormones on aortic elastin and collagen dynamics in castrated and intact male rats. Endocrinology 102:92–97
Fischer GM, Swain ML, Cherian K (1980) Increased vascular collagen and elastin synthesis in experimental atherosclerosis in the rabbit. Variation in synthesis among major vessels. Atherosclerosis 35:11–20
Fischer GM, Cherian K, Swain ML (1981) Increased synthesis of aortic collagen and elastin in experimental atherosclerosis. Inhibition by contraceptive steroids. Atherosclerosis 39:463–467
Fisher LW, Eanes ED, Denholm LJ, Heywood BR, Termine JD (1987) Two bovine models of osteogenesis imperfecta exhibit decreased apatite crystal size. Calcif Tissue Int 40:282–285
Fitch JM, Gibney E, Sanderson RD, Mayne R, Linsenmayer TF (1982) Domain and basement membrane specificity of a monoclonal antibody against chicken type IV collagen. J Cell Biol 95:641–647
Fitch JM, Mayne R, Linsenmayer TF (1983) Development acquisition of basement membrane heterogeneity: type IV collagen in the avian lens capsule. J Cell Biol 97:940–943
Fitch JM, Gross J, Mayne R, Johnson-Wint B, Linsenmayer TF (1984) Organization of collagen types I and V in the embryonic chicken cornea: monoclonal antibody studies. Proc Natl Acad Sci USA 81:2791–2795
Fitzpatrick RJ (1977) Changes in cervical function at parturition. Ann Rech Vet 8:438–439
Fitzpatrick RJ, Dobson H (1979) The cervix of the sheep and goat during parturition. Anim Rep Sci 2:209–224
Fitzpatrick RJ, Dobson H (1981) Softening of the ovine cervix at parturition. In: Ellwood DA, Anderson ABM (eds) The cervix in pregnancy and labour. Churchill Livingstone, Edinburgh, pp 40–56
Fitzsimmons CM, Barnes MJ (1985) The platelet reactivity of the $\alpha 2$(I) chain of type I collagen: platelet aggregation induced by polymers of the molecule $[\alpha 2(I)]_3$. Thromb Res 39:523–531
Fitzsimmons CM, Cawston TE, Barnes MJ (1986) The platelet reactivity of collagen type I: evidence for multiple platelet-reactive sites in the type I collagen molecule. Thromb Haemost 56:95–99
Fjolstad M, Helle O (1974) Hereditary dysplasia of collagen tissues in sheep. J Pathol 112:184 188
Fleckman PH, Jeffrey JJ, Eisen AZ (1973) A sensitive microassay for prolyl hydroxylase: activity in normal and psoriatic skin. J Invest Dermatol 60:46–52
Fleischmajer R (1964) The collagen in scleroderma. Arch Dermatol 89:437–442
Fleischmajer R, Krol S (1967) Chemical analysis of the dermis in scleroderma. Proc Soc Exp Biol Med 126:252–258
Fleischmajer R, Lara JV (1965) Scleroderma: a histochemical and biochemical study. Arch Dermatol 92:643–652

Fleischmajer R, Perlish JS (1980) Capillary alterations in scleroderma. J Am Acad Dermatol 2:161–170
Fleischmajer R, Faludi G, Krol S (1970) Scleroderma and diabetes mellitus. Arch Dermatol 101:21–26
Fleischmajer R, Samiano V, Nedwich A (1972) Alteration of subcutaneous tissue in systemic sclerosis. Arch Dermatol 105:59–66
Fleischmajer R, Perlish JS, Shaw KU, Prozzi DJ (1976) Skin capillary changes in early systemic scleroderma. Arch Dermatol 112:1553–1557
Fleischmajer R, Gay S, Meigel WN, Perlish JS (1978) Collagen in the cellular and fibrotic stages of scleroderma. Arthritis Rheum 21:418–428
Fleischmajer R, Gay S, Perlish JS, Cesarini JP (1980) Immunoelectron microscopy of type III collagen in normal and scleroderma skin. J Invest Dermatol 75:189–191
Fleischmajer R, Perlish JS, Krieg T, Timpl R (1981a) Variability in collagen and fibronectin synthesis by scleroderma fibroblasts in primary culture. J Invest Dermatol 76:400–403
Fleischmajer R, Timpl R, Tuderman L, Raisher L, Wiestner M, Perlish JS, Graves PN (1981b) Ultrastructural identification of extension aminopropeptides of type I and type III collagen in human skin. Proc Natl Acad Sci USA 78:7360–7364
Fleischmajer R, Perlish JS, Duncan M (1983) Scleroderma. A model for fibrosis. Arch Dermatol 119:957–962
Fleischmajer R, Krieg T, Dziadek M, Altchek D, Timpl R (1984) Ultrastructure and composition of connective tissue in hyalinosis cutis et mucosae skin. J Invest Dermatol 82:252–258
Fleischmajer R, Perlish JS, Olsen BR (1987) The carboxylpropeptide of type I procollagen in skin fibrillogenesis. J Invest Dermatol 89:212–215
Foidart JM, Rorive GL, Nusgens BV, Lapière CM (1978) The relationship between blood pressure and aortic collagen metabolism in renal hypertensive rats. Clin Sci Mol Med 55:27s–29s
Foidart JM, Berman JJ, Paglia L, Rennard S, Abe S, Perantoni A, Martin GK (1980) Synthesis of fibronectin, laminin and several collagens by a liver-derived epithelial line. Lab Invest 42:525–532
Folkhard W, Geercken W, Knörzer E, Mosler E, Nemetschek-Gansler H, Menetschek T, Koch MHJ (1987) Structural dynamic of native tendon collagen. J Mol Biol 193:405–407
Forbers RM, Cooper AR, Mitchell HH (1953) The composition of the adult body as determined by chemical analysis. J Biol Chem 203:359–366
Forbes RM, Mitchell HH, Cooper AR (1956) Further studies on the gross composition and mineral elements in the adult human body. J Biol Chem 223:969–975
Forest N, Boy-Lefevre ML, Duprey P, Grimaud JA, Jakob H, Paulin D (1982) Collagen synthesis in mouse embryonal carcinoma cells: effect of retinoid acid. Differentiation 23:153–163
Forster SJ, Talbot IC, Critchley DR (1984) Laminin and fibronectin in rectal adenocarcinoma: relationship to tumor grade, stage and metastasis. Br J Cancer 50:51–61
Fowler LJ, Peach CM, Bailey AJ (1970) In vitro studies on t he enzymatic biosynthesis of the collagen cross-links. Biochem Biophys Res Commun 41:251–259
Fox PK, White DD, Cavanagh M, Davies MG, Wusterman F (1982) Failure to demonstrate fibrotic changes in the skin of mice injected with glycosaminoglycan fractions from the urine of scleroderma patients. Dermatologica 164:90–94
Franc S (1984) Collagen of Coelenterates. In: Bairati A, Garrone R (eds) Biology of invertebrate and lower vertebrate collagens. Plenum, New York, pp 197–210
Franceschi RT, Romano PR, Park KY (1988) Regulation of type I collagen synthesis by 1,25-dihydroxyvitamin $D_3$ in human osteosarcoma cells. J. Biol Chem 263:18938–18945
Francis G, Donnelly PV, DiFerrante N (1976a) Abnormal soluble collagen produced in fibroblast cultures. Experientia 32:691–693
Francis MJO, Sanderson MC, Smith R (1976b) Skin collagen in idiopathic adolescent scoliosis and Marfan's syndrome. Clin Sci Mol Med 51:467–474
Francis MJO, Smith R, Sanderson MC (1977) Collagen abnormalities in idiopathic adolescent scoliosis. Calcif Tissue Res 22 [Suppl]:381–384
Franck WA, Bress NM, Singer FR, Krane SM (1974) Rheumatic manifestations of Paget's disease of bone. Am J Med 56:592–603

François J, Cambie E, Feher J (1973) Collagenase inhibition with penicillamine. Ophthalmologica 166:222–225
Francomano CA, Streeten EA, Meyers DA, Pyeritz RE (1988) Marfan syndrome: exclusion of genetic linkage to three major collagen genes. Am J Med Genet 29:457–462
Frankland DM, Wynn CH (1962a) The collagenolytic activity of rat liver lisosomes. Biochem J 84:20P
Frankland DM, Wynn CH (1962b) The degradation of acid-soluble collagen by rat-liver preparations. Biochem J 85:276–282
Franklin TJ, Hitchen M (1989) Inhibition of collagen hydroxylation by 2,7,8-trihydroxyanthraquinone in embryonic-chick tendon cells. Biochem J 261:127–130
Fredensborg N, Udén A (1976) Altered connective tissue in children with congenital dislocation of the hip. Arch Dis Child 51:887–889
Frederiksen DW, Hoffnung JM, Frederiksen RT, Williams RB (1978) The structural proteins of normal and diseased human myocardium. Circ Res 42:459–466
Freeman IL (1978) Collagen polymorphism in mature rabbit cornea. Invest Ophthalmol Vis Sci 17:171–177
Freeman IL (1980) Collagen biosynthesis in the healing corneal wound. In: Schachar RA, Levy NS, Schachar L (eds) Keratorefraction. LAL, Denison, pp 39–48
Freeman IL (1982) The eye. In: Weiss JB, Jayson MIV (eds) Collagen in health and disease. Churchill Livingstone, Edinburgh, pp 388–403
Freeman G, Crane SC, Stephens RJ, Furiosi NJ (1968a) Environmental factors in emphysema and a model system with $NO_2$. Yale J Biol Med 40:566–590
Freeman G, Crane SC, Stephens RJ, Furiosi NJ (1968b) Pathogenesis of the nitrogen dioxide-induced lesion in the rat lung: a review and presentation of new observations. Am Rev Respir Dis 98:429–443
Freeman G, Crane SC, Stephens RJ, Furiosi NJ (1969) The subacute nitrogen dioxide-induced lesion of the rat lung. Arch Environ Health 18:609–612
Frei A, Zimmermann A, Weigand K (1984) The N-terminal propeptide of collagen type III in serum reflects activity and degree of fibrosis in patients with chronic liver disease. Hepatology 4:830–834
Freihoffer U, Wellband WA (1963) Collagen formation in hypothyroid rats. Surg Forum 14:58–59
Frey J, Bayle JJ (1978) Le collagene du tissu hépatique. Lyon Med 239:87–93
Fricke R, Hartmann F (eds) (1974) Connective tissues. Biochemistry and pathophysiology. Springer, Berlin Heidelberg New York
Friedenstein AJ (1976) Precursors of mechanocytes. Int Rev Cytol 47:327–359
Friend J (1988) Physiology of the cornea. Metabolism and biochemistry. In: Smolin G, Thoft RA (eds) The cornea. Little Brown, Boston, pp 16–38
Frost HM (1987) Osteogenesis imperfecta. The set point proposal. A possible causative mechanism. Clin Orthop 216:280–297
Fry P, Harkness MLR, Harkness RD, Nightingale M (1962) Mechanical properties of tissues of lathyritic animals. J Physiol (Lond) 164:77–82
Fujimori E (1985) Changes induced by ozone and ultraviolet light in type I collagen. Bovine Achilles tendon collagen versus rat tail tendon collagen. Eur J Biochem 152:299–306
Fujimoto D (1980) Evidence for natural existence of pyredinoline crosslink in collagen. Biochem Biophys Res Commun 93:948–953
Fujimoto D (1984) Human tendon collagen: aging and crosslinking. Biomed Res 5:279–282
Fujimoto D, Fujie M, Abe E, Suda T (1979) Effect of vitamin D on the content of the stable crosslink, pyridinoline, in chick bone collagen. Biochem Biophys Res Commun 91:24–28
Fujita K, Hirano M, Ochiai J, Funabashi M, Nagatsu I, Nagatsu T, Sakakibara S (1978) Serum glycylproline p-nitroanilidase activity in rheumatoid arthritis and systemic lupus erythematosus. Clin Chim Acta 88:15–20
Fujiwara S, Nagai Y (1981) Basement membrane collagen from bovine lung: its chain associations as observed by two-dimensional electrophoresis. Coll Relat Res 1:491–504
Fujiwara K, Sakai T, Oda T, Igarashi S (1973) The presence of collagenase in Kupffer cells of the rat liver. Biochem Biophys Res Commun 54:531–537

Fukuhara M, Tsurufuji S (1969) The effect of locally injected anti-inflammatory drugs on the synthesis of collagen and non-collagen protein of carrageenan granuloma in rats. Biochem Pharmacol 18:2409–2414
Fuller GC (1981) Perspectives for the use of collagen synthesis inhibitors as antifibrotic agents. J Med Chem 24:651–658
Fuller GC, Matoney AL, Fisher DO, Fausto N, Cardinale GJ (1976) Increased collagen synthesis and the kinetic characteristics of prolylhydroxylase in tissues of rabbits with experimental arteriosclerosis. Atherosclerosis 24:483–490
Fullmer HM, Gibson WA (1966) Collagenolytic activity in the gingivae of man. Nature 209:728–729
Fulmer JD, Crystal RG (1976) The biochemical basis of pulmonary function. In: Crystal RG (ed) The biochemical basis of pulmonary function. Dekker, New York, pp 419–466
Fulmer JD, Elson N, Bradley K, Ferrans V, Crystal RG (1977) Comparison of type-specific collagens by lung epithelial and mesenchymal cells. Clin Res 25:503A
Fulmer JD, Bienkowski RS, Cowan MJ, Breul SD, Bradley KD, Ferrans VJ, Roberts WC, Crystal RG (1980) Collagen concentration and rates of synthesis in idiopathic pulmonary fibrosis. Am Rev Respir Dis 122:289–301
Furthmayr H, Timpl R (1976) Immunochemistry of collagens and procollagens. Int Rev Connect Tissue Res 7:61–103
Furthmayr H, Wiedemann H, Timpl R, Odermatt E, Engel J (1983) Electron-microscopical approach to a structural model of intima collagen. Biochem J 211:303–311
Gabbiani G, Majno G (1972) Dupuytren's contracture: fibroblast contraction? Am J Pathol 66:131–138
Gabrielli GB, Faccioli G, Casaril M, Capra F, Bonazzi L, Falezza G, Tomba A, Baracchino F, Corrocher R (1989) Procollagen III peptide and fibronectin in alcohol-related chronic liver disease: correlations with morphological features and biochemical tests. Clin Chim Acta 179:315–322
Gadek JE, Fells GA, Wright DG, Crystal RG (1980) Human neutrophil elastase functions as a type III collagen "collagenase". Biochem Biophys Res Commun 95:1815–1822
Gadek JE, Fells GA, Zimmerman RL, Crystal RG (1984) Role of connective tissue proteases in the pathogenesis of chronic inflammatory lung diseases. Environ Health Perspect 55:297–306
Gadher SJ, Eyre DR, Duance VC, Wotton SF, Heck LW, Schmid TM, Woolley DE (1988) Susceptibility of cartilage collagens type II, IX, X and XI to human synovial collagenase and neutrophil elastase. Eur J Biochem 175:1–7
Gadher SJ, Schmid TM, Heck LW, Woolley DE (1989) Cleavage of collagen type X by human synovial collagenase and neutrophil elastase. Matrix 9:109–115
Galeski A, Kastelić J, Baer E, Kohn RR (1977) Mechanical and structural changes in rat tail tendon induced by alloxan diabetes and ageing. J Biomech 10:775–782
Gallant C, Kenny P (1986) Oral glucocorticoids and their complications. J Am Acad Dermatol 14:161–177
Galloway D (1982) The primary structure. In: Weiss JB, Jayson MIV (eds) Collagen in health and disease. Churchill Livingstone, Edinburgh, pp 528–557
Garant PR, Cho MI (1979a) Cytoplasmic polarization of periodontal ligament fibroblasts. J Periodont Res 14:95–106
Garant PR, Cho MI (1979b) Autoradiographic evidence of the coordination of the genesis of Sharpey's fibers with new bone formation in the periodontium of the mouse. J Periodont Res 14:107–114
Garbisa S, De Giovanni C, Biagini G, Vasi V, Grigioni WF, D'Errico A, Mancini AM, Del Re B, Lollini PL, Nanni P, Nicoletti G, Prodi G (1988) Different metastatic aggressiveness by murine TS/A clones: ultrastructure, extracellular glycoproteins and type IV collagenolytic activity. Invasion Metastasis 8:177–192
Garron LK, Feeney ML (1959) Electron microscopic studies of the human eye. II. Study of the trabeculae by light and electron microscopy. Arch Ophthalmol 62:966–973
Garrone R (1978) Phylogenesis of connective tissue. Karper, Basel
Garrone R (1985) The collagen of the Porifera. In: Bairati A, Garrone R (eds) Biology of invertebrate and lower vertebrate collagens. Plenum, New York, pp 157–175

Garrone R, Huc A, Junqua S (1975) Fine structure and physicochemical studies on the collagen of the marine sponge Chondrosia reniformis Nardo. J Ultrastruct Res 52:261–275
Gascon-Barré M, Huet PM, Belgiorno J, Plourde V, Coulombe PA (1989) Estimation of collagen content of liver specimens. Variation among animals and among hepatic lobes in cirrhotic rats. J Histochem Cytochem 37:377–381
Gasser AB, Jeannet C, Depierre D, Courvoisier B (1981) L'hydroxyproline urinaire et sérique dans le diagnostic des métastases osseuses du cancer de la prostate. Schweiz Med Wochenschr 111:246–251
Gasset AR, Dohlman CH (1968) The tensile strength of corneal wounds. Arch Ophthalmol 79:595–602
Gastpar H, Kühn K, Marx R (eds) (1978) Collagen-platelet interaction. Schattauer, Stuttgart
Gay S, Kresina TF (1982) Immunological disorders of collagen. In: Weiss JB, Jayson MIV (eds) Collagen in health and disease. Churchill Livingstone, Edinburgh, pp 269–288
Gay S, Miller EJ (1978) Collagen in the physiology and pathology of connective tissue. Fischer, Stuttgart
Gay S, Balleisen L, Remberger K, Fietzek PP, Adelmann BC, Kühn K (1975) Immunohistochemical evidence for the presence of collagen type III in human arterial walls, arterial thrombi and in leukocytes, incubated with collagen in vitro. Klin Wochenschr 53:899–902
Gay S, Martin GR, Müller PK, Timpl R, Kühn K (1976) Simultaneous synthesis of types I and III collagen by fibroblasts in culture. Proc Natl Acad Sci USA 72:4037–4040
Gay S, Viljanto J, Raekallio J, Penttinen R (1978) Collagen types in early phases of wound healing in children. Acta Chir Scand 144:205–211
Gay RE, Buckingham RB, Prince RK, Gay S, Rodnan GP, Miller EJ (1980) Collagen types synthesized in dermal fibroblast cultures from patients with early progressive systemic sclerosis. Arthritis Rheum 23:190–196
Gay S, Martinez-Hernandez A, Rhodes RK, Miller EJ (1981) The collagenous exocytoskeleton of smooth muscle cells. Coll Relat Res 1:377–384
Geerts A, Vrijsen P, Rauterberg J, Burt A, Schellinck R, Wisse E (1989) In vitro differentiation of fat-storing cells parallels marked increase of collagen synthesis and secretion. J Hepatol 9:59–68
Geesin J, Murad S, Pinnell SR (1986) Ascorbic acid stimulates collagen production without altering intracellular degradation in cultured human skin fibroblasts. Biochim Biophys Acta 886:272–274
Gehlsen KR, Hendrix MJC (1986) In vitro assay demonstrates similar invasion profiles for B16 F1 and B16 F10 murine melanoma cells. Cancer Lett 30:207–212
Gelman RA, Blackwell J (1973) Interaction between collagen and chondroitin-6-sulphate. Connect Tissue Res 2:31–35
Gelman RA, Blackwell J, Kefalides NA, Tomichek E (1976) Thermal stability of basement membrane collagen. Biochim Biophys Acta 427:492–496
Gelman RA, Poppke DC, Piez KA (1979) Collagen fibril formation in vitro. The role of the nonhelical terminal regions. J Biol Chem 254:11741–11745
Genovese C, Rowe D, Kream B (1984) Construction of DNA sequences complementary to rat $\alpha_1$ and $\alpha_2$ collagen mRNA and their use in studying the regulation of type I collagen synthesis by 1,25-dihydroxyvitamin D. Biochemistry 23:6210–6216
Gerber G, Gerber G, Kurohara S, Altman KI, Hempelmann LH (1961) Urinary excretion of several metabolites in persons accidentally exposed to ionizing radiation. Radiat Res 15:314–318
Gershwin ME, Abplanalp H, Castles JJ, Ikeda RM, Van de Water J, Eklund J, Haynes D (1981) Characterization of a spontaneous disease of White Leghorn chickens resembling progressive systemic sclerosis (scleroderma). J Exp Med 153:1640–1659
Gershwin ME, Abplanalp H, Van de Water J, Haynes D (1984) An avian model for scleroderma. In: Black CM, Muers AR (eds) Systemic sclerosis (scleroderma). Gower Medical, New York, pp 151–156
Gerstenfeld LC, Finer MH, Boedtker H (1989) Quantitative analysis of collagen expression in embryonic chick chondrocytes having different developmental fates. J Biol Chem 264:5112–5120
Grosh P, Bushell GR, Taylor TKF, Pearce RH, Grimmer BJ (1980) Distribution of glycosaminoglycans across the normal and the scoliotic disc. Spine 5:310–317

Gibson MA, Cleary EG (1985) CL Glycoprotein is the tissue form of type VI Collagen. J Biol Chem 260:11149–11159

Gielen F, Dequeker J, Drochmans A, Wildiers J, Merlevede M (1976) Relevance of hydroxyproline excretion to bone metastasis in breast cancer. Br J Cancer 34:279–285

Gil J (1982) Alveolar wall relations. Ann NY Acad Sci 384:31–38

Gilbert HS (1980) The spectrum of myeloproliferative disorders. Med Clin North Am 57:355–393

Gilbertson TJ, Brunden MN, Gruszczyk SB, Whyte MP, Burnett MA (1983) Serum total hydroxyproline assay: effects of age, sex and Paget's bone disease. J Clin Chem Clin Biochem 21:129–132

Gillery P, Coustry F, Pujol JP, Borel JP (1989) Inhibition of collagen synthesis by interleukin-1 in three-dimensional collagen lattice cultures of fibroblasts. Experientia 45:98–101

Gioud M, Monier JC, Bonvoisin B, Meghlaoui A, Venet C, Lejeune E, Bouvier M (1983) Autoimmunité anticollagène humorale et cellulaire en pathologie rhumatismale. Rev Rhum Mal Osteo-Articulaires 50:579–583

Giraud-Guille MM (1988) Twisted plywood architecture of collagen fibrils in human compact bone osteons. Calcif Tissue Int 43:167–180

Glanville RW (1982) A comparison of models for the macromolecular structure of interstitial and basement membrane collagens. Arzneimittelforschung 32:1353–1357

Glanville RW (1987) Type IV collagen. In: Mayne R, Burgeson RE (eds) Structure and function of collagen types. Academic, Orlando, 43–79

Glanville RW, Qian RQ, Siebold B, Risteli J, Kühn K (1985) Amino acid sequence of the N-terminal aggregation and cross-linking region (7S domain) of the $\alpha 1$(IV) chain of human basement membrane collagen. Eur J Biochem 152:213–219

Glimcher MJ (1976) Composition, structure, and organization of bone and other mineralized tissues and the mechanism of calcification. In: Greep RO, Astwood EB (eds) Endocrinology. Williams and Wilkins, Baltimore, pp 25–116 (Handbook of physiology, vol VII)

Glimcher MJ, Katz EP (1965) The organization of collagen in bone: the role of noncovalent bonds in the relative insolubility of bone collagen. J Ultrastruct Res 12:705–729

Glimcher MJ, Krane SM (1968) The organization and structure of bone and the mechanism of calcification. In: Ramachandran GN, Gould BS (eds) Treatise on collagen. Academic, New York, pp 68–251

Glimcher MJ, Friberg UA, Levine PT (1964) The identification and characterization of a calcified layer of coronal cementum in erupted bovine teeth. J Ultrastruct Res 10:76–81

Glimelius B, Norling B, Westermark B, Wasteson A (1978) Composition and distribution of glycosaminoglycans in cultures of human and malignant glial cells. Biochem J 172:443–456

Gnädinger MC, Itoi M, Slansky HH, Dohlman CH (1969) The role of collagenase in the alkali-burned cornea. Am J Ophthalmol 68:478–483

Gnädinger MC, Refojo MF, Itoi M, Berman MB (1971) The epithelium in corneal wound healing. An assay with tritiated proline. Ophthalmic Res 2:65–76

Godfrey M, Hollister DW (1988) Type II achondrogenesis-hypochondrogenesis: identification of abnormal type II collagen. Am J Hum Genet 43:904–913

Godfrey M, Keene DR, Blank E, Hori H, Sakai LY, Sherwin LA, Hollister DW (1988) Type II achondrogenesis-hypochondrogenesis: morphologic and immunohistopathologic studies. Am J Hum Genet 43:894–903

Goidanich LF, Lenzi L, Silva E (1965) Urinary hydroxyproline excretion in normal subjects and in patients affected with primary diseases of bone. Clin Chim Acta 11:35–38

Goldberg RL, Parrott DP, Kaplan SR, Fuller GC (1980) Effect of gold sodium thiomalate on proliferation of human rheumatoid synovial cells and on collagen synthesis in tissue culture. Biochem Pharmacol 29:869–876

Goldberg RL, Parrott DP, Kaplan SR, Fuller GC (1981) A mechanism of action of gold sodium thiomalate in diseases characterized by a proliferative synovitis: reversible changes in collagen production in cultured human synovial cells. J Pharmacol Exp Ther 218:395–403

Goldberg RL, Kaplan SR, Fuller GC (1983) Effect of heavy metals on human rheumatoid synovial cell proliferation and collagen synthesis. Biochem Pharmacol 32:2763–2766

Goldrose MH, Maslow DE (1985) Relationship between the colonization potential of murine colon adenocarcinoma-38 (MCA-38) cells and in vitro motility. Proc Am Assoc Cancer Res 26:57

Golds EE, Poole AR (1984) Connective tissue antigens stimulate collagenase production in arthritic diseases. Cell Immunol 86:190–205
Goldsmith LA, Briggaman RA (1983) Monoclonal antibodies to anchoring fibrils for the diagnosis of epidermolysis bullosa. J Invest Dermatol 81:464–466
Goldstein I, Reybeyotte P, Parlebas J, Halpern B (1967) Isolation from heart valves of glycopeptides with shared immunological properties with Streptococcus hemolyticus group A polysaccharides. Nature 219:866–867
Goltz RW, Peterson WC, Gorlin RJ, Ravits HG (1962) Focal dermal hypoplasia. Arch Dermatol 86:708–717
Goltz RW, Henderson RR, Hitch JM, Ott JE (1970) Focal dermal hypoplasia syndrome. A review of the literature and report of two cases. Arch Dermatol 101:1–20
Golub LM, Kaplan R, Mulvihill JE, Ramamurthy NS (1979) Collagenolytic activity of crevicular fluid and of adjacent gingival tissue. J Dent Res 58:2132–2136
Golub LM, Lee HM, Lehrer G, Nemiroff A, McNamara TF, Kaplan R, Ramamurthy NS (1983) Minocycline reduces gingival collagenolytic activity during diabetes. Preliminary observations and a proposed new mechanism of action. J Periodont Res 18:516–526
Golub LM, Wolff M, Lee HM, McNamara TF, Ramamurthy NS, Zambon J, Ciancio S (1984a) Further evidence that tetracycline inhibit collagenase activity in human crevicular fluid and from other mammalian sources. J Periodont Res 19:205–215
Golub LM, Ramamurthy N, McNamara TF, Gomes B, Wolff M, Casino A, Kapoor A, Zambon J, Ciancio S, Schneir M, Perry H (1984b) Tetracyclines inhibit tissue collagenase activity. A new mechanism in the treatment of periodontal disease. J Periodont Res 19:373–377
Goodwin JS, Ceuppens JL (1983) Effect of nonsteroidal antiinflammatory drugs on immune function. Semin Arthritis Rheum 13:134–143
Gordon S, Werb Z (1976) Secretion of macrophage neutral proteinase is enhanced by colchicine. Proc Natl Acad Sci USA 73:872–876
Gordon MK, Gerecke DR, Olsen BR (1987) Type XII collagen: distinct extracellular matrix component discovered by cDNA cloning. Proc Natl Acad Sci USA 84:6040–6044
Goshowaki H, Ito A, Mori Y (1988) Effects of prostaglandins on the production of collagenase by rabbit uterine cervical fibroblasts. Prostaglandins 36:107–114
Gosline JM, Shadwick RE (1983a) Molluscan collagen and its mechanical organization in squid mantle. In: Mollusca. Metabolic biochemistry and molecular biomechanics, vol 1. Academic, Orlando, pp 371–398
Gosline JM, Shadwick RE (1983b) The role of elastic energy storage mechanisms in swimming: analysis of mantle elasticity in escape jetting in the squid Loligo opalescens. Can J Zool 61:1421–1435
Gosling JA, Dixon JS (1978) Functional obstruction of the ureter and renal pelvis. A histological and electron microscopic study. Br J Urol 50:145–152
Goto M, Yoshinoya S, Miyamoto T, Sasano M, Okamoto M, Nishioko K, Terato K, Nagai Y (1988) Stimulation of interleukin-1$\alpha$ and interleukin-1$\beta$ release from human monocytes by cyanogen bromide peptides of type II collagen. Arthritis Rheum 31:1508–1514
Gould BS (1960) Ascorbic acid and collagen fiber formation. Vitam Horm 18:89–120
Gould BS (ed) (1968) Biology of collagen. Academic Press, London
Gowen M, Wood DD, Ihrie EJ, McGuire MKB, Russell RGG (1983) An interleukin-1-like factor stimulates bone resorption in vitro. Nature 306:378–380
Grant RA (1967) Content and distribution of aortic collagen, elastin, and carbohydrate in different species. J Atheroscler Res 7:463–472
Grant ME, Prockop DJ (1972) The biosynthesis of collagen. N Engl J Med 286:194–199, 242–249, 291–300
Grasedyck K, Lindner J (1975) The behavior of ($^{14}$C)-D-penicillamine in collagen metabolism. Connect Tissue Res 3:171–176
Grasedyck K, Ropohl D, Szarvas F, Lindner J (1971) Kollagenpeptidaseaktivität in menschlichen Seren bei Lebercirrhose. Klin Wochenschr 49:163–164
Grassman W (1955) Unsere heutige Kenntnis des Kollagens. Leder 6:241–261
Grassman W (1960) Kollagen und Bindegewebe. Svensk Kem Tidskr 72:275–302

Graves PN, Weiss IK, Perlish JS, Fleischmajer R (1983) Increased procollagen mRNA levels in scleroderma skin fibroblasts. J Invest Dermatol 80:130–132

Green CM, Sweet HO, Bunker LE (1976) Tight-skin, a new mutation of the mouse causing excessive growth of connective tissue and skeleton. Am J Pathol 82:493–507

Gressner AM (1982) Connective tissue metabolism in liver diseases and its relevance in clinical-chemical diagnosis. In: Kaiser E, Gabl F, Müller MM, Bayer PM (eds) Proceedings of the XIth international congress of clinical chemistry. Walter de Gruyter, Berlin New York, pp 687–698

Gressner AM, Neu HH (1984) N-Terminal procollagen peptide and $\beta_2$-microglobulin in synovial fluids from inflammatory and non-inflammatory joint disease. Clin Chim Acta 141:241–245

Greve C, Opsahl W, Reiser K, Abbott U, Kenney C, Benson D, Rucker R (1988) Collagen crosslinking and cartilage glycosaminoglycan composition in normal and scoliotic chickens. Biochim Biophys Acta 967:275–283

Grevstad HJ (1987) Experimentally induced resorption cavities in rat molars. Scand J Dent Res 95:428–440

Grevstad HJ (1988) Collagen deposition during wound repair in rat gingiva. Scand J Dent Res 98:561–568

Griffin CA, Emanuel BS, Hansen JR, Cavence WK, Myers JC (1987) Human collagen genes encoding basement membrane $\alpha$1(IV) and $\alpha$2(IV) chain map to the distal long arm of chromosome 13. Proc Natl Acad Sci USA 84:512–516

Griffiths MM, DeWitt CW (1981) Immunogenetic control of experimental type II collagen-induced arthritis. J Immunogenet 8:463–470

Griffiths R, Tudball N, Thomas J (1976) Effect of induced elevated plasma levels of homocystine and methionine in rats on collagen and elastin structures. Connect Tissue Res 4:101–106

Griffiths MM, Eichwald EJ, Martin JH, Smith CB, DeWitt CW (1981) Immunogenetic control of experimental type II collagen-induced arthritis. Arthritis Rheum 24:781–789

Grimaud JA, Druguet M, Peyrol S, Chevalier O, Herbage D, El Badrawy N (1980) Collagen immunotyping in human liver: light and electron microscope study. J Histochem Cytochem 28:1145–1156

Grimaud JA, Boros DL, Takiya C, Mathew RC, Emonard H (1987) Collagen isotypes, laminin, and fibronectin in granulomas of the liver and intestines of Schistosoma mansoni-infected mice. Am J Trop Med Hyg 37:335–344

Grinnell F, Fukamizu H, Pawelek P, Nakagawa S (1989) Collagen processing, crosslinking, and fibril bundle assembly in matrix produced by fibroblasts in long-term cultures supplemented with ascorbic acid. Exp Cell Res 181:483–491

Grobelny D, Teater C, Galardy RE (1989) The ketone cinnamoyl-(1-$^{13}$C-Phe)-$^{c}$Gly-Pro-Pro is a tetrahedral transition state analog inhibitor of C. histolyticum collagenase. Biochem Biophys Res Commun 159:426–431

Groniowski J (1975) Electron microscopic observations on chronic aggressive hepatitis: participation of hepatocytes in liver fibrosis. Pathol Eur 10:37–50

Grosfeld JCM, Spaas J, Van de Staak WJBM, Stadhouders AM (1965) Hyalinosis cutis et mucosae. Dermatologia 130:239–266

Gross J (1961) Collagen. Sci Am 204:120–130

Gross J (1974) Collagen biology: structure, degradation and disease. Harvey Lect 68:351–432

Gross J, Lapière CM (1962) Collagenolytic activity in amphibian tissues: a tissue culture assay. Proc Natl Acad Sci USA 48:1014–1022

Gross J, Levene CI (1959) Effect of beta-aminopropionitrile on extractability of collagen from skin of mature guinea pigs. Am J Pathol 36:687–695

Gross J, Matoltsy AG, Cohen C (1955) Vitrosin: a member of the collagen class. J Biophys Biochem Cytol 1:215–223

Gross J, Highberger JH, Johnson-Wint B, Biswas C (1980) Mode of action and regulation of tissue collagenases. In: Woolley DE, Evanson JM (eds) Collagenase in normal and pathological connective tissues. Willey, Chichester, pp 11–35

Grotendorst GR, Martin GR (1986) Cell movements in wound-healing and fibrosis. In: Kühn K, Krieg T (eds) Connective tissue: biological and clinical aspects. Karger, Basel, pp 385–403

Grucka-Mamczar E, Dróżdż M, Kucharz EJ, Barańska-Gachowska M, Mamczar A (1982) Badania nad wpływem związków fluoru na metabolizm tkanki łącznej. I. Wpływ jonu fluorkowego na stężenie glikoproteidów i aktywność enzymów lizosomalnych w surowicy krwi szczurów w okresie rozwoju osobniczego (Studies on the influence of fluorine on the metabolism of connective tissue. I. Effect of fluoride ion on the concentration of glycoproteins and lysosomal enzyme activities in blood serum of growing rats). In: Machoy M (ed) Metabolizm fluoru (Metabolism of fluorine). Szczecińskie Towarzystwo Naukowe, Szczecin, pp 69–73 (in Polish)

Grunfeld JP (1985) The clinical spectrum of hereditary nephritis. Kidney Int 27:83–97

Guantieri V, Tamburro AM, Cabrol D, Vasilescu D (1987) Conformational studies on polypeptide models of collagen. Int J Pept Protein Res 29:216–230

Gudmundson C (1971a) Oxytetracycline-induced fragility of growing bones. An experimental study of rats. Clin Orthop 77:284–289

Gudmundson C (1971b) Oxytetracycline-induced disturbances of fracture healing. J Trauma 11:511–517

Gunson DE, Halliwell RA, Minor RR (1984) Dermal collagen degradation and phagocytosis. Occurrence in a horse with hyperextensible fragile skin. Arch Dermatol 120:599–604

Günther T, Carsten PM (1964) Über die Löslichkeit von Kollagen in Abhängigkeit von Nebennierenrindenhormonen. Naturwissenschaften 48:699

Günzler V, Hanauske-Abel HM, Myllylä R, Mohr J, Kivirikko KI (1987) Time-dependent inactivation of chick-embryo prolyl 4-hydroxylase by coumaric acid. Biochem J 242:163–169

Gustavson KH (1956) The chemistry and reactivity of collagen. Academic, New York, pp 1–342

Gusterson BA, Warburton MJ, Mitchell D, Ellison M, Neville AM, Rudland PS (1982) Distribution of myoepithelial cells and basement membrane proteins in the normal breast and malignant breast diseases. Cancer Res 42:4763–4770

Guttadauria M, Diamond H, Kaplan D (1977) Colchicine in the treatment of scleroderma. J Rheumatol 4:272–276

Hagihara Y, Kaminishi H, Cho T, Tanaka M, Kaita H (1988) Degradation of human dentine collagen by an enzyme produced by the yeast Candida albicans. Arch Oral Biol 33:617–619

Hahn EG (1984) Blood analysis for liver fibrosis. J Hepatol 1:67–73

Hahn EG, Martini GA (1980a) Clinical parameters of fibroplasia. Ital J Gastroenterol 12:41–46

Hahn EG, Martini GA (1980b) Diagnostische Parameter der Kollagenbiosynthese. Internist 21:195–201

Hahn EG, Schuppan D (1982) Collagen metabolism in liver disease. In: Bianchi L, Gerok W, Landman L, Sickinger K, Stalder GA (eds) Liver in metabolic diseases. MTP Press, Boston, pp 309–323 (Falk symposium, no 35)

Hahn EG, Timpl R, Miller EJ (1974) The production of specific antibodies to native collagens with the chain compositions $[\alpha 1(I)]_3$, $[\alpha 1(II)]_3$ and $[\alpha 1(I)]_2\alpha 2(I)$. J Immunol 113:421–423

Hakkinen PJ, Schmoyer RL, Witschi HP (1983) Potentiation of butylated-hydroxytoluene-induced acute lung damage by oxigen. Am Rev Respir Dis 128:648–651

Hall DA, Saxl H (1961) Studies of human and tunicate cellulose and of their relationship to reticulin. Proc R Soc Biol 155:202–210

Hall CE, Jakus MA, Schmitt FO (1942) Electron microscope observations of collagen. J Am Chem Soc 64:1234–1239

Halme J, Kivirikko KI, Kaitila I, Saxen L (1969) Effect of tetracycline on collagen biosynthesis in cultured embryonic bones. Biochem Pharmacol 18:827–838

Halme J, Kivirikko KI, Simons K (1970) Isolation and partial characterization of highly purified protocollagen proline hydroxylase. Biochim Biophys Acta 198:460–470

Halme J, Tyree BT, Jeffrey JJ (1980) Collagenase production by primary cultures of rat uterine cells. Partial purification and characterization of the enzyme. Arch Biochem Biophys 199:51–60

Halse J, Gordeladze JO (1981) Total and non-dialyzable urinary hydroxyproline in acromegalics and control subjects. Acta Endocrinol (Copenh) 96:451–457

Halwe T, Savunen T, Aho H, Vihersaari T, Penttinen R (1985) Elastin and collagen in the aortic wall: changes in the Marfan syndrome and annuloaortic ectasia. Exp Mol Pathol 43:1–12

Hamaguchi Y, Sakakura K, Majima Y, Sakakura Y (1987) Cathepsin B-like thiol proteases and collagenolytic proteases in middle ear effusion from acute and chronic otitis media with effusion. Acta Otolaryngol (Stockh) 104:119–124

Hämäläinen L, Oikarinen J, Kivirikko KI (1985) Synthesis and degradation of type I procollagen mRNAs in cultured human skin fibroblasts and the effect of cortisol. J Biol Chem 260:720–725

Hamlin CR, Kohn RR, Luschin JH (1975) Apparent accelerated ageing of human collagen in diabetes mellitus. Diabetes 24:902–904

Hance AJ, Crystal RG (1975) The connective tissue of lung. Am Rev Respir Dis 112:657–711

Handler CE, Child A, Light ND, Dorrance DE (1985) Mitral valve prolapse, aortic compliance, and skin collagen in joint hypermobility syndrome. Br Heart J 54:501–508

Hansen TM, Lorenzen I (1975) The effects of cyclophosphamide and azathioprine on collagen in skin and granulation tissue in rats, and the effects of cyclophosphamide on collagen in human skin. Acta Pharmacol Toxicol 36:448–461

Hanset R, Ansay M (1967) Dermatosparaxie (peau déchirée) chez le veau: un defaut general du tissue conjonctif, de nature hereditaire. Ann Med Vet 111:451–470

Hanset R, Lapière CM (1974) Inheritance of dermatosparaxis in the calf. A genetic defect of connective tissues. J Hered 65:356–358

Hanson AN, Bentley JP (1983) Quantitation of type I to type III collagen ratios in small samples of human tendon, blood vessels and atherosclerotic plaque. Anal Biochem 30:32–40

Hantaï D, Gautron J, Labat-Robert J (1983) Immunolocalization of fibronectin and other macromolecules of the intercellular matrix in the striated muscle fiber of the adult rat. Coll Relat Res 3:381–391

Haralson MA, Mitchell WM (1981) Cell-free synthesis of putative type V procollagen chains programmed by Chinese hamster lung cell mRNA. Coll Relat Res 1:309–325

Haralson MA, Mitchell WM, Rhodes RK, Kresina TF, Gay R, Miller EJ (1980) Chinese hamster lung cells synthesize and confine to the cellular domain a collagen composed solely of B chains. Proc Natl Acad Sci USA 77:5206–5210

Haralson MA, Mitchell WM, Rhodes RK, Miller EJ (1984) Evidence that the collagen in the culture medium of Chinese hamster lung cells contains components related at the primary structural level to the $\alpha 1(V)$ collagen chain. Arch Biochem Biophys 229:509–518

Haralson MA, Federspiel SJ, Martinez-Hernandez A, Rhodes RK, Miller EJ (1985) Synthesis of $[\mathrm{Pro}\,\alpha 1(IV)]_3$ collagen molecules by cultured embryo-derived parietal yolk sac cells. Biochemistry 24:5792–5795

Harbers K, Kühn M, Deluis H, Jaenisch R (1984) Insertion of retrovirus into the first intron of $\alpha 1(I)$ collagen gene leads to embryonic lethal mutation in mice. Proc Natl Acad Sci USA 81:1504–1508

Hardforth CP (1962) Isoproterenol-induced myocardial infarction in animals. Arch Pathol 73:161–165

Harkness RD (1952) Collagen in regenerating liver of the rat. J Physiol (Lond) 117:257–266

Harkness RD (1957) Regeneration of liver. Br Med Bull 13:87–93

Harkness RD (1961) Biological functions of collagen. Biol Rev 36:399–463

Harkness RD (1964) The physiology of the connective tissues of the reproductive tract. Int Rev Connect Tissue Res 2:155–211

Harkness MLR, Harkness RD (1954) The collagen content of the reproductive tract of the rat during pregnancy and lactation. J Physiol (Lond) 123:492–500

Harkness MLR, Harkness RD (1956) The distribution of the growth of collagen in the uterus of the pregnant rat. J Physiol (Lond) 132:492–501

Harkness MLR, Harkness RD (1959) Changes in the physical properties of the uterine cervix of the rat during pregnancy. J Physiol (Lond) 148:524–527

Harkness RD, Moralee BE (1956) The time course and route of loss of collagen from the rat's uterus during post partum involution. J Physiol (Lond) 132:502–508

Harkness MLR, Harkness RD, Moralee BE (1957) The effect of the oestrous cycle and of hormones on the collagen content of the uterus of the rat. J Physiol (Lond) 135:270–280

Harkness MLR, Harkness RD, James DW (1958) The effect of a protein-free diet on the collagen content of mice. J Physiol (Lond) 144:307–313

Harmon CE, Portanova JP (1982) Drug-induced lupus: clinical and serological studies. Clin Rheum Dis 8:121–135

Harnisch JP, Buchen R, Sinha PK, Barrach HJ (1978) Ultrastructural identification of type I and II collagen in the cornea of the mouse by means of enzyme labeled antibodies. Graefes Arch Clin Exp Ophthalmol 208:9–13

Harper E, Bloch KJ, Gross J (1971) The zymogen of tadpole collagenase. Biochemistry 10:3035–3041

Harris ED Jr (1984) Retinoid therapy for rheumatoid arthritis. Ann Intern Med 100:146–147

Harris ED Jr (1986) Regulation of collagenolysis in synovial cell systems. In: Kühn K, Krieg T (eds) Connective tissue: biological and clinical aspects. Karger, Basel, pp 197–215

Harris ED Jr, Krane SM (1971) Effects of colchicine on collagenase in cultures of rheumatoid synovium. Arthritis Rheum 14:669–684

Harris ED Jr, KRane SM (1974) Collagenases. N Engl J Med 291:557–563, 605–609, 652–661

Harris ED Jr, Millis M (1971) Treatment with colchicine of the periarticular inflammation associated with sarcoidosis. Arthritis Rheum 14:130

Harris ED Jr, Sjoerdsma A (1966a) Collagen profile in various clinical conditions. Lancet 2:707–709

Harris ED Jr, Sjoerdsma A (1966b) Effect of penicillamine on human collagen and its possible application to treatment of scleroderma. Lancet 2:996–999

Harris ED Jr, DiBona DR, Krane SM (1969) Collagenases in human synovial fluid. J Clin Invest 48:2104–2113

Harris ED Jr, Evanson JM, DiBona DR, Krane SM (1970) Collagenase and rheumatoid arthritis. Arthritis Rheum 13:83–94

Harris ED Jr, Gonnerman WA, Savage JE, O'Dell BL (1974) Connective tissue amine oxidase. II. Purification and partial characterization of lysyl oxidase from chick aorta. Biochim Biophys Acta 341:332–344

Harris ED Jr, Hoffman GS, McGuire JL, Strosberg JM (1975) Colchicine: effects upon urinary hydroxyproline excretion in patients with scleroderma. Metabolism 24:529–535

Harris ED Jr, Glauert AM, Murley AHG (1977) Intracellular collagen fibers at the pannus-cartilage junction in rheumatoid arthritis. Arthritis Rheum 20:657–665

Harris ED Jr, Welgus HG, Krane SM (1984) Regulation of the mammalian collagenases. Coll Relat Res 4:493–512

Harris JP, Woolf NK, Ryan AF (1986) A re-examination of experimental type II collagen autoimmunity: middle and inner ear morphology and function. Ann Otol Rhinol Laryngol 95:176–180

Harrison DE, Archer JR (1978) Measurement of changes in mouse tail collagen with age: temperature dependence and procedural details. Exp Gerontol 13:75–82

Harrison JE, McNeil KG, Sturtridge WC, Bayley TA, Murray TM, Williams C, Tam C, Fornasier V (1981) Three-year changes in bone mineral mass of osteoporotic patients based on neutron activation analysis of the central third of the skeleton. J Clin Endocrinol Metab 52:751–756

Hartman F (1966) Investigations on the pathogenesis of connective tissue damage in experimental lathyrism: urinary hydroxyproline and hexosamine determinations in lathyritic rats. Z Rheumaforsch 25:161–166

Hartmann DJ, Magloire H, Ricard-Blum S, Joffre A, Couble ML, Ville G, Herbage D (1983) Light and electron immunoperoxidase localization of minor disulfide-bonded collagens in fetal calf epiphyseal cartilage. Coll Relat Res 3:349–357

Haschek WM, Witschi H (1979) Pulmonary fibrosis – a possible mechanism. Toxicol Appl Pharmacol 50:123–136

Hashimoto K, Yamanishi Y, Dabbous MK (1972) Electron microscopic observations of collagenolytic activity of basal cell epithelioma of the skin in vivo and in vitro. Cancer Res 32:2561–2567

Hashimoto K, Yamanishi Y, Dabbous MK, Maeyens E (1973a) Collagenase activity in rheumatoid nodules. Ultrastructural in vivo and in vitro studies. Acta Derm Venerol (Stockh) 53:439–448

Hashimoto K, Yamanishi Y, Maeyens E, Dabbous MK, Kanzaki T (1973b) Collagenolytic activities of squamous-cell carcinoma of the skin. Cancer Res 33:2790–2801

Hassan G, Stefanini S, Bargagli AM, Autuori F (1989) Proline-incorporating cells in chronic active liver disease. Hepatology 9:37–49

Hassanein H, Herbage D, Chevalier O, Buffevant C, Grimaud JA (1983) Solubilization and charcterization of human liver collagens in schistosomiasis mansoni. Cell Mol Biol 29:139–148

Hasselbalch H, Junker P, Lisse I, Bentsen KD (1985) Serum procollagen III peptide in chronic myeloproliferative disorders. Scand J Haematol 35:550–557

Hasselbalch H, Junker P, Lisse I, Bentsen KD, Risteli L, Risteli J (1986) Serum markers for type IV collagen and type III procollagen in the myelofibrosis-osteomyelosclerosis syndrome and other chronic myeloproliferative disorders. Am J Hematol 23:101–111

Hassell TM, Page RC, Lindhe J (1978) Histologic evidence for impaired growth control in diphenylhydantoin gingival overgrowth in man. Arch Oral Biol 23:381–384

Hasslacher C, Brocks D, Mann J, Mall G, Waldherr R (1987) Influence of hypertension on serum concentration of type IV collagen antigens in streptozotocin-diabetic and non-diabetic rats. Diabetologia 30:344–347

Hasty KA, Hibbs MS, Kang AH, Mainardi CL (1986) Secreted forms of human neutrophil collagenase. J Biol Chem 261:5645–5650

Hata RI, Ninomiya Y (1984) Heptatocytes (hepatic parenchymal cells) produce a major part of liver collagen in vivo. Biochem Int 8:181–186

Hata RI, Hori H, Nagai Y, Tanaka S, Kondo M, Hiramatsu M, Utsumi N, Kumegawa M (1984) Selective inhibition of type I collagen synthesis in osteoblastic cells by epidermal growth factor. Endocrinology 115:867–876

Hatahara T, Seyer JM (1982) Isolation and characterization of a fibrogenic factor from $CCl_4$-damaged rat liver. Biochim Biophys Acta 716:377–382

Hatton DV, Leach CS, Nicogossian AE, DiFerrante N (1977) Collagen breakdown and nitrogen dioxide inhalation. Arch Environ Health 32:33–36

Haustein UF, Ziegler V (1985) Environmentally induced systemic sclerosis-like disorders. Int J Dermatol 24:147–151

Hauss WH, Losse H (eds) (1960) Struktur and Stoffwechsel des Bindegewebes. Thieme, Stuttgart, pp 1–182

Havenith MG, Arends JW, Simon R, Volovics A, Wiggers T, Bosman FT (1988) Type IV collagen immunoreactivity in colorectal cancer. Prognostic values of basement membrane deposition. Cancer 62:2207–2211

Hawgood S, Benson BJ, Hamilton RL Jr (1985) Effects of a surfactant-associated protein and calcium ions on the structure and surface activity of lung surfactant lipids. Biochemistry 25:184–190

Hayes RL, Rodnan GP (1971) The ultrastructure of skin in progressive systemic sclerosis (scleroderma). Dermal collagen fibres. Am J Pathol 63:433–442

Heath JK, Gowen M, Meikle MC, Reynolds JJ (1982) Human gingival tissues in culture synthesize three metalloproteinases and a metalloproteinase inhibitor. J Periodont Res 17:183–190

Heathcote JG (1982) Kidney. In: Weiss JB, Jayson MIV (eds) Collagen in health and disease. Churchill Livingstone, Edinburgh, pp 404–413

Heathcote JG, Bailey AJ, Grant ME (1980) Studies on the assembly of the rat lens capsule: biosynthesis of a cross-linked collagenous component of high molecular weight. Biochem J 190:229–237

Heckmann M, Aumailley M, Hatamochi A, Chu ML, Timpl R, Krieg T (1989) Down-regulation of $\alpha$ 3(VI) chain expression by $\gamma$-interferon decreases synthesis and deposition of collagen type VI. Eur J Biochem 182:719–726

Hedbom E, Heinegård D (1989) Interaction of a 59 kDa connective tissue matrix protein with collagen I and collagen II. J Biol Chem 264:6898–6905

Hedtmann A, Steffen R, Krämer J (1987) Prospective comparative study of intradiscal high-dose and low-dose collagenase versus chymopapain. Spine 12:388–392

Hegedus SI, Schorr WF (1972) Familial cutaneous collagenoma. Cutis 10:283–288

Hegreberg GA (1982) Animal models of collagen disease. Prog Clin Biol Res 94:229–244

Hegreberg GA, Padgett GA, Gorgam JR, Henson JB (1969) A connective tissue disease of dogs and mink resembling the Ehlers-Danlos syndrome. II. Mode of inheritance. J Hered 60:249–254

Hegreberg GA, Padgett GA, Ott RL, Henson JB (1970a) A hereditable connective tissue disease of dogs and mink resembling the Ehlers-Danlos syndrome of man. I. Skin tensile strength properties. J Invest 54:377–380

Hegreberg GA, Padgett GA, Henson JB (1970b) Connective tissue disease of dogs resembling Ehlers-Danlos syndrome of man. III. Histopathologic changes of the skin. Arch Pathol 90:159–166

Heilmann K, Suschke J, Kunze D, Murken JD (1972) Marfan-syndrome. Ophthalmologische, klinische, biochemische and genetische Untersuchungen. Klin Padiatr 185:43–51

Heino J, Kähäri VM, Jaakkola S, Peltonen J (1989) Collagen in the extracellular matrix of cultured scleroderma skin fibroblasts: changes related to ascorbic acid-treatment. Matrix 9:34–39

Heinrich D, Metz J, Kubli F (1977) Hydroxyprolin: ein neuer Parameter in der Diagnostik pathologischer Schwangerschaften. Arch Gynakol 224:130–132

Helgeland K (1977a) pH and the effect of fluoride and zinc on protein and collagen biosynthesis in rabbit dental pulp in vitro. Scand J Dent Res 85:407–413

Helgeland K (1977b) Effect of fluoride on protein and collagen biosynthesis in rabbit dental pulp in vitro. Scand J Dent Res 85:264–269

Helle O, Nes NN (1972) A hereditary skin defect in sheep. Acta Vet Scand 13:443–445

Hendel L, Ammitzbøll T, Dirksen K, Petri M (1984) Collagen in the esophageal mucosa of patients with progressive systemic sclerosis. Acta Derm Venerol (Stockh) 64:480–484

Henderson AS, Myers JC, Ramirez F (1983) Localization of the human $\alpha 2$(I) collagen gene (COL1A2) to chromosome 7q22. Cytogenet Cell Genet 36:586–587

Henderson RR, Wheeler CE Jr, Abele DC (1968) Familiar cutaneous collagenoma. Arch Dermatol 98:23–27

Henke E, Leader M, Tajima S, Pinnell S, Kaufman R (1985) A 38 base pair insertion in the pro $\alpha 2$(I) collagen gene of a patient with Marfan syndrome. J Cell Biochem 27:169–174

Henley KS, Laughrey EG, Appelman HD, Flecker K (1977) Effect of ethanol on collagen formation in dietary cirrhosis in the rat. Gastroenterology 72:502–506

Henney AM, Parker DJ, Davies MJ (1982) Collagen biosynthesis in normal and abnormal human valves. Cardiovasc Res 16:624–631

Herbage D, Huc A, Chabrand D, Chapuy MC (1972) Physicochemical study of articular cartilage from healthy and osteo-arthritic human hips. Orientation and thermal stability of collagen fibres. Biochim Biophys Acta 271:339–346

Herbage D, Lucas JM, Huc A (1974) Collagen and proteoglycan interactions in bovine articular cartilage. Biochim Biophys Acta 336:108–116

Herbert DM, Lindberg KA, Jayson MIV, Bailey AJ (1974) Biosynthesis and maturation of skin collagen in scleroderma, and effect of d-penicillamine. Lancet 1:187–192

Herbert CM, Lindberg KA, Jayson MIV, Bailey AJ (1975) Changes in the collagen of human intervertebral discs during ageing and degenerative disc disease. J Mol Med 1:79–91

Herring GM (1972) The organic matrix of bone. In: Bonne GH (ed) The biochemistry and physiology of bone. Academic, New York, pp 128–196

Hess EV (ed) (1981) Drug-induced lupus. Proceedings of the Kroc Foundation Conference. Arthritis Rheum 24:979–1108

Hess EV (1987) Drug-related lupus: the same or different? In: Lahita RG (ed) Systemic lupus eryhtematosus. Willey, New York, pp 869–880

Hess OM, Ritter M, Schneider J, Grimm J, Turina M, Krayenbühl HP (1984) Diastolic stiffness and myocardial structure in aortic valve disease and after valve replacement. Circulation 69:855–865

Hesterberg TW, Last JA (1981) Ozone-induced acute pulmonary fibrosis in rats. Prevention of increased rates of collagen synthesis by methylprednisolone. Am Rev Respir Dis 123:47–52

Higgins GM, Anderson RM (1931) Experimental pathology of the liver. I. Restoration of the liver of the white rat following partial surgical removal. Arch Pathol 12:186–202

Highberger JH, Gross J, Schmitt FO (1950) Electron-microscope observations of certain fibrous structures obtained from connective-tissue extracts. J Am Chem Soc 72:3321–3322

Hilfer SR (1972) Correlation between extracellular material and thyroid organogenesis. J Cell Biol 55 (2, 2):114

Hilfer SR, Pakstis GL (1977) Interference with thyroid histogenesis by inhibitors of collagen synthesis. J Cell Biol 75:446–463
Hinglais N, Grünfeld JP, Bois E (1972) Characteristic ultrastructural lesion of the glomerular basement membrane in progressive hereditary nephritis (Alport's syndrome). Lab Invest 27:473–487
Hino M, Nakano G, Harada M, Nagatsu T (1975) Distribution of PZ-peptidase and glycylprolyl β-naphthylamidase activities in oral tissues. Arch Orach Biol 20:19–22
Hioco D, Gruson M, Ryckewaert A, de Sèze S (1967) Intérêt et signififcation de l'hydroxyproline urinaire en pathologie ossuese. Ann Biol Clin (Paris) 25:725–734
Hirani S, Lambris JD, Müller-Eberhard HJ (1986) Structural analysis of the asparagine-linked oligosaccarides of human complement component C3. Biochem J 233:613–616
Hirayama C, Hiroshige K, Masuya T (1969) Hepatic collagenolytic activity in rats after carbon tetrachloride poisoning. Biochem J 115:843–847
Hirayama C, Morotomi I, Hiroshige K (1971) Corticosteroid effect on hepatic collagens. Experientia 27:893–984
Hirayama C, Kawabe M, Morotomi I, Kimura N (1972) Serum collagen-like protein in patients with chronic liver disease. Digestion 5:31–39
Hochweiss S, Fruchtman S, Hahn EG, Gilbert H, Donovan PB, Johnson J, Goldberg JD, Berk PD (1983) Increased serum procollagen III aminoterminal peptide in myelofibrosis. Am J Hematol 15:343–351
Hoehn H, Bryant EM, Karp LE, Martin GM (1974) Cultivated cells from diagnostic amniocentesis in second trimester pregnancies. I. Clonal morphology and growth potential. Pediatr Res 8:746–754
Hofer PA (1973) Urbach-Wiethe disease (lipoglycoproteinosis; lipoid proteinosis; hyalinosis cutis et mucosae). Acta Derm Venerol (Stockh) 53 [Suppl 71]:5–52
Högemann B, Balleisen L, Rauterberg J, Voss B, Gerlach U (1986) Basement membrane components (7S collagen, laminin P1) are increased in sera of diabetics and activate platelets in vitro. Haemostasis 16:428–432
Holbrook KA, Byers PH (1981) Ultrastructural characteristics of the skin in a form of Ehlers-Danlos syndrome type IV: storage in the rough endoplasmic reticulum. Lab Invest 44:342–350
Holbrook KA, Byers PH (1982) Structural abnormalities in dermal collagen and elastic matrix from the skin of patients with inherited connective tissue disorders. J Invest Dermatol 79:7s–16s
Holbrook KA, Byers PH, Counts DF, Hegreberg GA (1980) Dermatosparaxis in a Himalayan cat. II. Ultrastructural studies of dermal collagen. J Invest Dermatol 74:100–104
Holderbaum D, Ehrhardt LA, McCullagh KG (1975) Effects of hyperlipoproteinemic serum and exogenous proline concentration on collagen synthesis by isolated rabbit aortas. Proc Soc Exp Biol Med 150:363–367
Hollenberg MD (1977) Steroid-stimulated amino acid uptake in cultured human fibroblasts reflects glucocorticoid and anti-inflammatory potency. Mol Pharmacol 13:150–160
Hollinger MA, Chvapil M (1977) Effect of paraquat on rat lung prolyl hydroxylase. Res Commun Chem Pathol Pharmacol 16:159–162
Hollinger MA, Zuckermann JE, Giri SN (1978) Effect of acute and chronic paraquat on rat lung collagen content. Res Commun Chem Pathol Pharmacol 21:295–305
Hollister DW, Burgeson RW, Rimoin DL (1975) Abnormal cartilage collagen in thanatophoric dwarfism. Am J Hum Genet 27:783
Hollister DW, Byers PH, Holbroock KA (1982) Genetic disorders of collagen metabolism. Adv Hum Genet 12:1–87
Holmberg A (1965) Schlamm's canal and the trabecular meshwork: an electron microscopic study of the normal structure in man and monley (Cercopithecus ethiops). Doc Ophthalmol 19:339–373
Holmes DF, Capaldi MJ, Chapman JA (1986) Reconstitution of collagen fibrils in vitro; the assembly process depends on the initiating procedure. Int J Biol Macromolec 8:161–166
Holmgren E (1907) Über die Trophospongien der quergestreiften Muskelfasern, nebst Bemerkungen über den allgemeinen Bau dieser Fasern. Arch Mikrosk Anat 71:165–201
Holubarsch C, Holubarsch T, Jacob R, Medugorac I, Thiedemann KU (1983) Passive elastic properties of myocardium in different models and stages of hypertrophy: a study comparing mechanical, chemical and morphometric parameters. Perspect Cardiovasc Res 7:323–336

Hopkins SC, Palmieri GM, Niell HB, Moinuddin M, Soloway MS (1984) Total and nondialyzable hydroxyproline excretion in stage $D_2$ prostate cancer. Cancer 53:117–121

Hori Y, Hatakeyama H, Yamada K, Kurosawa A (1989) Effect of a novel thromboxane $A_2$ receptor antagonist, S-145, on collagen-induced ECG changes and thrombocytopenia in rodents. Jpn J Pharmacol 50:195–205

Hörlain D, McPherson J, Goh SH, Bornstein P (1981) Regulation of protein synthesis: translational control by procollagen-derived fragments. Proc Natl Acad Sci USA 78:6163–6167

Hørslev-Petersen K, Ammitzbøll T, Engstrom-Laurent A, Bentsen K, Junker P, Asboe-Hansen G, Lorenzen I (1988a) Serum and urinary aminoterminal type III procollagen peptide in progressive systemic sclerosis: relationship to sclerodermal involvement, serum hyaluronan and urinary collagen metabolites. J Rheumatol 15:460–467

Hørslev-Petersen K, Bentsen KD, Halberg P, Junker P, Kivirikko KI, Majamaa K, Risteli L, Risteli J, Lorenzen I (1988b) Connective tissue metabolites in serum as markers of disease activity in patients with rheumatoid arthritis. Clin Exp Rheumatol 6:129–134

Hørslev-Petersen K, Pedersen LR, Bentsen KD, Brocks D, Garbarsch C, Kim KY, Hahn EG, Schuppan D, Lorenzen I (1988c) Collagen type IV and procollagen type III during granulation tissue formation: a serological, biochemical, immunohistochemical and morphometrical study on the viscose cellulose sponge rat model. Eur J Clin Invest 18:352–359

Hørslev-Petersen K, Bentsen KD, Junker P, Mathiesen FK, Hansen TM, Lorenzen I (1988d) Serum aminoterminal type III procollagen peptide in inflammatory and degenerative rheumatic disorders. Clin Rheumatol 7:61–68

Horton WA, Rimoin DL (1970) Kniest dysplasia. A histochemical study of the growth plate. Pediatr Res 13:1266–1270

Hostikka SL, Tryggvason K (1988) The complete primary structure of the $\alpha 2$ chain of human type IV collagen and comparison with the $\alpha 1$(IV) chain. J Biol Chem 263:19488–19493

Hostikka SL, Kurkinen M, Tryggvason K (1987) Nucleotide sequence coding for the human type IV collagen $\alpha_2$ chain cDNA reveals extensive homology with the NC-1 domain of $\alpha_1$(IV) but not with the collagenous domain or 3'-untranslated region. FEBS Lett 216:281–286

Houck JC, Patel YM, Gladner J (1967) The effects of anti-inflammatory drugs upon the chemistry and enzymology of rat skin. Biochem Pharmacol 16:1099–1111

Housley TJ, Rowland FN, Ledger PW, Kaplan J, Tanzer ML (1980) Effects of tunicamycin on the biosynthesis of procollagen by human fibroblasts. J Biol Chem 255:121–128

Hovig T (1963) Aggregation of rabbit blood platelets produced in vitro by saline "extract" of tendons. Thromb Diath Haemorrh 9:248–263

Howard BV, Macarak EJ, Gunson D, Kefalides NA (1976) Characterization of the collagen synthesized by endothelial cells in culture. Proc Natl Acad Sci USA 73:2361–2364

Huang HC, Wu CH, Abramson M (1979) Collagenese activity in cultures of rat prostate carcinoma. Biochim Biophys Acta 570:149–156

Huang TW (1978) Composite epithelial and endothelial basal laminas in human lungs. Am J Pathol 93:681–688

Huerre C, Junien C, Weil D, Chu ML, Marabito M, Van Cong N, Myers JC, Foubert C, Gross MS, Prockop DJ, Bove A, Kaplan JC, de la Chappele A, Ramirez F (1982) Human type I procollagen genes are located on different chromosome. Proc Natl Acad Sci USA 79:6627–6630

Huerre-Jeanpierre C, Mattei MG, Weil D, Grzeschik KH, Chu ML, Sangiorgi FO, Sobel ME, Ramirez F, Junien C (1986) Further evidence for the dispersion of the human fibrillar collagen genes. Am J Hum Genet 38:26–37

Hugues J (1960) Accolement des plaquettes au collagene. C R Soc Biol (Paris) 154:866–868

Hugues J (1962) Accolement des plaquettes aux structures conjonctives périvasculaires. Thromb Diath Haemor 8:241–255

Hugues J, Herion F, Nusgens B, Lapière CM (1976) Type III collagen and probably not type I collagen aggregates platelets. Thromb Res 9:223–231

Hukins DWL (1982) Biomedical properties of collagen. In: Weiss JB, Jayson MIV (eds) Collagen in health and disease. Churchill Livingstone, Edinburgh, pp 49–72

Hulmes DJS (1983) A possible mechanism for the regulation of collagen fibril diameter in vivo. Coll Relat Res 3:317–321

Hulmes DJS, Holmes DF, Cummings C (1985) Crystalline regions in collagen fibrils. J Mol Biol 184:473–477

Hulmes DJS, Kadler KE, Mould AP, Hajima Y, Holmes DF, Cummings C, Chapman JA, Prockop DJ (1989) Pleomorphism in type I collagen fibrils produced by persistence of the N-propeptide. J Mol Biol 1989, 210, 337–345

Hung CH, Noelken Me, Hudson H (1981) Intestinal basement membrane of Ascaris suum. Physical properties of the collagenous domian. J Biol Chem 256:3822–3825

Hunt AH (1941) The role of vitamin C in wound healing. Br J Surg 28:436–461

Hunt S, Grant ME, Liebovich SJ (1970) Polymeric collagen isolated from squid (Loligo peallii) connective tissue. Experientia 26:1204–1205

Hutterer F, Rubin P, Popper H (1964) Mechanism of collagen resportion in reversible hepatic fibrosis. Exp Mol Pathol 3:215–223

Hutton JJ Jr, Marglin A, Witkop MB, Kurtz J, Berger A, Udenfriend S (1968) Synthetic polypeptides as substrates and inhibitors of collagen proline hydroxylase. Arch Biochem Biophys 125:779–785

Hyldstrup L, McNair P, Jensen GF, Nielsen HR, Yransbol I (1984) Bone mass as referent for urinary hydroxyproline excretion: age and sex-related changes in 125 normals and in primary hyperparathyroidism. Calcif Tissue Int 36:639–644

Iber FL, Rosen H, Levenson SM, Chalmers TC (1957) The plasma amino acids in patients with liver failure. J Lab Clin Med 50:417–423

Ibrahim J, Harding JJ (1989) Pinpointing the sites of hydroxylysine glycosides in peptide $\alpha$1-CB7 of bovine corneal collagen, and their possible role in determining fibril diameter and thus transparency. Biochim Biophys Acta 992:9–22

Iguchi M, Sano H (1974) Effect of cadmium on proline metabolism and its relation to the urinary amino acid in "Ouch-ouch disease". Jpn J Hyg 29:65–71

Iguchi M, Sano S (1985) Cadmium- or zinc-binding to bone lysyl oxidase and copper replacement. Connect Tissue Res 14:129–139

Ihme A, Risteli L, Krieg T, Risteli J, Kruse K, Müller PK (1983) Biochemical characterization of variants of the Ehlers-Danlos syndrome type VI. Eur J Clin Invest 13:357–362

Ihme A, Krieg T, Nerlich A, Feldmann U, Rautenberg J, Glanville RW, Edel V, Müller PK (1984) Ehlers-Danlos syndrome type VI: collagen type specificity of defective lysyl hydroxylation in various tissue. J Invest Dermatol 83:161–165

Iijima K, Ando K, Kishi M, Nakashizuka T, Hayakama T (1983) Collagenase activity in human saliva. J Dent Res 62:709–712

Ikeda F, Murawaki Y, Hirayama C (1983) Collagenase activity in the granulocytes of patients with various liver diseases. Clin Chim Acta 135:135–142

Inoue H (1981) Three-dimensional architecture of lumbar intervertebral discs. Spine 6:139–146

Iob V, Coon WW, Sloan M (1966) Altered clearance of free amino acids from plasma of patients with cirrhosis of the liver. J Surg Res 6:233–239

Ionasescu V, Zellweger H, Conway TW (1971) Ribosomal protein synthesis in Duchenne muscular dystrophy. Arch Biochem Biophys 144:51–58

Irving MG, Roll FJ, Huang S, Bissell DM (1984) Characterization and culture of sinusoidal endothelium from normal rat liver: lipoprotcin uptake and collagen phenotype. Gastroenterology 87:1233–1247

Ishikawa H, Suzuki S, Horiuchi R, Sato H (1975) An aproach to experimental scleroderma, using urinary glycosaminoglycans from patients with systemic scleroderma. Acta Derm Venerol (Stockh) 55:97–107

Ishikawa H, Saito Y, Yamakage A, Kitbatake M (1978) Scleroderma-inducing glycosaminoglycan in the urine of patients with systemic sclerosis. Dermatologica 156:193–204

Ishikawa H, Tamuta T, Kitabatake M, Yamakage A (1984) Experimental scleroderma in the mouse induced by urinary glycosaminoglycans isolated from patients with systemic sclerosis. In: Black CM, Muers AR (eds) Systmeic sclerosis (scleroderma). Gower Medical, New York, pp 157–163

Ishimaru T, Kanamaru T, Takahashi T, Ohta K, Okazaki H (1982) Inhibition of prolyl hydroxylase activity and collagen biosynthesis by the anthraquinone glycoside, P-1894B, an inhibitor produced by Streptomyces albogriceolus. Biochem Pharmacol 31:915–919

Ishimaru T, Kanamaru T, Ohta K, Okazaki H (1987) Fibrostatins, new inhibitors of prolyl hydroxylase. I. Taxonomy, isolation and characterization. J Antibiot 40:1231–1238

Ishimaru T, Kanamaru T, Takahashi T, Okazaki H (1988) Inhibition of prolyl hydroxylase activity and collgen biosynthesis by fibrostatin C, a novel inhibitor produced by Streptomyces catenulae subsp. griseospora No 23924. J Antibiot 41:1668–1674

Ishizeki K, Sakakura Y, Nawa T, Harada Y (1989) Morphological evidence of the formation of intracellular collagen fibrils in the embryonic mouse molar odontoblasts induced by colchicine administration. Acta Anat 134:133–140

Ishizuka S, Kiyoki M, Kurihara N, Hakeda Y, Ikeda K, Kumegawa M, Norman AW (1988) Effects of diastereoisomers of 1,25-dihyrdoxyvitamin $D_3$-26, 23-lactone on alkaline phosphatase and collagen synthesis in osteoblastic cells. Mol Cell Endocrinol 55:77–86

Isseroff H, Sawma JT, Reino D (1977) Fascioliasis: role of proline in bile duct hyperplasia. Science 198:1157–1159

Ito A, Naganeo K, Mori Y, Hirakawa S, Hayashi M (1977) PZ-peptidase activaty human uterine cervix in pregnancy at term. Clin Chim Acta 78:267–270

Ito A, Kitamura K, Mori Y, Hirakawa S (1979) The change in solubility of type I collagen in human uterine cervix in pregnancy at term. Biochem Med 21:262–270

Ito A, Sano H, Ikeuchi T, Sakyo K, Hirakawa S, Mori Y (1984) Effect of dehydroepiandrosterone sulphate on collagenase production in rabbit uterine cervix culture. Biochem Med 31:257–266

Itoi M (1969) Discussion on the collagenolytic activity in the cornea and its role in normal and pathological conditions. Arch Ophthalmol 81:147

Itoi M, Gnädiger M, Slansky H, Freeman H, Dohlman C (1969) Collagenase in the cornea. Exp Eye Res 8:369–373

Iwańska J, Barańska H, Najmowicz-Dąbrowa H, Baranowska E, Frank-Piskorska A, Dabrowa R (1975) Badania porównawcze wartości diagnostycznej oznaczania współczynnika efektywnej tyroksyny we krwi i hydroksyproliny wydalanej w moczu u osób z schorzeniami tarczycy (Comparison of diagnostic usefulness of the free thyroxin index in blood and the urinary hydroxyproline output in patients with disorders of the thyroid gland). Pol Tyg Lek 30:413–415 (in Polish)

Iwase-Okada K, Nagatsu T, Fujita K, Torikai K, Hamamoto T, Shibata T, Maeno Y, Sakakibara S (1985) Serum collagenase-like peptidase activity in rheumatoid arthritis and systemic lupus eryhtematosus. Clin Chim Acta 146:75–79

Iwatsuki K, Cardinale GJ, Spector S, Udenfried S (1977) Reduction of blood pressure and vascular collagen in hypertensive rats by $\beta$-aminopropionitrile. Proc Natl Acad Sci USA 74:360–362

Jablecki CK, Heuser JE, Kaufman S (1973) Autoradiographic localization of new RNA-synthesis in hypertrophying skeletal muscle. J Cell Biol 57:743–759

Jabłońska S (ed) (1975) Scleroderma and pseudoscleroderma. State Medical Publishers, Warsaw

Jackson DS (1957) Connective tissue stimulated by carrageenin. I. The formation and removal of collagen. Biochem J 65:277–284

Jackson DS (1979) Cell proliferation and collagen metabolism. Agents Actions [Suppl] 5:9–23

Jackson DS (1982) Dermal scar. In: Weiss JB, Jayson MIV (eds) Collagen in health and disease. Churchill Livingstone, Edinburgh, pp 466–471

Jackson DS, Bentley JP (1960) On the significance of the extractable collagens. J Biophys Biochem Cell 7:37–42

Jackson DS, Bentley JP (1968) Collagen-glycosaminoglycan interactions. In: Gould BS (ed) Treatise on collagen. Academic, New York, vol 2A, pp 189–214

Jacotot B (1983) Atherosclerose: mechanismes fondamentaux. Pathol Biol 31:607–612

Jaffe RM, Deykin D (1974) Evidence for a structural requirement for the aggregation of platelets by collagen. J Clin Invest 53:875–883

Jahnke A (1960) Electronenmikroskopische Untersuchungen über die Dupuytrensche Kontractur. Z Chir 85:2295–2303

Jain S, Scheuer PJ, McGee JOD, Sherlock S (1978) Hepatic collagen proline hydroxylase activity in primary biliary chirrhosis. Eur J Clin Invest 8:15–17

Jalil JE, Doering CW, Janicki JS, Pick R, Shroff SG, Weber KT (1989) Fibrillar collagen and myocardial stiffness in the intact hypertrophied rat left ventricle. Circ Res 64:1041–1050

Jalkanen M, Penttinen R (1982) Enhanced fibroblast collagen production by a macrophage-derived factor. Biochem Biophys Res Commun 108:447–453

Jander R, Rauterberg J, Voss B, von Bassewitz DB (1981) A cystein-rich collagenous protein from bovine placenta. Isolation of its constituent polypeptide chains and some properties of the non-denatured protein. Eur J Biochem 114:17–25

Janecki M, Drożdż M, Kucharz EJ, Barańska-Gachowska M, Grucka-Mamczar E, Mamczar A, Piwowarczyk B (1982) Badania na wpływem związków fluoru na metabolizm tkanki łącznej. II. Wpływ jonu fluorkowego na zawartość hydroksyproliny i hydroksylizyny w surowicy krwi i moczu szczurów w wieku rozwojowym (Studies on the influence of fluorine on the metabolism of connective tissue. II. Effect of fluoride ion on hydroxyproline and hydroxylysine level in blood serum and urine of growing rats). In: Machoy M (ed) Metabolizm fluoru (Metabolism of fluorine). Szczecińskie Towarzystwo Naukowe, Szczecin, pp 74–77 (in Polish)

Jaros E, Bradley WG (1979) Atypical axon-Schwnan cell relationships in the common peroneal nerve of the dystrophic mouse. An ultrastrutural study. Neuropathol Appl Neurobiol 5:133–147

Jasin HE, Ziff M (1962) Relationship between soluble collagen and urinary hydroxyproline in the lathyritic rats. Proc Soc Exp Med 110:837–841

Jasin HE, Fink CW, Wise W, Ziff M (1962) Relationship between urinary hydroxyproline and growth. J Clin Invest 41:1928–1935

Jayson MIV, Barks JS (1973) Structural changes in the intervertebral disc. Ann Rheum Dis 32:10–15

Jeffrey JJ (1986) The biological regulation of collagenase activity. In: Mecham R (ed) Regulation of matrix accumulation. Academic, London, pp 53–98

Jeffrey JJ, Gross J (1970) Collagenase from rat uterus. Isolation and partial characterization. Biochemistry 9:268–273

Jeffrey JJ, Martin GR (1966a) The role of ascorbic acid in the biosynthesis of collagen. I. Ascorbic acid requirement by embryonic chick tibia in tissue culture. Biochim Biophys Acta 121:269–280

Jeffrey JJ, Martin GR (1966b) The role of ascorbic acid in the biosynthesis of collagen. II. Site and nature of ascorbic acid participation. Biochim Biophys Acta 121:281–291

Jeffrey JJ, Coffey RJ, Eisen AZ (1971) Studies on uterine collagenase in tissue culture. II. Effect of steroid hormones on enzyme production. Biochem Biophys Acta 252:143–149

Jeffrey JJ, Koob TJ, Eisen AZ (1975) Hormonal regulation of mammalian collagenase. In: Burleigh PMC, Poole AR (eds) Dynamics of connective tissue macromolecules. North Holland, Amsterdam, pp 147–156

Jeffrey JJ, Di Petrillo T, Counts DF, Cutroneo KR (1985) Collagen accumulation in the neonatal rat skin absence of fibrillar collagen degradation during normal growth. Coll Relat Res 5:157–165

Jeleńska MM, Dancewicz AM (1969) Thermal aggregation of tropocollagen solutions irradiated with low doses of ionizing radiation. Int J Radiat Biol 16:193–196

Jeleńska MM, Dancewicz AM (1972) Effect of ionizing radiation on the reactivity of ε-amino groups in tropocollagen. Acta Biochim Pol 19:341–346

Jensen BA, Reinmann I, Fredensborg N (1986) Collagen type III predominance in newborns with congenital disolcation of the hip. Acta Orthop Scand 57:362–365

Jeraj K, Kim Y, Vernier RL, Fish AJ, Michael AF (1984) Absence of Goodpasture's antigen in the patients with familiar nephritis. Am J Kidney Dis 2:626–629

Jeremiah LE, Martin AH (1981) Intramuscular collagen content and solubility: their relationship to tenderness and alteration by postmorten aging. Can J Anim Sci 6:53–61

Jessen H (1967 The ultrastructure of odentoblasts in perfusion fixed, dermineralised incisors of adult rats. Acta Ondontol Scand 25:491–495

Jilek F, Hörmann H (1978) Cold insoluble globulin (fibronectin). IV. Affinity to soluble collagen of various types. Z Physiol Chem 359:247–250

Jilka RL (1989) Stimulation of collagenolytic enzyme release from cultured bone cells of normal and ostepetrotic mice by parathyroid hormone and lipopolysaccharide. Bone Miner 6:277–287

Jilka RL, Hamilton JW (1985) Evidence for two pathways for stimulation of collagenolysis in bone. Calcif Tissue Int 37:300–306

Jimenez SA, Bashey RI (1978) Solubilization of bovine heart-valve collagen. Biochem J 173:337–340

Jimenez SA, Rosenbloom J (1974) Decreased thermal stability of collagens containing analogs of proline or lysine. Arch Biochem Biophys 163:459–465

Jimenez SA, Millan A, Bashey RI, Schumacher HR (1984) The tight skin mouse: an experimental model resembling scleroderma. In: Black CM, Muers AR (eds) Systemic sclerosis (scleroderma). Gower Medical, New York, pp 145–150

Job JC, Sizonenko P, Pakey JP, Pakey P, Bourdon R, Rossier A (1966) L'hydroxyproline urinaire chez l'enfant normal et dans les retards de croissance. Arch Fr Pediatr 23:679–691

Johannessen JV, Bang G (1972) Transmission electron microscopy on sound demineralized guinea pig dentin. Scand J Dent Res 80:213–221

John HA, Purdom IF (1984) Myelin proteins and collagen in the spinal roots and sciatic nerves of muscular dystrophic mice. J Neurological Sci 65:69–80

Johson EF, Mitchell R, Berryman H, Cardoso S, Ueal O, Patterson D (1986) Secretory cells in the nucleus pulposus of the adult human intervertebral disc. Acta Anat 125:161–164

Johnson JR, Andrews FA (1970) Lung scleroproteins in age and emphysema. Chest 57:239–244

Johnson KJ, Oikarinen AI, Lowe NJ, Clark JG, Uitto J (1984) Ultraviolet radiation-induced connective tissue changes in the skin of hairless mice. J Invest Dermatol 82:587–590

Johnson ME, Jones GH (1978) Effects of marcaine, a myotoxic drug, on macromolecular synthesis in muscle. Med Pharmacol 27:1753

Johnson PC, Brendel K, Meezan E (1981) Human diabetic perineurial basement membrane thickening. Lab Invest 44:265–270

Johnson PC, Duhamel RC, Meezan E, Brendel K (1982) Preparation of cell-free extracellular matrix from human peripheral nerve. Muscle Nerve 5:335–344

Johnson RL, Ziff M (1976) Lymphokine stimulation of collagen accumulation. J Clin Invest 58:240–252

Johnson WH, Roberts NA, Borkakoti N (1987) Collagenase inhibitors: their design and potential therapeutic use. J Enzyme Inhib 2:1–22

Johnson-Muller B, Gross J (1978) Regulation of corneal collagenase production: epithelial-stromal cell interactions. Proc Natl Acad Sci USA 75:4417–4421

Joyce R, Garrett B (1978) Collagen and non-collagen protein synthesis in developing lung exposed to tobacco smoke. Environ Res 17:205–215

Jozsa L, Thöring J, Järvinen M, Kannus P, Lehto M, Kvist M (1988) Quantitative alterations in intramuscular connective tissue following immobilization: an experimental study in the rat calf muscles. Exp Mol Pathol 49:267–278

Judd JT, Wexler BC (1969) Myocardial connective tissue metabolism in response to injury. Histological and chemical studies of mucopolysaccharide and collagen in rat hearts after isoproterenol-induced infarction. Circ Res 25:201–214

Judd JT, Wexler BC (1970) Myocardial connective tissue metabolism in response ot injure. Investigation of the mucopolysaccharides involved in the isoproterenol-induced necrosis and repair in rat hearts. Circ Res 26:101–109

Junien C, Huerre C, Rathore MO (1983) Direct gene dosage determination in patients with unbalanced chromosomal abberations using cloned DNA sequences: application to ther regional assignment of the gene for $\alpha 2(I)$ procollagen (COL1A2). Am J Hum Genet 35:584–591

Junker P, Lorenzen I (1983) Reversibility of D-penicillamine induced collagen alterations in rat skin and granulation tissue. Biochem Pharmacol 32:1753–1757

Junker P, Helin G, Lorenzen I (1981) Effect of different doses of D-penicillamine and combined administration of D-penicillamine and methylprednisolone on collagen, glycosaminoglycans, DNA and RNA of granulation tissue, skin, bone and aorta in rats. Acta Pharmacol Toxicol 49:366–380

Junqueira LCU, Motes GS (1983) Biology of collagen-prteoglycan interaction. Arch Histol Jpn 46:589–595

Junqueira LCU, Toledo OMS, Montes GS (1981) Correlation of specific sulfated glycosaminoglycans with collagen type I, II, and III. Cell Tissue Res 217:171–175

Kadar A, Bihari-Varga M, Gero S, Jellinek H (1975) Composition and macromolecular structure of intima in normal and arteriosclerotic human aorta. Paroi Artérielle/Arterial Wall 3:3–15

Kadler KE, Hojima Y, Prockop DJ (1988) Assembly of type I collagen fibrils de novo. Between 37° and 41 °C the process is limited by micro-unfolding of monomers. J Biol Chem 263:10517–10523

Kagan HM, Raghavan J, Hollander W (1981) Changes in aortic lysyl oxidase activity in diet-induced atherosclerosis in the rabbit. Arteriosclerosis 1:287–291

Kahaleh MB, LeRoy EC (1983) Endothelial injury in scleroderma. A protease mechanism. J Lab Clin Med 101:553–560

Kahaleh MB, Sherer GK, LeRoy EC (1979) Endothelial injury in scleroderma. J Exp Med 149:1326–1335

Kähäri VM, Vuorio T, Nänto-Salonen K, Vuorio E (1984) Increased type I collagen mRNA levels in cultured scleroderma fibroblasts. Biochim Biophys Acta 781:183–186

Kähäri VM, Heino J, Larjava H, Vuorio E (1987) Alterations in scleroderma fibroblast surface glycoproteins associated with increased collagen synthesis. Acta Derm Venerol (Stockh) 67:199–205

Kähäri VM, Heino J, Vuorio T, Vuorio E (1988) Interferon-$\alpha$ and interferon-$\gamma$ reduce excessive collagen synthesis and procollagen mRNA levels of scleroderma fibroblasts in culture. Biochim Biophys Acta 968:45–50

Kahlson G, Zederfeldt B (1960) Association between histamine forming capacity and reparative growth. Acta Chir Scand 119:207–208

Kaibara N, Hotokebuchi T, Takagishi K, Katsuki I, Morinaga M, Arita C, Jingushi S (1984) Pathogenetic difference between collagen arthritis and adjuvant arthritis. J Exp Med 159:11388–1396

Kalenga M, Eeckhout Y (1989) Effects of protein deprivation from the neonatal period on lung collagen and elastin in the rat. Pediatr Res 26:125–127

Kaminishi H, Hagihara Y, Hayashi S, Cho T (1986) Isolation and characteristics of collagenolytic enzyme produced by Candida albicans. Infect Immun 53:312–316

Kang AH, Trelstad RL (1973) A collagen defect in homocystinuria. J Clin Invest 52:2571–2578

Kao KYT, Leslie JG (1977) Polymorphism in human uterine collagen. Connect Tissue Res 5:127–129

Kao KYT, McGavack TH (1959) Connective tissue. I. Age and sex influence on protein composition of rat tissues. Proc Soc Exp Biol Med 101:153–157

Kao KYT, Hilker DM, McGavack TH (1961 a) Connective tissue. IV. Synthesis and turnover of proteins in tissues of rat. Proc Soc Exp Biol Med 106:121–124

Kao KYT, Hilker DM, McGavack TH (1961 b) Connective tissue. V. Comparison of synthesis and turnover of collagen and elastin in tissues of rat at several ages. Proc Soc Exp Biol Med 106:335—338

Kao KYT, Hitt WE, Bush AT, McGavack TH (1964) Connective tissue. XII. Stimulating effects of estrogens on collagen synthesis in rat uterine slices. Proc Soc Exp Biol Med 117:86–93

Kao RT, Hall J, Stern R (1986) Collagen and elastin synthesis in human stroma and breast carcinoma cell lines: modulation by the extracellular matrix. Connect Tissue Res 14:245–255

Kao WWY, Foreman CA (1980) Chick corneal collagen. Eur J Biochem 106:41–47

Kapoor R, Bornstein P, Sage H (1986) Type VIII collagen from bovine Descemet's membrane: structural characterization of a triple-helical domain. Biochemistry 25:3930–3937

Karakiulakis G, Missirlis E, Maragoudakis ME (1989) Mode of action of razoxane: inhibition of basement membrane collagen-degradation by a malignant tumor enzyme. Methods Find (Exp) Clin Pharmacol 11:255–261

Karkavelas G, Kefalides NA, Amenta PS, Martinez-Hernandez A (1988) Comparative ultrastructural localization of collagen types III, IV, VI and laminin in rat uterus and kidney. J Ultrastruct Mol Struct Res 100:137–155

Karttunen T, Autio-Harmainen H, Räsänen O, Risteli J, Risteli L (1984) Immunohistochemical localization of epidermal basement membrane laminin and type IV collagen in bullous lessions of dermatitis herpetiformis. Br J Dermatol 111:389–394

Karttunen T, Alavaikko M, Apaja-Sarkkinen M, Autio-Harmainen H (1986) Distribution of basement membrane laminin and type IV collagen in human reactive lymph nodes. Histopathology 10:841–849

Karttunen T, Sormunen R, Risteli L, Risteli J, Autio-Harmainen H (1989) Immunoelectron microscopic localization of laminin, type IV collagen, and type III pN-collagen in reticular fibers of human lymph nodes. J Histochem Cytochem 37:279–286

Kastelic J, Galeski A, Baer E (1978) The multicomposite structure of tendon. Connect Tissue Res 6:11–23

Kastelic J, Bear E (1980) Deformation in tendon collagen. In: Vincent JFV, Currey JD (eds) The mechanical properties of biological materials. Symposia of the Society for Experimental Biology, vol XXXIV, Cambridge University Press, Cambridge, pp 397–435

Kato T, Kojima K, Imai K, Nagatsu T (1980) Changes in the activities of collagenase-like peptidase and dipeptidylaminopeptidase IV and hydroxyproline contents in developing rat salivary glands. Arch Oral Biol 25:181–185

Katsura M, Ito A, Hirakawa S, Mori Y (1989) Human recombinant interleukin-1$\alpha$ increases biosynthesis of collagenase and hyaluronic acid in cultured human chorionic cells. FEBS Lett 244:315–318

Katz EJ, Hasterlik RJ (1955) Aminoaciduria following total body irradiation in the human. J N C I 15:1085–1107

Katz EP, Li ST (1973a) The intermolecular space of reconstituted collagen fibrils. J Mol Biol 73:351–369

Katz EP, Li ST (1973b) Structure and function of bone collagen fibrils. J Mol Biol 80:1–15

Kawase T, Shiratori Y, Sugimoto T (1986) Collagen production by rat liver fat-storing cells in primary culture. Exp Cell Biol 54:183–192

Kay EP (1986) Rabbit corrneal endothelial cells modulated by polymorphonuclear leukocytes are fibroblasts: comparison with keratocytes. Invest Ophthalmol Vis Sci 27:891–897

Kay EP (1988) Molecular anatomy of the vertebrate eye: distribution of collagen in ocular tissues. In: Nimni ME (ed) Collagen, vol I. CRC Press, Boca Raton, pp 207–224

Kay EP (1989) Expression of types I and IV collagen genes in normal and in modulated corneal endothelial cells. Invest Ophthalmol Vis Sci 30:260–268

Kay EP, Cheung CC, Jester JV, Nimni ME, Smith RE (1982) Type I collagen and fibronectin synthesis by retrocorneal fibrous membrane. Invest Ophthalmol Vis Sci 22:200–212

Kay EP, Smith RE, Nimni ME (1985) Type Collagen synthesis by corneal endothelial cells modulated by polymorphonnuclear leukocytes. J Biol Chem 260:5139–5146

Kaytes P, Wood L, Theriault N, Kurkinen M, Vogeli G (1988) Head-to-head arrangement of murine type IV collagen genes. J Biol Chem 263:19274–19277

Kazmin AI, Merkurieva RV (1971) O roli narusheniia metabolizma glikozaminoglikanov w patogenze skolioza (The role of abnormal metabolism of glycosaminoglycans in pathogenesis of scoliosis). Orthop Travamtol Protez 32(11):87–91 (in Russian)

Kaznachev VP, Gichev YuP, Korobkova EN (1972) Vivedeniye oksiprolina s mochey u bolnikh khronicheskim gepatitom i cirrozom pecheni (Urinary output of hydroxyproline in patients with chronic hepatitis and cirrhosis of the liver). Ter Arkh 44(3):20–24 (in Russian)

Kedersha NL, Tkacz JS, Berg RA (1985a) Characterization of the oligosaccharides of prolyl hydroxylase, a microsomal glycoprotein. Biochemistry 24:5952–5960

Kedersha NL, Tkacz JS, Berg RA (1985b) Biosynthesis of prolyl hydroxylase: evidence for two separate dolicho-mediated pathways of glycosylation. Biochemistry 24:5960–5967

Keene DR, Sakai LY, Lunstrum GP, Morris NP, Burgeson RE (1987) Type VII collagen forms an extended network of anchoring fibrils. J Cell Biol 104:611–621

Keene DR, Engvall E, Glanville RW (1988) Ultrastructure of type VI collagen in human skin and cartilage suggests an anchoring function for this filamentous network. J Cell Biol 107:1995–2006

Kefalides NA (1971) Isolation of a collagen from basement membrane containing three identical $\alpha$ chains. Biochem Biophys Res Commun 45:226–234

Kefalides NA (1973) Structure and biosynthesis of basement membranes. Int Rev Connect Tissue Res 6:63–104

Kefalides NA (1974) Biochemical properties of human glomerular basement membrane in normal and diabetic kidneys. J Clin Invest 53:403–407

Kefalides NA (1982) Basement membrane collagen. In: Weiss JB, Jayson MIV (eds) Collagen in health and disease. Churchill Livingstone, Edinburgh, pp 313–332

Kefalides NA, Alper R (1988) Structure and organization of macromolecules in basement membranes. In: Nimni ME (ed) Collagen, vol II. CRC Press, Boca Raton, pp 73–94

Kefalides NA, Winzler RJ (1966) The chemistry of glomerular basement membrane and its relation to collagen. Biochemistry 5:702–712

Kehrer JP (1982) Collagen production rates following acute lung damage by butyrated hydroxytoluene. Biochem Pharmacol 31:2053–2058

Kehrer JP (1985) Collagen synthesis and degradation in acutely damaged mouse lung tissue following treatment with prednisolone. Biochem Pharmacol 34:2519–2524

Kehrer JP, Witschi HP (1981) The effect of indomethacin, prednisolone and cis-4-hydroxyproline on pulmonary fibrosis produced by butylated hydroxytoluene and oxygen. Toxicology 20:281–288

Kehrer JP, Klein-Szanto AJP, Sorensen EMB, Pearlman R, Rosner MH (1984) Enhanced lung damage following corticosteroid treatment. Am Rev Respir Dis 130:256–261

Keifer O (1973) Über die Niebenwirkungen der Bleomycintherapie auf der Hand. Dermatologica 146:229–235

Keil-Dlouha V, Keil B (1978) Subunit structure of Achromobacter collagenase. Biochim Biophys Acta 522:218–228

Keiser HR, Sjoerdsma A (1961) Effect of hormone on collagen metabolism. J Clin Invest 44:1371–1375

Keiser HR, Sjoerdsma A (1967) Studies on $\beta$-aminopropionitrile in patients with slceroderma. Pharmacol Ther 8:593–602

Keiser HR, Stein HD, Sjoerdsma A (1971) Increased protocollagen proline hydroxylase activity in sclerodermatous skin. Arch Dermatol 104:57–60

Keller F, Rehbein C, Schwarz A, Fleck M, Hayasaka A, Schuppan D, Offermann G, Hahn EG (1988) Increased procollagen III production in patients with kidney disease. Nephron 50:332–337

Kennedy L, Baynes JW (1984) Non-enzymatic glycosylation and the chronic complications of diabetes: an overview. Diabetologia 26:93–98

Kennedy A, Frank RN, Mancini MA, Lande M (1986) Collagens of the retinal microvascular basement membrane and of retinal microvascular cells in vitro. Exp Eye Res 42:177–199

Kent G, Fels GI, Dubin A, Popper H (1959) Collagen content based on hydroxyproline determination in human and rat livers: its relation to morphologically demonstrable reticulum and collagen fibers. Lab Invest 8:48–56

Kern P, Moczar M, Robert L (1979) Biosynthesis of skin collagens in normal and diabetic mice. Biochem J 182:337–345

Kern P, Sebert B, Robert L (1986) Increased type-III/type-I collagen ratios in diabetic human conjunctival biopsies. Clin Physiol Biochem 4:113–119

Kershenobich D, Fierro FJ, Rojkind M (1970) The relationship between free pool of proline and collagen content in human liber cirrhosis. J Clin Invest 49:2246–2249

Kerwar SS, Marcel RJM, Salvador RA (1976) Studies on the effect of L-3,4-dehydroproline on collagen synthesis by chick embryo polysomes. Arch Biochem Biophys 172:685–688

Kerwar SS, Bauman N, Oronsky AL, Sloboda AE (1982) Studies on type II collagen induced polyarthritis in rats. Effect of complement depletion. J Immunopharmacol 3:323–337

Kerwar SS, Englert ME, McReynolds RA, Landes MJ, Lloyd JM, Oronsky AL, Wilson FJ (1983) Type II collagen-induced arthritis: studies with purified anticollagen immunoglobulin. Arthritis Rheum 26:1120–1131

Kerwar SS, Ridge SC, Landes MJ, Nolan JC, Oronsky AL (1984) Induction of collagenase synthesis in chondrocytes by a factor synthesized by inflammed synovial tissue. Agents Actions 14:54–57

Keyser AJ, Nimni ME, Cooper SM (1984) Scleroderma serum stimulates collagen synthesis in normal human dermal fibroblasts. In: Black CM, Myers AR (eds) Systemic sclerosis (scleroderma). Gower Medical, New York, pp 215–219

Khilkin AM, Shekhter AB, Istranov LP, Lemenev VL (1976) Kollagen i iego primeneniye v meditsine (Collagen and its application in medicine). Izd Meditsina, Moskva

Kiang WL, Crockett CP, Margolis RK, Margolis RU (1978) Glycosaminoglycans and glycoproteins associated with microsomal subfractions of brain and liver. Biochemistry 17:3841–3848
Kibrick AC, Singh KD (1974) Hydroxyproline excreted in urine: its source from collagen of tissues after $^{14}$C-proline in rats with and without administration of prednisolone. J Clin Endocrinol Metab 38:594–601
Kibrick AC, Hashiro CQ, Walters MI, Milhorat AT (1964) Diketopiperazine of prolylhydroxyproline in normal human urine. Clin Chim Acta 10:344–351
Kidawa Z (1978) Wpływ leczenia immunosupresyjnego na zachowanie się hydroksyproliny w krwi i moczu u chorych z kłębkowym zapaleniem nerek (Influence of immunosupressive treatment on serum and urine levels of hydroxyproline in patiens with glomerulonephritis). Reumatologia 16:91–95 (in Polish)
Kikuchi Y, Fujimoto D, Tamiya N (1969) Enzymic hydroxylation of protocollagen models. Biochem J 115:569–574
Killackey JJ, Roughley PJ, Mort JS (1983) Proteinase inhibitors of human articular cartilage. Coll Relat Res 3:419–430
Killar LM, Dunn CJ (1989) Interleukin-1 potentiates the development of collagen-induced arthritis in mice. Clin Sci 76:535–538
Kimata K, Barrach HJ, Brown KS, Pennypacker JP (1979) Absence of proteoglycan core protein in cartilage from the cmd/cmd (cartilage matrix deficient) mouse. J Biol Chem 256:6961–6968
Kimura S (1985) The interstitial collagens of the fishes. In: Bairati A, Garrone R (eds) Biology of invertebrate and lower vertebrate collagens. Plenum, New York, pp 397–408
Kimura S, Matsuura F (1974) The chain composition of several invertebrate collagens. J Biochem (Tokyo) 75:1231–1238
Kimura S, Takema Y, Kubota M (1981) Octopus skin collagen: isolation and characterization of collagen comprising three distinct $\alpha$ chains. J Biol Chem 256:13230–13234
Kimura S, Miura S, Park YH (1983) Collagen as the major edible component of jellyfish (Stomolophus nomurai). J Food Sci 48:1758–1760
Kimura T, Mattei MG, Stevens JW, Goldring MB, Ninomiya Y, Olsen BR (1989) Molecular cloning of rat and human type IX cDNA and localization of the $\alpha$1(IX) gene on the human chromosome 6. Eur J Biochem 179:71–78
King GS, Starcher BC (1979) Elastin catabolism: measurement of urine desmosine by radioimmunoassay. Clin Res 27:705A (abstract)
Kirk JME, Heard BE, Kerr I, Turner-Warwick M, Laurent GJ (1984) Quantitation of types I and III collagen in biopsy lung samples from patients with cryptogenic fibrosing alveolitis. Coll Relat Res 4:169–182
Kirrane JA, Glynn LE (1968) Immunology of collagen. Int Rev Connect Tissue Res 4:1–34
Kirschke H, Kembhavi AA, Bohley P, Barrett AJ (1982) Action of rat liver cathepsin L on collagen and other substrates. Biochem J 201:367–370
Kischer CW, Shetlar MR (1979) Electron microscopic studies of connective tissue repair after myocardial injury. Tex Rep Biol Med 39:357–361
Kishi JI, Hayakawa T (1984) Purification and characterization of bovine dental pulp collagenase inhibitor. J Biochem 96:395–404
Kishi JI, Iijima KI, Hayakawa T (1979) Dental pulp collagenase: initial demonstration and characterization. Biochem Biophys Res Commun 86:27–31
Kishi JI, Kawai T, Hayakawa T (1984) Dual regulation of collagenase activity in bovine dental pulps. Biochem Int 9:343–350
Kitamura K, Ito A, Mori Y, Hirakawa S (1979) Changes in the human uterine cervical collagenase with special reference to cervical ripening. Biochem Med 22:332–338
Kittelberger R, Davis PF, Greenhill NS (1989) Immunolocalization of type VIII collagen in vascular tissue. Biochem Biophys Res Commun 159:414–419
Kivirikko KI (1970) Urinary excretion of hydroxyproline in health and disease. Int Rev Connect Tissue Res 5:93–163
Kivirikko KI, Kuivaniemi H (1987) Posttranslational modifications of collagen and their alterations in heritable diseases. In: Uitto J, Perejda AJ (eds) Connective tissue disease. Molecular pathology of the extracellular matrix. Marcel, New York, pp 263–292

Kivirikko KI, Laitinen O (1965) Effect of cortisone on the hydroxyproline in the serum and urine of young rats. Acta Physiol Scand 64:356–360

Kivirikko KI, Myllylä R (1979) Collagen glycosyltransferases. Int Rev Connect Tissue Res 8:23–72

Kivirikko KI, Myllylä R (1980) The hydroxylation of prolyl and lysyl residues. In: Freedman RB, Hawkins HC (eds) The enzymology of post-translational modifications of proteins, vol 1. Academic, London, pp 53–104

Kivirkko KI, Myllylä R (1982a) Posttranslational enzymes in the biosynthesis of collagen: intracellular enzymes. Methods Enzymol 82:245–304

Kivirikko KI, Myllylä R (1982b) Post-translational modifications. In: Weiss JB, Jayson MIV (eds) Collagen in health and disease. Churchill Livingstone, Edinburgh, pp 101–120

Kivirikko Ki, Myllylä R (1984) Biosynthesis of collagen. In: Piez KA, Reddi AH (eds) Extracellular matrix biochemistry. Elsevier, New York, pp 83–118

Kivirikko KI, Myllylä R (1987) Recent developments in posttranslational modifications: intracellular processing. Methods Enzymol 144:96–114

Kivirikko KI, Prockop DJ (1967a) Purification and partial characterization of the enzyme for the hydroxylation of proline in protocollagen. Arch Biochem Biophys 118:611–618

Kivirikko KI, Prockop DJ (1967b) Enzymic hydroxylation of proline and lysine in protocollagen. Proc Natl Acad Sci USA 57:782–789

Kivirikko KI, Prockop DJ (1967c) Hydroxylation of proline in synthetic polypeptides with purified protocollagen hydroxylase. J Biol Chem 242:4007–4012

Kivirikko KI, Risteli L (1976) Biosynthesis of collagen and its alterations in pathological states. Med Biol 54:159–186

Kivirikko KI, Savolainen ER (1988) Hepatic collagen metabolism and its modification by drugs. In: Testa B, Perrissonol D (eds) Liver drugs: from experimental pharmacology to therapeutic applications. CRC Press, Boca Raton, pp 193–222

Kivirikko KI, Liesmaa M, Luukainen T (1958) Effect of growth hormone and pregnancy on the hydroxyproline concentration of the blood of rats. Acta Endocrinol (Copenh) 27:118–122

Kivirikko KI, Koivusalo M, Laitinen O, Liesmaa M (1963) Effect of thyroxine on the hydroxyproline in rat urine and skin. Acta Physiol Scand 57:462–467

Kivirikko KI, Koivusalo M, Laitinen O, Lamberg BA (1964) Hydroxyproline in the serum and urine of patients with hyperthyroidism. J Clin Endocrinol 24:222–223

Kivirikko KI, Laitinen O, Lamberg BA (1965) Value of urine and serum hydroxyproline in the diagnosis of thyroid disease J. Clin Endocrinol 25:1347–1352

Kivirikko KI, Laitinen O, Aer J, Halme J (1967) Metabolism of collagen in experimental hyperthyroidism and hypothyroidism in the rat. Endocrinology 80:1051–1061

Kivirikko KI, Ganser V, Engel J, Prockop DJ (1967) Comparison of poly-L-proline I and II as inhibitors of procollagen hydroxylase. Z Physiol Chem 348:131–134

Kivirikko KI, Bright HJ, Prockop DJ (1968) Kinetic patterns of procollagen hydroxylase and further studies on the polypeptide substrate. Biochim Biophys Acta 151:558–567

Kivirikko KI, Prockop DJ, Lorenzi GP, Blout ER (1969) Oligopeptides with the sequences alanine-proline-glycine and glycine-proline-glycine as substrates or inhibitors for protocollagen proline hydroxylase. J Biol Chem 244:2755–2760

Kivirikko KI, Kishida Y, Sakakibara S, Prockop DJ (1972) Hydroxylation of $(X\text{-}Pro\text{-}Gly)_n$ by protocollagen proline hydroxylase. Effect of chain lenght, helical conformation, and amino acid sequence in the substrate. Biochim Biophys Acta 271:347–356

Kjellström T (1986) Influence of glucose on collagen and protein production in cultured human skin fibroblasts from diabetic and non-diabetic subjects. Diabetes Res 3:77–82

Kjellström T, Malmquist J (1984) Insulin effects on collagen and protein production in cultured human skin fibroblasts from diabetic and non-diabetic subjects. Horm Metab Res 16:168–171

Klein JA, Hukins DWL (1982) Collagen fibre orientation in the annulus fibrosus of intervertebral disc during bending and torsion measured by x-ray diffraction Biochim Biophys Acta 719:98–101

Klein L, Yen SSC (1970) Urinary peptide hydroxyproline before and during postpartum involution of human uterus. Metabolism 19:19–23

Kleinerman J (1979) Effects of nitrogen dioxide on elastin and collagen contents of lungs. Arch Environ Health 34:228–232

Kleinman HK, Wilkes CM (1982) Interaction of fibronectin with collagen. In: Weiss JB, Jayson MIV (eds) Collagen in health and disease. Churchill Livingstone, Edinburgh, pp 198–205

Kleinman HK, Wilkes CM, Martin GR (1981) Interaction of fibronectin with collagen fibrils. Biochemistry 20:2325–2530

Kleissl HP, Van der Rest M, Naftolin F, Glorieux FH, De Leon A (1978) Collagen changes in the human uterine cervix at parturition. Am J Obstet Gynecol 130:748–753

Klemperer P, Pollack AD, Baehr G (1942) Diffuse collagen disease. Acute disseminated lupus erythematosus and diffuse scleroderma. JAMA 119:331–332

Kleppel MM, Kashtan CE; Butkowski RJ, Fish AJ, Michael AF (1987) Alport familial nephritis. Absence of 28 kDa non-collagenous monomers of type IV collagen in glomerular basement membrane. J Clin Invest 80:263–266

Kligman LH, Akin FJ, Kligman AM (1982) Prevention of ultraviolet damage to the dermis of hairless mice by sunscreens. J Invest Dermatol 78:181–189

Kligman LH, Akin FJ, Kligman AM (1983) Sunscreens promote repair of ultraviolet radiation-induced dermal damage. J Invest Dermatol 81:98–102

Kligman LH, Gebre M, Alper R, Kefalides NA (1989) Collagen metabolism in ultraviolet irradiated hairless mouse skin and its correlation to histochemical observations. J Invest Dermatol 93:210–214

Knese KH (1978) Kristallisation und Auflösung von Kollagen Fibrillen während der Histogenese der Zwischenwirbelscheibe. Acta Anat 100:328–346

Knox JM, Cockerell EG, Freeman RG (1962) Eitological factors and premature aging. JAMA 179:136–142

Kobayasi T, Asboe-Hansen G (1972) Ultrastructure of generalized scleroderma. Acta Derm Venerol (Stockh) 52:81–93

Kobayashi M, Watanabe A, Nakatsukasa H (1987) Elevation of type V collagen-degrading enzyme activity in invasive hepatocellular carcinoma. Res Commun Chem Pathol Pharmacol 58:143–144

Kocher P, Vuille R, Rovarino D, Courvoisier B (1965) Etude de l'hydroxyprolinurie dans les ostéopathies. Helv Med Acta 32:480–486

Kohn LD (1987) Connective tissue. In: Ingbar SH, Bravermann LE (eds) The thyroid. A fundamental and clinical text. Lippincott, London. pp 816–839, 1128–1130

Kohn LD, Winand RJ (1975) Structure of an exophthalmos-producing factor derived from thyrotropin by partial pepsin digestion. J Biol Chem 250:6503–6508

Kohn LD, Isersky C, Zupnik J, Lenaers A, Lee G, Lapière CM (1974) Calf tendon procollagen peptidase: its purification and endopeptidase mode of action. Proc Natl Acad Sci USA 71:40–44

Kohn RR, Cerami A, Monnier VM (1984) Collagen aging in vitro by nonenzymatic glycosylation and browning. Diabetes 33:57–59

Koivu J, Myllylä R (1987) Interchain disulfide bond formation in types I and II procollagen. Evidence for a protein disulfide isomerase catalyzing bond formation. J Biol Chem 262:6159–6164

Koivu J, Myllylä R, Helaakoski T, Pihlajaniemi T, Tasanen K, Kivirikko KI (1987) A single polypeptide acts both as the $\beta$-subunit of prolyl 4-hydroxylase and as a protein disulfide-isomerase. J Biol Chem 262:6447–6449

Koller E, Winterhalter KH, Trueb B (1989) The globular domains o type VI collagen are related to the collagen-binding domains of cartilage matrix protein and von Willebrand factor. EMBO J 8:1073–1077

Komarov FI, Sutchkov AV, Pogramov AP (1977) Issledovaniya metabolitov kollagena pri khronitcheskikh diffuznikh zabolevaniyakh petcheni (Studies on collagen metabolities in patients with chronic diffuse disorders of the liver). Klin Med (Mosk) 42(11):62–67 (in Russian)

Kondo A, Sussett JG (1973) Collagen content in detrusor muscle and ratio of bladder weight to body weight. Acta Urol Jpn 19:683–691

Kondo A, Susset JG (1974) Viscoelastic properties of bladder. Invest Urol 11:459–465

Konno K, Tetsuka T (1964) Studies on the connective tissue. III. Biosynthesis of collagen by the granuloma tissue in vitro. J Biochem (Tokyo) 56:581–590

Konno K, Traelnes KR, Altman KI (1984) The effect of whole- and partial-body X-irradiation on the metabolism of collagen in polyethylene sponges and the urinary excretion of pyrrole-2-carboxylic acid. Int J Radiat Biol 8:367–372

Konohana A, Kawakubo Y, Tajima S, Kitamura K, Nishikawa T (1985) Glycosaminoglycans and collagen in skin of a patient with diabetic sclerederma. Keio J Med 34:221–226

Konomi H, Sano J, Nagai Y (1981) Immunohistochemical localization of types I, III and IV (basement membrane) collagens in the lymph node: co-distribution of type I and III collagen in the reticular fibers. Biomed Res 2:536–541

Koob TJ, Jeffrey JJ (1974) Hormonal regulation of collagen degradation in the uterus: inhibition of collagenase expression by progesterone and cyclic AMP. Biochim Biophys Acta 354:61–70

Koob TJ, Jeffrey JJ, Eisen AZ (1974) Regulation of human skin collagenase activity by hydrocortisone and dexamethasone in organ culture. Biochem Biophys Res Commun 61:1083–1088

Korting GW, Holzmann H, Kühn K (1964) Biochemische Bindegewebsanalysen bei progressiver Sklerodermie. Klin Wochenschr 42:247–249

Kosher RA, Church RL (1975) Stimulation of in vitro somite chondrogenesis by procollagen and collagen. Nature 258:327–330

Kosher RA, Lash JW (1975) Notochord stimulation of in vitro somite chondrogenesis before and after enzymatic removal of perinotochordal materials. Dev Biol 42:362–378

Kostrzyńska M, Schalén C, Wadström T (1989) Specific binding of collagen type IV to Streptococcus pyogenes. FEMS Microbiol Lett 59:229–234

Kotek J, Deyl Z, Rosmus J, Krajicek M (1964) Reconstitution of soluble collagen by irradiation. Int J Appl Radiat Isotop 15:551–553

Kovanen V (1989) Effects of ageing and physical training on rat skeletal muscle. Acta Physiol Scand 135 [Suppl 577]:1–56

Kovanen V, Suominen H (1988) Effects of age and life-long endurance training on the passive mechanical properties of rat skeletal muscle. Compr Gerontol [A]2:18–23

Kovanen V, Suominen H, Heikkinen E (1980) Connective tissue of "fast" and "slow" skeletal muscle in rats-effects of endurance training. Acta Physiol Scand 108:173–180

Kovanen V, Suominen H, Heikkinen E (1984) Collagen of slow twitch and fast twitch muscle fibres in different types of rat skeletal muscle. Eur J Appl Physiol 52:235–242

Kovanen V, Suominen H, Peltonen L (1987) Effects of aging and life-long physical training on collagen in slow and fast skeletal muscle in rats. A morphometric and immunohisto-chemical study. Cell Tissue Res 248:247–255

Kovanen V, Suominen H, Risteli J, Risteli L (1988) Type IV collagen and laminin in slow and fast skeletal muscle in rats–effects of age and life-time endurance training. Coll Relat Res 8:145–153

Kowalewski K (1969) Changes in the growth and body collagen of sucling rats caused by the treatment of lactating mothers with cortisone. Endocrinology 84:432–434

Kowalewski K, Yong S (1968) Effect of growth hormone and an anabolic steroid on hydroxyproline in healing dermal wounds in rats. Acta Endocrinol (Copenh) 59:53–66

Kowalska M (1988) The effect of vanadium on lung collagen content and composition in two successive generations of rats. Toxicol Lett 41:203–208

Kowalska M (1989) Changes in rat lung collagen after life-time treatment with vanadium. Toxicol Lett 47:185–190

Kowashi Y, Jaccard F, Cimasoni G (1979) Increase on free collagenase and neutral protease activities in the gingival crevice during experimental gingivitis in man. Arch Oral Biol 24:645–650

Kowashi Y, Cimasoni G, Matter J (1980) Collagen breakdown by gingival collagenase and elastase. Experientia 36:395–396

Kozlovskis PL, Fieber LA, Pruitt DK, Bailey BK, Smets MJD, Bassett AL, Kimura S, Myerburg RJ (1987) Myocardial changes during the progression of left ventricular pressure-overload by renal hypertension or aortic constriction: myosin, myosin ATPase and collagen. J Mol Cell Cardiol 19:105–114

Kraemer HP, Nemetschek T, Gross F (1979) Effect of hydralazine on the elasticity of collagen. Experientia 35:527–528

Kramer RM, Fuh GM, Karasek MA (1985) Type IV collagen synthesis by cultured human microvascular endothelial cells and its deposition into the subendothelial basement membrane. Biochemistry 24:7423–7427

Kramsch D, Chan CT (1976) Increased in vivo synthesis of elastin and collagen in atherosclerotic arteries and its suppression by drugs. Fed Proc 35:578

Krane SM (1981) Aspects of the cell biology of the rheumatoid synovial lesion. Ann Rheum Dis 40:433–448

Krane SM (1982) Collagenases and collagen degradation. J Invest Dermatol 79 [Suppl 1]:83s–86s

Krane SM, Pinnell SR, Erbe RW (1972) Lysyl-protocollagen hydroxylase deficiency in fibroblasts from siblings with hydroxylysine-deficient collagen. Proc Natl Acad Sci USA 69:2899–2903

Kratzsch KH (1969a) Urinary hydroxyproline excretion in cirrhosis of the liver. Acta Hepatosplenol 16:2–8

Kratzsch KH (1969b) The urinary excretion of hydroxyproline in cirrhosis of the liver. Germ Med Monatsschr 14:457–460

Krieg T, Müller PK (1979) The Marfan's syndrome. In vitro study of collagen metabolism in tissue specimens of the aorta. Exp Cell Biol 45:207–221

Krieg T, Müller PK, Goerz G (1977) Fibroblasts from a patient with scleroderma reveal abnormal metabolism. Arch Dermatol Res 259:105–107

Krieg T, Hörlein D, Wiestner M, Müller PK (1978) Amino terminal extension peptides from type I procollagen normalize excessive collagen synthesis of scleroderma fibroblasts. Arch Dermatol Res 263:171–180

Krieg T, Feldmann U, Kessler W, Müller PK (1979) Biochemical characterization of Ehlers-Danlos syndrome type VI in a family with one affected infant. Hum Genet 46:41–49

Krieg T, Luderschmidt C, Weber L, Müller PK, Braun-Falco O (1981) Scleroderma fibroblasts: some aspects of in vitro assessment of collagen synthesis. Arch Dermatol Res 270:263–272

Krieg T, Braun-Falco O, Perlish JS, Fleischmajer R (1983) Collagen synthesis in generalized morphea. Arch Dermatol Res 275:393–396

Krieg T, Perlish JS, Fleischmajer R, Braun-Falco O (1985) Collagen synthesis in scleroderma: selection of fibroblast populations during subcultures. Arch Dermatol Res 277:373–376

Krieg T, Langer I, Gerstmeier H, Keller J, Mensing H, Goerz G, Timpl R (1986) Type III collagen aminopropeptide levels in serum of patients with progressive systemic scleroderma. J Invest Dermatol 87:788–791

Krieg T, Hein R, Hatamochi A, Aumailley M (1988) Molecular and clinical aspects of connective tissue Eur J Clin Invest 18:105–123

Kruze D, Wojtecka E (1972) Activation of leucocyte collagenase proenzyme by rheumatoid synovial fluid. Biochim Biophys Acta 285:436–446

Kruze D, Iwańska J, Stanowski E (1973) Hydroksyprolina w moczu chorych z nadczynnością tarczycy leczonych chirurgicznie (Urinary excretion of hydroxyproline in patients with hyperthyroidism before and after surgery). Pol Arch Med Wewn 50:251–257 (in Polish)

Kruze D, Salgam P, Cohen G, Fehr K, Böni A (1978) Purification and some properties of collagenase proenzyme activator from rheumatoid synovial fluid. Z Rheumatol 37:355–365

Kryschtalskyj E, Sodek J, Ferrier JM (1986) Correlation of collagenolytic enzymes and inhibitors in gingival crevicular fluid with clinical and microscopic changes in experimental periodontitis in the dog. Arch Oral Biol 31:21–31

Kuboki Y, Takagi T, Shimokawa H, Oguchi H, Sasaki S, Mechanic GL (1981a) Location of an intermolecular crosslink in bovine collagen. Connect Tissue Res 9:107–114

Kuboki Y, Tsazaki M, Sasaki S, Liu CF, Mechanic GL (1981b)Location of the intermolecular cross-links in bovine dentin collagen, solubilization with trypsin and isolation of cross-link peptides containing dihydroxylysinonorleucine and pyridinoline. Biochem Biophys Res Commun 102:119–126

Kubota Y, Shoji S, Funakoshi T, Shionaga K, Ueki H (1983) Hydroxyproline-containing citrus esterase. I. Purification and properties of the enzyme from Citrus natsudaidai Hayata. Yakugaku Zasshi 103:655–661

Kubota Y, Fujiwara K, Ogata I, Hori Y, Oka H (1985) Collagen prolyl hydroxylase inhibitor and reduced collagen formation in cotton pellet-induced granuloma and skin wound healing in rats. Br J Dermatol 113:559–563

Kucharz EJ (1976a) Conditions of the high-effective production of LE-positive, experimental hydralazine-induced collagen disease-like syndrome in guinea pigs. Exp Pathol 12:121–122

Kucharz EJ (1976b) Współczesne poglądy na patogenezę zespołu Ehlersa-Danlosa (Modern concepts of the pathogenesis of the Ehlers-Danlos syndrome). Wiad Lek 29:609—614 (in Polish)

Kucharz EJ (1978) Regulacja metabolizmu kolagenu w procesie włóknienia (Regulation of collagen metabolism in fibrosis). Pol Tyg Lek 32:1807–1810 (in Polish)

Kucharz EJ (1980) Collagen metabolism in liver disorders. VIIth European symposium on connective tissue research, Prague, abstracts, pp 115–116

Kucharz EJ (1981) La influo de antiinflamaj steroidoj je metabolismo de konjunktiva histo en fibrozigita hepato (Influence of anti-inflammatory steroids upon metabolism of the connective tissue in the fibrotic liver). In: Popov K (ed) Sciencaj prelego. III internacia medicina esperanto konferenco (Scientific Proceedings. The 3rd International Medical Esperanto Meeting). Universala Medicina Esperanto Asocio, Ruse, pp 68–76 (in Esperanto)

Kucharz EJ (1984) Clinical and experimental studies on collagen metabolism in hepatic disorders. Med Intern 22:129–140

Kucharz EJ (1985) Inhibitors of collagenolytic enzymes in the serum of patients with chronic liver disorders. Gut 26:1364–1366

Kucharz EJ (1986a) Indices of collagen metabolism in pregnancy and postpartum period. Gynecol Obstet Invest 22:169–171

Kucharz EJ (1986b) Effect of lead intoxication on collagen content in the liver of rats. Rev Roum Biochim 23:315–317

Kucharz EJ (1986c) Metabolity kolagenu w surowicy krwi i płynie stawowym u chorych na reumatoidalne zapalenie stawów z dodatnim lub ujemnym odczynem Waalera-Rosego (Collagen metabolites in serum and synovial fluid in patients with seropositive or seronegative rheumatoid arthritis). Wiad Lek 39:602–606 (in Polish)

Kucharz EJ (1986d) Kinetics of glycosaminoglycan accumulation in the liver of rats with carbon tetrachloride-induced hepatic fibrosis. Arch Roum Pathol Exp Microbiol 45:129–136

Kucharz EJ (1987a) Patogeneza polekowego zespołu kolagenozopodobnego (Pathogenesis of drug-induced lupus-like syndrome). Przegl Lek 44:443–449 (in Polish)

Kucharz EJ (1987b) Metabolizm kolagenu w procesie włóknienia wątroby (Collagen metabolism in hepatic fibrosis). Postępy Hig Med Dośw 41:302–330 (in Polish)

Kucharz EJ (1987c) Dynamics of collagen accumulation and activity of collagen-degrading enzymes in the liver of rats with carbon tetrachloride-induced hepatic fibrosis. Connect Tissue Res 16:143–151

Kucharz EJ (1988a) Effect of cadmium intoxication on collagen and elastin content in tissues of the rat. Bull Environ Contam Toxicol 40:273–279

Kucharz EJ (1988b) Hormonal control of collagen metabolism. Endocrinologie 26:69–79, 229–237

Kucharz EJ, Drożdż M (1975) Comparative studies on collagen metabolism in patients with collagen diseases and experimental drug-induced collagen-like syndrome in guinea pigs. II. Congressus cěchoslovacus de biochimia clinica, Bratislava, abstracta, p 36

Kucharz EJ, Dróżdż M (1977a) Influence of a single administration of hydralazine on collagen content in liver injured with carbon tetrachloride. Rev Roum Biochim 14:27–29

Kucharz EJ, Dróżdż M (1977b) Wydalanie hydroksylizyny jako wskaźnik zaburzeń przemian kolagenu (Urinary excretion of hydroxylysine as an index of abnormal collagen metabolism). Wiad Lek 30:695–700 (in Polish)

Kucharz EJ, Dróżdż M (1987a) Increased solubility of collagen. A biochemical defect in the experimental hydralazine-induced collagen disease-like syndrome. Exp Pathol 15:63–65

Kucharz EJ Dróżdż M (1978b) Wpływ leków cytostatycznych na metabolizm kolagenu u świnek morskich (Influence of cytostatic drugs upon collagen metabolism in guinea pigs). Reumatologia 16:85–89 (in Polish)

Kucharz EJ, Dróżdż M (1978c) Collagenolytic activity in leukocytes isolated from patients with leukemias and Hodgkin's disease. Neoplasma 25:621–624

Kucharz EJ, Hawryluk D (1987) Effect of acetylsalicylic acid and naproxen on collagen solubility and heterogeneity in the rat liver. Pharmazie 42:117–118

Kucharz EJ, Jabłońska D (1978) Porazheniye legkikh pri sklerodermii (Lung involvement in systemic sclerosis). Klin Med (Mosk) 61(4):32–37 (in Russian)

Kucharz EJ, Nowak M (1986) Circadian changes of collagen peptidase activity in the serum and the liver of rat. Rev Roum Biochim 23:115–118

Kucharz EJ, Sierakowski SJ (1987) Zespoły przypominające twardzinę układową wywołane przez czynniki chemiczne (Chemically induced scleroderma-like syndromes). Przegl Lek 44:666–671

Kucharz EJ, Stawiarska B (1981) Udział kolagenu płuc w procesach fizjologicznych i patologicznych (Pathophysiology of the lung collagen). Pol Tyg Lek 36:69–71 (in Polish)

Kucharz EJ, Stawiarska-Pięta B (1986) The effect of lead on protein and lactate dehydrogenase activity in hepatic slices cultured in vitro. Arh Hig Rada Toksikol 37:225–229

Kucharz EJ, Miodońska G, Kozłowski A (1972) Wpływ tlenków azotu na metabolizm kolagenu w płucach świnek morskich (Effect of nitrogen oxides on collagen metabolism in the lungs of guinea pigs). Informator XII Uczelnianej konferencji naukowej Studenckiego Koła Naukowego Śląskiej Akademii Medycznej, Zabrze, pp 30–31 (in Polish)

Kucharz EJ, Dróżdż M, Grucka-Mamczar E, Mamczar A (1978) Effect of long-term exposure to hydroxyfluoride on collagen metabolism in growing rats. 12th Federation of European Biochemical Societies meeting, Dresden, abstracts, vol II, p 159

Kucharz EJ, Dróżdż M, Obuchowicz E (1981) Collagen peptidase in chronic liver disorders. Z Med Lab Diagn 22:307–313

Kucharz EJ, Dróżdż M, Cerazy B, Jendryczko A (1982a) Collagen metabolism in patients with myocardial infarction. Cor Vasa 24:339–344

Kucharz EJ, Stawiarska B, Dróżdż M, (1982b) Influence of anti-rheumatic drugs on the activity of collagenolytic cathepsin in hepatic culture. Acta Biol Med Germ 41:39–46

Kucharz EJ, Dróżdż M, Jendryczko A (1986a) Influence of fluoride compounds on lysosomal enzymes activity in blood serum of growing rats. Rev Roum Biochim 23:115–118

Kucharz EJ, Olczyk K, Dróżdż M, Wieczorek M (1986b) Influence of thyroid hormones and of methylthiouracil upon collagen content in the liver of rats with hepatic fibrosis. Endokrinologie 24:21–25

Kucharz EJ, Nowak M, Rzepecka G, Kocot E (1987) Collagen peptidase in serum of patients with myocardial infarction. Cor Vasa 29:98–101

Kuettner KE, Hiti J, Eisenstein R, Harper E (1976) Collagenase inhibition by cationic proteins derived from cartilage and aorta. Biochem Biophys Res Commun 72:40–46

Kuettner KE, Soble L, Croxen RL, Marczynska B, Hiti J, Harper E (1977) Tumour cell collagenase and its inhibition by a cartilage-derived protease inhibitor. Science 196:653–654

Kühn K (1982a) Chemical properties of collagen. In: Furtmayr H (ed) Immunochemistry of the extracellular matrix. CRC Press, Boca Raton, pp 1–29

Kühn K (1982b) Relationship between amino acid sequence and higher structures of collagen. Connect Tissue Res 10:5–10

Kühn K (1986) The collagen family – variations in the molecular and supramolecular structure. In: Kühn K, Krieg T (eds) Connective tissue: biological and clinical aspects, vol I. Karger, Basel, pp 29–69

Kühn K (1987) The classical collagens: types I, II, and III. In: Mayne R, Burgeson RE (eds) Structure and function of collagen types. Academic, Orlando, pp 1–42

Kühn K, Krieg T (eds) (1986) Connective tissue: biological and clinical aspects. Karger, Basel

Kühn K, Iwangoff P, Hammerstein F, Stecher K, Durruti M, Holzmann H, Korting GW (1964) Untersuchungen über den Stoffwechsel des Kollagens. II. Der Eibau von [$^{14}$C]Glycin in Kollagen bei mit Prednison behandelten Ratten. Z Physiol Chem 337:249–256

Kühn K, Wiedermann H, Timpl R, Risteli J, Dieringer H, Voss T, Glanville RW (1981) Macromolecular structure of basement membrane collagens. FEBS Lett 125:123–128

Kühn K, Wiestner M, Krieg T, Muller PK (1982) Structure and function of the amino terminal propeptide of type I and III collagen. Connect Tissue Res 10:43–50

Kühn K, Glanville RW, Babel W, Qian RQ, Dieringer H, Voss T, Siebold B, Oberbaumer I, Schwarz U, Yamada Y (1985) The structure of type IV collagen. Ann NY Acad Sci 460:14–24

Kuivaniemi H, Peltonen L, Palotie A, Kaitila I, Kivirikko KI (1982) Abnormal copper metabolism and deficient lysyl oxidase activity in a heritable connective tissue disorder. J Clin Invest 69:730–733

Kuivaniemi H, Peltonen L, Kivirikko KI (1985) Type IX Ehlers-Danlos syndrome and Menke's syndrome: the decrease in lysyl oxidase activity is associated with a corresponding deficiency in the enzyme protein. Am J Hum Genet 37:798–808

Kuntz E, White E (1961) Effect of electron beam irradiation on collagen. Fed Proc 20:376

Kurelec B, Rijavec M (1966) Amino acid pool of the liver fluke (Fasciola hepatica L. ). Comp Biochem Physiol 19:525–531

Kuutti ER, Tuderman L, Kivirikko KI (1975) Human prolyl hydroxylase. Purification, partial characterization, and preparation of antiserum to the enzyme. Eur J Biochem 57:181–188

Kuutti-Savolainen ER (1979) Enzymes of collagen biosynthesis in skin and serum in dermatological diseases. II. Serum enzymes. Clin Chim Acta 96:53–58

Kuutti-Savolainen ER, Kero M (1979) Enzymes of collagen biosynthesis in skin and serum in dermatological diseases. I. Enzymes of the skin. Clin Chim Acta 96:43–51

Kuutti-Savolainen ER, Risteli J, Miettinen TA, Kivirikko KI (1979a) Collagen biosynthesis enzymes in serum and hepatic tissue in liver disease. I. Prolyl hydroxylase. Eur J Clin Invest 9:89–95

Kuutti-Savolainen ER, Anttinen H, Miettinen TA, Kivirikko KI (1979b) Collagen biosynthesis enzymes in serum and hepatic tissue in liver disease. II. Galactosylhydroxylysyl glucosyltransferase. Eur J Clin Invest 9:97–101

Laato M, Heino J (1988) Interleukin 1 modulates collagen accumulation by rat granulation tissue both in vivo and in vitro. Experientia 44:32–34

Laato M, Kähäri VM, Niinikoski J, Vuorio E (1987) Epidermal growth factor increases collagen production in granulation tissue by stimulation of fibroblast proliferation and not by activation of collagen genes. Biochem J 247:385–388

Labat-Robert J, Robert L (1988) Interactions between structural glycoproteins and collagen. In: Nimni ME (ed) Collagen, vol I. CRC Press Inc. Boca Raton, pp 173–186

Labat-Robert J, Szendrol J, Godeau G, Robert L (1985) Comparative distribution patterns of type I and III collagens in fibronectin in human artherosclerotic aorta. Pathol Biol 33:261–266

Laitinen O (1977) Bone, calcium, and hydroxyproline metabolism in hyperparathyroidism and after removal of parathyroid adenoma. Acta Med Scand 202:39–42

Laitinen O, Nikkil EA, Kivirikko KI (1966a) Hydroxyproline in serum and urine. Normal values and clinical significance. Acta Med Scand 179:275

Laitinen O, Uitto J, Hannuksela M, Mustakallio KK (1966b) Increased soluble collagen content of affected and normal-looking skin in dermatomyositis, lupus erythematosus, and scleroderma. Ann Med Exp Biol Fenn 44:507–509

Laitinen O, Uitto J, Ilvainen M, Hannuksela M, Kivirikko KI (1968) Collagen metabolism of the skin in Marfan's syndrome. Clin Chim Acta 21:320–326

Laitinen O, Uitto J, Hannuksela M, Mustakallio KK (1969) Solubility and turnover of collagen in collagen disease. Ann Clin Res 1:64–73

Lamberg SI, Dorfman A (1973) Synthesis and degradation of hyaluronic acid in the cultured fibroblasts of Marfan's syndrome. J Clin Invest 52:2428–2433

Lane JM, Parkes LJ, Prockop DJ (1971) Effect of the proline analogue azetidine-2-carboxylic acid on collagen synthesis in vivo. II. Morphological and physical properties of collagen containing the analogue. Biochim Biophys Acta 236:528–541

Lane JM, Bora FW, Prockop DJ, Heppenstall RB, Black J (1972) Inhibition of scar formation by the proline analog cis-hydroxyproline. J Surg Res 13:135–137

Langer LO, Spranger JW, Greinacher I, Herdman RC (1969) Thanatopharic dwarfism. A condition manifested with achondroplasia in the neonate, with brief comments on achondrogenesis and homozygous achondroplasia. Radiology 92:285–294

Langner RO, Modrak JB (1977) Aortic collagen synthesis in rabbits following removal of atherogenic diet. Exp Mol Pathol 26:310–317

Lanman TH, Ingalls TH (1937) Vitamin C deficiency and wound healing. An experimental and clinical study. Am J Surg 105:616–623

Lapière CM Nusgens B, Hugues J (1978) Collagen type I and type III polymerization and platelet aggregation. In: Gastpar H, Kühn K, Marx R (eds) Collagen-platelet interaction. Schattaner, Stuttgart, pp 227–232
Laqua H (1981) Collagen formation by periretinal cellular membranes. Dev Ophthalmol 2:396–406
Larjava H, Sandberg M, Vuorio E (1989) Altered distribution of type I collagen mRNA in periodontal disease. J Periodont Res 24:171–177
Lasek W (1978) Kolagen. Chemia i wykorzystanie (Collagen. Chemistry and application). Wydawnictwo Naukowo Techniczne, Warszawa
Lash JW, Vasan NS (1978) Somite chondrogenesis in vitro. Stimulation by exogenous matrix components. Dev Biol 66:151–171
Last JA, Greenberg DB, Castleman WL (1979) Ozone-induced alterations in collagen metabolism of rat lungs. Toxicol Appl Pharmacol 51:247–258
Last JA, Reiser KM, Tyler WS, Rucker RB (1984) Long-term consequences of exposure to ozone. I. Lung collagen content. Toxicol Appl Pharmacol 72:111–118
Last JA, Summers P, Reiser KM (1989) Biosynthesis of collagen cross-links II. In vivo labelling and stability of lung collagen in rats. Biochim Biophys Acta 990:182–189
Laszlo J (1975) Myeloproliferative disorders: myelofibrosis, myelosclerosis, extramedullary hematopoiesis, undifferentiated MPD and hemorrhagic thrombocythemia. Semin Hematol 12:409–432
Laurent GJ (1982) Rates of collagen synthesis in lung, skin and muscle obtained in vivo by a simplified method using [$^3$H]-proline. Biochem J 206:535–544
Laurent GJ (1986) Lung collagen: more than scaffolding. Thorax 41:418–428
Laurent GJ, McAnulty RJ, (1983) Protein metabolism during bleomycin-induced pulmonary fibrosis in rabbits. Am Rev Respir Dis 128:82–88
Lautsch EV (1979) Morphological factors of clinical significance in myocardial infarction. Tex Rep Biol Med 39:371–377
Lawrie RA (1979) Meat science, 3rd edn. Pergamon, Oxford, pp 75–131
Layman DL, Narayanan AS, Martin GR (1972) The production of lysyl oxidase by human fibroblasts in culture. Arch Biochem Biophys 149:97–101
Lazarus GS (1972) Collagenase and connective tissue metabolism in epidermolysis bullosa. J Invest Dermatol 58:242–248
Leach CS, Rambaut PC (1977) Biomedical responses of the skylab crewmen. In: Johnson RS, Dietlein LF (eds) Biomedical results from skylab. NASA SP-377, Washington DC, pp 204–216
Leblond CP (1989) Synthesis and secretion of collagen by cells of connective tissue, bone, and dentin. Anat Rec 224:123–138
Lee RE, Davison PF (1984) The collagens of the developing bovine cornea. Exp Eye Res 39:639–652
Lee CA, Lloyd HM (1964) Urinary hydroxyproline in diseases involving bone and calcium metabolism. Med J Aust 25:992–995
Lee SL, Siegel M (1968) Drug-induced systemic lupus erythematosus. In: Meyler L, Peck HM (eds) Drug-induced disease, vol 3. Excerpta Medica, Amsterdam, pp 239–248
Lee Y, Lee F, Lu A, Chang C, Chen H, Liang K, Chang C (1983) Biochemical analysis and electron microscopy of human mitral valve collagen in patients with various etiologies of mitral valve disease. Jpn Heart J 24:529–538
Lees S (1987) Considerations regarding the structure of the mammalian mineralized osteoid from viewpoint of the generalized packing model. Connect Tissue Res 16:281–303
Leeson TS, Speakman JS (1961) The fine structure of extracellular material in the pectinate ligament (trabecular meshwork) of the human iris. Acta Anat (Basel) 46:363–379
Leheup BP, Federspiel SJ, Guerry-Force ML, Wetherall NT, Commers PA, DiMari SJ, Haralson MA (1989) Extracellular matrix biosynthesis by cultured fetal rat lung epithelial cells. I. Characterization of the clone and the major genetic types of collagen produced. Lab Invest 60:791–807
Lehtinen P, Aho S, Kulonen E (1983) Effect of silica on the rat lung with special reference to RNA. Ann Occup Hyp 27:81–87

Lelièvre Y, Boubouton R, Boiziau J, Cartwright T (1989) Inhibition de la collagènase synoviale par l'actinonine, étude de relations structure/activité. Pathol Biol (Paris) 37:43–46
Lenco W, McKnight M, MacDonald AS (1975) Effects of cortisone acetate, methylprednisolone and medroxyprogesterone on wound contracture and epithelization in rabbits. Ann Surg 181:67–73
Lenkiewicz JE, Davies MJ, Rosen D (1972) Collagen in human myocardium as a function of age. Cardiovasc Res 6:549–555
LeRoy EC (1972) Connective tissue synthesis by scleroderma fibroblasts in cell culture. J Exp Med 135:1351–1362
LeRoy EC (1974) Increased collagen synthesis by scleroderma skin fibroblasts in vitro: a possible defect in the regulation or activation of the scleroderma fibroblast. J Clin Invest 54:880–889
LeRoy EC (1981) The connective tissue in scleroderma. Coll Relat Res 1:301–308
LeRoy EC, Kahaleh MB, Mercurio S (1983) A fibroblast mitogen present in scleroderma but not control sera: inhibition by proteinase inhibitors. Rheumatol Int 3:35–38
Lesot H, Ruch JV (1979) Analyse des types de collagene synthétisés par l'ébauche dentaire et ses constituants dissociés chez l'embryon de Souris. Biol Cell 34:23–38
Lesot H, von der Mark K, Ruch JV (1978) Localisation par immunofluorescence des types de collagène synthetises par l'ébauche dentaire chez l'embryon de Souris. Acad Sci [III] 286:765–768
Leung K, Munck A (1975) Peripheral actions of glucocorticoids. Annu Rev Physiol 37:245–272
Leushner JRA, Clarson CL (1986) Analysis of the collagens from the fetal membranes of diabetic mothers. Placenta 7:65–72
Leushner JRA, Tevaarwerk GJM, Clarson CL, Harding PGR, Chance GW, Haust MD (1986) Analysis of the collagens of diabetic placental villi. Cell Mol Biol 32:27–35
Levene CI (1961) Structural requirements for lathyrogenic agents. J Exp Med 114:295–310
Levene CI, Gross J (1959) Alterations in the state of molecular aggregation of collagen induced in chick embryos by $\beta$-aminopropionitrile (lathyrus factor). J Exp Med 110:771–789
Levene CI, Poole JCF (1962) The collagen content of the normal and atherosclerotic human aortic intima. Br J Exp Pathol 43:469–476
Levin LS, Brady JM, Meinick M (1980) Scanning electron microscopy of teeth in dominant osteogenesis imperfecta: support for genetic heterogeneity. Am J Med Genet 5:189–199
Li W, Stramm LE, Aguirre GD, Rockey JH (1984) Extracellular matrix production by cat retinal pigment epithelium in vitro: characterization of type IV collagen synthesis. Exp Eye Res 38:291–304
Liakakos D, Vlachos P, Anoussakis C (1971) The effect of acetylsalicylic acid on bone collagen metabolism in children. Iatr Athinai 19:207–212
Liakakos D, Anoussakis C, Dalles K, Vlachos P, Platakis I, Ikkos D (1973) On the mode of action of acetylosalicylic acid (aspirin) on collagen. Helv Paediatr Acta 28:109–116
Liakakos D, Vlachos P, Anoussakis C, Doulas NL (1975) Urinary hydroxyproline values following continuous and intermittent administration of prednisolone in children. Helv Paediatr Acta 30:495–499
Liang JN, Chakrabarti B (1981) Spectroscopic studies on pepsin-solubilized vitreous and cartilage collagens. Curr Eye Res 1:175–181
Lichtenstein JR, Martin GR, Kohn LD, Byers PH, McKusick VA (1973) Defect in conversion of procollagen to collagen in a form of Ehlers-Danlos syndrome. Science 182:298–300
Lichtler A, Stover ML, Angilly J, Kream B, Rowe DW (1989) Isolation and characterization of the rat $\alpha 1$(I) collagen promoter. Regulation by 1,25-dihydroxyvitamin D. J Biol Chem 264:3072–3077
Liener IE (1967) Lathyrogens. Ind J Gen Plant Breed 27:34
Light N, Champion AE (1984) Characterization of muscle epimysium, perimysium and endomysium collagens. Biochem J 219:1017–1026
Light ND (1979) Bovine type I collagen. A study of cross-linking in various mature tissues. Biochim Biophys Acta 581:96–105
Light ND, Bailey AJ (1979) Changes in crosslinking during aging in bovine tendon collagen. FEBS Lett 97:183–188

Lim KO, Boughner DR, Perkins DG (1983) Ultrastructure and mechanical properties of chordae tendineae from a myxomatous tricuspid valve. Jpn Heart J 24:539–548

Limeback HF, Sodek J (1979) Procollagen synthesis and processing in periodontal ligament in vivo and in vitro. Eur J Biochem 100:541–546

Linch DC, Aston DHC (1979) Ehlers-Danlos syndrome presenting with juvenile destruction periodontitis. Br Dent J 147:95–97

Lindahl U, Höök M (1978) Glycosaminoglycans and their binding to biological macromolecules. Annu Rev Biochem 47:385–417

Lindberg KA, Hassert MA, Pinnell SR (1976) Inhibition of lysyl oxidase by homocysteine. A proposed connective tissue defect in homocystinuria. Clin Res 24:265 A (abstract)

Linde A, Robins SP (1988) Quantitative assessment of collagen cross-links in dissected predentin and dentin. Coll Relat Res 8:443–450

Lindner J, Grasedyck K, Prinz G, Grade J, Kolln H (1974) Studies on collagen peptidases. In: Fricke R, Hartmann F (eds) Connective tissues: biochemistry and pathophysiology. Springer, Berlin Heidelberg New York, pp 178–190

Lindsay JR (1980) Otosclerosis. In: Paparella MM, Schumrick DA (eds) Otolaryngology, vol 2. Saunders, Philadelphia, pp 1617–1643

Lindy S, Sorsa T, Suomalainen K, Turto H (1986) Gold sodium thiomalate activates latent human leukocyte collagenase. FEBS Lett 208:23–25

Lindy S, Sorsa T, Suomalainen K, Turto H (1988) Effects of gold (I) compounds on latent human leucocyte collagenase and gelatinase. Scand J Rheumatol [Suppl] 67:5–9

Linsenmayer TF (1982) Immunology of purified collagens and their use in localization of collagen types in tissues. In: Weiss JB, Jayson MIV (eds) Collagen in health and disease. Churchill Livingstone, Edinburgh, pp 244–268

Linsenmayer TF, Smith GN Jr, Hay ED (1977) Synthesis of two collagen types by embryonic chick corneal epithelium in vitro. Proc Natl Acad Sci USA 74:39–43

Linsenmayer TF, Fitch JM, Schmid TM, Zak NB, Gibney E, Sanderson RD, Mayne R (1983) Monoclonal antibodies against chicken type V collagen: production, specificity, and use for immunocytochemical localization in embryonic cornea and other organs. J Cell Biol 96:124–132

Linsenmayer TF, Gibney E, Fitch JM, Gross J, Mayne R (1984) Thermal stability of the helical structure of type IV collagen within basement membrane in situ: determination with a conformation-dependent monoclonal antibody. J Cell Biol 99:1405–1409

Linsenmayer TF, Fitch JM, Gross J, Mayne R (1985) Are collagen fibrils in the developing avian cornea composed of two different collagen types? Evidence from monoclonal antibody studies. Ann NY Acad Sci 460:232–245

Liotta LA, Rao CN (1985) Role of the extracellular matrix in cancer. Ann NY Acad Sci 460:333–344

Liotta LA, Kleinerman J, Catanzaro P, Rynbrandt D (1977) Degradation of basement membrane collagen by murine tumor cells. JNCI 58:1427–1431

Liotta LA, Foidart SM, Robey PG, Martin G, Gullino PM (1979) Identification of micrometastasis of breast carcinomas by presence of basement membrane collagen. Lancet 2:146–147

Liotta LA, Tryggvason K, Garbisa S, Hart IR, Foltz CM, Shafie S (1980) Metastatic potential correlates with enzymatic degradation of basement membrane collagen. Nature 284–67–68

Liotta LA, Kalebic T, Reese CA, Mayne R (1982a) Protease susceptibility of HMW $1\alpha 2\alpha$, but not $3\alpha$ cartilage collagens are similar to type V collagen. Biochem Biophys Res Commun 104:500–506

Liotta LA, Thorgeirsson UP, Garbisa S (1982b) Role of collagenases in tumor cell invasion. Cancer Metastasis Rev 1:277–288

Liotta LA, Rao CN, Wewer UM (1986) Biochemical interactions of tumor cells with the basement membrane. Annu Rev Biochem 55:1037–1057

Lipner H (1973) Mechanism of mammalian ovulation. In: Greep RO (ed) Handbook of physiology: endocrinology. American Physiological Society, Washington, pp 409–437

Lipton BH (1977) Collagen synthesis by normal and bromodeoxyuridine-modulated cells in myogenic culture. Dev Biol 61:153–165

Listgarten MA, Shapiro IM (1974) Fine structure and composition of coronal cementum in guinea pig molars. Arch Oral Biol 19:679–696

Little CD, Church RL, Miller RA, Ruddle FH (1977) Procollagen and collagen produced by a teratocarcinoma-derived cell line, TSD4: evidence for a new molecular from of collagen. Cell 10:287–295

Liu TZ, Bhatnagar RS (1973) Inhibition of protocollagen proline hydroxylase by dilantin. Proc Soc Exp Biol Med 142:253–260

Loeb JN (1976) Corticosteroids and growth. N Engl J Med 295: 547–552

Löhler J, Timpl R, Jaenisch R (1984) Embryonic lethal mutation in mouse collagen I gene causes rupture of blood vessels and is associated with erythopoietic and mesenchymal cell death. Cell 38:597–607

Lonati-Galligani M, Galligani L, Fuller GC (1979) Effect of (+)-catechin on purified prolyl hydroxylase and collagen synthesis in skin fibroblasts in culture. Biochem Pharmacol 28:2573–2577

López-Escalera R, Pardo A (1987) Carrageenin-stimulated peritoneal macrophages release in vitro collagenase and gelatinase. Coll Relat Res 7:249–257

Lorenzen I (1969) The effects of the glucocorticoids on connective tissue. Acta Med Scand [Suppl] 500:17–21

Lovell CR, Nicholls AC, Duance VC, Bailey AJ (1979) Characterization of dermal collagen in systemic sclerosis. Br J Dermatol 100:359–369

Lubec G (1977) Collagenase activity of rat kidney with glomerulonephritis during the heterologous phase. Clin Chim Acta 76:89–94

Luck JV (1959) Dupuytren's contracture: a new concept of the pathogenesis correlated with surgical management. J Bone Joint Surg 41:635–664

Ludowieg JJ, Adams J, Wang AC, Parker J, Fudenberg HH (1973) The mammalian intervertebral disc. The collagen of whale fetal nucleus pulposus. Connect Tissue Res 2:21–29

Lugano EM Dauber JH, Daniele RP (1982) Acute experimental silicosis. A J Pathol 109:27–36

Lukinmoa PL (1988) Immunofluorescent localization of type III collagen and the N-terminal propeptide of type III procollagen in dentin matrix in osteogenesis imperfecta. J Craniofac Genet Dev Biol 8:235–243

Luparello C (1987) Desmoplasia in a non-tumoral disease: determination of collagen content in human mammary dysplasia. Bull Mol Biol Med 12:49–57

Lütjen-Drecoll E, Shimizu T, Rohrbach M, Rohen JW (1986) Quantitative analysis of "plaque material" in the inner and outer wall of Schlemm's canal in normal and glaucomatous eyes. Exp Eye Res 42:443–455

Lütjen-Drecoll E, Rittig M, Rauterberg J, Jander R, Mollenhauer J (1989) Immunomicroscopical study of type VI collagen in the trabecular meshwork of normal and glaucomatous eyes. Exp Eye Res 48:139–147

Lynas NA, Merritt AD (1958) Marfan's syndrome in Northern Ireland. Ann Hum Genet 22:289–309

Lyons BL, Schwarz RI (1984) Ascorbate stimulation of PAT cells causes an increase in transcription rates and a decrease in degradation rates of procollagen mRNA. Nucleic Acids Res 12:2569–2579

Lyons H, Jones E, Quinn FE, Sprunt DH (1964) Protein-polysaccharide complexes of normal and herniated human intervertebral discs. Proc Soc Exp Biol Med 115:610–614

MacCartney HW, Tschesche H (1983) Latent and active human polymorphonuclear leukocyte collagenase. Isolation, purification, and characterization. Eur J Biochem 130:71–77

MacDonald AS (1971) Topical corticosteroid preparations, hazard and size-effects. Br J Clin Pract 25:421–425

Macek M, Hurych J, Chvapil M, Kadlecova V (1966) Study on fibroblasts in Marfan's syndrome. Humangenetik 3:87–97

Maciewicz RA, Etherington DJ (1985) Separation of cathepsins B, L, N, and S from rabbit spleen. Biochem Soc Trans 13:1169–1170

Maciewicz RA, Etherington DJ, Kos J, Turk V (1987) Collagenolytic cathepsins of rabbit spleen: a kinetic analysis of collagen degradation and inhibition by chicken cystatin. Coll Relat Res 7:295–304

Mack PB, LaChance PC (1967) Effects of recumbency and space flight on bone density. Am J Clin Nutr 30:1194–1205
Mackel AM, DeLustro F, Harper FE, LeRoy EC (1982) Antibodies to collagen in scleroderma. Arthritis Rheum 25:522–531
MacKenzie AR, Pick CR, Sibley PR, White BP (1978) Suppression of rat adjuvant disease by cyclophosphamide pretreatment: evidence for an antibody mediated component in the pathogenesis of the disease. Clin Exp Immunol 32:86–96
Madden JW, Chvapil M, Carlson EC, Ryan JN (1973) Toxicity and metabolic effects of 3,4-dehydroproline in mice. Toxicol Appl Pharmacol 26:426–437
Madri JA, Dreyer B, Pittick FA, Furtmayr H (1980) The collagenous components of the subendothelium: correlation of structure and function. Lab Invest 43:303–315
Madri JA, Foellmer HG, Furtmayr H (1982) Type V collagen of the human placenta: trimer $\alpha$-chain composition, ultrastructural morphology and peptide analysis. Coll Relat Res 2:19–29
Madrid RE, Jaros E, Cullen MJ, Bradley WG (1975) Genetically determined defect of Schwann cell basement membrane in dystrophic mouse. Nature 257:319–321
Maher JJ, Bissell DM, Friedman SL, Roll FJ (1988) Collagen measured in primary cultures of normal rat hepatocytes derives from lipocytes within the monolayer. J Clin Invest 82:450–459
Mainardi CL (1985) Collagenase in rheumatoid arthritis. Ann NY Acad Sci 460:345–354
Mainardi CL (1987) The role of connective tissue degrading enzymes in human pathology. In: Uitto J, Perejda AJ (eds) Connective tissue disease. Molecular pathology of the extracellular matrix. Dekker, New York, pp 523–542
Mainardi CL, Dixit SN, Kang AH (1980) Degradation of type IV (basement membrane) collagen by a proteinase isolated from human polymorphonuclear leukocyte granules. J Biol Chem 225:5435–5441
Majamaa K, Kuutii-Savolainen ER, Tuderman L, Kivirikko KI (1979) Turnover of prolyl hydroxylase tetramers and the monomer-size protein in chick-embryo cartilaginous bone and lung in vivo. Biochem J 178:313–322
Majamaa K, Hanauske-Abel HM, Günzler V, Kivirikko KI (1984) The 2-oxyglutarate binding site of prolyl 4-hydroxylase. Identification of distinct subsites and evidence for 2-oxyglutarate decarboxylation in a ligand reaction at the enzyme-bound ferrous ion. Eur J Biochem 138:239–245
Majewska MR, Dancewicz AM (1976) Fluorescence of radiation-induced collagen aggregates. Acta Biochim Pol 23:353–355
Majewski S, Skiendzielewska A, Makieła B, Jabłońska S, Błaszczyk M (1987) Serum levels of type III collagen aminopeptide in patients with systemic scleroderma. Arch Dermatol Res 279:484–486
Mäkelä JK, Raassina M, Virta A, Vuorio E (1988) Human pro $\alpha 1$(I) collagen: cDNA sequence for the C-propeptide domain. Nucleic Acids Res 16:349
Małdyk E (ed) Patomorfologia chorób tkanki łącznej (Pathomorphology of the connective tissue disorders). Państwowy Zakład Wydawnictw Lekarskich, Warszawa
Mallya SK, Van Wart HE (1987) Inhibition of human neutrophil collagenase by gold (I) salts used in chrysotherapy. Biochem Biophys Res Commun 144:101–108
Mallya SK, Van Wart HE (1989) Mechanism of inhibition of human neutrophil collagenase by gold (I) chrysotherapeutical compounds. Interaction at a heavy metal binding site. J Biol Chem 264:1594–1601
Mammo W, Singh G, Dolby AE (1982) Enhanced cellular immune response to type I collagen in patients with periodontal disease. Int Arch Allergy Appl Immunol 67:149–154
Man M, Adams E (1975) Basement membrane and interstitial collagen content of whole animals and tissues. Biochem Biophys Res Commun 66:9–16
Man M, Nagle RB, Adams E (1978) Chemical estimation of interstitial and basement membrane collagen in obstructive nephropathy. Exp Mol Pathol 29:144–148
Mandl I (1961) Collagenases and elastases. Adv Enzymol 23:163–264
Mani I, Mani UV (1986) Urinary excretion of hydroxyproline in diabetics under different modes of treatment. Indian J Biochem Biophys 23:297–298
Manicourt D, Orloff S, Rao VH (1979) Synovial fluid and serum hydroxyproline fractions in rheumatoid arthritis. Scand J Rheumatol 8:161–167

Manicourt D, Brauman H, Orloff S (1980) Synovial fluid $\beta_2$ microglobulin and hydroxyproline fractions in rheumatoid arthritis and nonautoimmune arthropathies. Ann Rheum Dis 39:207–216

Manicourt D, Orloff S, Rao VH (1981) Synovial fluid hydroxyproline fractions before and after osmic acid treatment in rheumatoid arthritis. Scand J Rheumatol 10:43–48

Mannschott P, Herbage D, Weiss M, Buffevant C (1976) Collagen heterogeneity in pig heart valves. Biochim Biophys Acta 434:177–183

Mansour MM, Dunn MA, Salah LA (1988) Effect of colchicine on collagen synthesis by liver fibroblasts in murine schistosomiasis. Clin Chim Acta 177:11–20

Manzke E, Rawley R, Vose G, Roginsky M, Rader JI, Baylink DJ (1977) Effect of fluoride therapy on nondialyzable urinary hydroxyproline, serum alkaline phosphatase, parathyroid hormone, and 25-hydroxyvitamin D. Metabolism 26:1005–1010

Maquart FX, Bellon G, Cornillet-Stoupy J, Randoux A, Triller R, Kalis B, Borel JP (1985) Inhibition of collagen production in scleroderma fibroblast cultures by a connective tissue glycoprotein extracted from normal dermis. J Invest Dermatol 85:156–160

Marchini M, Morocutti M, Ruggeri A, Koch MHJ, Bigi A, Roveri N (1986) Differences in the fibril structure of corneal and tendon collagen. An electron microscopy and x-ray diffraction investigation. Connect Tissue Res 15:269–281

Marfan ABJ (1896) Un cas de deformation congenitale des quatre membres plus prononcece aux extremites characterisée par l'allongement des os avec un certain degré d'amincissement. Bull Mem Soc Med Hôp Paris 13:220–226

Margolis RK, Margolis RU (1979) Structure and distribution of glycoproteins and glycosaminoglycans. In: Margolis RU, Margolis RK (eds) Complex carbohydrates of nervous tissue. Plenum, New York, pp 45–51

Margolis RU, Lalley K, Kiang WL, Crockett C, Margolis RK (1976) Isolation and properties of a soluble chondrioitin sulfate proteoglycans from brain. Biochem Biophys Res Commun 73:1018–1024

Maricq HR, LeRoy EC (1973) Patterns of finger capillary abnormalities in connective tissue disease by "wide field" microscopy. Arthritis Rheum 16:619–628

Marini JC, Grange DK, Gottesman GS, Lewis MB, Koeplin DA (1989) Osteogenesis imperfecta type IV. Detection of a point mutation in one $\alpha$1(I) collagen allele (COL1A1) by RNA/RNA hybrid analysis. J Biol Chem 264:11893–11900

Mark LG, Isseroff H (1983) Levels of type I and type III collagen in the bile duct of rats infected with Fasciola hepatica. Molec Biochem Parasitol 8:253–262

Maroudas A, Stockwell RA, Nachemson A, Urban J (1975) Factors involved in the nutrition of the human lumbar intervertebral disc: cellularity and diffusion of glucose in vitro. J Anat 111:219–227

Marshall PA, Williams PE, Goldspink G (1989) Accumulation of collagen and altered fiber-type ratios as indicators of abnormal muscle gene expression in the MDX dystrophic mouse. Muscle Nerve 12:528–537

Martin HB, Boatman ES (1965) Electron microscopy of human pulmonary emphysema. Am Rev Respir Dis 91:206–212

Martin GR, Byers PH, Piez KA (1975) Procollagen. Adv Enzymol 42:167–191

Martin TJ, Ng KW, Nicholson GC (1988) Cell biology of bone. Baillieres Clin Endocrinol Metab 2:1–30

Martinez-Hernandez A (1985) The hepatic extracellular matrix. II. Electron immunohistochemical studies in rats with $CCl_4$-induced cirrhosis. Lab Invest 53:166–186

Martinez-Hernandez A (1988) The extracellular matrix and neoplasia. Lab Invest 58:609–612

Martinez-Hernandez A, Amenta PS (1983) The basement membrane in pathology. Lab Invest 48:656–677

Martinez-Hernandez A, Catalano E (1980) Stromal reaction in neoplasia: colonic caracinomas. Ultrastruct Pathol 1:403–410

Martinez-Hernandez A, Gay S, Miller EJ (1982) Ultrastructural localization of type V collagen in rat kidney. J Cell Biol 92:343–349

Maruyama K, Feinman L, Okazaki I, Lieber CS (1981) Direct measurement of neutral collagenase activity in homogenates from baboon and human liver. Biochim Biophys Acta 658:124–131

Maruyama K, Okazaki I, Kobayashi T, Suzuki H, Kashiwazaki K, Tsuchiya M (1983) Collagenase production by rabbit liver cells in monolayer culture. J Lab Clin Med 102:543–550

Maśliński C, Sternik K, Magrowicz J, Bilski R, Kafliński W, Markowiak W (1965) Sur le rôle de l'histamine dans la cicatrisation. In: Ppoisseu A (ed) La cicatrisation CNRS, Paris 1965, p 297

Maslow DE (1987) Collagenase effects on cancer cell invasiveness and motility. Invasion Metastasis 7:297–310

Mason RM, Palfrey AJ (1977) Some aspects of hereditary kyphoscoliosis in mice. In: Zorab PA (ed) Scoliosis. Academic, London, pp 349–367

Massaro D, Handler A, Katz S, Young R Jr (1966) Excretion of hydroxyproline in patients with sarcoidosis. Am Rev Resp Dis 93:929–933

Mata JM, Kershenobich D, Villarreal E, Rojkind M (1975) Serum free proline and free hydroxyproline in patients with chronic liver disease. Gastroenterology 68:1265–1269

Matalon R, Dorfman A (1968) The accumulation of hyaluronic acid in cultured fibroblasts of the Marfan syndrome. Biochem Biophys Res Commun 32:150–154

Mathews MB (1965) The interaction of collagen and acid mucopolysaccharides. A model for connective tissue. Biochem J 96:710–716

Mathews MB (1975) Connective tissue. Macromolecular structure and evolution. Springer, Berlin Heidelberg New York

Matsumoto E, Muragaki Y, Ooshima A, Terasawa M, Imayoshi T (1988) Increased serum immureactive prolyl hydroxylase in rats with carrageenan-induced granuloma and adjuvant arthritis. Exp Mol Pathol 49:243–253

Matsumura T (1974) Collagen fibrils of the sea cucumber, Stichopus japonicus: purification and morphological study. Connect Tissue Res 2:117–125

Matsumura T, Hasegawa M, Shigei M (1979) Collagen biochemistry and phylogeny of echinodermis. Comp Biochem Physiol 62B:101–109

Matsumura Y, Shibata Y, Tamaki T, Iguchi M (1967) Effect of chronic administration of tyrosine on urinary hydroxyproline in rats. Metabolism 16:957–959

Matsuo S, Brentjens JR, Andres G, Foidart JM, Martin GR, Martinez-Hernandez A (1986) Distribution of basement membrane antigens in glomeruli of mice with autoimmune glomerulonephritis. Am J Pathol 122:36–49

Matsuoka M, Pham NT, Tsukamoto H (1989) Differential effects of interleukin-1 $\alpha$, tumor necrosis factor $\alpha$, and transforming growth factor $\beta_1$ on cell proliferation and collagen formation by cultured fat-storing cells. Liver 9:71–78

Matsushima K, Bano M, Kidwell WR, Oppenheim JJ (1985) Interleukin 1 increases collagen type IV production by murine mammary epithelial cells. J Immunol 134:904–909

Mauch C, Krieg T (1986) Pathogenesis of fibrosis – introduction and general aspects. In: Kühn K, Krieg T (eds) Connective tissue: biological and clinical aspects, Karger, Basel,, pp 372–384

Maumenee IH, Trauboulsi EI (1985) The ocular findings in Kniest dysplasia. Am J Opthalmol 100:155–160

Maurice DM (1969) The cornea and sclera. In: Davson H (ed) The eye, vol 1. Academic, New York, p 489

Maurice-Williams RS (1981) Spinal degenerative disease. Wright, Bristol, pp 32–46

Maxwell WA, Spicer SS, Miller RL, Halushka PV, Westphal MC, Setser ME (1977) Histochemical and ultrastructural studies in fibrodysplasia ossificans progressiva (myositis ossificans progressiva). Am J Pathol 87:483–492

Maxwell WL, Duance VC, Lehto M, Ashurst DE, Berry M (1984) The distribution of types I, III, IV and V collagens in penetrant lesions of the central nervous system of the rat. Histochem J 16: 1219–1219-1229

May MA, Beauchamp GR (1987) Collagen maturation defects in Ehlers-Danlos keratopathy. J Pediatr Opthalmol Strabismus 24:78–82

Mayne R (1986) Collagenous proteins of blood vessels. Arteriosclerosis 6:585–593

Mayne R (1987) Vascular conective tissue: normal biology and derangement in human diseases. In: Uitto J, Perejda AJ (eds) Connective tissue disease. Molecular pathology of the extracellular matrix. Dekker, New York, pp 163–183

Mayne R (1989) Cartilage collagens. What is their function, and are they involved in articular disease? Arthritis Rheum 32:241–246

Mayne R, Burgeson RE (1987) Structure and function of collagen types. Academic, New York
Mayne R, Irwin MH (1986) Collagen types in cartilage. In: Kuettner KE, Schleyerbach R, Haseall VC (eds) Articular cartilage biochemistry. Raven, New York, pp 23–38
Mayne R, Vail MS, Miller EJ (1978) Characterization of the collagen chains synthesized by cultured smooth muscle cells derived from rhesus monkey thoracic aorta. Biochemistry 17:446–452
Mayo ME, Lloyd-Davies RW, Shuttleworth KED, Tighe JR (1973) The damaged human detrusor: functional and electron microscopic changes in disease. Br J Urol 45:116–125
Mays PK, Bishop JE, Laurent GJ (1988) Age-related changes in the proportion of types I and III collagen. Mech Ageing Dev 45:203–212
Mays PK, McAnulty RJ, Laurent GJ (1989) Age-related changes in lung collagen metabolism. A role for degradation in regulating lung collagen production. Am Rev Respir Dis 140:410–416
Mbuyi-Muamba JM, Dequeker J, Gevers G (1988) Biochemistry of bone. Baillieres Clin Rheumatol 2:63–102
McAdam KPWJ; Fudenberg HH, Michaeli D (1978) Antibodies to collagen in patients with leprosy. Clin Immunol Immunopathol 9:16–21
McAnulty RJ, Laurent GJ (1987) Collagen synthesis and degradation in vivo. Evidence for rapid rates of collagen turnover with extensive degradation of newly synthesized collagen in tissues of the adult rat. Coll Relat Res 7:93–104
McAnulty RJ, Staple LH, Guerreiro D, Laurent GJ (1988) Extensive changes in collagen synthesis and degradation during compensatory lung growth. Am J Physiol 255:C754–C759
McCarthy JB, Wahl SM, Rees JC, Olsen CE, Sandberg AL, Wahl LM (1980) Mediation of macrophage collagenase production by 3'–5' cyclic adenosine monophosphate. J Immunol 124:2405–2409
McCarthy TL, Centrella M, Canalis E (1989) Regulatory effects of insulin-like growth factors I and II on bone collagen synthesis in rat calvarial cultures. Endocrinology 124:301–309
McCay CM (1947) Effect of restricted feeding upon aging and chronic diseases in rats and dogs. Am J Publ Health 37:521–528
McCloskey DI, Cleary EG (1974) Chemical composition of the rabbit aorta during development. Circ Res 34:828–835
McConkey B, Walton KW, Carney SA, Lawrence JC, Ricketts CR (1967) Significance of the occurence of transparent skin. A study of histological characteristics and biosynthesis of dermal collagen. Ann Rheum Dis 26:219–225
McCroskery PA, Richards JF, Harris ED Jr (1975) Purification and characterization of a collagenase extracted from rabbit tumours. Biochem J 152:131–142
McCullagh KA, Balian G (1975) Collagen characterization and cell transformation in human atherosclerosis. Nature 258:73–75
McCullagh KG, Ehrhart LA (1974) Increased arterial collagen synthesis in experimental canine atherosclerosis. Atherosclerosis 19:13–28
McCullagh KG, Ehrhart LA (1977) Enhanced synthesis and accumulation of collagen in cholesterol-aggravated pigeon atherosclerosis. Atherosclerosis 26:341–352
McCullagh KG, Duance VC, Bishop KA (1980) The distribution of collagen types I, III and V (AB) in normal and atherosclerotic human aorta. J Pathol 130:45–55
McCune WJ, Buckley JA, Trentham DE (1982) Immunosupression by fractionated total lymphoid irradiation in collagen arthritis. Arthritis Rheum 25:532–539
McCurley TL, Gay RE, Gay S, Glick AD, Haralson MA, Collins RD (1986) The extracellular matrix in "sclerosing" follicular center cell lymphomas: an immunohistochemical and ultrastructural study. Hum Pathol 17:930–938
McDonald JA (1988) Fibronectin. A primitive matrix. In: Clark RAF, Henson PM (eds) The molecular and cellular biology of wound repair. Plenum, New York, pp 405–435
McGee JOD (1982) Liver. In: Weiss JB, Jayson MIV (eds) Collagen in health and disease. Churchill Livingstone, Edinburgh, pp 414–423
McGee JOD, Patrick RS (1972) The role of perisinusoidal cells in hepatic fibrogenesis. Lab Invest 26:429–440
McGee JOD, O'Hare RP, Patrick RS (1973) Stimulation of the collagen biosynthetic pathway by factors isolated from experimentally injured liver. Nature New Biol 243:121–123

McKusick VA (1972) Heritable disorders of connective tissue, 4th edn. Mosby, St Louis, pp 1–878
McKusick VA (1983) Mendelian inheritance in man: catalogs of autosomal dominant, autosomal recessive, and X-linked phenotypes, 6th edn. John Hopkins University Press, Baltimore, pp 1–1378
McPherson JM, Piez KA (1988) Collagen in dermal wound repair. In: Clark RAF, Henson PM (eds) The molecular and cellular biology of wound repair. Plenum, New York, pp 471–496
Mechanic G, Gallop PM, Tanzer ML (1971) The nature of crosslinking in collagens from mineralized tissues. Biochem Biophys Res Commun 45:644–653
Mechanic GL, Toverud SU, Ramp WK (1972) Quantitative changes of bone collagen crosslinks and precursors in vitamin D deficiency. Biochem Biophys Res Commun 47:760–765
Mechanic GL, Banes AJ, Henmi M, Yamauchi M (1985) Possible collagen structural control of mineralization. In: Butler WT (ed) The chemistry and biology of mineralized tissue. Ebsco Media, Birmingham, pp 98–102
Mechanic GL, Young DR, Banes AJ, Yamauchi M (1986) Nonmineralized and mineralized bone collagen in bone of immobilized monkeys. Calcif Tissue Int 39:63–68
Megaw JM, Priest JH, Priest RE, Johnson LD (1977) Differentiation in human amniotic fluid cell cultures. II. Secretion of an epithelial basement membrane glycoprotein. J Med Genet 14:163–167
Mei Liu H (1988) Collagens in nervous tissue. In: Nimni ME (ed) Collagen, vol I. CRC Press, Boca Raton, pp 243–256
Meilman E, Urivetzky MM, Rapoport CM (1963) Urinary hydroxyproline peptides. J Clin Invest 42:40–44
Melin M, Saeland S, Magloire H, Hartmann DJ, Guerret S, Blanchart D, Banchereau J, Grimand JA (1987) Supernatant from an activated human CD4+ T-cell clone modulates the proliferation and collagen synthesis of human dental pulp fibroblasts. Coll Relat Res 7:371–381
Melin M, Hartmann DJ, Magloire H, Falcoff E, Aurialt C, Grimand JA (1989) Human recombinant gamma-interferon stimulates proliferation and inhibits collagen and fibronectin production by human dental pulp fibroblasts. Cell Mol Biol 35:97–110
Mendenhall CL, Chedid A, Kromme C (1984) Altered proline uptake by mouse liver cells after chronic exposure to ethanol and its metabolites. Gut 25:138–144
Menton DN, Hess RA (1980) The ultrastructure of collagen in the dermis of tight-skin mutant mice. J Invest Dermatol 74:139–147
Menton DN, Hess RA, Lichtenstein JR, Eisen AZ (1980) The structure and tensile properties of the skin of tight skin mutant mice. J Invest Dermatol 70:4–10
Menzel DB (1967) Reaction of oxidizing lipids with ribonuclease. Lipids 2:83–84
Menzel J, Steffen C (1977) Detection of collagenase-activity in RA synovial fluids using soluble $^{14}$C-labelled collagen. Z Rheumatol 36:364–377
Menzel J, Steffen C, Kolarz G, Eberl R, Frank O, Thumb N (1976) Demonstration oof antibodies to collagen and of collagen-anticollagen immune complexes in rheumatoid arthritis synovial fluids. Ann Rheum Dis 35:446–450
Menzel J, Steffen C, Kolarz G, Kojer M, Smolen J (1978) Demonstration of anticollagen antibodies in rheumatoid arthritis synovial fluids by $^{14}$C-radioimmunoassay. Arthritis Rheum 21:243–248
Merrilees MJ, Tiang KM, Scott L (1987) Changes in collagen fibril diameters across artery walls including a correlation with glycosaminoglycan content. Connect Tissue Res 16:237–257
Métivier H, Richebé R, Bernaudin JF, Chrétien J (1978) Le collagène pulmonaire. Rev Fr Mal Respir 6:529–546
Meyer O, Semmache M, Cyna J, Mitrovic D, Ryckewaert A (1983) Les anticorps anticollagène détection au cours de la polyarthrite rhumatoide, de la polychondrite chronique atrophiante et de divers rhumatismes inflammatoires chroniques. Rev Rhum Mal Osteoartic 50:493–499
Michaeli D (1977) Immunochemistry of collagen. In: Atassi MZ (ed) Immunochemistry of proteins, vol. 1. Plenum, New York, pp 371–399
Michaeli D, Fudenberg HH (1974) The incidence and antigenic specificity of antibodies against denatured human collagen in rheumatoid arthritis. Clin Immunol Immunopathol 2:153–159

Michel JB, Salzmann JL, Ossondo Nlom M, Bruneval P, Barres D, Camilleri JP (1986) Morphometric analysis of collagen network and plasma perfused capillary bed in the myocardium or rats during evolution of cardiac hypertrophy. Basic Res Cardiol 81:142–154

Miech G, Myara I Mangeot M, Lemonnier A (1988) Activity of the two prolidase isoforms in rat liver after chronic $CCl_4$ intoxication. Biomed Biochim Acta 47:1073–1075

Mikkonen L, Touminen T, Kulonen E (1960) Collagen fractions in lathyritic rats. Biochem Pharmacol 3:181–186

Mikkonen L, Lampiaho K, Kulonen E (1966) Effect of thyroid hormones, somatotrophin, insulin and corticosteroids on synthesis of collagen in granulation tissue both in vivo and in vitro. Acta Endocrinol 51:23–31

Mikuliková D, Trnavský K (1980) Influence of colchicine derivates on lysosomal enzyme release from polymorphonuclear leukocytes and intracellular levels of cAMP after phagocytosis of monosodium urate crystals. Biochem Pharmacol 29:2146–2148

Miller A, Wray TS (1971) Molecular packing in collagen. Nature 230:437–439

Miller EJ (1985) The structure of fibril-forming collagens. Ann NY Acad Sci 460: 1–13

Miller EJ (1988) Collagen types: structure, distribution, and functions. In: Nimni ME (ed) Collagen, vol I. CRC Press, Boca Raton, pp 139–156

Miller EJ, Gay S (1987) The collagens: an overview and update. Methods Enzymol 144:3–41

Miller EJ, Matukas VJ (1969) Chick cartilage collagen: a new type $\alpha 1$ chain not present in bone or skin of the species. Proc Natl Acad Sci USA 64:1264–1268

Miller RL, Elsas LJ, Priest RE (1979) Ascorbate action in normal and mutant lysyl hydroxylase from cultured dermal fibroblasts. J Invest Dermatol 72:241–247

Millesi H (1981) Collagen grafts. In: Gorio A, Millesi H, Mingrino S (eds) Posttraumatic peripheral nerve regeneration. Raven, New York, pp 215–226

Minafra S, Luparello C, Rallo F, Pucci-Minafra I (1988) Collagen biosynthesis by a breast carcinoma cell strain and biopsy fragments of the primary tumor. Cell Biol Int Rep 12:895–905

Minisola S, Antonelli R, Mazzuoli G (1985) Clinical significance of free plasma hydroxyproline measurement in metabolic disease. J Clin Chem Clin Biochem 23:515–519

Minor RR (1980) Collagen metabolism: a comparison of diseases of collagen and diseases affecting collagen. Am J Pathol 98:225–280

Minor RR (1983) Animal models of heretable diseases of skin. In: Goldsmith LA (ed) Biochemistry and physiology of the skin. Oxford University Press, New York, pp 1139–1152

Minor RR, Lein DH, Patterson DF, Krook L, Porter TG, Kane AC (1983) Defects in collagen fibrillogenesis causing hyperextensible, fragile skin of dogs. J Am Vet Med Assoc 182:142–148

Minor RR, Wootton JAM Patterson DF, Uitto J, Bartel D (1984a) Genetic diseases of collagen in animals. In: Uitto J, Perejda AJ (eds) Diseases of connective tissues: molecular pathology of the extracellular matrix. Dekker, New York, pp 293–319

Minor RR, Pihlajaniemi T, Prockop DJ, Denholm LJ, Wootton JAM (1984b) Post-translational overmodification of type I, type III and type V collagens in bovine and human osteogenesis imperfecta. J Cell Biol 99:405a

Minor RR, Strausse EL, Koszalka TR, Brent RL, Kefalides NA (1986) Organ cultures of the embryonic rat parietal yolk sac. II. Synthesis, accumulation and turnover of collagen and non-collagen basement membrane glycoproteins. Dev Biol 48:365–371

Minor RR, Wootton JAM, Prockop DJ, Patterson DF (1987) Genetic diseases of connective tissues in animals. Curr Probl Dermatol 17:199–215

Miranda AF, Nette EG, Khan S, Brockbank K, Schonberg M (1978) Alteration of myoblast phenotype by dimethyl sulfoxide. Proc Natl Acad Sci USA 75:3826–3830

Mishra G, Behera HN (1986) Alloxan-induced changes in the collagen characteristics in the skin of male garden lizards, Calotes versicolor, from three age groups. Arch Gerontol Geriatr 5:11–19

Mitomo Y, Nakao K, Angrist AA (1968) Incidental virus particles in chicken heart valve. Arch Pathol 86:508–515

Mitomo Y, Nakao K, Angrist AA (1969) The fine structure of the heart valves in the chicken. I. Mitral valve. Am J Anat 125:147–153

Mizel SB, Dayer JM, Krane SM, Mergenhagen SE (1981) Stimulation of rheumatoid synovial cell collagenase and prostaglandin by partially purified lymphocyte-activating factor (interleukin-1). Proc Natl Acad Sci USA 78:2474–2478

Modavi S, Isseroff H (1984) Fasciola hepatica: collagen deposition and other histopathology in the rat host's bile duct caused by the parasite and by proline infusion. Exp Parasitol 58:239–244

Modesti A, Kalebic T, Scarpa S, Togo S, Grotendorst G, Liotta LA, Triche TJ (1984) Type V collagen in human amnion is a 12 nm fibrillar component of the pericellular interstitium. Eur J Cell Biol 35:246–255

Modrak JB, Langner RO (1980) Possible relationship of cholesterol accumulation and collagen synthesis in rabbit aortic tissues. Atherosclerosis 37:211–218

Mogensen CE, Osterby R, Gundersen HJG (1979) Early functional and morphological vascular renal consequences of the diabetic state. Diabetologia 17:71–76

Monnier VM, Cerami A (1982) Non-enzymatic glycosylation and browning of proteins in diabetes. Clin Endocrinol Metab 10:432–452

Monnier VM, Kohn RR, Cerami A (1984) Accelerated age-related browning of human collagen in diabetes mellitus. Proc Natl Acad Sci USA 81:583–587

Monnier VM, Sell DR, Abdul-Karim FW, Emancipator SN (1988) Collagen browning and crosslinking are increased in chronic experiemental hyperglycemia. Relevance to diabetes and aging. Diabetes 37:867–872

Montfort I, Pérez-Tamayo R (1962) The muscle-collagen ratio in normal and hypertrophic human hearts. Lab Invest 11:463–470

Mooppan MMU, Wax SH, Kim H, Wang JC, Tobin MS (1980) Urinary hydroxyproline excretion as a marker of osseous metastasis in carcinoma of the prostate. J Urol 123:694–696

Morales TI, Woessner JF Jr, Howell DS, Marsh JM, LeMaire WJ (1978a) A microassay for the direct demonstration of collagenolytic activity in Graafian follicles of the rat. Biochim Biophys Acta 524:428–434

Morales TI, Woessner JF Jr, Marsh JM, LeMaire WJ (1978b) Collagenase in Graafian follicles. 60th Meeting of the Endocrine Society, abstracts, p 501

Morales TI, Woessner JF Jr, Marsh JM, LeMaire WJ (1983) Collagen, collagenase and collagenolytic activity in rat Graafian follicles during follicular growth and ovulation. Biochim Biophys Acta 756:119–122

Morgan CF (1963) A study of estrogenic action on the collagen, hexosamine and nitrogen content of skin, uterus and vagina. Endocrinology 73:11–16

Mori Y, Bashey RI, Angrist A (1967) The in vitro biosynthesis of collagen in bovine heart valves. Biochem Med 1:295–303

Moriguchi T, Fujimoto D (1978) Age-related changes in the content of the collagen crosslink, pyridinoline. J Biochem 84:933–935

Moro L, Modricky C, Rovis L, de Bernard B (1988a) Determination of galactosyl hydroxylysine in urine as a means for the identification of osteoporotic women. Bone Miner 3:271–276

Moro L, Mucelli SP, Gazzarrini C, Modricky C, Marotti F, de Bernard B (1988b) Urinary $\beta$-1-galactosyl-0-hydroxylysine as a marker of collagen turnover of bone. Calcif Tissue Int 42:87–90

Morrione TG (1947) Quantitative study of collagen content in experimental cirrhosis. J Exp Med 85:217–226

Morrione TG (1949) Factors influencing collagen content in experimental cirrhosis. Am J Pathol 25:273–285

Morrione TG, Levine J (1967) Collagenolytic activity and collagen resorption in experimental cirrhosis. Arch Pathol 84:59–63

Morris NP, Bächinger HP (1987) Type XI collagen is a heterotrimer with the composition ($1\alpha 2\alpha 3\alpha$) retaining non-triple-helical domains. J Biol Chem 262:11345–11350

Morris SC, McClain PE (1972) Heterogeneity in the cyanogen bromide peptides from striated muscle and heart valve collagen. Biochem Biophys Res Commun 47:27–34

Morton LF, Barnes MJ (1983) Analysis of the collagenous products synthesized in vitro by chick blood vessels: alterations during development and effects of serum on the pattern of synthesis. Artery 11:361–383

Morton LF, Fitzsimmons CM, Rauterberg J, Barnes MJ (1987) Platelet-reactive sites in collagen. Collagens I and III possess different aggregatory sites. Biochem J 248:483–487

Morton LF, Peachey AR, Barnes MJ (1989) Platelet-reactive sites in collagens type I and type III. Evidence for separate adhesion and aggregatory sites. Biochem J 258:157–163

Motohashi T, Okamura H, Okazaki T, Morikawa H, Nishimura T (1977) A study on ovarian lysosomal enzyme activities during ovulation in rodents. Proceedings of the 2nd Asian congress of obstetrics and gynecology, Bankok, pp 860–872

Motz W, Strauer BE (1989) Left ventricular function and collagen content after regression of hypertensive hypertrophy. Hypertension 13:43–50

Müller PK, Nerlich AG, Böhm J, Phan-Than L, Krieg T (1986) Feedback regulation of collagen synthesis. In: Mecham RP (ed) Regulation of matrix accumulation. Academic, Orlando, pp 100–118

Müller KP, Neglich AG, Kunze D, Müller PK (1987) Studies on collagen metabolism in the Marfan syndrome. Eur J Clin Invest 17:218–225

Munksgaard EC (1979) Collagen in dentin. J Biol Buccale 7:131–135

Munksgaard EC, Rhodes M, Mayne R, Butler WT (1978) Collagen synthesis and secretion by rat incisor odontoblasts in organ culture. Eur J Biochem 82:609–617

Muona P, Jaakkola S, Salonen V, Peltonen J (1989) Diabetes induces the formation of large diameter collagen fibrils in the sciatic nerves of BB rats. Matrix 9:62–67

Murata K, Motayama T, Kotake C (1986) Collagen types in various layers of the human aorta and their changes with the atherosclerotic process. Atherosclerosis 60:251–262

Murawaki Y, Hirayama C (1980) Hepatic collagenolytic cathepsin in patients with chronic liver disease. Clin Chim Acta 108:121–128

Murdoch JH, Walker BA, Halpern BL, Juzma JM, McKusick VA (1972) Life expectancy and causes of death in the Marfan syndrome. N Engl J Med 286:804–808

Murphy G, Cartwright EC, Sellers A, Reynolds JJ (1977) The detection and characterisation of collagenase inhibitors from rabbit tissues in culture. Biochim Biophys Acta 483:493–498

Murphy G, Cawston TE, Reynolds JJ (1981) An inhibitor of collagenase from human amniotic fluid. Biochem J 195:167–170

Murphy G, Nagase H, Brinckerhoff CE (1988) Relationship of procollagenase activator, stromelysin and matrix metalloproteinase 3. Coll Relat Res 8:389–391

Murray LW, Tanzer ML (1983) Characterization of a large fragment from annelid cuticle collagen and its relationship to the intact molecule. Coll Relat Res 3:445–458

Murray LW, Tanzer ML (1985) The collagens of the Annelida. In: Bairati A, Garrone R (eds) Biology of invertebrate and lower vertebrate collagens. Plenum, New York, pp 243–258

Murray JC, Lindberg KA, Pinnell SR (1977) In vitro inhibition of chick embryo lysyl hydroxylase by homogentisic acid. A proposed connective tissue defect in alkaptonuria. J Clin Invest 59:1071–1079

Murray JC, Fraser DR, Levene CI (1978) The effect of pyridoxine deficiency on lysyl oxidase activity in the chick. Exp Mol Pathol 28:301–308

Murray LW, Bautista J, James P, Rimoin DL (1987) Normal type II collagen is not detected in cartilage of patients with achondrogenesis II-hypochondrogenesis. Am J Hum Genet 41 [Suppl]:A12

Myara I, Cosson C (1988) Fibrose hépatique: modifications du collagène et marqueurs sérique lies a son metabolisme. Gastroenterol Clin Biol 12:99–106

Myara I, Myara A, Mangeot M, Fabre M, Charpentier C, Lemonnier A (1984) Plasma prolidase activity: a possible index of collagen catabolism in chronic liver disease. Clin Chem 30:211–215

Myers JC, Emanuel BS (1987) Chromosomal localization of human collagen genes. Coll Relat Res 7:149–159

Myllylä R, Kuutti-Savolainen ER, Kivirikko KI (1978) The role of ascorbate in the prolyl hydroxylase reaction. Biochem Biophys Res Commun 83:411–448

Myllylä R, Anttinen H, Kivirikko KI (1979) Metal activation of galactosyl-hydroxylysyl glucotransferase, an intracellular enzyme of collagen biosynthesis. Eur J Biochem 101:261–269

Na GC (1988) UV spectroscopic characterization of type I collagen. Coll Relat Res 8:315–330

Nachemson AL, Sahlstrand T (1977) Etiologic factors in adolescent idiopathic scoliosis. Spine 2:176–184

Nagai M (1989) The effects of prostaglandin $E_2$ on DNA and collagen synthesis in osteoblasts in vitro. Calcif Tissue Int 44:411–420

Nagai Y, Sakakibara S, Noda H, Akabori S (1960) Hydrolysis of synthetic peptide by collagenase. Biochim Biophys Acta 37:567–569

Nagai Y, Sato M, Sasaki M (1982) Effect of cadmium administration upon urinary excretion of hydroxylysine and hydroxyproline in the rat. Toxicol Appl Pharm 63:188–193

Nagai Y, Sunada H, Sano J, Onodera S, Aroi K, Konomi H, Minamoto T, Hata R, Hori H, Nakanishi I, Kitaoka H, Sakamoto G (1985) Biochemical and immunohistochemical studies on the scirrhous carcinoma of human stomach. Ann NY Acad Sci 460:321–332

Nagase H, Jackson RC, Brinckerhoff CE, Vater CA, Harris ED Jr (1981) A precursor form of latent collagenase produced in a cell-free system with mRNA from rabbit synovial cells. J Biol Chem 256:11951–11954

Nagayama M, Sakamoto S, Sakamoto M (1984) Mouse bone collagenase inhibitor: purification and partial characterization of the inhibitor from mouse calvaria cultures. Arch Biochem Biophys 228:653–659

Nagent de Deuxchaisnes C, Krane SM (1964) Paget's disease of bone: clinical and metabolic observations. Medicine (Baltimore) 43:233–266

Nageotte J (1927) Coagulation fibrillaire in vitro due collagène dissous dans un acide dilué. Acad Sci 184:115–126

Nagle RB, Bulger RE (1978) Unilateral obstructive nephropathy in the rabbit. II. Late morphologic changes. Lab Invest 38:270–278

Nagle RB, Bulger RE, Cutler RE, Jervis HR, Benditt EP (1973) Unilateral obstructive nephropathy in the rabbit. I. Early morphologic, physiologic and histochemical changes. Lab Invest 28:456–467

Nagle RB, Johnson ME, Jervis HR (1976) Proliferation of renal interstitial cells following injury induced by ureteral obstruction. Lab Invest 35:18–22

Nagy IZ, Tóth VN, Verzár F (1974) High-resolution electron microscopy of thermal collagen denaturation in tail tendons of young, adult and old rats. Connect Tissue Res 2:265–272

Naito Y, Kino I, Horiuchi K, Fujimoto D (1984) Promotion of collagen production by human fibroblasts with gastric cancer cells in vitro. Virchows Arch [A]46:145–154

Nakagawa H, Tsurufuji S (1972) Action of betamethasone disodium phosphate on the metabolism of collagen and noncollagen protein in rat carrageenin granuloma. Biochem Pharmacol 21:839–846

Nakagawa H, Fukuhara M, Tsurufuji S (1971) Effect of a single injection of betamethasone disodium phosphate on the synthesis of collagen and noncollagen protein of carrageenin granuloma in rats. Biochem Pharmacol 20:2253–2261

Nakagawa H, Aihara H, Tsurufuji S (1977) The role of cathepsin B1 in the collagen breakdown of carrageenin granuloma in rats. J Biochem (Tokyo) 81:801–804

Nakajima T, Oda H, Kusumoto S, Nogami H (1980) Biological effects of nitrogen dioxide and nitric oxide. In: Lee SD (ed) Nitrogen oxides and their effects on health. Ann Arbor Science, Ann Arbor, pp 121–141

Nakano T, Thompson JR, Aherne FX, Christian RG (1984) Non-reducible crosslink in normal and degenerative joint cartilage. Biomed Res 5:371–374

Nakao K, Mao P, Ghidoni J, Angrist AA (1966) An electron microscopic study of the aging process in the rat heart valve. J Gerontol 21:72–79

Nakayasu K, Tanaka M, Konomi H, Hayashi T (1986) Distribution of types I, II, III, IV and V collagen in normal and keratoconus corneas Ophthalmic Res 18:1–8

Nandi PK, Grant ME, Robinson DR (1985) Destabilization of collagen structure by amides and detergents in solution. Int J Pept Protein Res 25:206–212

Näntö-Salonen K, Penttinen R (1982) Metabolism of collagen in aspartylglycosaminuria: decreased synthesis by cultured fibroblasts. J Inherited Metab Dis 5:197–203

Näntö-Salonen K, Autio S, Käro E, Kivamäki T, Koskela SL, Näntö V, Penttinen R (1984a) Metabolism of collagen in aspartylglycosaminuria: urinary excretion of hydroxyproline. J Interited Metab Dis 7:117–121

Nänto-Salonen K, Pelliniemi L, Autio S, Kivamäki T, Rapola J, Penttinen R (1984b) Disturbed collagen fibril formation in aspartylglycosaminuria: altered ultrastructure in a glycoprotein storage disorder. Lab Invest 51:464–468

Nänto-Salonen K, Larjava H, Aalto M, Kivamäki T (1985) Urinary glycosaminoglycans in aspartylglycosaminuria: evidence for disturbed proteoglycan metabolism. Clin Chim Acta 146:111–118

Nänto-Salonen K, Larjava H, Saamanen AM, Heino J, Penttinen R, Pelliniemi LJ, Tammi M (1987) Abnormal dermal proteoglycan in aspartylglycosaminuria: a possible mechanism for ultrastructural changes of collagen fibrils in a glycoprotein storage disorder. Connect Tissue Res 16:367–376

Narayan SN, Page R (1983) Connective tissues of the periodontium. Coll Relat Res 3:33–64

Narayan AS, Page RC (1976) Biochemical characterization of collagens synthesized by fibroblasts derived from normal and diseased human gingiva. J Biol Chem 251:5464–5469

Narayanan AS, Meyers DF, Page RC, Welgus HG (1984) Action of mammalian collagenases on type I trimer collagen. Coll Relat Res 4:289–296

Naylor A (1970) The structure and function of the intervertebral disc. Orthopaedics (Oxf) 3:7–22

Neblock DS, Berg RA (1987) Intracellular degradation as a modulator of collagen production. In: Uitto J, Perejda AJ (eds) Connective tissue diseases. Molecular pathology of the extracellular matrix. Dekker, New York, pp 233–246

Nefussi JR, Baron R (1985) $PGE_2$ stimulates both resorption and formation of bone in vitro: differential responses of the periosteum and the endosteum in fetal rat long bone cultures. Anat Rec 211:9–16

Neldner KH, Jones JD, Winkelmann RK (1966) Scleroderma: dermal amino acid composition with particular reference to hydroxyproline. Proc Soc Exp Biol Med 122:39–45

Nelson DL, King RA (1981) Ehlers-Danlos syndrome type VIII. J Am Acad Dermatol 5:297–303

Németh-Csóka M, Mészáros T (1984) The importance of minor collagenous chains in the articular cartilage. Acta Biol Hung 35:163–180

Németh-Csóka M (1974) The effect of acid mucopolysaccharides on the activation energy of collagen fibril-formation. Exp Pathol 9:256–262

Németh-Csòka M, Kovácsay A (1979a) The effect of glycosaminoglycans on the in vitro fibril formation of collagen type I and type III. Exp Pathol 17:82–87

Németh-Csóka M, Kovácsay A (1979b) The effect of glycosaminoglycans on the intramolecular binding of collagen. Actal Biol Hung 30:303–308

Némethy G, Scheraga HA (1986) Stabilization of collagen fibrils by hydroxyproline. Biochemistry 25:3184–3188

Nerlich AG, Pöschl E, Voss T, Müller PK (1986) Biosynthesis of collagen and its control. In: Kühn K, Krieg T (eds) Connective tissue: biological and clinical aspects. Karger, Basel, pp 70–90

Nesterova VI, Barabanova AB, Petuskhov VN, Krylov VM (1980) Giperoksiprolinuria pri ostroy sotshetannoy radiacionnoy travme (Increased excretion of hydroxyproline in urine in acute incidental radiational injury). Med Radiol 25(11):50–53 (in Russian)

Nethery A, O'Grady RL (1989) Identification of a metalloproteinase co-purifying with rat tumour collagenase and the characteristics of fragments of both enzymes. Biochim Biophys Acta 994:149–160

Neuberger A, Perrone JC, Slack HGB (1951) The relative metabolic inertia of tendon collagne in the rat. Biochem J 49:199–204

Newman RA, Cutroneo KR (1978) Glucocorticoids selectively decrease the synthesis of hydroxylated collagen peptides. Mol Pharmacol 14:185–198

Newsome Da, Kenyon KR (1973) Collagen production in vitro by the retinal pigmented epithelium of the chick embryo. Dev Biol 32:387–400

Newsome DA, Gross J, Hassell JR (1982) Human corneal stroma contains three distinct collagens. Invest Ophthalmol Vis Sci 22:376–381

Niebes P (1977) Action of (+)-cathechin on connective tissue. Proceeding of the 5th Hungarian bioflavonoid symposium, Mátrafüred. Akadémiai Kiado, Budapest, pp 347–361

Niedermüller H, Skalicky M, Hofecker G, Kment A (1977) Investigations on the kinetics of collagen metabolism in young and old rats. Exp Gerontol 12:159–168

Niemelä O, Risteli L, Sotaniemi EA, Risteli J (1983) Aminoterminal propeptide of type III procollagen in serum in alcoholic liver disease. Gastroenterology 85:254–259

Niemelä O, Risteli L, Sotaniemi EA, Risteli J (1985) Type IV collagen and laminin-related antigens in human serum in alcoholic liver disease. Eur J Clin Invest 15:132–137

Niewiarowski S, Stuart RK, Thomas DP (1966) Activation of intravascular coagulation by collagen. Proc Soc Exp Biol Med 123:196–202

Nimni ME (1965) Accumulation of a collagen precursor in the skin of penicillamine-treated rats. Biochem Biophys Acta 111:576–581

Nimni ME (1968) A defect in the intramolecular and intermolecular crosslinks of collagen caused by penicillamine. I. Metabolic and functional abnormalities in soft tissues. J Biol Chem 243:1457–1466

Nimni ME (1974) Collagen: its structure and function in normal and pathological connective tissues. Semin Arthritis Rheum 4:95–150

Nimni ME (ed) (1988) Collagen, vol I–III. CRC Press, Boca Raton

Nimni ME, Bavetta LA (1965) Collagen defect induced by penicillamine. Science 150:905–907

Nimni ME, Deshmukh K (1973) Differences in collagen metabolism between normal and osteoarthritic human articular cartilage. Science 181:751–752

Nimni ME, Harkness RD (1988) Molecular structures and functions of collagen. In: Nimni ME (ed) Collagen, vol. I. CRC Press, Boca Raton, pp 1–78

Nimni ME, Deshmukh K, Bavetta LA (1967) Turnover and age distribution of a collagen fraction extractable from rat skin by mercaptoethylamine. Arch Biochem Biophys 122:292–298

Nimni ME, Deshmukh K, Gerth N (1972) Tissue lysyl oxidase activity and the nature of the collagen defect induced by penicillamine. Nature (New Biol) 240:220–222

Nissen R, Cardinale GJ, Udenfriend S (1978) Increased turnover of arterial collagen in hypertensive rats. Proc Natl Acad Sci USA 75:451–453

Niyibizi C, Eyre DR (1989) Identification of the cartilage $\alpha 1$(XI) chain in type V collagen from bovine bone. FEBS Lett 242:314–318

Niyibizi C, Fietzek PP, van der Rest M (1984) Human placenta type V collagens. Evidence for the existence of an alpha-1(V), alpha-2(V), alpha-3(V) collagen molecule. J Biol Chem 259:4170–4174

Nordwall A, Waldeström J (1976) Metachromasia of fibroblasts from patients with idiopathic scoliosis. Spine 1:97–98

Nordwig A (1971) Collagenolytic enzymes. Adv Enzymol 34:155–205

Nordwig A, Hayduk U (1969) Invertebrate collagens: isolation, characterization and phylogenetic aspects. J Mol Biol 44:161–172

Nordwig A, Nowack H, Hieber-Rogall E (1973) Sea anemone collagen: further evidence for the existence of only one $\alpha$-chain type. J Mol Evol 2:175–179

Norling B, Glimelius B, Westermark B, Watersin A (1978) A chondroitin sulfate proteoglycans from human cultured glial cells, aggregated with hyaluronic acid. Biochem Biophys Res Commun 84:914

Noro A, Kimata K, Oike A, Shinomura T, Maeda N, Yano S, Takahashi N, Suzuki S (1983) Isolation and characterization of a third proteoglycan (PG-Lt) from chick embryo cartilage which contains disulfide-bounded collagenous polypeptide. J Biol Chem 258:9323–9331

Norström A, Bryman I, Lindblom B, Christensen NJ (1985) Effects of 9-deoxo-16,16-dimethyl-9-methylene $PGE_2$ on muscle contractile activity and collagen synthesis in the human cervix. Prostaglandins 29:337–346

Norton WL, Nardo JM (1970) Vascular disease in progressive systemic sclerosis. Ann Intern Med 73:317–324

Norton WL, Hurd ER, Lewis DC, Ziff M (1968) Evidence of microvascular injury in scleroderma and sytemic lupus erythematosus: quantitative study of the microvascular bed. J Lab Clin Med 71:919–933

Notley RG (1968) Electron microscopy of the upper ureter and the pelvi-ureteric junction. Br J Urol 40:37–52

Notley RG (1970) The musculature of the human ureter. Br J Urol 42:724–727

Notley RG (1971) The structural basis for normal and abnormal ureteric motility. Ann R Col Surg Engl 49:250–267

Notley RG (1972) Electron microscopy of the primary obstructive megaureter. Br J Urol 44:229–234

Novikov PV, Shilov AV, Daikhin EI, Storozhev VL (1980) Potzetznaya ekskreciya oksiliz-inglikozidov u detey s rehkitopodobnimi zabolevaniyami (Urinary excretion of hydroxylysine glycosides in children with rachitis-like disorders). Vopr Okhr Materin Det 25(9):53–57 (in Russian)

Novotny GEK, Pau H (1984) Myofibroblast-like cells in human anterior capsular cataract. Virchows Arch [A] 404:393–401

Novotny GEK, Pau H, Arnold G (1989) Organization of collagen and other extracellular material in anterior capsular cataract. Anat Anz 168:127–133

Nowack H, Hahn E, Timpl R (1976) Requirement for T cells in the antibody response of mice to calf skin collagen. Immunology 30:29–32

Numata Y, Takei T, Hayakawa T (1981) Hydralazine as an inhibitor of lysyl oxidase activity. Biochem Pharmacol 30:3125–3126

Öbrink B, Laurent TC, Carlsson B (1975) The binding of chondroitin sulphate to collagen. FEBS Lett 56:166–169

Oda Y, Kawahara E, Minamoto T, Ueda Y, Ikeda K, Nagai Y, Nakanishi I (1988) Immunohistochemical studies on the tissue localization of collagen types I, III, IV, V and VI in schwannomas. Correlation with ultrastructural features of the extracellular matrix. Virchows Arch [B] 56:153–163

Odermatt E, Risteli J, Van Delden V, Timpl R (1983) Structural diversity and domain composition of a unique collagenous fragment (intima collagen) obtained from human placenta. Biochem J 211:295–303

Oegema T Jr, Laidlaw J, Hascall VC, Dziewiatkowski DD (1975) The effect of proteoglycans on the formation of fibrils from collagen solutions. Arch Biochem Biophys 170:698–709

O'Hara PJ, Read WK, Romane WM, Bridges CH (1970) A collagenous tissue dysplasia of calves. Lab Invest 23:307–314

Ohno T, Tsurufuji S (1970) Mode of action of hydrocortisone on the protein metabolism in rat carrageenin granuloma. Biochem Pharmacol 19:1–8

Ohta A, Louie JS, Uitto J (1987a) Retinoid modulation of collagenase production by adherent human mononuclear cells in culture. Ann Rheum Dis 46:357–362

Ohta K, Kasahara F, Ishimaru T, Wada Y, Kanamaru T, Okazaki H (1987b) Structures of fibrostatins, new inhibitors of prolyl hydroxylase. J Antibiot (Tokyo) 40:1239–1248

Ohtani O (1988) Three-dimensional organization of the collagen fibrillar framework of the human and rat livers. Arch Histol Cytol 51:473–488

Ohuzi K, Tsurufuji S (1972) Protocollagen proline hydroxylase in isolated rat liver cells. Biochim Biophys Acta 258:731–740

Ohyama H, Hashimoto K (1977) Collagenase of human skin basal cell epithelioma. J Biochem 82:175–184

Oikarinen A (1977) Effect of betamethasone-17-valerate on synthesis of collagen and prolyl hydroxylase activity in chick-embryo tendon cells. Biochem Pharmacol 26:875–879

Oikarinen A, Hannuksela M (1980) Effect of hydrocortisone-17-butyrate, hydrocortisone and clobetasol-17-propionate on prolyl hydroxylase activity in human skin. Arch Dermatol Res 267:79–82

Oikarinen J, Ryhänen L (1981) Cortisol decreases the concentration of translatable type-I procollagen mRNA species in the developing chick-embryo calvaria. Biochem J 198:519–524

Oikarinen J, Uitto J (1987) Molecular mechanisms of glucocorticoid action on connective tissue metbolism. In: Uitto J, Perejda AJ (eds) Connective tissue disease. Molecular pathology of the extracellular matrix. Dekker, New York, pp 385–398

Oikarinen J, Pihlajaniemi T, Hämäläinen L, Kivirikko KI (1983a) Cortisol decreases the cellular concentration of translatable procollagen mRNA species in cultured human skin fibroblasts. Biochim Biophys Acta 741:297–302

Oikarinen AI, Meeker C, Oikarinen H, Tan E, Uitto J (1983b) Glucocorticoid receptors in human skin fibroblasts in culture: correlation of receptor density with changes in collagen metabolism. J Invest Dermatol 80:325–326

Oikarinen AI, Uitto J, Oikarinen J (1986) Glucocorticoid action on connective tissue: from molecular mechanisms to clinical practice. Med Biol 64:221–230

Oikarinen AI, Mörtenhumer M, Kallioinen M, Savolainen ER (1987) Necrobiosis lipoidica: ultrastructural and biochemical demonstration of a collagen defect. J Invest Dermatol 88:227–232

Oikarinen AI, Vuorio EI, Zaragoza EJ, Palotie A, Chu ML, Uitto J (1988) Modulation of collagen metabolism by glucocorticoids. Receptor-mediated effects of dexamethasone on collagen-biosynthesis in chick embryo fibroblasts and chondriocytes. Biochem Pharmacol 37:1451–1462

Oikarinen AI, Kinnunen T, Kallioinen M (1989) Biochemical and immunohistochemical comparison of collagen in granuloma annulare and skin sarcoidosis. Acta Derm Venereol (Stockh) 69:277–283

Ojima Y, Ito A, Nagas H, Mori Y (1989) Calmodulin regulates the interleukin 1-induced procollagenase production in human uterine cervical fibroblasts. Biochim Biophys Acta 1011:61–66

Oka M, Angrist AA (1961) Fibrous thickening with billowing sails distortion of aging heart valve. Proc NY State Assoc Public Health Lab 16:21–26

Oka M, Shirota A, Angrist AA (1966) Experimental endocarditis. Arch Pathol 82:85–92

Okada K, Kikuchi Y, Kawashiri Y, Hiramoto M (1972) Syntheses and enzymatic hydroxylation of protocollagen model of peptides containing glutamyl or leucyl residue. FEBS Lett 28:226–230

Okada Y, Nagase H, Harris ED Jr (1986) A metalloproteinase from human rheumatoid synovial fibroblasts that digests connective tissue matrix components. Purification and characterization. J Biol Chem 261:14245–14255

Okamura H, Takenaka A, Yajima Y, Nishimura T (1980) Ovulatory changes in the wall at the apex of the human Graafian follicle. J Reprod Fertil 58:153–155

Okamura H, Fukumoto M, Mori T (1985) Prostaglandin-mediated changes of vasculature and collagen degradation/synthesis in the follicle wall during ovulation. Adv Prostaglandin Thromboxane Leukotriene Res 15:597–599

Okazaki I, Maruyama K (1974) Collagenase activity in experimental hepatic fibrosis. Nature 252:49–50

Okazaki I, Oda M, Maruyama K, Funatsu K, Matsuzaki S, Kamegaya K, Tsuchiya M (1974) Mechanism of collagen resorption in experimental hepatic fibrosis, with special reference to the activity of lysosomal enzymes. Biochem Exp Biol 11:15–28

Okazaki H, Ohta K, Kanamaru T, Ishimaru T, Kishi T (1981) A potent prolyl hydroxylase inhibitor, P-1894B, produced by a strain of Streptomyces. J Antibiot (Tokyo) 34:1355–1356

Okazaki I, Maruuama K, Okuno F, Suzuki H, Aoyagi K, Sera Y (1983) Serum type III procollagen peptide in patients with pneumoconiosis. J Univ Occup Environ Health 5:461–467

Okazaki R, Matsuoka K, Horiuchi A, Maruyama K, Okazaki I (1988) Assays of serum laminin and type III procollagen peptide for monitoring the clinical course of diabetic microangiopathy. Diabet Res Clin Pract 5:163–170

Oken DE, Boucek RJ (1957) Quantitation of collagen in human myocardium. Circ Res 5:357–361

Okuno F, Marayama K, Okazaki I, Arai M, Suzuki H (1985) Serum type III procollagen peptide diagnosis of lung fibrosis due to silicosis and bleomycin toxicity. Sanyo Ika Daigaku Zaisshi 7:291–297

Olczyk K, Kucharz EJ, Wieczorek M, Szczębara M, Sonecki P (1990) Collagen and elastin in the liver of rats intoxicated with mercuric chloride. Arh Hig Rada Toksikol 41:1–6

Old LJ (1987) Polypeptide mediator network. Nature 326:330–332

Oldberg Å, Antonsson P, Lindblom K, Heinegård D (1989) A collagen-binding 59-kd protein (fibromodulin) is structurally related to the small interstitial proteoglycans PG-S1 and PG-S2 (decorin). EMBO J 8:2601–2604

Olden K, Parent JW, White SL (1982) Carbohydrate moieties of glycoproteins. A re-evaluation of their function. Biochim Biophys Acta 650:209–232

Olmarker K, Rydevik B, Dahlin LB, Danielsen N, Nordborg C (1987) Effects of epidural and intrathecal application of collagenase in the lumbar spine: an experimental study in rabbits. Spine 12:477–482

Olsen BR (1965) Electron microscope studies on collagen. IV. Structure of vitrosin fibrils and interaction properties of vitrosin molecules. J Ultrastruct Res 13:172–178

Olsen AS, Prockop DJ (1989) Transcription of human type I collagen genes. Variation in the relative rates of transcription of the Pro $\alpha 1$ and Pro $\alpha 2$ genes. Matrix 9:73–81
Olsen BR, Alper R, Kefalides NA (1973a) Structural characterization of a soluble fraction from lens-capsule basement membrane. Eur J Biochem 38:220–228
Olsen BR, Berg RA, Kishida Y, Prockop DJ (1973b) Collagen synthesis. Localization of prolyl hydroxylase in tendon cells detected with ferritin-labeled antibodies. Science 182:825–827
Olsen BR, Berg RA, Kivirikko KI, Prockop DJ (1973c) Structure of protocollagen proline hydroxylase from chick embryos. Eur J Biochem 35:137–147
Olsen RF, Wangsness PJ, Patton WH, Martin RJ (1980) Relationship of serum somatomedin-like activity and fibroblast proliferative activity with age and growth in the rat. Growth 44:19–28
Ooshima A, Fuller G, Cardinale G, Spector S, Udenfriend S (1975) Collagen biosynthesis in blood vessels of brain and other tissues of the hypertensive rat. Science 190:898–900
Orbison JL, McCrary C, Callahan M (1965) Urinary excretion and serum composition of lathyritic rats. Arch Pathol 79:292–298
Orekhovich VN (1952) Prokollageny ikh khimicheskii sostav, svoistva i biologicheskaia rol (Procollagens: their chemical composition, properties and biological role). Izd Akademii Nauk SSSR, Moskva (in Russian)
Orlowski WA (1974) Analysis of collagen, glycoproteins and acid mucpolysaccharides in the bovine and porcine dental pulp. Arch Oral Biol 19:255–258
Orlowski WA (1976) The incorporation of $^3$H-proline into the collagen of the periodontium of a rat. J Periodont Res 11:96–102
Orlowski WA, Doyle JL (1976) Collagen metabolism in the pulps of rat teeth. Arch Oral Biol 21:391–392
Osterud B, Rapoport SI, Lavine KK (1977) Factor V activity of platelets: evidence for an activated factor V molecule and for a platelet activator. Blood 49:819–834
Otaka Y (ed) (1974) Biochemistry and pathology of connective tissue. Thieme, Stuttgart
Otouka K, Murota S, Mori Y (1976) Stimulatory effect of bleomycin on the synthesis of acidic glycosaminoglycans in cultured fibroblasts derived from rat carrageenan granuloma. Biochem Biophys Acta 444:359–368
Ouazana R (1985) The collagen of Aschelminthes. In: Bairati A, Garrone R (eds) Biology of invertebrate and lower vertebrate collagens. Plenum, New York, pp 217–236
Overall CM, Wrana JL, Sodek J (1989) Independent regulation of collagenase, 72-kDa progelatinase, and metalloendoproteinase inhibitor expression in human fibroblasts by transforming growth factor-$\beta$. J Biol Chem 264:1860–1869
Owen M (1970) The origin of bone cells. Int Rev Cytol 28:215–238
Owen M (1980) The origin of bone cells in the postnatal organism. Arthritis Rheum 23:1073–1079
Owen M (1985) Lineage of osteogenic cells and their relationship to the stromal system. Bone Miner Res 3:28–46
Oxlund H (1986) Relationships between the biomechanical properties, composition and molecular structure of connective tissues. Connect Tissue Res 15:65–72
Pääkö P, Sormunen R, Risteli L,Risteli J, Ala-Kokko L, Ryhänen L (1989) Malotilate prevents accumulation of type III pN-collagen, type IV collagen, and laminin in carbon tetrachloride-induced pulmonary fibrosis in rats. Am Rev Respir Dis 139:1105–1111
Pacificic M, Iozzo RV (1988) Remodeling of the rough endoplasmic reticulum during stimulation of procollagen secretion by ascorbic acid in cultured chondrocytes. A biochemical and morphological study. J Biol Chem 263:2483–2492
Page RC, Narayanan AS, Schroeder HE (1980) Connective tissue composition and collagen synthesis in diseased and normal gingiva of adult dogs with spontaneous periodontitis. Arch Oral Biol 25:727–736
Paglia L, Wolczek J, Diaz De Leon L, Martin GR, Hörlein D, Müller P (1979) Inhibition of procollagen cell-free synthesis by aminoterminal extension peptides. Biochemistry 18:5030–5034
Paglia LM, Wiestner M, Duchene M, Ouellette LA, Horlain D, Martin GR, Müller PK (1981) Effects of procollagen peptides on the translation of type II collagen messenger ribonucleic acid and on collagen biosynthesis in condrocytes. Biochemistry 20:3523–3527
Palade PE, Farquhar MG (1965) A special fibril of the dermis. J Cell Biol 27:215–224

Pałka J, Galewska Z (1985) Zmiany relacji ilościowych kolagenu typu I i III w skórze szczurów traktowanych niektórymi lekami przeciwzapalnymi (Quantitative changes in type I and type III collagens in skin of rats treated with some anti-inflammatory drugs). XXI Zjazd Polskiego Towarzystwa Biochemicznego (Proceedings of the 21st Meeting of the Polish Biochemical Society), Kraków, abstracts, p 142

Palmer RM, Robins SP, Lobley GE (1980) Measurement of the synthesis rates of collagens and total protein in rabbit muscle. Biochem J 192:631–636

Palotie A, Tryggvason K, Peltonen L, Seppä H (1983) Components of subendothelial aorta basement membrane. Immunohistochemical localization and role in cell attachement. Lab Invest 49:362–370

Palotie A, Ott J, Elima K, Cheah K, Väisänen P, Ryhänen L, Vikkula M, Vuorio E, Peltonen L (1989) Predisposition to familial osteoarthrosis linked to type II collagen gene. Lancet 1:924–927

Panduro A, Shalaby F, Biempica L, Shafritz DA (1988) Changes in ablumin, $\alpha$-fetoprotein and collagen gene transcription in $CCl_4$-induced hepatic fibrosis. Hepatology 8:259–266

Pänkäläinen M, Aro H, Simons K, Kivirikko KI (1970) Procollagen proline hydroxylase: molecular weight, subunits, and isoelectric point. Biochim Biophys Acta 221:559–565

Parke WW, Schiff DCM (1971) The applied anatomy of the intervertebral disc. Orthop Clin 2:309–324

Parker MI, Gevers W (1984) Demethylation of the type I procollagen genes in transformed fibroblasts treated with 5-azacytidine. Biochem Biophys Res Commun 124:236–243

Parry DAD (1988) The molecular and fibrillar structure of collagen and its relationship to the mechanical properties of connective tissue. Biophys Chem 29:195–209

Parry DAD, Craig AS, Barnes GRG (1978) Tendon and ligament from the horse: an ultrastructural study of collagen fibrils and elastic fibres as a function of age. Proc R Soc London 203B:293–303

Parry DAD, Flint MH, Gillard GC, Craig AS (1982) A role for glycosaminoglycans in the development of collagen fibrils. FEBS Lett 149:1–7

Parsons TJ, Haycraft DL, Hoak JC, Sage H (1983) Diminished platelet adherence to type V collagen. Artherosclerosis 3:589–598

Pascalis W, Pia G, Aresu G, Rosetti L, Cherchi R (1986) Compartamento del rame e del peptide amino-terminale del procollageno III nel liquido di lavaggio broncoalveolare e nei sangue di soggetti affetti da silicosi. In: Atti del 40^e^ Congresso Nazionale della Societa Italiana di Medicina del Lavoro e Igiene Industriale. Monduzzi, Bologna, pp 769–778

Pasquali-Ronchetti I, Fornieri C, Castellani I, Bressan GM, Volpin D (1981) Alterations of the connective tissue components induced by beta-aminopropionitrile. Exp Mol Pathol 35:42–56

Pasternak RD, Hubbs SJ, Caccese RG, Marks J, Conaty M, DiPasquale G (1986) Interleukin-1 stimulates the secretion of proteoglycan and collagen-degrading proteases by rabbit articular chondrocytes. Clin Immunol Immunopathol 41:351–367

Pasternak RD, Hubbs SJ, Caccese RG, Marks RL, Conaty JM, DiPasquale G (1987) Interleukin-1 induces chondrocyte protease production: the development of collagenase inhibitors. Agents Actions 21:328–330

Patterson DF, Minor RR (1977) Hereditary fragility and hyperextensibility of the skin of cats: a defect of collagen fibrillogenesis. Lab Invest 37:170–179

Patterson DF, Haskins ME, Jezyk PF (1982) Models of human genetic disease in domestic animals. Adv Hum Genet 12:263–339

Pauli BU, Memoli VH, Kuettner KE (1981) In vitro determination of tumor invasiveness using extracted hyaline cartilage. Cancer Res 41:2084–2091

Pawelec D (1972a) Wydalanie hydroksyproliny w moczu u chorych z gruźlicą płuc (Urinary excretion of hydroxyproline in tuberculous patients). Gruźl Chor Płuc 40:913–918 (in Polish)

Pawelec D (1972b) Wydalanie hydroksyproliny w moczu u chorych z gruźlicą płuc przewlekłą włóknisto-jamistą (Urinary excretion of hydroxyproline in patients with chronic fibro-cavernous pulmonary tuberculosis). Gruźl Chor Płuc 40:919–923 (in Polish)

Pawelec D (1973) Wydalanie hydroksyproliny w moczu w sarkoidozie (Urinary excretion of hydroxyproline in patients with sacroidosis). Pol Tyg Lek 28:1412–1414 (in Polish)

Pawelec D (1974) Wydalanie hydroksyproliny w moczu ludzi z przebytą, nieczynną gruźlicą płuc (Urinary excretion of hydroxyproline in patients with inactive pulmonary tuberculosis). Gruźl Chor Płuc 42:1163–1166 (in Polish)
Pawelec D (1975) Hydroxyprolina a choroby płuc (Hydroxyproline in disorders of the lungs). Pol Tyg Lek 30:1081–1082 (in Polish)
Pawlicka E, Bańkowski E, Chrostek L, Szmitkowski M (1988) An increase of collagen biosynthesis precedes other symptoms of ethanol-induced liver damage in rats. Drug Alcohol Depend 22:113–116
Peacock EE Jr (1973) Biological frontiers in the control of healing. Am J Surg 126:708–713
Peacock EE Jr, Madden JW (1969) Some studies on the effects of $\beta$-aminopropionitrile in patients with injured flexor tendons. Surgery 66:215–223
Pearson RW (1962) Studies on the pathogenesis of epidermolysis bullosa. J Invest Dermatol 39:551–575
Pearson CM, Wood FD (1964) Passive transfer of adjuvant arthritis by lymph node or spleen cells. J Exp Med 120:547–560
Pedrini VA, Ponseti IV, Dohrman SC (1973) Glycosaminoglycans of intervertebral disc in idiopathic scoliosis. J Lab Clin Med 82:938–950
Peltonen L, Palotie A, Prockop DJ (1980) A defect in the structure of type I procollagen in a patient who had osteogenesis imperfecta: excess mannose in the COOH-terminal propeptide. Proc Natl Acad Sci USA 77:6179–6183
Peltonen L, Palotie A; Myllylä R, Krieg T, Oikarinen A (1985) Collagen biosynthesis in systemic scleroderma: regulation of posttranslational modifications and synthesis of procollagen in cultured fibroblasts. J Invest Dermatol 84:14–18
Perejda AJ (1987) Vascular and cutaneous complications of diabetes mellitus. The role of nonenzymatic glucosylation of collagens. In: Uitto J, Perejda AJ (eds) Connective tissue disease. Molecular pathology of the extracellular matrix. Dekker, New York, pp 475–490
Perejda AJ, Abraham PA, Carnes WH, Coulson WF, Uitto J (1985) Marfan's syndrome: structural, biochemical, and mechanical studies of the aortic media. J Lab Clin Med 106:376–383
Pérez-Tamayo R (1982) Degradation of collagen: pathology. In: Weiss JB, Jayson MIV (eds) Collagen in health and disease. Churchill Livingstone, Edinburgh, pp 135–159
Pérez-Tamayo R, Monfort I, Gonzalez E (1987) Collagenolytic activity in experimental cirrhosis of the liver. Exp Mol Pathol 47:300–308
Perier C, Gautier M, Baril A, Bayle JJ, Patouillard G, Frey J (1984) Measurement of changes in amino acids related to total collagen in fibrotic human liver. Clin Physiol Biochem 2:279–286
Perlish JS, Timpl R, Fleischmajer R (1985) Collagen synthesis regulation by the aminopeptide of procollagen I in normal and scleroderma fibroblasts. Arthritis Rheum 28:647–651
Perron RR, Wright BA (1950) Alteration of collagen structure by irradiation with electrons. Nature 166:863–864
Person P, Philpott DE (1969) The nature and significance of invertebrate cartilage. Biol Rev 44:1–16
Peters KM, Snyder KF, Rush BD, Ruwart MJ, Henley KS (1989) 16,16-Dimethyl prostaglandin $E_2$ decreases the formation of collagen in fibrotic rat liver slices. Prostaglandins 37:445–456
Petkov R (1978a) Ultrastructure of the collagen fibril. I. Some features of the structure of the collagen fibril. Anat Anz 144:301–318
Petkov R (1978b) Ultrastructure of the collagen fibril. II. Evidence of the spiral organization of the fibril. Anat Anz 144:485–501
Phalen G (1966) The carpal tunnel syndrome. J Bone Joint Surg 48:211–228
Phan SH, Thrall RS, Williams C (1981) Bleomycin-induced pulmonary fibrosis. Effects of steroid on lung collagen metabolism. Am Rev Respir Dis 124:428–434
Phang JM, Finerman GAM, Singh B, Rosenberg LE, Berman M (1971) Compartmental analysis of collagen synthesis in fetal rat calvaria. I. Perturbation of proline transport. Biochim Biophys Acta 230:146–159
Phillips LS, Vassilopoulou-Sellin R, Reichard LA (1979) Nutrition and somatomedin. VIII. The somatomedin inhibitor in diabetic rat serum is a general inhibitor of growing cartilage. Diabetes 28:919–924

Picou D, Alleyne GAO, Waterlow JC, Seakins A (1965) Hyroxyproline and creatinine excretion in protein-depleted infants. Biochem J 95:18P

Piedra de la C, Toural V, Rapado A (1987) Osteocalcin and urinary hydroxyproline/creatinine ratio in the differential diagnosis of primary hyperparathyroidism and hypercalcaemia of malignancy. Scand J Lab Invest 47:587–592

Pietilä K, Nikkari T (1980) Enhanced synthesis of collagen and total protein by smooth muscle cells from atherosclerotic rabbit aortas in culture. Atherosclerosis 37:11–19

Piez JA (1968) Molecular weight determination of random coil polypeptides from collagen by molecular sieve chromatography. Anal Biochem 26:305–312

Piez KA (1984) Molecular and aggregate structure of the collagens. In: Piez KA, Reddi AH (eds) Extracellular matrix biochemistry. Elsevier, New York, p 1–39

Piez KA, Reddi AH (eds) (1984) Extracellular matrix biochemistry. Elsevier, New York

Pignatelli M, Bodmer WF (1988) Genetics and biochemistry of collagen binding-triggered glandular differentiation in a human colon carcinoma cell line. Proc Natl Acad Sci USA 85:5561–5565

Pihlajaniemi T, Dickson LA, Pope FM, Korhonen VR, Nicholls A, Prockop DJ, Myers JC (1984) Osteogenesis imperfecta: cloning of a pro-$\alpha$2(I) collagen gene with a frameshift mutation. J Biol Chem 259:12941–12944

Pihlajaniemi T, Myllylä R, Seyer J, Kurkinen M, Prockop DJ (1987) Partial characterization of a low molecular weight human collagen that undergoes alternative splicing. Proc Natl Acad Sci USA 84:940–944

Pikkarainen J (1968) The molecular structure of vertebrate skin collagens. Acta Physiol Scand [Suppl] 309, 1–60

Pikkarainen J, Rantanen J, Vastamaki M, Lampiaho K, Kari A, Kulonen E (1968) On collagens of invertebrates with special reference to Mytilus edulis. Eur J Biochem 4:555–561

Pinnell SR (1983) Disorders of collagen. In: Stanbury JB, Wyngaarden JB, Friedrickson DS (eds) The metabolic basis of inherited diseases. McGraw-Hill, New York, pp 1366–1394

Pinnell SR (1985) Regulation of collagen biosynthesis by ascoribc acid: a review. Yale J Biol Med 58:553–559

Pinnell SR, Fox R, Krane SM (1971) Human collagens: differences in glycosylated hydroxylsines in skin and bone. Biochim Biophys Acta 229:119–124

Pinnell SR, Krane SM, Kenzora JE, Glimcher MJ (1972) A hereditable disorder of connective tissue. Hydroxylysine deficient collagen. N Engl J Med 289:1013–1020

Pinnell SR, Murad S, Darr D (1987) Induction of collagen synthesis by ascorbic acid. A possible mechanism. Arch Dermatol 123:1684–1686

Pirie A (1951) Composition of ox lens capsule. Biochemistry 48:368–371

Piukovich I, Morvay J (1973) Urinary excretion of hydroxyproline in pregnancy. Acta Med Acad Sci Hung 30:27–32

Planas-Bohne F (1973) Pharmakokinetische Untersuchungen an $^{14}$C-markiertem Penicillamin. Arzneimittelforschung 22:1426–1433

Plastow SR, Lovell CR, Young AR (1987) UVB-induced collagen changes in the skin of the hairless albino mouse. J Invest Dermatol 88:145–148

Plastow SR, Harrison JA, Young AR (1988) Early changes in dermal collagen of mice exposed to chronic UVB irradiation and the effects of a UVB sunscreen. J Invest Dermatol 91:590–592

Policard A, Collet A (1961) Physiologie du tissue conjonctif normal et pathologique. Masson, Paris, pp 1–258

Poole AR (1986) Changes in the collagen and proteoglycan of articular cartilage in arthritis. In: Kühn K, Krieg T (eds) Connective tissue: biological and clinical aspects. Karger, Basel, pp 316–371

Poole A (1987) Collagen synthesis in rats with silica-induced pulmonary fibrosis. Arch Toxicol [Suppl]11:285–287

Poole CA, Ayad S, Schofield JR (1988a) Chondrons from articular cartilage. I. Immunolocalization of type VI collagen in the pericellular capsule of isolated canine tibial chondrons. J Cell Sci 90:635–643

Poole AR, Pidoux I, Reiner A, Rosenberg L, Hollister D, Murray L, Rimoin D (1988b) Kniest dysplasia is characterized by an apparent abnormal processing of the C-propeptide of type II cartilage collagen resulting in imperfect fibril assembly. J Clin Invest 81:579–589

Poole A, Myllylä R, Davies BH (1989) Activities of enzymes of collagen biosynthesis and levels of type III procollagen peptide in the serum of patients with sarcodiosis. Life Sci 45:319–326

Pope FM, Martin GR, Lichtenstein JR, Penttinen RP, Gerson G, Rowe DW, McKusick VA (1975) Patients with Ehlers-Danlos syndrome type IV lack type III collagen. Proc Natl Acad Sci USA 72:1314–1316

Pope FM, Jones PM, Wells RS; Lawrence D (1980) Ehlers-Danlos syndrome IV (Acrogeria): new autosomal dominant and recessive types. J R Soc Med 73:180–186

Pope FM, Nicholls AC, Lewkonia RM, Halme T, Dorrance DE, Pomerancem A (1987) Clinical and genetic heterogeneity of the Marfan syndrome. Curr Probl Dermatol 17:95–110

Popper H, Udenfriend S (1970) Hepatic fibrosis. Correlation of biochemical and morphological investigations. Am J Med 49:707–721

Porter J (1896) Further researches on the closure of the coronary arteries. J Exp Med 1:46–70

Postlethwaite EE, Seyer JM, Kang AH (1978) Chemotactic attraction of human fibroblasts to type I, II and III collagens and collagen derived peptides. Proc Natl Acad Sci USA 75:871–875

Postlethwaite AE, Lachman LB, Mainardi CL, Kang AH (1983) Interleukin-1 stimulation of collagenase production by cultured fibroblasts. J Exp Med 157:801–806

Potter SR, Bienenstock J, Goldstein S, Buchanan WW (1985) Fibroblast growth factors in scleroderma. J Rheumatol 12:1129–1135

Powis G (1973) Binding of catecholamines to connective tissue and the effect upon the response of blood vessels to noradrenaline and nerve stimulation. J Physiol (Lond) 234:145–162

Praus R, Brettschneider I, Adam M (1979) Heterogeneity of the bovine corneal collagen. Exp Eye Res 29:469–475

Priest RE, Moinuddin JF, Priest JH (1973) Collagen of Marfan syndrome is abnormally soluble. Nature 245:264–266

Priest RE, Priest JH, Moinuddin JF, Keyser AJ (1977) Differentiation in human amniotic fluid cell cultures. I. Collagen production. J Med Genet 14:157–162

Prockop DJ (1985) Mutations of collagen genes. Consequences for rare and common diseases. J Clin Invest 75:783–787

Prockop DJ (1988) Osteogenesis imperfecta. A model for genetic causes of osteoporosis and perhaps several other common diseases of connective tissue. Arthritis Rheum 31:1–8

Prockop DJ, Juva K (1965) Hydroxylation of proline in particulate fractions from cartilage. Biochem Biophys Res Commun 18:54–59

Prockop DJ, Kivirikko KI (1967) Relationship of hydroxyproline excretion in urine to collagen metabolism. Biochemistry and clinical applications. Ann Intern Med 66:1243–1266

Prockop DJ, Kivirikko KI (1984) Heretable diseases of collagen. N Engl J Med 311:376–386

Prockop DJ, Kuivaniemi H (1986) Inborn errors of collagen. In: Kühn K, Krieg T (eds) Connective tissue: biological and clinical aspects. Karger, Basel, pp 246–271 (Rheumatology. An annual review, vol 10)

Prockop DJ, Sjoerdsma A (1961) Significance of urinary hydroxyproline in man. J Clin Invest 40:843–848

Prockop DJ, Tuderman L (1982) Posttranslational enzymes in the biosynthesis of collagen: extracellular enzymes. Methods Enzymol 82:305–319

Prockop DJ, Berg RA, Kivirikko KI, Uitto J (1976) Intracellular steps in the synthesis of collagen. In: Ramachandran GN, Reddi AH (eds) Biochemistry of collagen. Plenum, New York, pp 163–254

Prockop DJ, Kivirikko KI, Tuderman L, Guzman NA (1979) The biosynthesis of collagen and its disorders. N Engl J Med 301:13–23, 77–85

Proye MP, Polson AM (1982) Repair in different zones of the periodontium after tooth reimplantation. J Periodont Res 15:43–52

Puistola U, Rönnberg L, Martikainen H, Turpeeniemi-Hujanen T (1989) The human embryo produces basement membrane collagen (type IV collagen) – degrading protease activity. Hum Reprod 4:309–311

Pycock C, Blaschke E, Bergvist U, Uvnäs B (1975) On the possible involvement of sulfo-mucopolysaccharides in the storage of catecholamine within the central nervous system. Acta Physiol Scand 95:373–382

Pyeritz RE (1983) Marfan syndrome. In: Emery AE, Rimon DL (eds) Principles and practice of medical genetics, vol 2. Churchill Livingstone, Edinburgh, pp 820–835

Pyeritz RE (1986) The Marfan syndrome. Am Fam Physician 34(6):83–94

Pyeritz RE, McKusick VA (1979) The Marfan syndrome: diagnosis and management. N Engl J Med 300:772–775

Qian RG, Glanville RW (1984) Separation and characterization of two polypeptide chains from the 7S cross-linking domain of basement membrane (type IV) collagen. Biochem J 222:447–452

Radom S, Sienczewska-Burzyńska J, Jaruga-Dolińska E, Żuławski M (1971) Współczynnik kreatyninowo-hydroksyprolinowy u zdrowych, w naczynności tarczycy i w nowotworach kości (Creatinine-hydroxyproline index in healthy individuals and in patients with hyperthyroidism and bone neoplasms). Pol Arch Med Wewn 47:257–264 (in Polish)

Raghow R, Gossage D, Seyer JM, Kang AH (1984) Transcriptional regulation of type I collagen genes in cultured fibroblasts by a factor isolated from thioacetamide-induced fibrotic rat liver. J Biol Chem 259:12718–12723

Raghu G, Masta S, Meyers D, Narayanan AS (1989) Collagen synthesis by normal and fibrotic human lung fibroblasts and the effect of transforming growth factor-$\beta$. Am Rev Respir Dis 140:95–100

Raisz LG (1988) Bone metabolism and its hormonal regulation: an update. Triangle 27:5–10

Rajabi M, Woessner JF (1984) Rise of serum levels of PZ-peptidase, an enzyme involved in collagen breakdown in human pregnancy and labor. Am J Obstet Gynecol 150:821–826

Ramachandran GN (1963) Molecular structure of collagen. Int Rev Connect Tissue Res 1:127–182

Ramachandran GN (ed) (1967a) Treatise on collagen. Academic, London

Ramachandran GN (1967b) Structure of collagen at the molecular level. In: Ramachandran GN (ed) Treatise on collagen, vol 1. Academic, London, pp 103–179

Ramachandran GN (1988) Stereochemistry of collagen. Int J Pept Protein Res 31:1–16

Ramachandran GN, Kartha G (1954) Structure of collagen. Nature 174:269–270

Ramachandran GN, Reddi AH (eds) (1976) Biochemistry of collagen. Plenum, New York

Ramirez F, Sangiorgi FO, Tsipouras P (1986) Human collagens: biochemical, molecular and genetic features in normal and diseased states. Horiz Biochem Biophys 8:341–375

Ramon Y, Cajal S (1968) Degeneration and regeneration of the nervous system. Haffner, London

Rana SVS, Prakash R (1986) Collagen in the liver of metal fed rats. Exp Pathol 29:193–195

Rao NV, Adams E (1978) Partial reaction of prolyl hydroxylase (Gly-Pro-Ala)$_n$ stimulates $\alpha$-ketoglutarate decarboxylation without prolyl hydroxylation. J Biol Chem 253:6327–6330

Rao LG, Wang HM, Kalliecharan R, Heersche JNM, Sodek J (1979) Specific immunohistochemical localization of type I collagen in porcine periodontal tissues using the peroxidase-labelled antibody technique. Histochem J 11:73–79

Rapaka RS, Parr RW, Tsan-Zon Liu, Bhatnagar RS (1977) Biochemical basis of skeletal defects induced by hydralazine: inhibition of collagen synthesis and secretion in embryonic cartilage in vitro. Teratology 15:185–194

Rapaka RS, Renugopalakrishnan V, Urry DW, Bhatnagar RS (1978) Hydroxylation of proline in polytripeptide models of collagen: stereochemistry of polytripeptide-prolyl hydroxylase interaction. Biochemistry 17:2892–2898

Rauterberg J, Allam S, Brehmer U, Wirth W, Hauss WH (1977) Characterization of the collagen synthesized by cultured human smooth muscle cells from fetal and adult aorta. Z Physiol Chem 358:401–407

Raynes JG, Anderson JC, Fitzpatrick RJ, Dobson H (1988) Increased collagenase activity is not detectable in cervical softening in the ewe. Coll Relat Res 8:461–469

Recchia O, Sereno L, Isidori A (1967) Iddrossiprolina urinaria ed HGH circolante. Primi rilievi in alcune condizioni disendocrine. Folia Endocrinol (Roma) 20:420–441

Reddi AH (1984) Extracellular matrix and development. In: Piez KA, Reddi AH (eds) Extracellular matrix biochemistry. Elsevier, New York, pp 375–412

Reddy GS, Srikantia SG (1971) Effect of dietary calcium, vitamin C and protein in development of experimental skeletal fluorosis. I. Growth, serum chemistry, and changes in composition and radiological appearance of bones. Metabolism 20:642–649

Redini F, Galera P, Mauviel A, Loyau G, Pujol JP (1988) Transforming growth factor $\beta$ stimulates collagen and glycosaminoglycans biosynthesis in cultured rabbit articular condrocytes. FEBS Lett 234:172–176

Reeser FH, Aarberg TM (1979) Vitreous humor. In: Record RE (ed) Physiology of the human eye and visual system. Harper and Row, Hagerstown, p 261

Refetoff S, Matalon R, Bigazzi M (1972) Metabolism of L-thyroxine ($T_4$) and L-triiodothyronine ($T_3$) by human fibroblasts in tissue culture: evidence for cellular binding proteins and conversion of $T_4$ to $T_3$. Endocrinology 91:934–947

Reid KBM (1982) Proteins containing collagen sequences. In: Weiss JB, Jayson MIV (eds) Collagen in health and disease. Churchill Livingstone, Edinburgh, pp 18–27

Reid KBM (1989) Chemistry and molecular genetics of C1q. Behring Inst Mitt 84:8–19

Reid GC, Woods DR, Robb FT (1980) Peptone induction and rifampin-insensitive collagenase production by Vibrio alginolyticus. J Bacteriol 142:447–454

Reiser KM, Last JA (1981) Pulmonary fibrosis in experimental acute respiratory disease. Am Rev Respir Dis 123:58–63

Reiser KM, Last JA (1983) Type V collagen. Quantitation in normal lungs and in lungs of rats with bleomycin-induced pulmonary fibrosis. J Biol Chem 258:269–275

Reiser KM, Hascheck WM, Hesterberg TW, Last JA (1983) Experimental silicosis. II. Long-term effects of intratracheally instilled quartz on collagen metabolism and morphologic characteristics of rat lungs. Am J Pathol 110:30–41

Reiser KM, Tyler WS, Hennessy SM, Diminiguez JJ, Last JA (1987a) Long-term consequences of exposure to ozone. II. Structural alterations in lung collagen of monkeys. Toxicol Appl Pharmacol 89:314–322

Reiser KM, Hennessy SM, Last JA (1987b) Analysis of age-associated changes in collagen crosslinking in the skin and lung in monkeys and rats. Biochim Biophys Acta 926:339–348

Remberger K, Gay S, Adelmann BC (1976) Immunhistochemische Charakterisierung und Lokalisation unterschiedlicher Kollagentypen bei chronischen Nierenerkrankungen. Verh Dtsch Ges Pathol 60:314

Rennard SI, Crystal RG (1982) Lung. In: Weiss JB, Jayson MIV (eds) Collagen in health and disease. Churchill Livingstone, Edinburgh, pp 424–444

Resnick RH, Cerda JC, Boitnott J, Aron J, Iber FL (1973) Urinary hydroxyproline excretion in hepatic disorders. Am J Gastroenterol 60:576–584

Resnick RH, Boitnott J, Iber FL, Makipour H, Cerda JJ (1975) Penicillamine therapy in acute alcoholic liver disease. In: Popper H, Becker K (eds) Collagen metabolism in the liver. Stratton Intercontinental Medical Book, New York, pp 207–218

Retief E, Parker MI, Retief AE (1985) Regional chromosome mapping of human collagen genes alpha 2(I) and alpha 1(I) (COL1A2 and COL1A1). Hum Genet 69:304–308

Revell PA (1986) Pathology of bone. Springer, Berlin Heidelberg New York, pp 1–302

Rhoads RE, Udenfriend S (1970) Purification and properties of collagen proline hydroxylase from newborn rat skin. Arch Biochem Biophys 139:329–339

Rhodes RK (1982) Blood vessel. In: Weiss JB, Jayson MIV (eds) Collagen in health and disease. Churchill Livingstone, Edinburgh, pp 376–787

Rhodes RK, Miller EJ (1978) Physicochemical characterization and molecular organization of the collagen A and B chains. Biochemistry 17:3442–3448

Rhodes RK, Miller EJ (1981) Evidence for the existence of an alpha-1(V), alpha-2 (V), alpha-3 (V) collagen molecule in human placental tissue. Coll Relat Res 1:337–343

Ribeiro P, Walesby R, Edmondson S, Jadhav AV, Trayner I, Oaklex CM, Thompson GR (1983) Collagen content of atherosclerotic arteries is higher in smokers than in non-smokers. Lancent 1:1070–1073

Ricard-Blum S, Ville G (1988) Collagen cross-linking. Cell Mol Biol 34:581–590

Rich A, Crick FHC (1955) Thr structure of collagen. Nature 176:915–916

Riley DJ, Berg RA, Edelman NH, Prockop DJ (1980) Prevention of collagen deposition following pulmonary oxygen toxicity in the rat by cis-4-hydroxy-L-proline. J Clin Invest 65:643–651

Rimoin DL (1975) The chondrodystrophias. Adv Hum Genet 5:1–116

Rimoin DL, Lachman DS (1983) The chondrodysplasias. In: Emery AEH, Rimoin DL (eds) Principles and practice of medical genetics, vol 2. Churchill Livingstone, New York, pp 703–735

Risteli J (1977a) Intracellular enzymes of collagen biosynthesis in the liver. Effects of age, hepatic injury and prednisolone treatment in the rat, and the influence of liver disease on serum immunoreactive prolyl 4-hydroxylase in rats and humans. Acta Univ Ouluensis [A] 55:1–44

Risteli J (1977b) Effect of prednisolone on the activities of the intracellular enzymes of collagen biosynthesis in rat liver and skin. Biochem Pharmacol 26:11295–1298

Risteli J, Kivirikko KI (1974) Activities of prolyl hydroxylase, lysyl hydroxylase, collagen galactosyltransferase and collagen glucosyltransferase in the liver of rats with hepatic injury. Biochem J 144:115–122

Risteli J, Kivirikko KI (1976) Intracellular enzymes of collagen biosynthesis in rat liver as a function of age and in hepatic injury induced by dimethylnitrosamine. Changes in prolyl hydroxylase, lysyl hydroxylase, collagen galactosyltransferase and collagen glucosyltransferase activities. Biochem J 158:361–367

Risteli J, Tuderman L, Kivirikko KI (1976) Intracellular enzymes of collagen biosynthesis in rat liver as a function of age and in hepatic injury induced by dimethylnitrosamine. Purification of rat prolyl hydroxylase and comparison of changes in prolyl hydroxylase activity with changes in immunoreactive prolyl hydroxylase. Biochem J 158:369–376

Risteli J, Tryggvason K, Kivirikko KI (1977) Prolyl 3-hydroxylase: partial characterization of the enzyme from rat kidney cortex. Eur J Biochem 73:485–492

Risteli J, Tuderman L, Tryggvason K, Kivirikko KI (1978) Effect of hepatic injury on prolyl 3-hydroxylase and 4-hydroxylase activities in rat liver and on immunoreactive prolyl 4-hydroxylase concentrations in the liver and serum. Biochem J 170:129–135

Risteli J, Schuppan D, Glanville RW, Timpl R (1980) Immunochemical distinction between two different chains of type IV collagen. Biochem J 191:517–522

Risteli J, Wick G, Timpl R (1981) Immunological characterization of the 7S domain of type IV collagens. Coll Relat Res 1:419–432

Risteli J, Draeger KE, Regitz G, Neubauer HP (1982) Increase in circulating basement membrane antigens in diabetic rats and effects of insulin treatment. Diabetologia 23:266–269

Risteli L, Puistola U, Hohtari H, Kauppila A, Risteli J (1987) Collagen metabolism in normal and complicated pregnancy: changes in the aminoterminal propeptide of type III procollagen in serum. Eur J Clin Invest 17:81–86

Rizzo R, Contri MB, Micali G, Quaglino D, Pavone L, Ronchetti IP (1987) Familiar Ehlers-Danlos syndrome type II: abnormal fibrillogenesis of dermal collagen. Pediatr Dermatol 4:197–204

Roberts S (1985) Collagen of the calcified layer of human articular cartilage. Experientia 41:1138–1139

Robertson PB (1976) The periodontium. In: Lazzari EP (ed) Dental biochemistry. Lea and Febiger, Philadelphia, pp 187–200

Robertson PB, Simpson J (1976) Collagenase: current concepts and relevance to periodontal disease. J Periodontol 47:29–35

Robertson PB, Lantz M, Marucha PT, Kornman KS, Trummel CL, Holt SC (1982) Collagenolytic activity associated with Bacteroides species and Actinobacillus actinomycetencomitans. J Periodont Res 17:275–283

Robins SP (1982) Turnover and cross-linking of collagen. In: Weiss JB, Jayson MIV (eds) Collagen in health and disease. Churchill Livingstone, Edinburgh, pp 160–178

Robins SP, Duncan A (1988) Pyridinium crosslinks of bone and their location in peptides isolated from rat femur. Biochim Biophys Acta 914–233–239

Robinson JD, Green JP (1962) Sulfomucopolysaccharides in brain. Yale J Biol Med 35:248–254

Robinson TF, Cohen-Gould L, Factor SM (1983) Skeletal framework of mammalian heart muscle. Arrangement of inter- and pericellular connective tissue structures. Lab Invest 49:482–298

Robinson TF, Cohen-Gould L, Remily RM, Capasso JM, Factor SM (1984) Extracellular structures in heart muscle. Adv Myocardiol 5:243–255

Rodnan GP, Luksick J (1969) Plasma hypro-protein level and urinary hydroxyproline excretion in progressive systemic sclerosis (scleroderma). Arthritis Rheum 12:693–699

Rodnan GP, Lipinski E, Luksick J (1979) Skin thickness and collagen content in progresssive systemic sclerosis and localized scleroderma. Arthritis Rheum 22:1301–1304

Rogers HJ, Weidmann SM, Parkinson A (1952) Studies on the skeletal tissues. The collagen content of bones from rabbits, oxen and humans. Biochem J 50:537–542

Rohde H, Hahn E, Timpl R (1978) Radioimmunoassay for aminoterminal procollagen peptide in liver disease. Fresenius Z Analyt Chem 290:151–152

Rohde H, Vargas M, Hahn E, Kalbfleisch H, Bruguerra M, Timpl R (1979) Radioimmunoassay for type III procollagen peptide and its application to human liver disease. Eur J Clin Invest 9:451–459

Rohen JW (1962) Über das Ligamentum pectinatum der Primaten. Z Zellforsch 58:403–421

Rojkind M (1973) Inhibition of liver fibrosis by L-azetidine-2-carboxylic acid in rats treated with carbon tetrachloride. J Clin Invest 52:2451–2456

Rojkind M (1984) The blue glass and the predictive value of serum amino-terminal propeptide of type III procollagen as a marker of liver fibrosis. Hepatology 4:977–978

Rojkind M, Diaz de Leon L (1970) Collagen biosynthesis in cirrhosis rat liver slices. A regulatory mechanism. Biochim Biophys Acta 217:512–522

Rojkind M, Dunn MA (1979) Hepatic fibrosis. Gastroenterology 76:849–863

Rojkind M, Kershenobich D (1975) Regulation of collagen synthesis in liver cirrhosis. In: Popper H (ed) Collagen metabolism in the liver. Stratton Intercontinental Medical Book, New York, pp 129–138

Rojkind M, Mourelle M (1988) The liver as a bioecological system: modifications during regeneration and repair. In: Nimni ME (ed) Collagen, vol II. CRC Press, Boca Raton, pp 137–159

Rojkind M, Pérez-Tamayo R (1983) Liver fibrosis. Int Rev Connect Tissue Res 10:333–393

Rojkind M, Ponce-Noyola P (1982) The extracellular matrix of the liver. Coll Relat Res 2:151–175

Rojkind M, Giambrone MA, Biempica L (1979) Collagen types in normal and cirrhotic liver. Gastroenterology 76:710–719

Rojkind M, Giambrone MA, Takahashi S (1982) Collagen polymorphism in normal and fibrotic liver. J UOEH Jpn [Suppl] 4:157–168

Rojkind M, Rojkind MH, Cordero-Hernández J (1983) In vivo collagen synthesis and deposition in fibrotic and regenerating rat livers. Coll Relat Res 3:335–347

Rokowski RJ, Sheehy J, Cutroneo KR (1981) Glucocorticoid-mediated selective reduction of functioning collagen messenger ribonucleic acid. Arch Biochem Biophys 210:74–81

Roll FJ, Madri AJ, Albert J, Furthmayr H (1980) Codistribution of collagen types IV and $AB_2$ in basement immunoferritin study of ultrathin frozen sections. J Cell Biol 85:597–616

Rona G, Cappel EI, Balazs T, Gaundry R (1958) An infarct-like myocardial lesion and other toxic manifestations produced by isoproterenol in the rat. Arch Pathol 67:443–455

Rönnemaa T, Doherty NS (1977) Effect of serum and liver extracts from hypercholesterolemic rats on the synthesis of collagen by isolated aortas and cultured aortic smooth muscle cells. Atherosclerosis 26:261–272

Rorie DK, Newton M (1967) Histologic and chemical studies of the smooth muscle in the human cervix and uterus. Am J Obstet Gynecol 99:466–469

Rosenberg H, Modrak JB, Hassing JM, Al-Turk WA, Stohs SJ (1979) Glycosylated collagen. Biochem Biophys Res Commun 91:498–501

Rosenberg L, Schubert M, Sandson J (1967) The protein-polysaccharides of bovine nucleus pulposus. J Biol Chem 242:4691–4701

Rosenberry TL (1975) Acetyl-cholinesterase. Adv Enzymol 43:103–125

Rosenberry TL, Barnett P, Mays C (1982) Acetylcholinesterase. Methods Enzymol 82:325–338

Rosenbloom J (1971) Trans-hydroxyproline is not incorporated into collagen. Arch Biochem Biophys 142:718–719

Rosenbloom J, Prockop DJ (1970) Incorporation of 3,4-dehydroproline into procollagen and collagen. J Biol Chem 245:3361–3368

Rosenbloom AL, Silverstein JH, Lezotte DC, Richardson K, McCallum M (1981) Limited joint mobility in childhood diabetes mellitus indicates increased risk for microvascular disease. N Engl J Med 305:191–194

Rosenbloom J, Feldman G, Freundlich B, Jimenez SA (1984) Transcriptional control of human diploid fibroblast collagen synthesis by gammainterferon. Biochem Biophys Res Commun 123:365–372
Ross R, Benditt EP (1961) Wound healing and collagen formation. I. Sequential changes in components of guinea pig skin wounds observed in the electron microscope. J Biophys Biochem Cytol 11:677–700
Ross R, Glomset JA (1976) The pathogenesis of atherosclerosis. N Engl J Med 295:368–377, 420–425
Ross R, Klebanoff SJ (1971) The smooth muscle cell. I. In vivo synthesis of connective tissue proteins. J Cell Biol 50:159–172
Roswit WT, Halme J, Jeffrey JJ (1983) Purification and properties of rat uterine procollagenase. Arch Biochem Biophys 255:285–292
Roumestand C, Yiotakis A, Dive V, Morgat JL, Fromageot P, Toma F, Hammadi A, Poulin JC, Kagan HB (1989) Tritium and deuterium selective stereospecific labelling of peptide inhibitors of bacterial collagenases. In: Aubry A, Marraud M, Vittoux B (eds) Secon forum on peptides. Libbey Eurotext, Paris, pp 243–245
Roupe G, Laurent TC, Malmström A, Suurküla M, Särnstrand B (1987) Biochemical characterization and tissue distribution of the scleredema in a case of Buschke's disease. Acta Derm Venereol (Stockh) 67:193–198
Rowe RWD (1985a) The structure of rat tail tendon. Connect Tissue Res 14:9–20
Rowe RWD (1985b) The structure of rat tail tendon fascicles. Connect Tissue Res 14:21–30
Rowe DW, McGoodwin EB, Martin GR, Sussman MD, Grahn D, Faris B, Franzblau C (1974) A sex-linked defect in the cross-linking of collagen and elastin associated with the mottled locus in mice. J Exp Med 139:180–192
Rowe DW, Starman BJ, Fujimoto WY, Williams RH (1977) Abnormalities in proliferation and protein synthesis in skin fibroblast cultures from patients with diabetes mellitus. Diabetes 26:284–290
Rowe DW, Shapiro JR, Poirier M, Schlesinger S (1985) Diminished type I collagen synthesis and reduced $\alpha$ 1(I) collagen messenger RNA in cultured fibroblasts from patients with dominantly inherited (type 1) osteogenesis imperfecta. J Clin Invest 76:604–611
Rowley MJ, Williamson DJ, Mackay IR (1987) Evidence for local synthesis of antibodies to denatured collagen in the synovium in rheumatoid arthritis. Arthritis Rheum 30:1420–1425
Royce PM, Barnes MJ (1977) Comparative studies on collagen glycosylation in chick skin and bone. Biochim Biophys Acta 498:132–142
Royce PM, Danks DM (1982) Normal lysyl oxidase activity in skin fibroblasts from patients with Marfan's syndrome. IRCS Med Sci Biochem 10:41
Rozanis J, Slots J (1982) Collagenolytic activity of Actinobacillus actinomycetemcomitans and black-pigmented Bacteroides. IADR Prog 61:870
Rubegni M, Gennari C, Ravenni C, Forconi S, Bencini M (1964) L'idrossiprolinuria nelle malattie della tiroide. Boll Soc Ital Biol Sper 41:355–357
Rudnicki M, Wojtyczka A (1984) Biochemical and mechanical aspects of the healing of experimental gastric wound after antibiotic prophylaxis. 9th meeting of the Federation of European Connective Tissue Societes, Budapest, 1984, abstracts p 168
Rusenko KW, Gammon WR, Fine JD, Briggaman RA (1989) The carboxylterminal domain of type VII collagen is present at the basement membrane in recessive dystrophic epidermolysis bullosa. J Invest Dermatol 92:623–627
Ryan WL (1964) Regulation of the free amino acids of skin by hydrocortisone. J Invest Dermatol 43:121–124
Saarni H (1977) The effect of certain anti-inflammatory steroids on collagen synthesis in vitro. Biochem Pharmacol 26:1961–1966
Sage H, Bornstein P (1982) Endothelial cells from umbilical vein and a haemangioendothelioma secrete basement membrane largely to the exclusion of interstitial procollagens. Arteriosclerosis 2:27–36
Sage H, Bornstein P (1987) Type VIII collagen. In: Mayne R, Burgeson RE (eds) Structure and function of collagen types. Academic, Orlando, pp 173–194

Sage H, Pritzl P, Bornstein P (1980) A unique, pepsin-sensitive collagen synthesized by aortic endothelial cells in culture. Biochemistry 19:5747–5755

Sage H, Pritzl P, Bornstein P (1981) Susceptibility of type V collagen to neutral proteases: evidence that the major molecular species is a thrombin-sensitive heteropolymer, $[\alpha 1(V)]_2\alpha 2(V)$. Biochemistry 20:3778–3784

Sage H, Trüeb B, Bornstein P (1983) Biosynthetic and structural properties of endothelial cell type VIII Collagen. J Biol Chem 258:13391–13401

Sakakibara K, Takaoka T, Katsuta H, Umeda M, Tsukada Y (1978) Collagen fiber formation as a common property of epithelial liver lines in cultures. Exp Cell Res 111:63–71

Sakamoto S (1982) Bone. In: Weiss JB, Jayson MIV (eds) Collagen in health and disease. Churchill Livingstone, Edinburgh, pp 362–375

Sakamoto S, Sakamoto M (1981) Heparin and bone metabolism. In: Brown WV, Mann KG, Robert HR, Lundblad RL (eds) The chemistry and biology of heparin. Elsevier-North Holland, New York, pp 133–156

Sakamoto S, Sakamoto M, Goldhaber P, Glimcher MJ (1975) Collagenase and bone resorption: isolation of collagenase from culture medium containing serum after stimulation of bone resorption by addition of parathyroid hormone extract. Biochem Biophys Res Commun 63:172–178

Sakamoto S, Sakamoto M, Goldhaber P, Glimcher MJ (1978) Mouse bone collagenase. Purification of the enzyme by heparin-substituted Sepharose 4B affinity chromatography and preparation of specific antibody to the enzyme. Arch Biochem Biophys 188:438–449

Salamon DS, Liotta LA, Kidwell WR (1981) Differential response to growth factor by rat mammary epithelium plated on different collagen substrata in serum-free medium. Proc Natl Acad Sci USA 78:382–386

Saldino RM (1971) Lethal short-limbed dwarfism: achondrogenesis and thanatophoric dwarfism. Am J Roentgenol Radium Ther Nucl Med 112:185–197

Salle V (1912) Über einen Fall von angeborener abnormen Größe der Extremitäten mit einem an Akromegalia erinnerden Symptomenkomplex. Jahrb Kinderheilkd 75:540–552

Salo T, Liotta LA, Tryggvason K (1983) Purification and characterization of a murine basement membrane collagen-degrading enzyme secreted by metastatic tumor cells. J Biol Chem 258:3058–3063

Salvador RA, Tsai I (1973) Collagen proline hydroxylase activity in the uterus of the rat during rapid collagen synthesis in vivo. Arch Biochem Biophys 154:583–592

Salvador RA, Tsai I, Marcel RJ, Felix AM, Kerwar SS (1976) The in vivo inhibition of collagen synthesis and the reduction of prolyl hydroxylase activity by 3,4-dehydroproline. Arch Biochem Biophys 174:379–390

Sandberg M, Mäkelä JK, Multimäki P, Vuorio T, Vuorio E (1989a) Construction of a human pro $\alpha 1$(III) collagen cDNA clone and localization of type III collagen expression in human fetal tissues. Matrix 9:82–91

Sandberg M, Tamminen M, Hirvonen H, Vuorio E, Pihlajaniemi T (1989b) Expression of mRNAs coding for the $\alpha 1$ chain of type XIII collagen in human fetal tissues: comparison with expression of mRNAs for collagen types I, II, and III. J Cell Biol 109:1371–1379

Sandberg N (1962) Accelerated collagen formation and histamine. Nature 94:183–185

Sandberg N (1964a) Enhanced rate of healing in rats with an increased rate of histamine formation. Acta Chir Scand 127:9–21

Sandberg N (1964b) Time relationship between administration of cortisone and wound healing in rats. Acta Chir Scand 127:446–455

Sanes JR (1982) Laminin, fibronectin and collagen in synaptic and extrasynaptic portions of muscle fiber basement membrane. J Cell Biol 93:442–451

Sartoris DJ, Luzzati L, Weaver DD, MacFarlane JD, Hollister DW, Parker BR (1984) Type IX Ehlers-Danlos syndrome. A new variant with pathognomonic radiographic features. Radiology 152:665–670

Sasaki T, Arai K, Ono M, Yamaguchi T, Furata S, Nagai Y (1987a) Ehlers-Danlos syndrome. A variant characterized by the deficiency of pro $\alpha 2$ chain of type I procollagen. Arch Dermatol 123:76–79

Sasaki T, Majamaa K, Uitto J (1987b) Reduction of collagen production in keloid fibroblast cultures by ethyl-3,4-dihydroxybenzoate. Inhibition of prolyl hydroxylase activity as a mechanism of action. J Biol Chem 262:9397–9403

Sato K, Yoshinaka R, Itoh Y, Sato M (1989) Molecular species of collagen in the intramuscular connective tissue of fish. Comp Biochem Physiol 92B:87–91

Sato T, Saito T, Kokubun M, Ito M, Inoue M, Saito K, Yoshinaga K (1980) Urinary excretion of O-hydroxylysylglycosides in diabetes mellitus. Tohoku J Exp Med 131:97–98

Satwekar K, Radhakrishnan AN (1964) Urinary excretion of hydroxyproline in tropical sprue. Clin Chim Acta 10:284–285

Sauer A, Robinson DG (1985) Intracellular localization of posttranslational modifications in the synthesis of hydroxyproline-rich glycoproteins. Peptydyl proline hydroxylation in maize roots. Planta Med 164:287–294

Savolainen ER (1979) Enzymes of collagen biosynthesis in diseases of the liver and connective tissues. Changes in prolyl hydroxylase and galactosylhydroxylysyl glucosyltransferase in serum and tissues. Acta Univ Ouluensis [A] 85:1–46

Savolainen ER, Miettinen TA, Pikkarainen P, Salaspuro M, Kivirikko KI (1983) Enzymes of collagen synthesis and type III procollagen aminopropeptide in the evoluation of D-penicillamine and medroxyprogesterone treatments of primary biliary cirrhosis. Gut 24:136–142

Savolainen ER, Goldberg B, Leo MA, Velez M, Lieber CS (1984) Diagnostic value of serum procollagen peptide measurements in alcoholic liver disease. Alcoholism Clin Exp Res 8:384–389

Savolainen ER, Brocks D, Ala-Kokko L, Kivirikko KI (1988) Serum concentrations of the N-terminal propeptide of type III procollagen and two type IV fragments and gene expression of the respective collagen types in liver in rats with dimethylnitrosamine-induced hepatic fibrosis. Biochem J 249:753–757

Scharffetter K, Lankat-Buttgereit B, Krieg T (1988) Localization of collagen mRNA in normal and scleroderma skin by in-situ hybridization. Eur J Clin Invest 18:9–17

Schaub MC (1963) Qualitative and quantitative changes in parenchymatous organs of the rat during ageing. Gerontologia 8:114–122

Schaub MC (1964/65) The ageing of collagen in the heart muscle. Gerontologia 10:38–41

Scheck M, Siegel RC, Parker J, Chang YH, Fu JCC (1979) Aortic aneurism in Marfan's syndrome: changes in the ultrastructure and composition of collagen. J Anat 129:645–657

Scheven BAA, Hamilton NJ, Farquharson C, Rucklidge GJ, Robins SP (1988) Immunohistochemical localization of native and denatured collagen types I and II in fetal and adult rat long bones. Bone 9:407–414

Schmid TM, Conrad HE (1982) A unique low molecular weight collagen secreted by cultured chick embryo chondrocytes. J Biol Chem 257:12444–12450

Schmid TM, Linsenmayer TF (1987) Type X collagen. In: Mayne R, Burgeson RE (eds) Structure and function of collagen types. Academic, Orlando, pp 223–259

Schmid TM, Mayne R, Bruns RR, Linsenmayer TF (1984) Molecular structure of short-chain cartilage collagen by electron microscopy. J Ultrastruct Res 86:186–191

Schmid TM, Mayne R, Jeffrey JJ, Linsenmayer TF (1986) Type X collagen contains two cleavage sites for a vertebrate collagenase. J Biol Chem 261:4184–4189

Schmitt FO, Hall CE, Jakson MA (1942) Electron microscope investigation of the structure of collagen. J Cell Comp Physiol 20:11–33

Schmut O (1977) The identification of type III collagen in calf and bovine cornea and sclera. Exp Eye Res 25:505–511

Schmut O (1978) The organization of tissues of the eye by different collagen types. Arch Clin Exp Ophthalmol 207:189–195

Schmut O, Reich ME, Hofmann H (1979) Isolation of different hydroxyproline containing proteins from bovine vitreous body collagen. Arch Clin Exp Ophthalmol 211:329–336

Schmut O, Mallinger R, Paschke E (1984) Studies on a distinct fraction of bovine vitreous body collagen. Arch Clin Exp Ophthalmol 221:286–292

Schneir M, Vogan I, Yu H, Yavelow J, Liu-Montealm A, Furuto D (1975) Ability of salt and acetic acid to extract human and animal gingival collagen. J Dent Res 54:1095–1101

Schneir M, Ogata S, Fine A (1978) Confirmation that neither phenotype nor hydroxylation of collagen is altered in overgrown gingiva from diphenyl-hydantoin-treated patients. J Dent Res 57:506–512

Schneir M, Ramamurthy NS, Golub LM (1982) Skin collagen metabolism in the streptozotocin-induced diabetic rat. Enhanced catabolism of collagen formed both before and during the diabetic state. Diabetes 31:426–431

Schneir M, Ramamurthy NS, Golub LM (1984) Extensive degradation of recently synthesized collagen in gingiva of normal and streptozotocin-induced diabetic rats. J Dent Res 63:23–27

Schneir M, Imberman M, Ramamurthy N, Golub L (1986) The in vivo fractional rate of gingival collagen production in non-diabetic and diabetic rats. J Periodont Res 21:56–63

Schöcket SS (1916) A suggestion as to the process of ovulation and ovarian cyst formation. Anat Rec 10:447–457

Schröder CH, Monnens LAH, van Lith-Zanders HMA, Trijbels JMF, Veerkamp JH, Langeveld JPM (1986) Urinary excretion of hydroxylysine and its glycosides in Alport's syndrome and several other glomerulopathies. Nephron 44:103–107

Schwartz D, Veis A (1978) Characterization of basement membrane collagen of bovine anterior lens capsule via segment-long-spacing crystallites and the specific cleavage of the collagen by pepsin. FEBS Lett 85:326–332

Schwartz D, Veis A (1980) Characterization of bovine anterior-lenscapsule basement-membrane collagen. Eur J Biochem 103:29–35

Schwartz SM, Gajdusek CM, Owens GK (1982) Vessel wall growth control. In: Nossel HL, Vogel HJ (eds) Pathobiology of the endothelial cell. Academic, New York, pp 63–78

Schwartz E, Cruickshank FA, Perlish JS, Fleischmajer R (1989) Alterations in dermal collagen in ultraviolet irradiated hairless mice. J Invest Dermatol 93:142–146

Scott DM, Harwood R, Grant ME, Jackson DS (1977) Characterisation of the major collagen species present in porcine aortae and the synthesis of their precursors by smooth muscle cells in culture. Connect Tissue Res 5:7–13

Scott JE (1980) Collagen-proteoglycan interactions. Localisation of proteoglycans in tendon by electron microscopy. Biochem J 187:887–891

Scott JE (1986) Proteoglycan-collagen interactions. Ciba Found Symp 124:104–124

Scott JE, Orford CR, Hughes EW (1981) Proteoglycan-collagen arrangements in developing rat tail tendon. An electron-microscopical and biochemical investigation. Biochem J 195:573–581

Scow RO (1951) Development of obesity in force fed young thyroidectomized rats. Endocrinology 49:522–529

Scubert M, Hamerman D (1988) A primer on connective tissue biochemistry. Lea and Febiger, Philadelphia

Sedowofia KI, Tomlinson IW, Weiss JB, Hilton RC, Jayson MIV (1982) Collagenolytic enzyme systems in human intervertebral disc. Their control, mechanism, and their possible role in the initiation of biomechanical failure. Spine 7:213–222

Seegmiller RE, Myers RA, Dorfman A, Horwitz AL (1981) Structural and associative properties of cartilage matrix constituents in mice with hereditary chondrodysplasia. Connect Tissue Res 9:69–77

Segrest JP, Cunningham LW (1970) Variations in human urinary O-hydroxylysyl glycoside levels and their relationship to collagen metabolism. J Clin Invest 49:1497–1509

Sela M, Schechter B, Schechter M, Borek F (1967) Antibodies to sequential and conformational antigenic determinants. Cold Spring Harbor Symp Quant Biol 32:537–545

Sellers A, Murphy G (1981) Collagenolytic enzymes and their naturally occurring inhibitors. Int Rev Connect Tissue Res 9:151–182

Sellers A, Reynolds JJ (1977) Identification and partial characterization of an inhibitor of collagenase form rabbit bone. Biochem J 167:353–360

Sellers A, Cartwright E, Murphy C, Reynolds JJ (1977) Evidence that latent collagenases are enzyme-inhibitor complexes. Biochem J 163:303–307

Seltzer JL, Welgus HG, Jeffrey JJ, Eisen AZ (1976) The function of calcium ion in the action of mammalian collagenase. Arch Biochem Biophys 173:355–361

Seltzer JL, Jeffrey JJ, Eisen AZ (1977) Evidence for mammalian collagenases as zinc ion metalloenzymes. Biochim Biophys Acta 484:179–187

Seltzer JL, Eschbach ML, Winberg JO, Bauer EA, Eisen AZ, Weingarten H (1987) Eriochrome black T inhibition of human skin collagenase, but not gelatinase, using both protein and synthetic substrates. Coll Relat Res 7:399–407

Selye H (1957) Lathyrism. Rev Can Biol 16:1–82

Selye H (1970) Prevention of various forms of metabolic myocardial necrosis by catatoxic steroids. J Mol Cell Cardiol 1:91–99

Seyer M (1980) Interstitial collagen polymorphism in rat liver with $CCl_4$-induced cirrhosis. Biochim Biophys Acta 629:490–498

Seyer J, Hasty KA, Kang AH (1989) Covalent structure of collagen. Amino acid sequence of an arthritogenic cyanogen bromide peptide from type II collagen of bovine cartilage. Eur J Biochem 181:159–173

Seyer JM, Kang AH (1989) Covalent structure of collagen: amino acid sequence of three cyanogen bromide-derived peptides from human $\alpha$1(V) collagen chain. Arch Biochem Biophys 271:120–129

Seyer JM, Kang AJ, Rodnan G (1981) Investigation of type III collagen in the lung in progressive systemic sclerosis (scleroderma). Arthritis Rheum 24:625–631

Shain WG, Hilfer SR, Fonte VG (1972) Early organogenesis of the embryonic chick thyroid. I. Morphology and biochemistry. Dev Biol 28:202–218

Shambaugh GE III (1986) Thyroid hormone action. Biologic and cellular effects. In: Ingbar SH, Braverman LE (eds) Werner's the thyroid: a fundamental and clinical text. Lippincott, Philadelphia, pp 201–219

Sharimianov YG, Buishvili LL, Mrevlishvili GM (1979) Izmeneniya gidratacii kollagena pri denaturacii (Changes of collagen hydratation after denaturation). Biofizika 24:606–610 (in Russian)

Sharma YD (1982) Variations in the metabolism and maturation of collagen after fluoride ingestion. Biochem Biophys Acta 715:137–141

Sheffield LG, Anderson RR (1984) Effect of estradiol and relaxin on collagen and non-collagen protein synthesis by mammary fibroblasts. Life Sci 35:2199–2203

Shekhonin BV, Domogatsky SP, Idelson GL, Koteliansky VE, Rukosuev VS (1987) Relative distribution of fibronectin and type I, III, IV, V collagens in normal and atherosclerotic intima of human arteries. Atherosclerosis 67:9–16

Shekhonin BV, Domogatsky SP, Idelson GL, Koteliansky VE (1988) Participance of fibronectin and various collagen types in the formation of fibrous extracellular matrix in cardiosclerosis. J Mol Cell Cardiol 20:501–508

Shellswell GB, Restall DJ, Duance VC, Bailey AJ (1979) Identification and differential distribution of collagen types in the central and peripheral nervous systems. FEBS Lett 106:305–308

Shen S, Stramm LE, Li W, Robertson GA, Aguirre CD, Rockey JH (1985) Translation of type IV procollagen messenger RNA from cultured cat retinal pigment epithelial cells. Ophthalmic Res 17:216–221

Shen V, Kohler G, Jeffrey JJ, Peck WA (1988) Bone-resorbing agents promote and interferon-$\gamma$ inhibits bone cell collagenase production. J Bone Miner Res 3:657–666

Sherman MI, Gay R, Gay S, Miller EJ (1980) Association of collagen with preimplantation and peri-implantation mouse embryos. Dev Biol 74:470–476

Shetlar MR, Shetlar CL, Kischer CW (1979) Healing of myocardial infarction in animal models. Tex Rep Biol Med 39:339–355

Shinkai H (1984) pN-Collagen peptides in sera of patients with progressive systemic sclerosis. In: Black CM, Muers AR (eds) Systemic sclerosis (scleroderma). Gower, New York, pp 209–214

Shinkai H, Hirabayashi O, Tamaki A, Matsubayashi S, Sano S (1976) Connective tissue metabolism in culture fibroblasts of a patient with Ehlers-Danlos syndrome type I. Arch Dermatol Res 257:113–123

Shiota G, Murawaki Y, Hirayama C (1987) Hepatic collagen content and lysyl oxidase activity in rats fed a low protein-ethanol diet. Res Commun Chem Pathol Pharmacol 58:115–127

Shiratori Y, Geerts A, Ichida T, Kawase T, Wisse E (1986) Kupffer cells from $CCl_4$-induced fibrotic livers stimulate proliferation of fat-storing cells. J Hepatol 3:294–303

Shoshan S, Finkelstein S, Kushner W, Weinreb M (1972) Maturation and crosslinking of collagen implants in hypophysectomized rats in vivo. Connect Tiss Res 1:47–53

Shows TB, Tikka L, Byers MG, Eddy RL, Haley LL, Henry WM, Prockop DJ, Tryggvason K (1989) Assignment of the human collagen α 1(XIII) chain gene (COL 13 A 1) to the q22 region of chromosome 10. Genomics 5:128–133

Shtacher G, Maayan R, Feinstein G (1973) Proteinase inhibitors in human synovial fluids. Biochem Biophys Acta 303:138–147

Shull S, Cutroneo KR (1983) Glucocorticoids coordinately regulate procollagens type I and type III synthesis. J Biol Chem 258:3364–3369

Shupp-Byrne DE, Church RL (1982) "Embryonic" collagen (type I trimer) α 1-chains are genetically distinct from type I collagens α 1-chains. Coll Relat Res 2:481–494

Shuttleworth CA, Smalley JW (1983) Periodontal ligament. Int Rev Connect Tissue Res 10:211–247

Shuttleworth CA, Berry L, Wilson N (1980) Collagen synthesis in rabbit dental pulp fibroblast cultures. Arch Oral Biol 25:201–208

Shvelidze TI, Tsagarelli ZG (1978) Electron-microscopic study of collagenous fibers of cultures of bone marrow from healthy children. Acta Anat (Basel) 100:111–113

Shvelidze TI, Tsagarelli ZG (1980) The ultrastructure of collagenous fibers in bone marrow long-term cultures. Acta Anat (Basel) 108:153–155

Siegel RC (1979) Lysyl oxidase. Int Rev Connect Tissue Res 8:73–118

Siegel RC, Chen KH, Greenspan JS, Aguiar JM (1978) Biochemical and immunochemical study of lysyl oxidase in experimental hepatic fibrosis in the rat. Proc Natl Acad Sci USA 75:2945–2949

Siegel RC, Black CM, Bailey AJ (1979) Crosslinking of collagen in the X-linked Ehlers-Danlos syndrome type V. Biochem Biophys Res Commun 88:281–287

Sierakowski SJ, Kucharz EJ (1988) Doświadczalne modele twardziny uogólnionej (Experimental animal models of systemic sclerosis). Postępy Hig Med Dośw 42:619–640

Siggers DC, Rimoin DL, Dorst JP, Dotz SB, Williams BR, Hollister DW, Silberberg R, Cranley RE, Kaufman RL, McKusick VA (1974) The Kniest syndrome. Birth Defects 10:193–208

Silbert CK, Kleinman HK (1979) Studies of cultured human fibroblasts in diabetes mellitus. Changes in heparon sulfate. Diabetes 28:61–64

Sillence DO (1981) Osteogenesis imperfecta: an expanding panorama of variants. Clin Orthop 159:11–25

Sillence DO, Senn A, Danks DM (1979a) Genetic heterogeneity in osteogenesis imperfecta. J Med Genet 16:101–116

Sillence DO, Horton WA, Rimoin DL (1979b) Morphologic studies in the skeletal dysplasias. A review. Am J Pathol 96:811–870

Sillence DO (1983) Disorders of bone density, volume and mineralisation. In: Emery AEH, Rimoin DL (eds) Principles and practice of medical genetics. Churchill Livingstone, Edinburgh, vol 2, pp 736–751

Sillness J, Gustavsen F, Fejerskov O, Karring T, Loe H (1976) Cellular, afibrillar coronal cementum in human teeth. J Periodont Res 11:331–336

Simila S, Vera L, Wasz-Höckert O (1970) Hereditary onychoosterodysplasia (the nail-patella syndrome) with nephrosis-like renal disease in a newborn boy. Pediatrics 46:61–65

Simon LS, Krane SM, Wortman PD, Krane IM, Kovitz KL (1984) Serum levels of type I and III procollagen fragments in Paget's disease of bone. J Clin Endocrinol Metab 58:110–120

Simons ER, Chesney CM, Colman RW, Harper E, Samberg E (1975) The effect of the conformation of collagen on its ability to aggregate platelets. Thromb Res 7:123–139

Simpson CF, Taylor WJ (1982) Effect of hydralazine on aortic rupture induced by beta-aminopropionitrile in turkeys. Circulation 65:704–708

Singh M, Bachhwat BK (1965) The distribution and variation with age of different uronic acid-containing mucopolysaccharides in brain. J Neurochem 12:519–525

Sippola M, Kaffe S, Prockop DJ (1984) A heterozygous defect for structurally altered pro-α2 chain of type I procollagen in a mild variant of osteogenesis imperfecta. J Biol Chem 259:14094–14100

Siuko H, Sävelä J, Kulonen E (1959) Effect of the hydrocortisone on the formation of collagen in guinea pig skin. Acta Endocrinol 31:113–116

Sjoerdsma A, Davidson JD, Udenfriend S, Mitoma C (1958) Increased excretion of hydroxyproline in Marfan's syndrome. Lancet 2:994

Sjoerdsma A, Udenfriend S, Keiser H, LeRoy EC (1965) Hydroxyproline and collagen metabolism. Clinical implications. Ann Intern Med 63:672–694

Skinner SJM, Campos GA, Liggins GC (1981) The collagen content of human amniotic membranes: effect of gestation lenght and premature rupture. Obstet Gynecol 57:487–489

Skirving AP, Sims TJ, Bailey AJ (1984) Congenital dislocation of the hip: a possible inborn error of collagen metabolism. J Inherited Metab Dis 7:27–31

Skosey JL, Zak R, Martin AF, Aschenbrenner V, Rabinowitz M (1972) Biochemical correlates of cardiac hypertrophy. V. Labeling of collagen, myosin, and nuclear DNA during experimental myocardial hypertrophy in the rat. Circ Res 31:145–157

Slansky HH, Berman MB, Dohlman CH, Rose J (1970) Cysteine and acetylcysteine in the prevention of corneal ulcerations. Ann Ophthalmol 2:488–491

Slansky HH, Dohlman CH, Berman MB (1971) Prevention of corneal ulcers. Trans Am Acad Ophthalmol Otolaryngol 75:1208–1211

Slavkin HC (ed) (1972) The comparative molecular biology of extracellular matrices. Academic, New York

Slutskii LI (1984) Current concepts about the collagen components of cartilage tissue. Acta Biol Hung 35:151–161

Smedsrød B (1988) Aminoterminal propeptide of type III procollagen is cleared from the circulation by receptor-mediated endocytosis in liver. Coll Relat Res 8:375–388

Smelser GK (1937) A comparative study of experimental and clinical exophthalmos. Am J Ophthalmol 20:1189–1193

Smelser GK, Polack FM, Ozanics V (1965) Persistence of donor collagen in corneal transplants. Exp Eye Res 4:349–354

Smiley JD, Ziff M (1964) Urinary hydroxyproline excretion and growth. Physiol Rev 44:30–36

Smith BH (1966) Peyronie's disease. Am J Clin Pathol 45:670–678

Smith DM, Sommers Smith SK (1988) Morphology of the fibrogenic response of adult rat lung to constinuous beta-blockade. Histol Histopathol 3:291–295

Smith FM (1918) The ligation of coronary arteries with electrocardiographic study. Arch Intern Med 22:8–27

Smith GN Jr, Williams JM, Brandt KD (1985) Interaction of proteoglycans with the pericellular ($1\alpha$, $2\alpha$, $3\alpha$) collagens of cartilage. J Biol Chem 260:10761–10767

Smith JG, Davidson EA, Sams WM, Clark RD (1962) Alterations in human dermal connective tissue with age and chronic sun exposure. J Invest Dermatol 39:347–350

Smith JW, Serafini-Fracassini A (1968) The distribution of the protein-polysaccharide complex in the nucleus pulposus matrix in young rabbits. J Cell Sci 3:33–40

Smith OW, Kaltreider NB (1963) Collagen content of the non-pregnant rat uterus as related to the functional responses to estrogen and progesterone. Endocrinology 73:619–624

Smith P, Heath D, Kay JM (1974) The pathogenesis and structure of paraquat-induced pulmonary fibrosis in rats. J Pathol 114:57–67

Smith QT, Allison DJ (1965) Skin and femur collagens and urinary hydroxyproline of cortisone-treated rats. Endocrinology 77:785–791

Smith QT, Allison DJ (1966) Changes of collagen content in skin, femur and uterus of 17-estradiol benzoate-treated rats. Endocrinology 79:486–491

Smith QT, Rukavina GT, Haaland EM (1965) Urinary hydroxproline in various diseases. Acta Derm Venereol (Stockh) 45:44–48

Smith R (1967) Total urinary hydroxyproline in primary hyperparathyroidism. An assessment of its clinical significance. Clin Chim Acta 18:47–50

Smith R (1975) Myositis ossificans progressiva: a review of current problems. Sem Arthritis Rheum 4:369–380

Smith R (1980) Collagen and disorders of bone. Clin Sci 59:215–223

Smith R (1986) Osteogenesis imperfecta. Clin Rheum Dis 12:655–690

Smith R, Sykes B (1985) Osteogenesis imperfecta (the brittle bone syndrome) advances and controversies. Calcif Tissue Int 37:107–111

Smith R, Francis MJO, Houghton GR (1983) The brittle bone syndrome: osteogensis imperfecta. Butterworth, London
Smith RA, Stehbens WE, Weber P (1976) Hemodynamically-induced increase in soluble collagen in the anastomosed veins of experimental arteriovenous fistulae. Atherosclerosis 23:429–436
Snowden J McK, Swann DA (1980) Effects of glycosaminoglycans and proteoglycans on the in vitro assembly and thermal stability of collagen fibrils. Biopolymers 19:767–780
Snowden JM, Eyre DR, Swann DA (1982) Vitreous structure. VI. Age-related changes in the thermal stability and crosslinks of vitreous, articular cartilage and tendon collagens. Biochim Biophys Acta 706:153–162
Sodek J (1977) A comparison of the rates of synthesis and turnover of collagen and non-collagen proteins in adult rat periodontal tissues and skin using a microassay. Arch Oral Biol 22:655–665
Sodek J, Limeback HF (1979) Comparison of the rates of synthesis, conversion, and maturation of type I and type III collagens in rat periodontal tissues. J Biol Chem 254:20496–20501
Soininen R, Huotari M, Ganguly A, Prockop DJ, Tryggvason K (1989) Structural organization of the gene for the $\alpha 1$ chain of human type IV collagen. J Biol Chem 264:13565–13571
Solis-Herruzo JA, Brenners DA, Chojkier M (1988) Tumor necrosis factor a inhibits collagen gene transcription and collagen synthesis in cultured human fibroblasts. J Biol Chem 263:5841–5845
Solomon E, Hioras L, Dalgleish R, Tolstoshev P, Crystal R, Sykes B (1983) Regional localization of the human $\alpha 2$(I) collagen gene on chromosome 7 by molecular hybridization. Cytogenet Cell Genet 35:64–66
Solomon E, Hiorns L, Sheer D, Rowe D (1984) Conformation that the type I collagen gene on chromosome 17 is COL1A1 ($\alpha 1$(I)), using a human genomic probe. Ann Hum Genet 48:39–42
Sølvsten-Sørensen M, Nielsen LP, Bretlau P, Jørgensen HB (1988) The role of type II collagen autoimmunity in otosclerosis revisited. Acta Otolaryngol (Stockh) 105:242–247
Somers KD, Sismour EN, Wright GL Jr, Devine CJ Jr, Gilbert DA, Horton CE (1989) Isolation and characterization of collagen in Peyronie's disease. J Urol 141:629–631
Sommers Smith SK, Smith DM (1989) Effect of continuous $\beta$-blockade on collagen synthesis in interstitial fibroblasts isolated from adult rat lung. An in vitro model of progressive pulmonary fibrogenesis. Toxicol In Vitro 3:129–136
Song ZX, Quesenberry P (1984) Radioresistant murine marrow stromal cells: a morphologic and functional characterization. Exp Hematol 12:523–533
Sorgente N, Brownell AG, Slavkin HC (1977) Basal lamina degradation: the identification of mammalian-like collagenase activity in mesenchymal-derived matrix vesicles. Biochem Biophys Res Commun 74:448–454
Sorsa T, Uitto VJ, Suomalainen K, Turto H, Lindy S (1987) Human leukocyte collagenase: recent biochemical findings. Proc Finn Dent Soc 83:111–117
Soskel NT, Sandberg LB (1986) Pulmonary emphysema. From animal models to human diseases. In: Uitto J, Perejda AJ (eds) Connective tissue disease: molecular pathology of the extracellular matrix. Dekker, New York, pp 423–453
Sottrup-Jensen L, Birkedal-Hansen H (1989) Human fibroblast collagenase-$\alpha$-macroglobulin interactions. Localization of cleavage sites in the bait regions of five mammalian $\alpha$-macroglobulins. J Biol Chem 264:393–401
Spanheimer RG (1988) Direct inhibition of collagen production in vitro by diabetic rat serum. Metabolism 37:479–485
Spelsberg WW, Chapman GB (1962) Fine structure of human trabeculae. Arch Ophthalmol 67:773–784
Speranza ML, Valentini G, Calligaro A (1987) Influence of fibronectin on the fibrillogenesis of type I and type III collagen. Col Relat Res 7:115–123
Spiro RG, Spiro MJ (1971) Studies on the biosynthesis of the hydroxylysine-linked disaccharide unit of basement membranes and collagen. J Biol Chem 246:4919–4925
Sporn MB, Roberts AB (1986) Peptide growth factors and inflammation, tissue repair, and cancer. J Clin Invest 78:329–332

Spranger JW, Langer LO, Wiedemann HR (1974) Bone dysplasias: an atlas of constitutional disorders of skeletal development. Saunders, Philadelphia

Squier CA, Bausch WH (1984) Three-dimensional organization of fibroblasts and collagen fibrils in rat tail tendon. Cell Tissue Res 238:319–327

Ssadikow WS (1927) Über ein neues Kollagenlösendes Ferment (Kollagenase). Biochem Z 181:267–283

Stachow A (1976) The role of disorders of collagen in scleroderma. In: Jabłońska S (ed) Scleroderma and pseudoscleroderma. State Medical, Warsaw, pp 86–95

Stack MS, Gray RD (1989) Comparison of vertebrate collagenase and gelatinase using a new fluorogenic substrate peptide. J Biol Chem 264:4277–4281

Stančiková M, Trnavský K, Keilová H (1977) The effect of antirheumatic drugs on collagenolytic activity of cathepsin B1. Biochem Pharmacol 26:2121–2124

Stančiková M, Fryšák Z, Trnavský K (1987) Effect of colchicine on the activity of cathepsin B and D in human liver cirrhosis. Acta Med Hung 44:181–188

Stanescu V, Maroteaux P, Stanescu R (1976) Etude par electrophorese sur gel des chaines a et des CNBr peptides du collagene du cartilage de croissance dans les chondriodysplasies. Ann Genet (Paris) 19:119–125

Staubesand J, Fischer N (1980) The ultrastructural characteristics of abnormal collagen fibrils in various organs. Connect Tissue Res 7:213–217

St Clair RW, Toma JJ Jr, Lofland HB (1975) Proline hydroxylase activity and collagen content of pigeon aortas with naturally-occurring and cholesterol-aggrevated atherosclerosis. Atherosclerosis 21:155–165

Steffen C (1969) Tissue antibodies in rheumatoid arthritis and other connective tissue diseases. Ann Immunol 1:47–58

Steffen C, Timpl R (1963) Antigenicity of collagen and its application in the serological investigation of rheumatoid arthritis sera. Int Arch Allergy Appl Immunol 22:333–349

Steffen C, Ludwig H, Knapp W (1974) Collagen-anticollagen immune complexes in rheumatoid arthritis synovial fluid cells. Z Immunitaetsforsch 147:229–235

Steffen C, Zeitlhofer J, Zielinski C, Menzel J, Smolen J (1979) Collagenase-induced experimental arthritis. Z Rheumatol 38:1–10

Steinbronn D, Carlson E, Chvapil M (1974) Antifibroblast serum: a new method of controlling collagen synthesis. Surg Forum 25:47–49

Steinetz BG, Beach VL, Elden HR (1966) Some effects of hormones on contractile properties of rat tail tendon collagen. Endocrinology 79:1047–1052

Steinman B, Gitzelmann R, Vogel A, Grant ME, Harwood R (1975) Ehlers-Danlos syndrome in two sybblings with deficient lysyl hydroxylase in skin. Helv Paediatr Acta 30:255–274

Steinmann BU, Martin GR, Baum BI, Crystal RG (1979) Synthesis and degradation of collagen by skin fibroblasts from controls and from patients with osteogenesis imperfecta. FEBS Lett 101:269–272

Steinmann B, Tuderman L, Peltonen L, Martin GR, McKusick VA, Prockop DJ (1980) Evidence for a structural mutation of procollagen type I in a patient with Ehlers-Danlos syndrome type VII. J Biol Chem 255:8887–8893

Steinmann B, Nicholls A, Pope FM (1986) Clinical variability of osteogenesis imperfecta reflecting molecular heterogeneity: cysteine substitution in the $\alpha 1$(I) collagen chain producing lethal and mild forms. J Biol Chem 261:8958–8964

Stemerman MB (1974) Vascular intimal components: precursors of thrombosis. Prog Hemost Thromb 2:1–47

Stemerman MB, Baumgartner HR, Spaet TH (1971) Subendothelial microfibril and platelet adhesion. Lab Invest 24:179–186

Stenbäck F, Risteli J, Risteli L, Wasenius VM (1985) Basement membrane laminin and type IV collagen in endometrial adenocarcinoma: relation to differentiation and treatment. Oncology 32:370–376

Sterling KM Jr, DiPetrillo TA, Cutroneo KR, Prestayko A (1982) Inhibition of collagen accumulation by glucocorticoids in rat lung after intratracheal bleomycin instillation. Cancer Res 42:405–408

Sterling KM Jr, Harris MJ, Mitchell JJ, DiPetrillo TA, Delaney GL, Cutroneo KR (1983a) Dexamethasone decreases the amounts of type I procollagen mRNAs in vivo and in fibroblast cell cultures. J Biol Chem 258:7644–7647
Sterling KM Jr, Harris MJ, Mitchell JJ, Cutroneo KR (1983b) Bleomycin treatment of chick fibroblasts causes an increase of polysomal type I procollagen mRNAs. Reversal of the bleomycin effect by dexamethasone. J Biol Chem 258:14438–14444
Sternlieb I (1975) The beneficial and adverse effects of penicillamine. In: Popper H, Becker K (eds) Collagen metabolism in the liver. Stratton Interconticontinental Medical Book, New York, pp 183–190
Sterzel RB, Lovett DH, Foellmer HG, Perfetto M, Biemesderfer D, Kashgarian M (1986) Mesangial cell hillocks. Nodular foci of exaggerated growth of cells and matrix in prolonged culture. Am J Pathol 125:130–140
Stetler-Stevenson WG, Krutzsch HC, Wacher MP, Margulies IMK, Liotta LA (1989) The activation of human type IV collagenase proenzyme. Sequence identification of the major conversion product following organomercurial activation. J Biol Chem 264:1353–1356
Stetten MR, Schoenheimer R (1944) The metabolism of L(–) proline studied with the aid of deuterium and isotopic nitrogen. J Biol Chem 153:113–132
Steven FS (1966) The depolymerising action of pepsin on collagen. Molecular weights of the component polypeptide chains. Biochim Biophys Acta 130:190–195
Steven FS, Knott J, Jackson DS, Podrazky V (1969) Collagen-protein-polysaccharide interactions in human intervertebral disc. Biochim Biophys Acta 188:307–313
Stevenson JC (1988) Pathophysiology of osteoporosis. Triangle 27:47–52
Stewart RE, Hollister DW, Rimoin DL (1977) A new variant of Ehlers-Danlos syndrome: an autosomal dominant disorder of fragile skin, abnormal scarring and generalized periodontosis. Birth Defects 38:88–93
Stockman A, Rowley MJ, Emery P, Muirden KD (1989) Activity of rheumatoid arthritis and levels of collagen antibodies: a prospective study. Rheumatol Int 8:239–243
Stoerk HC, Bielinski TC, Budzilovich T (1954) Chronic polyarthritis in rats injected with spleen adjuvants. Am J Pathol 30:616–621
Stoianova IG, Nekrasova TA, Zaides AL (1960) Issledovaniye kollagena vo vlazhnom sostoyanii v gazovoi mikrokamere elektronnogo mikroskopa. Deistvie na kollagen ionizuruyushtzego izlutzeniya (Investigation of collagen under native conditions of gasous microchamber in electron microscopy. Effect of ionizing radiation on collagen). Dokl Akad Nauk SSSR 130:1366–1369 (in Russian)
Stolle CA, Pyeritz RE, Myers JC, Prockop DJ (1985) Synthesis of an altered type III procollagen in a patient with type IV Ehlers-Danlos syndrome. J Biol Chem 260:1937–1944
Stone N, Meister A (1962) Function of ascorbic acid in the conversion of proline to collagen hydroxyproline. Nature 194:555–557
Strang CJ, Slayter HS, Lachmann PJ, Davis AE (1986) Ultrastructure and composition of bovine conglutinin. Biochem J 234:381–389
Stransky G, Weis S, Neumüller J, Hakimzadeh A, Firneis F, Ammer K, Partsch G, Eberl R (1987) Morphometric analysis of collagen fibrils in idiopathic carpal tunnel syndrome. Exp Cell Biol 55:57–62
Stratton MA (1985) Drug-induced systemic lupus erythematosus. Clin Pharm 4:657–663
Strauch L, Vencelj H (1967) Collagenases in mammalian cells. Z Physiol Chem 348:465–468
Strickland S, Beers WH (1976) Studies on the role of plasminogen activator in ovulation. In vitro response of granulosa cells to gonadotropins, cyclic nucleotides and prostaglandins. J Biol Chem 251:5694–5699
Stricklin GP, Hibbs MS (1988) Biochemistry and physiology of mammalian collagenases. In: Nimni ME (ed) Collagen, vol I. CRC Press, Boca Raton, pp 187–206
Stricklin GP, Welgus HG (1983) Human skin fibroblast collagenase inhibitor. Purification and biochemical characterization. J Biol Chem 258:12252–12257
Stricklin GP, Bauer EA, Jeffrey JJ, Eisen AZ (1977) Human skin collagenase: isolation of precursor and active forms from both fibroblast and organ cultures. Biochemistry 16:1607–1615
Stricklin GP, Eisen AZ, Bauer EA, Jeffrey JJ (1978) Human skin fibroblast collagenase: chemical properties of precursor and active forms. Biochemistry 17:2331–2337

Stricklin GP, Jeffrey JJ, Roswitt WT, Eisen AZ (1983) Human skin fibroblast procollagenase: mechanisms of activation by organomercurials and trypsin. Biochemistry 22:61–68
Strom CM, Eddy RL, Shows TB (1984) Localization of human type II procollagen gene (COL2A1) to chromosome 12. Somatic Cell Mol Genet 10:651–655
Stuart JM, Dixon FJ (1983) Serum transfer of collagen-induced arthritis in mice. J Exp Med 158:378–392
Stuart JM, Postlethwaite AE, Kang AH (1976) Evidence for cell-mediated immunity to collagen in progressive systemic sclerosis. J Lab Clin Med 88:601–607
Stuart JM, Cremer MA, Dixit SN, Kang AH, Townes AS (1979) Collagen-induced arthritis in rats. Comparision of vitreous and cartilage-derived collagens. Arthritis Rheum 22:347–352
Stuart JM, Postlethwaite AE, Kang AH, Townes AS (1980) Cell-mediated immunity to collagen in rheumatoid arthritis and other rheumatic diseases. Am J Med 69:13–18
Stuart JM, Cremer MA, Townes AS, Kang AH (1982) Type II Collagen-induced arthritis in rats: passive transfer with serum and evidence that IgG anticollagen antibodies can cause arthritis. J Exp Med 155:1–12
Stuart JM, Huffstutter EH, Townes AS, Kang AH (1983) Incidence and specificity of antibodies to types I, II, III, IV, and V collagen in rheumatoid arthritis and other rheumatic diseases as measured by $^{125}$I-radioimmunoassay. Arthritis Rheum 26:832–840
Stys SJ, Clewell WH, Meschia G (1978) Changes in cervical compliance at parturition independent of uterine activity. Am J Obstet Gynecol 130:414–419
Suárez G, Rajaram R, Bhuyan KC, Oronsky AL, Goidl JA (1988) Administration of an aldolase reductase inhibitor induces a decrease of collagen fluorescence in diabetic rats. J Clin Invest 82:624–627
Suki WN, Caskey CT (1979) Hereditary chronic nephropathies. In: Earley LE, Gottschalk CW (eds) Strauss and Welt's diseases of the kidney vol 2. Little and Brown, Boston, pp 1167–1195
Sultan C, Loire C, Kern P, Fenard O, Terraza A (1986) Collagène et hormones stéroïdes. Ann Biol Clin (Paris) 44:285–288
Sun JD, Pickrell JA, Harkema JR, McLaughlin SI, Hahn FF, Henderson RF (1988) Effects of buthionine sulfoximine on the development of ozone-induced pulmonary fibrosis. Exp Mol Pathol 49:254–266
Sundar-Raj CV, Freeman IL, Brown SI (1980) Selective growth of rabbit corneal epithelial cells in culture and basement membrane collagen synthesis. Invest Ophthalmol Vis Sci 19:1222–1228
Sundar-Raj N, Freeman I, Buckingham RB, Prince RK, Rodnan GP (1984) Surface proteins of scleroderma fibroblasts in culture. J Rheumatol 11:53–55
Sunderland S (1965) The connective tissue of peripheral nerves. Brain 88:841–846
Superti-Furga A, Steinmann B (1988) Impaired secretion of type III procollagen in Ehlers-Danlos syndrome type IV fibroblasts: correction of the defect by incubation at reduced temperature and demonstration of subtle alterations in the triple-helical region of the molecule. Biochem Biophys Res Commun 150:140–147
Superti-Furga A, Steinmann B, Ramirez F, Byers PH (1989) Molecular defects of type III procollagen in Ehlers-Danlos syndrome type IV. Hum Genet 82:104–108
Surrenti C, Casini A, Milani S, Ambu S, Ceccatelli P, D'Agata A (1987) Is determination of serum N-terminal procollagen type III peptide a marker of hepatic fibrosis? Dig Dis Sic 32:705–709
Sussman BJ (1968) Intervertebral discolysis with collagenase. J Natl Med Assoc 60:184–187
Sussman BJ, Mann M (1969) Experimental intervertebral discolysis with collagenase. J Neurosurg 31:628–632
Suzuki F, Koyama E (1969) Hydroxylation of proline in collagen model peptide. Biochim Biophys Acta 177:154–156
Suzuki Y, Ito M, Hamaguchi Y, Yamagami I (1974) Biochemical studies on the mechanism of action of a new anti-inflammatory agent, naproxen. I. Effects of naproxen on the changes in connective tissue components due to inflammation. Folia Pharmacol Jpn 70:465–477
Suzuki Y, Ito M, Yamagami I (1976) Mechanism of action of a new antiinflammatory agent, naproxen. II. Effects of naproxen on activities of mucopolysaccharase, acid proteinase and collagenolytic enzymes in inflammed tissues. Jpn J Pharmacol 26:91–103

Svercar J (1975) Biochemical analysis of epiphyseal cartilage during growth and in some chondriodyslpasias. Birth Defects 11:227–230
Svojtková E, Deyl Z, Rosmus J, Adam M (1972) Aging of connective tissue. The effect of diet and x-irradiation. Exp Gerontol 7:157–167
Svojetková E, Deyl Z, Šmíd A, Adam M (1977) The occurrence of collagen type II in bronchogenic carcinoma. Neoplasma 24:437–443
Swann DA, Sotman SS (1980) The chemical composition of bovine vitreous-humour collagen fibres. Biochem J 185:545–554
Swann JC, Reynolds JJ, Galloway WA (1981) Zinc metalloenzyme properties of active and latent collagenase from rabbit bone. Biochem J 195:41–49
Swarz W, Graf-Kaysenlingk D (1969) Electron microscopy of normal and opaque human cornea. In: Langham M (ed) The cornea. John Hopkins, Baltimore, p 123
Sykes B, Smith R (1985) Collagen and collagen gene disorders. Q J Med 56:533–547
Sylvén B (1949) Ester sulfuric acids in stroma connective tissue. Acta Radiol 32:11–16
Sylvén B (1958) The biochemical mechanism underlying the destructive growth of tumors. Acta Union Int Contra Cancrum 14:61–62
Sylvester MF, Yannas IV, Salzman EW, Forbes MJ (1989) Collagen banded fibril structure and the collagen-platelet reaction. Thromb Res 55:135–148
Taffel M, Harvey SC (1938) Effect of absolute and partial vitamin C deficiency on healing of wounds. Proc Soc Exp Biol Med 38:518–525
Taïeb A, Aumailley M, Courouge-Dorcier D, Rabaud M, Bioulac-Sage P, Surléve-Bazeille JE, Maleville J (1987) Collagen studies in congenital cutis laxa. Arch Dermatol Res 279:308–314
Tajima S, Pinnell SR (1982) Regulation of collagen synthesis by ascorbic acid: ascorbic acid increases type I procollagen mRNA. Biochem Biophys Res Commun 106:632–637
Takács I, Verzár F (1968) Macromolecular ageing of collagen. III. Stimulation of collagen production in the skin and uterus. Gerontologia 14:126–132
Takahashi S (1979) A study on myocardial fibrosis in myocardial infarction and in idiopathic cardiomyopathy: a measurement of hydroxyproline level in plasma and in myocardium. Jpn Circ J 43:913–921
Takehara K, Grotendorst GR, Trojanowska M, LeRoy EC (1986) Ascorbate effects on type I procollagen synthesis by human skin fibroblasts: different migration positions of type I procollagen chains on SDS polyacrylamide gel after incubation with ascorbate. Coll Relat Res 6:455–466
Takeuchi T, Prockop DJ (1969) Protocollagen proline hydroxylase in normal liver and in hepatic fibrosis. Gastroenterology 56:744–750
Tamburro AM, Guantieri V, Cabrol D, Broch H, Vasilescu D (1984) Experimental and theoretical conformational studies on polypeptide models of collagen. Int J Pept Protein Res 24:627–635
Tan EML, Uitto J, Bauer EA, Eisen AZ (1981) Human skin fibroblasts in culture: procollagen synthesis in the presence of sera from normal human subjects and from patients with dermal fibrosis. J Invest Dermatol 76:462–467
Tan EML, Ryhänen L, Uitto J (1983) Proline analogues inhibit human skin fibroblast growth and collagen production in culture. J Invest Dermatol 80:261–267
Tane N, Hashimoto K, Kanzaki T, Ohyama H (1978) Collagenolytic activities of cultured human malignant melanoma cells. J Biochem (Tokyo) 84:1171–1176
Tanzer ML (1965) Experimental lathyrism. Int Rev Connect Tissue Res 3:91–112
Tanzer ML (1976) Cross-linking In: Ramachandran GN, Reddi AH (eds) Biochemistry of collagen. Plenum, New York, pp 137–162
Tanzer ML, Gross J (1964) Collagen metabolism in the normal and lathyritic chick. J Exp Med 119:275–284
Tanzer ML, Kimura S (1988) Phylogenetic aspects of collagen structure and function. In: Nimni ME (ed) Collagen, vol. II. CRC Press, Boca Raton, pp 25–40
Tarin D, Hoyt BJ, Evans DJ (1982) Correlation of collagenase secretion with metastatic-colonization potential in naturally occurring murine mammary tumors. Br J Cancer 46:266–278
Taubenhaus M, Amromin GD (1950) The effects on the hypophysis, thyroid, sex steroids, and the adrenal cortex upon granulation tissue. J Lab Clin Med 36:7–18

Taylor TKF, Ghosh P, Bushell GR, Sutherland JM (1976a) Disc metabolism in scoliosis. In: Zorab PA (ed) Scoliosis. Academic, London, pp 231–246

Taylor TKF, Ghosh P, Braund KG, Sutherland JM, Shewood AA (1976b) The effect of spinal fusion on intervertebral disc composition: an experimental study. J Surg Res 21:91–104

Termine JD, Gehron Robey P, Fisher LW, Shimokawa MA, Drum MA, Conn KM, Hawkins GR, Cruz JB, Thompson KG (1984) Osteonectin bone proteoglycan, and phosphophoryn defects in a form of bovine osteogenesis imperfecta. Proc Natl Acad Sci USA 81:2213–2217

Thesleff I, Lehtonen E, Saxen L (1978) Basement membrane formation in transfilter tooth culture and its relation to odontoblast differentiation. Differentiation 10:71–75

Thesleff I, Stenman S, Vaheri Å, Timpl R (1979) Changes in the matrix proteins, fibronectin and collagen, during differentiation of mouse tooth germ. Dev Biol 70:116–126

Thiedemann KU, Holubarsch CH, Medugorac I, Jacob R (1983) Connective tissue content and myocardial stiffness in pressure overload hypertropy; a combined study of morphologic, morphometric, biochemical, and mechanical parameters. Basic Res Cardiol 78:140–155

Thiel S, Reid KBM (1989) Structures and functions associated with the group of mammalian lectins containing collagen-like sequences. FEBS Lett 250:78–84

Thilander H (1968) Epithelial changes in gingivitis. J Periodont Res 3:303–308

Thomas HF (1979) The extent of the odontoblast process in human dentin. J Dent Res 58:2207–2218

Thorgeisson UP, Liotta LA, Kalebic T, Margulies IM, Thomas K, Rios-Candelore M, Russo RG (1982) Effect of natural protease inhibitors and a chemoattractant on tumor cell invasion in vitro. J N C I 69:1049–1054

Thorner PS, Baumal R, Eddy A, Marrano P (1989) Characterization of the NC1 domain of collagen type IV in glomerular basement membranes (GBM) and of antibodies to GBM in a patient with anti-GBM nephritis. Clin Nephrol 31:160–168

Tidman MJ, Eady RA (1985) Evaluation of anchoring fibrils and other components of the dermal-epidermal junction in dystrophic epidermolysis bullosa by a quantitative ultrastructural technique. J Invest Dermatol 84:374–376

Tikka L, Pihlajaniami T, Hentle P, Prockop DJ, Tryggvason K (1988) Gene structure for the $\alpha 1$ chain of a human short chain collagen (type XIII) with alternatively spliced transcripts and translation termination codon at the 5′ end of the last exon. Proc Natl Acad Sci USA 85:7491–7495

Timpl R (1982) Antibodies to collagens and procollagens. Methods Enzymol 82:472–498

Timpl R (1984) Immunology of the collagens. In: Piez KA, Reddi AH (eds) Extracellular matrix biochemistry. Elsevier, New York, pp 159–190

Timpl R (1986) Recent advances in the biochemistry of glomerular basement membrane. Kidney Int 30:293–298

Timpl R, Engel J (1987) Type VI collagen. In: Mayne R, Burgeson RE (eds) Structure and function of collagen types. Academic, Orlando, pp 105–143

Timpl R, Furthmayr H, Hahn E, Becker U, Stoltz M (1973) Immunochemistry of collagen. Behring Inst Mitt 53:66–79

Timpl R, Wick G, Gay S (1977) Antibodies to distinct types of collagens and procollagens and their application in immunohistology. J Immunol Methods 18:165–171

Timpl R, Wiedemann H, Van Delden V, Furtmayr H, Kühn K (1981) A network model for the organization of type IV collagen molecules in basement membranes. Eur J Biochem 120:203–211

Tkocz C, Kühn K (1969) The formation of triple-helical collagen molecules from $\alpha 1$ and $\alpha 2$ polypeptide chains. Eur J Biochem 7:454–462

Tobin G, Aronson A, Chvapil M (1974) Effect of Ca EDTA administration on urinary hydroxyproline excretion and skin wound healing in the rat. J Surg Res 17:346–351

Todorovich-Hunter L, Johnson DJ, Ranger P, Keeley FW, Rabinovitch M (1988) Altered elastin and collagen synthesis associated with progressive pulmonary hypertension induced by monocrotaline. A biochemical and ultrastructural study. Lab Invest 58:184–195

Tonnesen MG, Worthen S, Johnston RB Jr (1988) Neutrophil emigration, activation, and tissue damage. In: Clark RAF, Henson PM (eds) The molecular and cellular biology of wound repair. Plenum, New York, pp 149–184

Too CKL, Bryant-Greenwood GD, Greenwood FC (1984) Relaxin increases the relases of plasminogen activator, collagenase, and proteoglycanase from rat granulosa cells in vitro. Endocrinology 115:1043–1050

Toole BP, Kang AH, Trelstad RL, Gross J (1972) Collagen heterogeneity within different growth regions of long bones of rachitic and non-rachitic chicks. Biochem J 127:715–720

Townes AS (1984) Antibodies to type II collagen. Mayo Clin Proc 59:791–796

Traub W, Piez KA (1971) The chemistry and structur of collagen. Adv Protein Chem 25:243–352

Trelstad RL, Hayashi K (1979) Tendon collagen fibrillogenesis: intracellular subassemblies and cell surface changes associated with fibril growth. Dev Biol 71:228–242

Trelstad RL, Kang AH (1974) Collagen heterogeneity in the avian eye: lens, vitreous body, cornea and sclera. Exp Eye Res 18:395–406

Trelstad RL, Hayashi K, Toole BP (1974) Epithelial collagens and glycosaminoglycans in the embryonic cornea: macromolecular order and morphogenesis in the basement membrane. J Cell Biol 62:875–880

Tremblay G (1976) Ultrastructure of elastosis in scirrhous carcinoma of the breast. Cancer 38:307–316

Trentham DE, Brinckerhoff CE (1982) Augmentation of collagen arthritis by synthetic analogues of retinoic acid. J Immunol 129:2668–2672

Trentham De, Townes AS, Kang AH (1977) Autoimmunity to type II collagen: an experimental model of arthritis. J Exp Med 146:857–868

Trentham DE, Dynesius RA, Rocklin RE, David JR (1978) Cellular sensitivity to collagen in rheumatoid arthritis. N Engl J Med 299:327–332

Trentham DE, McCune WJ, Susman P, David JR (1980) Autoimmunity to collagen in adjuvant arthritis of rats. J Clin Invest 66:1109–1117

Trentham DE, Kammer GM, McCune WJ, David JR (1981) Autoimmunity to collagen. A shared feature of psoriatic and rheumatoid arthritis. Arthritis Rheum 24:1363–1369

Trnavská Z, Trnavský K (1974) Influence of nonsteroidal antirheumatic drugs on collagen metabolism in rats with adjuvant-induced arthritis. Pharmacology 12:110–116

Trnavská Z, Mikulíková D, Trnavský K, Rovenský J (1978) Wirkung von Levamisol auf die Kollagensynthese in vitro. Z Rheumatol 37:221–224

Trnavský K, Trnavská Z (1971) The influence of phenylbutazone on collagen metabolism in vivo. Pharmacology 6:9–16

Trnavský K, Trnavská Z (1973) Vplyv nesteroidnych antireumatik na metabolizmus kolagenu (Effect of non-steroid anti-rheumatic drugs on metabolism of collagen). Vydavatelstvo Slovenskej Akademie Ved, Bratislava (in Slovak)

Trnavský K, Trnavská Z (1974) Pathophysiological mechanisms in the action of antirheumatics. Cesk Fysiol 23:423–435

Trnavský K, Vykydal M (1976) Farmakoterapie revmatických chorob (Pharmacotherapy of rheumatic diseases). Avicenum Zdravotnické Nakladatelství, Praha, pp 1–200 (in Slovak)

Trnavský K, Trnavska Z, Cebecauer L (1965) The influence of sodium salicylate on the increased solubility of collagen in lathyrism. Med Pharmacol Exp 13:98–102

Trout JJ, Buckwalter JA, Moore KC (1982) Ultrastructure of the human intervertebral disc. II. Cells of the nucleus pulposus. Anat Rec 204:307–314

Trüeb B, Winterhalter K (1986) Type VI collagen is composed of a 200 Kd subunit and two 140 kd subunits. EMBO J 5:2815–2819

Trüeb B, Gröbli B, Spiess M, Odermatt BF, Winterhalter KH (1982) Basement membrane (type IV) collagen is a heteropolymer. J Biol Chem 257:5239–5245

Trüeb B, Schaeren-Wiemers N, Schreier T, Winterhalter KH (1989) Molecular cloning of chicken type VI collagen. Primary structure of the subunit $\alpha$2(VI)-pepsin. J Biol Chem 264:136–140

Trupin JS, Russell SB, Russell JD (1983) Variation in prolyl hydroxylase activity in keloid-derived and normal human fibroblasts in response to hydrocortisone and ascorbic acid. Coll Relat Res 3:13–23

Tryggvason K, Risteli J, Kivirikko KI (1978) Glomerular basement membrane collagen and activities of the intracellular enzymes of collagen biosynthesis in congenital nephrotic syndrome of the Finnish type. Clin Chim Acta 82:233–240

Tseng SCG, Smuckler D, Stern R (1982) Comparison of collagen types in adult and fetal bovine corneas. J Biol Chem 257:2627–2631

Tsipouras P, Byers PH, Schwartz RC, Mon-Li Chu, Weil D, Pepe G, Cassidy SB, Ramirez F (1986) Ehlers-Danlos syndrome type IV: cosegregation of the phenotype to a COL3A1 allele of type III procollagen. Hum Genet 74:41–46

Tsipouras P, Schwartz RC, Liddell AC, Salkeld CS, Weil D, Ramirez F (1988) Genetic distance of two fibrillar collagen loci, COL3A1 and COL5A2, located on the long arm of human chromosome 2. Genomic 3:275–277

Tsubura A, Shikata N, Inui R, Morii S, Hatano T, Oikawa T, Matsuzawa A (1988) Immunohistochemical localization of myoepithelial cells and basement membrane in normal, benign and malignant human breast lesions. Virchows Arch Anet Pathol [A] 413:133–139

Tuderman L, Kivirikko KI (1977) Immunoreactive prolyl hydroxylase in human skin, serum and synovial fluid: changes in the content and components with age. Eur J Clin Invest 7:295–299

Tuderman L, Prockop DJ (1982) Procollagen N-proteinase. Properties of the enzyme purified from chick embryo tendons. Eur J Biochem 125:545–549

Tuderman L, Kuutti ER, Kivirikko KI (1975) An affinity-column procedure using poly(L-proline) for the purification of prolyl hydroxylase. Purification of the enzyme from chick embryos. Eur J Biochem 52:9–16

Tuderman L, Myllylä R, Kivirikko K I(1977a) Mechanism of the prolylhydroxylase reaction. I. Role of co-substrates. Eur J Biochem 80:340–348

Tuderman L, Risteli J, Miettinen TA, Kivirikko KI (1977b) Serum immunoreactive prolyl hydroxylase in liver disease. Eur J Clin Invest 7:537–541

Tünte W, Becker PE, Knorre GV (1967) Zur Genetik der Myositis ossificans progressiva. Humangenetik 4:320–351

Turner-Warwick M, Burrows B, Johnson A (1980) Cryptogenic fibrosing alveolitis: clinica features and their influence on survival. Thorax 35:171–180

Turpeenniemi-Hujanen T, Thorgeirsson UP, Liotta LA (1984) Collagenases in tumor cell extravasation. Ann Rep Med Chem 19:231–239

Turpeenniemi-Hujanen T, Thorgeirsson UP, Hart IR, Grant SS, Liotta LA (1985) Expression of collagenase IV (basement membrane collagenase) activity in murine tumor cell hybrids that differ in metastatic potential. J N C I 75:99–103

Tyopponen J, Forsander OA, Kulonen E (1980) Influence of extracellular proline on collagen synthesis in rat liver slices. Scand J Gastroenterol 15:373–376

Tyree B, Halme J, Jeffrey JJ (1980) Latent and active forms of collagenase in rat uterine explant cultures: regulation of conversion by progestatioal steroids. Arch Biochem Biopyhs 202:314–317

Udén A, Nilsson IM, Willner S (1980) Collagen changes in congenital and idiopathic scoliosis. Acta Orthop Scand 51:271–274

Uitto J (1975) Effect of tetracyclines on collagen biosynthesis in the dental pulp. Acta Odontol Scand 33:279–285

Uitto J, Bauer EA (1982) Diseases associated with collagen abnormalities. In: Weiss JB, Jayson MIV (eds) Collagen in health and disease. Churchill Livingstone, Edinburgh, pp 289–312

Uitto J, Mustakallio KK (1971) Effect of hydrocortisone acetate, fluorinolone acetonide, floclorolone acetonide, bemethasone-17-valerate and fluprednyliden-21-acetate on collagen biosynthesis. Biochem Pharmacol 20:2495–2503

Uitto J, Perejda AJ (1987) Connective tissue disease. Molecular pathology of the extracellular matrix. Dekker, New York

Uitto J, Prockop DJ (1974) Incorporation of proline analogues into collagen polypeptides. Effect on the production of extracellular procollagen and on the stability of the triple helical structure of the molecule. Biochim Biophys Acta 336:234–251

Uitto J, Laitinen O, Lamberg BA, Kivirikko KI (1968) Further evaluation of the significance of urinary hydroxyproline determinations in the diagnosis of thyroid disorders. Clin Chim Acta 22:583–591

Uitto J, Halme J, Hannuksela M, Peltokallio P, Kivirikko KI (1969) Procollagen proline hydroxylase activity in the skin of normal subjects and of patients with scleroderma. Scand J Lab Invest 23:241–247

Uitto J, Hannuksela M, Rasmussen O (1970a) Protocollagen proline hydroxylase activity in scleroderma and other connective tissue disorders. Ann Clin Res 2:235–239
Uitto J, Helin P, Rasmussen O, Lorenzen I (1970b) Skin collagen in patients with scleroderma: biosynthesis and maturation in vitro and the effect of D-penicillamine. Ann Clin Res 2:228–234
Uitto J, Helin G, Helin P, Lorenzen H (1971a) Connective tissue in scleroderma. A biochemical study on the correlation of fractioned glycosaminoglycans and collagen in human skin. Acta Derm Venereol (Stockh) 51:401–407
Uitto J, Ohlenschager K, Lorenzen I (1971b) Solubility of skin collagen in normal subjects and in patients with generalized scleroderma. Clin Chim Acta 31:13–18
Uitto J, Teir H, Mustakallio KK (1972) Corticosteroid-induced inhibition of the biosynthesis of human skin collagen. Biochem Pharmacol 21:2161–2167
Uitto J, Bauer EA, Santa-Cruz DJ, Loewinger RJ, Eisen AZ (1979a) Skin fibroblasts in focal dermal hypoplasia: growth characteristics and collagen metabolism in culture. J Invest Dermatol 72:199–205
Uitto J, Bauer EA, Eisen AZ (1979b) Scleroderma. Increased biosynthesis of triple-helical type I and type III procollagens associated with unaltered expression of collagenase by skin fibroblasts in culture. J Clin Invest 64:921–930
Uitto J, Santa-Cruz DJ, Eisen AZ (1979c) Familial cutaneous collagenoma: genetic studies on a family. Br J Dermatol 101:185–195
Uitto J, Santa-Cruz DJ, Eisen AZ (1980a) Connective tissue nevi of the skin. Clinical, genetic, and histopathological classification of hamartomas of the collagen, elastin, and proteoglycan type. J Am Acad Dermatol 3:741–761
Uitto J, Bauer EA, Santa-Cruz DJ, Loewinger RJ, Eisen AZ (1980b) Focal dermal hypoplasia: abnormal growth characteristics of skin fibroblasts in culture. J Invest Dermatol 75:170–175
Uitto VJ, Raeste AM (1978) Activation of latent collagenase of human leukocytes and gingival fluid by bacterial plaque. J Dent Res 57:844–851
Uitto VJ, Turto H, Saxén L (1978) Extraction of collagenase from human gingiva. J Periodont Res 13:207–214
Uitto VJ, Appelgren R, Robinson PJ (1981) Collagenase and neutral metalloproteinase activity in extracts of inflammed human gingiva. J Periodont Res 16:417–427
Uitto VJ, Elliot I, Robinson PJ (1984) Activation of human gingival collagenase. J Oral Pathol 13:412–418
Uitto VJ, Chan ECS, Chin-Quee T (1986) Initial characterization of neutral proteases from oral spirochetes. J Periodont Res 21:95–100
Uitto VJ, Larjava H, Heino J, Sorsa T (1989) A protease of Bacteroides gingivalis degrades cell surface and matrix glycoproteins of cultured gingival fibroblasts and induces secretion of collagenase and plasminogen activator. Infect Immun 57:213–218
Uldbjerg N, Ulmsten U, Ekman G (1983) The ripening of the human uterine cervix in terms of connective tissue biochemistry. Clin Obstet Gynecol 26:14–26
Uvelius B, Mattiasson A (1984) Collagen content in the rat urinary bladder subjected to intravesical outflow obstruction. J Urol 132:587–590
Uvelius B, Mattiasson A (1986) Detrusor collagen content in the denervated rat urinary bladder. J Urol 136:1110–1116
Vacanti JP, Folkman J (1979) Bile duct enlargement by infusion of L-proline: potential significance in biliary atreasia. J Pediatr Surg 14:814–818
Vacelet J, Garrone R (1985) Two distinct populations of collagen fibrils in a "sclerosponge". In: Bairati A, Garrone R (eds) Biology of intervertebrate and lower vertebrate collagens. Plenum, New York, pp 183–189
Vadilo-Ortega F, González-Avila G, Chevez P, Abraham CR, Montaño M, Selman-Lama M (1989) A latent collagenase in human aqueous humor. Invest Ophthalmol Vis Sci 30:332–335
Vaes G (1971) A latent collagenase released by bone and skin explaints in culture. Biochem J 123:23P–24P
Vaes G (1972) Multiple steps in the activation of the inactive precursor of bone collagenase by trypsin. FEBS Lett 28:198–205

Vaes G (1980) Collagenase, lysosomes and osteoclastic bone resorption. In: Woolley DE, Evanson JM (eds) Collagenases in normal and pathological connective tissues. Wiley, Chichester, pp 185–207

Vaes G, Eeckhout Y, Druetz JE (1976) A latent neutral protease released by bone in culture. Arch Int Physiol Biochim 84:666–668

Vaes G, Eeckhout Y, Lenaers-Claeys G, Francois-Gillet C, Druetz JE (1978) The simultaneous release by bone explants in culture and the parellel activation of procollagenase and of a latent neutral proteinase that degrades cartilage proteoglycans and denatured collagen. Biochem J 172:261–274

Vaes GM, Nichols G Jr (1962) Metabolism of glycine-1-$C^{14}$ by bone in vitro: effects of hormones and other factors. Endocrinology 70:890–901

Valle KJ, Bauer EA (1980) Enhanced biosynthesis of human skin collagenase in fibroblast cultures from recessive dystrophic epidermolysis bullosa. J Clin Invest 66:176–187

Van Arman CG (1976) Pathway to adjuvant arthritis. Fed Proc 35:2442–2446

Van Blerkom J, Motta P (1978) A scanning electron microscopic study of the luteo-follicular complex. III. Repair of ovulated follicle and formation of the corpus luteum. Cell Tissue Res 189:131–153

Van den Hooff A (1983) Connective tissue changes in cancer. Int Rev Connect Tissue Res 10:395–431

Van der Rest M, Mayne R (1987) Type IX collagen. In: Mayne R, Burgeson RE (eds) Structure and function of collagen types. Academic, Orlando, pp 195–221

Van der Rest M, Mayne R (1988) Type IX collagen proteoglycan from cartilage is covalently cross-linked to type II collagen. J Biol Chem 263:1615–1618

Van der Rest M, Mayne R, Ninomiya Y, Seidah NG, Chretien M, Olsen BR (1985) The structure of type IX collagen. J Biol Chem 260:220–225

Van der Rest M, Rosenberg LC, Olsen BR, Poole AR (1986) Chondrocalcin is identical with the C-propeptide of type II procollagen. Biochem J 237:923–925

Van de Water J, Gershwin ME, Abplanalp H, Wick G, Von der Mark K (1984) Serial observations and definition of mononuclear cell infiltrates in avian scleroderma, an inherited fibrotic disease in chickens. Arthritis Rheum 27:807–815

Van Gool J, Van Vugt H, De Nie I (1986a) Acute phase reactants enhance $CCl_4$-induced liver cirrhosis in the rat. Exp Mol Pathol 44:157–168

Van Gool J, De Nie I, Smit HJ Zuyderhoudt FMJ (1986b) Mechanisms by which acute phase proteins enhance development of liver fibrosis: effects on collagenase and prolyl-4-hydroxylase activity in the rat liver. Exp Mol Pathol 45:160–170

Van Holst GJ, Klis FM (1981) Hydroxyproline glycosides in secretory arabinogalactan-protein of Phaseolus vulgaris L. Plant Physiol 68:979–980

Van Holst GJ, Klis FM, Bouman F, Stegwee D (1980) Changing cell-wall compositions in hypocotyls of dark-grown bean seedlings. Planta 149:209–212

Van Holst GJ, Martin SR, Allen AK, Ashford D, Desai NN, Neuberger A (1986) Protein conformation of potato (Solanum tuberosum) lectin determined by circular dichroism. Biochem J 233:731–736

Van Hussen N, Fegeler K, Eberhardt G, Gerlach U (1971) "Collagen-like Protein" im Blutserum von Patienten mit chronischen Leberkrankheiten. Med Welt 22:1877–1878

Van Ness KP, Koob TJ, Eyre DR (1988) Collagen cross-linking: distribution of hydroxypyridinium cross-links among invertebrate phyla and tissues. Comp Biochem Physiol 91B:531–534

Vassan N (1983) Effects of physical stress on the synthesis and degradation of cartilage matrix. Connect Tissue Res 12:49–58

Vater CA, Nagase H, Harris ED Jr (1983) Purification of an endogenous activator of procollagenase from rabbit synovial fibroblast culture medium. J Biol Chem 258:9374–9382

Vater CA, Nagase H, Harris ED Jr (1986) Activator-dependent activation of procollagenase induced by treatment with EGTA. Biochem J 237:853–858

Vaughan L, Winterhalter KB, Bruckner P (1985) Proteoglycan Lt from chicken embryo sternum identified as type IX collagen. J Biol Chem 260:4758–4763

Vaughan L, Mendel M, Huber S, Bruckner P, Winterhalter KH, Irwin MI, Mayne R (1988) D-periodic distribution of collagen type IX along cartilage fibrils. J Cell Biol 106:991–997

Vazquez D (1974) Inhibitors of protein synthesis. FEBS Lett 40 [Suppl]: 63–84

Vegge T, Ringvold A (1971) The ultrastructure of the extracellular components of the trabecular meshwork in the human eye. Z Zellforsch 115:364–376

Veis A (1984) The macromolecular chemistry of gellatin. Academic, New York

Veis A, Payne K (1988) Collagen fibrillogenesis. In: Nimni ME (ed) Collagen, vol. CRC Press, Boca Raton, pp 113:138

Vellenga E, Mulder NH, Van Zanten AK, Nieweg HO, Woldring MG (1983) The significance of the aminoterminal propeptide of type III procollagen in paroxysmal nocturnal haemoglobinuria and myelofibrosis. Eur J Nucl Med 8:499–501

Verbruggen LA, Geerts A, Wisse E, Lapiere C (1986) Serum and synovial fluid antibodies to collagen in rheumatic diseases: a review. Clin Rheumatol 5:440–444

Vernon-Roberts B (1987) Pathology of intervertebral discs and apophyseal joints. In: Jayson MIV (ed) The lumbar spine and back pain. Churchill Livingstone, Edinburgh, pp 37–55

Vernon-Roberts B, Pirie CJ (1977) Degenerative changes in the intervertebral discs of the lumbar spine and their sequelae. Rheumatol Rehabil 16:13–21

Versi E, Cardozo L, Brincat M, Cooper D, Montgomery J, Studd J (1988) Correlation of urethal physiology and skin collagen in postmenopausal women. Br. J Obstet Gynecol 95:147–152

Verzár F (1956) Das Altern des Kollagens. Helv Physiol Pharmacol Acta 14:207–221

Verzár F (1964) The aging of collagen fibre. Int Rev connect Tissue Res 2:243–300

Viidik A, Vuust J (eds) (1980) Biology of collagen. Academic, New York

Vikkula M, Peltonen L (1989) Structural analyses of the polymorphic area in type II collagen gene. FEBS Lett 250:171–174

Viljanto J (1964) Biochemical basis of tensile strength in wound healing. Acta Chir Scand [Suppl] 333:1–101

Viljoen D, Goldblatt J, Thompson D, Beighton P (1987) Ehlers-Danlos syndrome: yet another type? Clin Genet 32:196–201

Ville DB, Powers M (1977) Effect of glucose and insulin on collagen secretion by human skin fibroblasts in vitro. Nature 268:156–158

Villela B, Birkedal-Hansen H (1985) Collagenolytic activity as an indicator of periodontal disease activity. J Dent Res 64:374 (special issue)

Vitellaro-Zuccarello L, Cheli F, Cetta G (1985) The interstitial collagen of Lumbricus sp (Annelida). In: Bairati A, Garrone R (eds) Biology of intervertebrate and lower vertebrate collagens. Plenum, New York, pp 259–266

Vitello L, Breen M, Weinstein HG, Sittig RA, Blacik LJ (1978) Keratan sulfate-like glycosaminoglycan in the cerebral cortex of the brain and its variation with age. Biochem Biophys Acta 539:305–314

Vlădescu C (1975) Researches regarding the effect of total x-ray irradiation on the collagen metabolism in rats. Rev Roum Morphol Embryol Physiol Physiol 12:73–75

Vlădescu C, Ganea E, Petrescu L (1975) The effect of ionizing irradiation on the collagen general metabolism in normal and adrenelectomized rats. Rev Roum Morphol Embryol Physiol Physiol 12:307–311

Vlădescu C, Gheorghe N, Ganea E, Petrescu L (1976) Recherches concernant le métabolisme du collagène hépatique et global chez les rats irradiés. Pathol Biol (Paris) 24:671–676

Vinogradova ZA (1979) Soderzhanye nerastvórimogo kollagena v tkanyakh sobak posle deistviya khronitsheskogo gamma-izlutsheniya v malikh dozakh (Insoluble collagen content in tissues of dogs after a long-term exposure to low doses of gamma irradiation). Kosm Biol Aviakosm Med 13(6):80–82 (in Russian)

Vogel BE, Minor RR, Freund M, Prockop DJ (1987) A point mutation in a type I procollagen gene converts glycine 748 of the $\alpha 1$ chain to cysteine and destabilizes the triple helix in a lethal variant of osteogenesis imperfecta. J Biol Chem 262:14737–14744

Vogel HG (1973) Connective tissue and ageing. Excerpta Medica, Amsterdam, pp 1–272

Vogel HG (1974) Correlation between tensile strength and collagen content in rat skin. Effect of age and cortisol treatment. Connect Tissue Res 2:177–182

Vogel HG (1975) Collagen and mechanical strength in various organs of rats treated with D-penicillamine or amino-acetonitrile. Connect Tissue Res 3:237–244

Vogel KG, Trotter JA (1987) The effect of proteoglycans on the morphology of collagen fibrils formed in vitro. Coll Relat Res 7:105–114

Vogel KG, Paulsson M, Heinegård D (1984) Specific inhibition of type I and type II collagen fibrillogenesis by the small proteoglycan of tendon. Biochem J 223:587–597

Vogel VHG (1970) Beeinflussung der mechanischen Eigenschaften der Haut von Ratten durch Hormone. Arzneimittelforschung 20:1849–1857

Völkl KP (1989) A sensitive in vivo platelet function test in rats based on intravenously injected collagenase. Haemostasis 19:174–179

Volpin D, Veis A (1973) Cyanogen bromide peptides from insoluble skin and dentin bovine collagens. Biochemistry 12:1452–1457

Von der Mark K (1981) Localization of collagen types in tissues. Int Rev Connect Tissue Res 9:265–324

Von der Mark K (1986) Differentiation, modulation and dedifferentiation of chondrocytes. In: Kühn K, Krieg T (eds) Connective tissue: biological and clinical aspects. Karger, Basel, pp 272–315

Von der Mark K, Von der Mark H, Gay S (1976) Study of differential collagen synthesis during development of the chick embryo by immunofluorescence. Dev Biol 48:237–248

Von der Mark K, Von der Mark H, Timpl R, Trelstad R (1977) Immunofluorescent localization of collagen types I, II, and III in the embryonic chick eye. Dev Biol 59:75–79

Von der Mark H, Aumailley M, Wick G, Fleischmajer R, Timpl R (1984) Immunochemistry, genuine size and tissue localization of collagen VI. Eur J Biochem 142:493–502

Von Hahn O, Fricke RJ, Hoffmann-Blume E, Predel K (1974) Die Bedeutung der Kollagenpeptidase-Aktivitätsbestimmung in der Differentialdiagnose chronischer Lebererkrankungen. Z Inn Med 29:347–351

Von Knorring J (1970) Effect of age on the collagen content of the normal rat myocardium. Acta Physiol Scand 79:216–225

Von Maillot K, Zimmermann BK (1976) Solubility of collagen of uterine cervix during pregnancy and labour. Arch Gynäkol 220:275–280

Voss T, Eistetter H, Schafer KP, Engel J (1988) Macromolecular organisation of natural and recombinant lung surfactant protein SP 28–36. Structural homology with complement factor Clq. J Mol Biol 201:219–227

Vrany B, Hnátková Z, Lettl A (1988) Occurrence of collagen-degrading microorganisms in associations of mesophilic heterotrophic bacteria from various soils. Folia Microbiol (Praha) 33:458–461

Vuorio EI, Mäkela JK, Vuorio TK, Poole A, Wagner JC (1989) Characterization of excessive collagen production during development of pulmonary fibrosis induced by chronic silica inhalation in rats. Br J Exp Pathol 70:305–315

Wahl SM (1986) Inflammatory cell regulation of connective tissue metabolism. In: Kühn K, Krieg T (eds) Connective tissue: biological and clinical aspects. Karger, Basel, pp 404–429

Wahl LM, Winter CC (1984) Regulation of guinea pig macrophage collagenase production by dexamethasone and colchicine. Arch Biochem Biophys 230:661–667

Wahl LM, Olsen CE, Sandberg AL, Mergenhagen SE (1977) Prostaglandin regulation of macrophage collagenase production. Proc Natl Acad Sci USA 74:4955–4958

Waksman BH, Wennersten C (1963) Passive transfer of adjuvant arthritis in rats with living lymphoid cells of sensitized donors. Int Arch Allergy Appl Immunol 23:129–139

Walsh PN (1972) The effects of collagen and kaolin on the intrinsic coagulant activity of platelets: evidence for an alternative pathway in intrinsic coagulation not requiring factor XII. Br J Haematol 22:393–405

Walters C, Eyre DR (1983) Collagen crosslinks in human dentin: increasing content of hydroxypyridinium residues with age. Calcif Tissue Int 35:401–405

Wang HM, Nando V, Rao LG, Melcher AH, Heersche JNM, Sodek J (1980) Specific immunohistochemical localization of type III collagen in porcine periodontal tissues using the peroxidase-antiperoxidase method. J Histochem Cytochem 28:1211–1215

Wang HM, Hurum S, Sodek J (1983) Con A stimulation of collagenase synthesis by human gingival fibroblasts. J Periodont Res 18:149–155

Ward PA (1967) A plasmin-split fragment of C'3 as a new chemotactic factor. J Exp Med 126:189–206
Ward HP, Block MH (1971) The natural history of agnogenic myeloid metaplasia and a critical evaluation of its relationship with the myeloproliferative syndrome. Medicine (Baltimore) 50:357–420
Waterhouse EJ, Quesenberry PJ, Balian G (1986) Collagen synthesis by murine bone marrow cell culture. J Cell Physiol 127:397–402
Watson RF, Rothbard S, Vanamee P (1954) The antigenicity of rat collagen. J Exp Med 99:535–549
Weaver AL (1967) Lathyrism: a review. Arthritis Rheum 10:470–478
Weber KT (1989) Cardiac interstitium in health and disease: the fibrillar collagen network. J Am Coll Cardiol 13:1637–1652
Weber KT, Janicki JS, Shroff SG, Pick R, Chen RM, Bahey RI (1988) Collagen remodeling of the pressure-overloaded, hypertrophied non-human primate myocardium. Circ Res 62:757–765
Wegelius O, Von Knorring J (1964) The hydroxyproline and hexosamine content in human myocardium at different ages. Acta Med Scand [Suppl] 412:233–237
Weidman ER, Bala RM (1980) Direct mitogenic effects of human somatomedin on human embryonic lung fibroblasts. Biochem Biophys Res Commun 92:577–585
Weigand Z, Zaugg PY, Frei A, Zimmermann A (1984) Long-term follow-up of serum N-terminal propeptide of collagen type III levels in patients with chronic liver disease. Hepatology 4:835–838
Weil D, Mattei MG, Passage E, Van Long N, Pribula-Conway D, Mann K, Deutzmann R, Timpl R, Chu ML (1988) Cloning and chromosomal localization of human genes encoding the three chains of type VI collagen. Am J Hum Genet 42:435–445
Weil D, D'Alessio M, Ramirez F, de Wet W, Cole WG, Chan D, Baterman JF (1989) A base substitution in the exon of a collagen gene causes alternative splicing and generates a structurally abnormal polypeptide in a patient weith Ehlers-Danlos syndrome type VII. EMBO J 8:1705–1710
Weimar VL (1959) Activation of corneal stromal cells to take up the vital dye neutral red. Exp Cell Res 18:1–14
Weiner FR, Czaja MJ, Jefferson DM, Giambrone MA, Tur-Kaspa R, Reid LM, Zern MA (1987) The effects of dexamethasone on in vitro collagen gene expression. J Biol Chem 262:6955–6958
Weinstein A (1980) Drug-induced systemic lupus erythematosus. Prog Clin Immunol 4:1–21
Weinstein JN, Lehmann TR, Hejna W, McNeil T, Spratt K (1986) Chemonucleolysis versus open discectomy. A ten-year follow-up study. Clin Orthop 206:50–55
Weinstrup RJ, Hunter AGW, Byers PH (1986a) Osteogenesis imperfecta type IV: evidence of abnormal triple helical structure of type I collagen. Hum Genet 74:47–53
Weinstrup RJ, Tsipouras P, Byers PH (1986b) Osteogenesis imperfecta type IV. Biochemical confirmation of genetic linkage to the pro $\alpha$2(I) gene of type I collagen. J Clin Invest 78:1449–1455
Weintraub LR, Goral A, Grasso J, Franzblau C, Sullivan A, Sullivan S (1988) Collagen biosynthesis in iron overload. Ann NY Acad Sci 526:179–184
Weis S, Stransky G, Dimitrov L, Wenger E, Neumüller J, Hakimzadeh A, Firneis F, Partsch G, Eberl R (1987) Morphometric analysis of collagen fibrils in idiopathic carpal tunnel syndrome, part 2. Exp Cell Biol 55:179–182
Weiss A, Livne E, Brandeis E, Silbermann M (1988) Triamcinolone impairs the synthesis of collagen and noncollagen proteins in condylar cartilage of newborn mice. Calcif Tissue Int 42:63–69
Weiss JB (1976) Enzymatic degradation of collagen. Int Rev Connect Tissue Res 7:101–157
Weiss JB, Jayson MIV (eds) (1982) Collagen in health and disease. Churchill Livingstone, Edinburgh, pp 1–571
Weiss JB, Shuttleworth CA, Brown R, Hunter JA (1975a) Polymeric type III collagen in inflamed human synovia. Lancet 1:85

Weiss JB, Shuttleworth CA, Brown R, Sedowfia K, Baildam A, Hunter JA (1975b) Occurrence of type III collagen in inflamed synovial membranes: a comparison between non rheumatic, rheumatoid and normal synovial collagens. Biochem Biophys Res Commun 65:907–912

Weiss PH, Klein L (1969) The quantitative relationship of urinary peptide hydroxyproline excretion to collagen degradation. J Clin Invest 48:1–10

Welgus HG, Stricklin GP, Eisen AZ, Bauer EA, Cooney RV, Jeffrey JJ (1979) A specific inhibitor of vertebrate collagenase produced by human skin fibroblasts. J Biol Chem 254:1938–1943

Welgus HG, Jeffrey JJ, Eisen AZ, Roswitt WT, Stricklin GP (1985) Human skin fibroblast collagenase: interaction with substrate and inhibitor. Coll Relat Res 5:167–179

Wells BB, Kendell EC (1940) The influence of corticosterone and $C_{17}$ hydroxydehydrocorticosterone (compound E) on somatic growth. Proc Staff Meet Mayo Clin 15:324–328

Wener MH, Uwatoko S, Mannik M (1989) Antibodies to the collagen-like region of Clq in sera of patients with autoimmune rheumatic diseases. Arthritis Rheum 32:544–551

Werb Z (1978) Biochemical actions of glucocorticoids on macrophages in culture. Specific inhibition of elastase, collagenase and plasminogen activator secreted and effects on other metabolic functions. J Exp Med 147:1695–1712

Wessells NK (1977) Tissue interactions and development. Benjamin, Menlo Park

Westberg NG, Michael AF (1973) Human glomerular basement membrane: chemical composition in diabetes mellitus. Acta Med Scand 194:39–47

Weve H (1931) Ueber Arachnodaktylie (Dystrophia mesenchymalis congenita, typus Marfanis). Arch Augenheilkd 104:1–46

Wewer UM, Albrechtsen R, Rao CN, Liotta LA (1986) The extracellular matrix in malignancy. In: Kühn K, Krieg T (eds) Connective tissue: biological and clinical aspects. Karger, Basel, pp 451–478

Wheat MR, McCoy SL, Barton ED, Starcher BM, Schwane JA (1989) Hydroxylysine excretion does not indicate collagen damage with downhill running in young men. Int J Sports Med 10:155–160

Whitehead RG (1969) Urinary hydroxyproline and protein nutrition. Lancet 1:103

Whiteside TL, Worrall JG, Prince RK, Buckingham RB, Rodnan GP (1985) Soluble mediators from mononuclear cells increase the synthesis of glycosaminoglycan by dermal fibroblast cultures derived from normal subjects and progressive systemic sclerosis patients. Arthritis Rheum 28:188–197

Whitson TC, Peacock EE Jr (1969) Effect of $\alpha,\alpha'$-dipyridil on collagen synthesis in healing wounds. Surg Gynecol Obstet 128:1061–1064

Wick G, Brunner H, Penner E, Timpl R (1978) The diagnostic application of specific antiprocollagen sera. II. Analysis of liver biopsies. Int Arch Allergy Appl Immunol 56:316–324

Wick G, Glanville RW, Timpl R (1979) Characterization of antibodies to basement membrane (type IV) collagen in immunohistological studies. Immunobiology 156:372–381

Widomska-Czekajska T, Miturzyńska-Stryjecka P, Bielak Z (1968) Hydroksyprolina w zawale i dusznicy bolesnej (Hydroxyproline in patients with myocardial infarction and coronary insufficiency). Pol Tyg Lek 23:1764–1768 (in Polish)

Wie H, Beck EI (1981a) Effects of cyclophosphamide on the formation and solubility of collagen. Acta Pharmacol Toxicol 48:289–293

Wie H, Beck EI (1981b) Synthesis and solubility of collagen in rats during recovery after high-dose cyclophosphamide administration. Acta Pharmacol Toxicol 48:294–299

Wieczorek M (1981) Histopatologia ogólna i podstawy cytodiagnostyki (General histopathology and basic cytodiagnostics). Śląska Akademia Medyczna, Katowice (in Polish)

Wienecke K, Glas R, Robinson DG (1982) Organelles involved in the synthesis and transport of hydroxyproline-containing glycoproteins in carrot root discs. Planta 155:58–63

Wieslander J, Barr JF, Butkowski RJ, Edwards SJ, Bygren P, Heinegård D, Hudson BG (1984) Goodpasture antigen of the glomerular basement membrane: localization to noncollagenous regions of type IV collagen. Proc Natl Acad Sci USA 81:3838–3842

Wiestner M, Krieg T, Hörlein D, Glanville RW, Fietzek P, Müller PK (1979) Inhibiting effect of procollagen peptides on collagen biosynthesis in fibroblast cultures. J Biol Chem 254:7016–7023

Wildhirt E (1974) Therapie chronischer Leberkrankheiten mit D-Penicillamin. Münch Med Wochenschr 116:217–220
Wilhelm SM, Eisen AZ (1985) Induction of collagenase expression in human fetal skin fibroblasts and certain other human cells by phorbol diester. Clin Res 33:532A (abstract)
Williams E (1873–1879) Rare cases, with practical remarks. Trans Am Ophthalmol Soc 2:291–298
Wilner GD, Nossel HL, LeRoy EC (1968) Activation of Hageman factor by collagen. J Clin Invest 47:2608–2615
Winand RJ, Kohn LD (1970) Relationship of thyrotropin to exophthalmic-producing substance: purification of homogeneous glycoproteins containing both activities from $^3$H-labeled pituitary extracts. J Biol Chem 245:967–975
Winberg JO, Gedde-Dahr T Jr, Bauer EA (1989) Collagenase expression in skin fibroblasts from families with recessive dystrophic epidermolysis bullosa. J Invest Dermatol 92:82–85
Windsor ACW, Williams CB (1970) Urinary hydroxyproline in the elderly with low leucocyte ascorbic acid levels. Br Med J 1:732–733
Wintermantel E, Emde H, Loew F (1985) Intradiscal collagenase for treatment of lumbar disc herniations. A comparison of clinical results and computed tomography follow-up. Acta Neurochir (Wien) 78:98–104
Wiontzek H (1970) Behandlung der chronischen aggressiven Hepatitis mit D-Penicillamin. Med Welt 21:1419–1423
Wirl G (1977) Extractable collagenase and carcinogenesis of the mouse skin. Connect Tissue Res 5:171–178
Wirl G (1984) Biological significance of interstitial collagenase in DMBA-induced mammary tumors of the rat. Cancer Metastasis Rev 3:237–248
Wirtschafter ZT, Bentley JP (1962) The extractable collagen of lathyritic rats with relation to age. Lab Invest 11:365–367
Wirtz MK, Glanville RW, Steinmann B, Rao VH, Hollister DW (1987) Ehlers-Danlos syndrome type VIIB. Deletion of 18 amino acids comprising the N-telopeptide region of a pro-$\alpha$ 2(I) chain. J Biol Chem 262:16376–16385
Witschel H (1981) Intraretinal collagen formation in lattice degeneration. Dev Ophthalmol 2:369–384
Wize J, Sopata I, Gietka J, Jakubowski S, Kruze D (1975) Hydroxyproline levels and collagenolytic activity in synovial fluids of patients with rheumatic diseases. Scand J Rheumatol 4:65–72
Woessner JF Jr (1962) Catabolism of collagen and non-collagen protein in the rat uterus during post partum involution. Biochem J 83:304–314
Woessner JF Jr (1973) Mammalian collagenases. Clin Orthop 96:310–326
Woessner JF Jr (1977) A latent form of collagenase in the involuting rat uterus and its activation by a serine proteinase. Biochem J 161:535–542
Woessner JF Jr (1979) Separation of collagenase and a metal-dependent endopeptidase of rat uterus that hydrolyzes a heptapeptide related to collagen. Biochim Biophys Acta 571:313–320
Woessner JF Jr (1982) Uterus, cervix and ovary. In: Weiss JB, Jayson MIV (eds) Collagen in health and disease. Churchill Livingstone, Edinburgh, pp 506–526
Woessner JF Jr, Ryan JN (1973) Collagenase activity in homogenates of the involuting rat uterus. Biochim Biophys Acta 309:397–405
Wohllebe M, Carmichael DJ (1978) Type-I trimer and type-I collagen in neutral salt-soluble lathyritic-rat dentine. Eur J Biochem 92:183–188
Wohllebe M, Carmichael DJ (1979) Biochemical characterization of guanidinium chloride-soluble dentine collagen from lathyritic-rat incisors. Biochem J 181:667–676
Wojtecka-Łukasik E, Dancewicz AM (1974) Inhibition of human leucocyte collagenase by some drugs used in the therapy of rheumatic diseases. Biochem Pharmacol 23:2077–2081
Wojtecka-Łukasik E, Maśliński S (1984) Histamine, 5-hydroxytryptamine and compound 48/80 activate PMN-leucocyte collagenase of the rat. Agents Actions 14:451–453
Wojtecka-Łukasik E, Sopata I, Maśliński S (1986) Aurarafin modulates mast cell histamine and polymorphonuclear leukocyte collagenase release. Agents Actions 18:68–70
Wolf-Spengler ML, Isseroff H (1983) Fascioliasis: bile duct collagen induced by proline from the worm. J Parasitol 69:290–294

Wolinsky H (1972) Long-term effects of hypertension on the rat aortic wall and their relation to concurrent aging changes. Circ Res 30:301–309

Wolpers C (1943) Kollagenquerstreifung und Grundsubstanz. Klin Wochenschr 22:624

Woo SLY, Sites TJ (1988) Current advances on the study of the biomechanical properties of tendons and ligaments. In: Nimni ME (ed) Collagen, vol II. CRC Press, Boca Raton, pp 223–241

Wood GG (1960) The formation of fibrils from collagen solutions. 3. Effect of chondroitin sulphate and some other naturally occurring polyanions on the rate of formation. Biochem J 75:605–612

Woodbury D, Benson-Chanda V, Ramirez F (1989) Amino-terminal propeptide of human pro-$\alpha$ 2 (V) collagen conforms to the structural criteria of a fibrillar procollagen molecule. J Biol Chem 264:2735–2738

Woodhead-Galloway J (1982) Structure of the collagen fibril: an interpretation. In: Weiss JB, Jayson MIV (eds) Collagen in health and disease. Churchill Livingstone, Edinburgh, pp 28–48

Woodhouse NJY (1972) Paget's disease of bone. Clin Endocrinol Metab 1:125–141

Woodley DT, O'Keefe EJ, Prunieras M (1985) Cutaneous wound healing: a model for cell-matrix interactions. J Am Acad Dermatol 12:420–433

Wooley PH, Luthra HS, Stuart JM, David CS (1981) Type II collagen-induced arthritis in mice. I. Major histocompatibility complex (I-region) linkage and antibody correlates. J Exp Med 154:688–700

Wooley PH, Dillon AM, Luthra HS, Stuart JM, David CS (1983) Genetic control of type II collagen-induced arthritis in mice: factors influencing disease susceptibility and evidence for multiple MHC-associated gene control. Transplant Proc 15:180–185

Wooley PH, Luthra HS, O'Duffy JD, Bunch TW, Moore SB, Stuart JM (1984) Anti-type II collagen antibodies in rheumatoid arthritis. The influence of HLA phenotype. Tissue Antigens 23:263–269

Woolley DE (1982) Collagenase immunolocalization studies of human tumors. In: Liotta LA, Hart I (eds) Tumor invasion and metastasis. Nijhoff, Boston, pp 391–404

Woolley DE (1984a) Collagenolytic mechanisms in tumor cell invasion. Cancer Metastasis Rev 3:361–372

Woolley DE (1984b) Mammalian collagenases. In: Piez KA, Reddi AH (eds) Extracellular matrix biochemistry. Elsevier, New York, pp 119–158

Woolley DE, Davies RM (1981) Immunolocalization of collagenase in periodontal disease. J Periodont Res 16:292–297

Woolley DE, Evanson JM (eds) (1980) Collagenase in normal and pathological connective tissues. Willey, Chichester, pp 1–285

Woolley DE, Roberts DR, Evanson JM (1976) Small molecular-weight serum protein which specifically inhibits human collagenases. Nature 261:325–327

Woolley DE, Akroyd C, Evanson JM, Soames JV, Davies RM (1978a) Characterization and serum inhibition of neutral collagenase from cultured dog gingival tissue. Biochim Biophys Acta 522:205–217

Woolley DE, Glanville RW, Roberts DR, Evanson JM (1978b) Purification, characterization and inhibition of human skin collagenase. Biochem J 169:265–276

Woolley DE, Tetlow LC, Mooney CJ, Evanson JM (1980) Human collagenase and its extracellular inhibitors in relation to tumor invasiveness. In: Strand P, Barrett AJ, Baiei A (eds) Proteinases and tumor invasion. Raven, New York, pp 97–113

Wordsworth P, Ogilvie D, Smith R, Sykes B (1985) Exclusion of the $\alpha 1$(II) collagen structural gene as the mutant locus in type II Ehlers-Danlos syndrome. Ann Rheum Dis 44:431–433

Wotton SF, Duance VC, Fryer PR (1988) Type IX collagen: a possible function in articular cartilage. FEBS Lett 234:79–82

Wright GW, Kleinerman J, Zorn EM (1960) The elastin and collagen content of normal and emphysematous human lungs. Am Rev Respir Dis 81:938–943

Wu A, Levi AJ (1975) Effect of alcohol on total urinary hydroxyproline excretion. Am J Gastroenterol 64:217–220

Wu JJ, Eyre DR, Slayter HS (1987) Type VI collagen of the intervertebral disc. Biochemical and electron-microscopic characterization of the native protein. Biochem J 248:373–381

Wünsch E, Heidrich HG (1963) Darstellung von Prolinpeptiden. III. Ein neues Substrat zur Bestimmung der Kollagenase. Z Physiol Chem 332:300–304

Wyckoff RWG, Corey RB (1936) X-ray diffraction patterns from reprecipitated connective tissue. Proc Soc Exp Biol Med 34:285–287

Wyler DJ (1983) Regulation of fibroblast functions by products of schistosomal egg granulomas: potential role of the pathogenesis of hepatic fibrosis. Ciba Found Symp 99:190–206

Wyler DJ, Rosenwasser LJ (1982) Fibroblast stimulation in schistosomiasis. II. Functional and biochemical characteristics of egg granuloma-derived fibroblast-stimulating factor. J Immunol 129:1706–1710

Wyler DJ, Stadecker MJ, Dinarello CA, O'Dea JF (1984) Fibroblast stimulation in schistosomiasis. V. Egg granuloma macrophage spontaneously secrete a fibroblast stimulating factor. J Immunol 132:3142–3148

Wynne-Davies R (1970) Acetabular dysplasia and familial joint laxity: two etiological factors in congenital dislocation of the hip. J Bone Joint Surg 52B:704–716

Yamada Y, Mudryj M, de Crombrugghe B (1983) A uniquely conserved regulatory signal is found around the translation initiation site in three different collagenous genes. J Biol Chem 258:14914–14919

Yamagata S, Yamagata T (1984) FBJ Virus-induced osteosarcoma contains type I, type I trimer, type III as well as type V collagens. J Biochem (Tokyo) 96:17–26

Yamaguchi K, Yasumasu I (1977a) Effects of thyroxine and prolactin on the rates of protein synthesis in the thigh bones of the tadpole of Rana catesbeiana. Dev Growth Differ 19:161–169

Yamaguchi K, Yasumasu I (1977b) Hormones controlling collagen sythesis during metamophosis in the tigh bone of the tadpole of Rana catesbeiana. Dev Growth Differ 19:149–159

Yamamoto K, Maeyama I, Kishimoto H, Morio Y, Harada Y, Ishitobi K, Ishikawa J (1983) Suppressive effect of elcatonin, an eel calcitonin analogue, on excessive urinary hydroxyproline excretion in polyostotic fibrous dysplasia (McCune-Albright's syndrome). Endocrinol Jpn 30:651–656

Yamanishi Y, Dabbous MK, Hashimoto K (1972) Effect of collagenolytic activity in basal cell epithelioma of the skin on reconstituted collagen and physical properties and kinetics of the crude enzyme. Cancer Res 32:2551–2560

Yamanishi Y, Maeyens E, Dabbous MK, Ohyama H, Hashimoto K (1973) Collagenolytic activity in a malignant melanoma: physicochemical studies. Cancer Res 33:2507–2512

Yamauchi M, Mechanic G (1988) Cross-linking in collagen. In: Nimni ME (ed) Collagen, vol I. CRC Press, Boca Raton, pp 157–172

Yamauchi M, Woodley DT, Mechanic GL (1988a) Aging and cross-linking of skin collagen. Biochem Biophys Res Commun 152:898–903

Yamauchi M, Young DR, Chandler GS, Mechanic GL (1988b) Cross-linking and new bone collagen synthesis in immobilized and recovering primate osteoporosis. Bone 9:415–418

Yanagisawa Y, Nishimura H, Matsuki H, Osaka F, Kasuga H (1986) Personal exposure and health effect relationship for $NO_2$ with urinary hydroxyproline to creatinine ratio as indicator. Arch Environ Health 41:41–48

Yasui N, Nimni ME (1988) Cartilage collagens. In: Nimni ME (ed) Collagen, vol I. CRC Press, Boca Raton, pp 225–242

Yochim JM, Blahna DG (1976) Effects of estrone and progesterone on collagen and ascorbic acid content in the endometrium and myometrium of the rat. J Reprod Fertil 47:79–82

Yonath A, Traub W (1969) Polymers of tripeptides as collagen models. IV. Structure analysis of poly(L-prolyl-glycyl-L-proline). J Mol Biol 43:461–477

Yoo TJ (1984a) Etiopathogenesis of otosclerosis: a hypothesis. Ann Otol Rhinol Laryngol 93:28–33

Yoo TJ (1984b) Etiopathogenesis of Meniere's disease: a hypothesis. Ann Otol Rhinol Laryngol 93:6–12

Yoo TJ, Tomoda K (1988) Type II collagen distribution in rodents. Laryngoscope 98:1255–1260

Yoo TJ, Stuart JM, Kang AH, Townes A, Tomoda K, Dixit SN (1982) Type II collagen autoimmunity in otosclerosis and Meniere's disease. Science 217:1153–1155

Yoo TJ, Tomoda K, Stuart JM, Kang AH, Townes AS, Cremer MA (1983) Type II collagen-induced autoimmune sensorineural hearing loss and vestibular dysfunction in rats. Ann Otol Rhinol Laryngol 92:267–271

Yoon K, Davidson JM, Boyd C, May M, LuValle P, Ornstein-Goldstein N, Smoth J, Indik Z, Ross A, Golub E, Rosenbloom J (1985) Analysis of the 3′ region of the sheep elastin gene. Arch Biochem Biophys 241:684–691

Yoshiki S, Yanagisawa T, Kimura M, Otaki N, Suzuki M, Suda T (1978) Bone and kidney lesions in experimental cadmium intoxication. Arch Environ Health 30:559–562

Yu SY, Keller NR (1978) Synthesis of lung collagen in hamsters with elastase-induced emphysema. Exp Mol Pathol 29:37–43

Yu SY, Sun CN, Still MF (1977) Ultrastructural changes of elastic tissue in hamster lung during elastase-emphysema. In: Sandberg LB, Gray WR, Franzblau C (eds) Elastin and elastic tissue. Plenum, New York, pp 39–56

Yue DK, McLennan S, Delbridge L, Handelsman DJ, Reeve T, Turtle JR (1983) The thermal stability of collagen in diabetic rats: correlation with severity of diabetes and non-enzymatic glycosylation. Diabetologia 24:282–285

Yue DK, McLennan S, Handelsman DJ, Delbridge L, Reeve T, Turtle JR (1984) The effect of salicylates on nonenzymatic glycosylation and thermal stability of collagen in diabetic rats. Diabetes 33:745–751

Zaccharidės MPA (1900) Des actions diverses des acides sur la substance conjonctive. CR Soc Biol (Paris) 52:1127–1129

Zachariae H, Halkier-Sorensen L, Heickendorff L (1989) Serum aminoterminal propeptide of type III procollagen in progressive systemic sclerosis. Acta Derm Venereol (Stockh) 69:66–70

Zatońska I, Widomska-Czekajska T (1970) Hydroksyprolina we krwi i moczu chorych z nadczynnością i niedoczynnością tarczycy (Serum and urine hydroxyproline in patients with hyper- and hypothyroidism). Pol Tyg Lek 25:1353–1355 (in Polish)

Zimmer F (1965) Die Uterusvergrösserung in der Schwangerschaft. Arch Gynekol 202:31–40

Zipori D, Duksin D, Tamir M, Argaman A, Toledo J, Malik Z (1985) Cultured mouse marrow stromal cell lines. II. Distinct subtypes differing in morphology, collagen types, myelopoietic factors, and leukemic cell growth modulating activities. J Cell Physiol 122:81–90

Zorab PA (1969) Normal creatinine and hydroxyproline excretion in young persons. Lancet 2:1164–1165

Zorab PA (1971) Urinary total hydroxyproline excretion in normal and scoliotic children. In: Zorab PA (ed) Scoliosis and growth. Churchill Livingstone, Edinburgh, pp 112–116

Zorab PA, Clark S, Cotrel Y, Harrison A (1971) Bone collagen turnover in idiopathic scoliosis estimated from total hydroxyproline excretion. Arch Dis Child 46:828–832

Zucker MB, Borelli J (1962) Platelet clumping produced by connective tissue suspensions and by collagen. Proc Soc Exp Biol Med 109:779

Zuckerman KS, Wicha MS (1982) Long term murine hematopoietic cell production in vitro is dependent on deposition of collagen in the extracellular matrix. Blood 60 [Suppl 1]:106a

Zuckerman KS, Wicha MS (1983) Extracellular matrix production by the adherent cells of long-term murine bone marrow cultures. Blood 61:540–547

# Subject Index


acetylcholinesterase 26, 27
N-acetyl-L-cysteine 239
acetylsalicylic acid 304
achondrogenesis 127
Achromobacter iophagus 66
Actinia equina 83
Actinobacillus actinomycetemcomitans 224
acute otitis media 243
adhesion of platelets 265
adjuvant arthritis 173
aggregation of platelets 266
aging 279–80
alkaptonuria 132
allysine 46
Alport's syndrome, see hereditary progressive glomeruropathy
alveolitis 200
Amadori rearrangement 47
β-aminopropionitrile 193, 296
aminoterminal propeptide of type III procollagen 215, 247, 254, 269
amniotic fluid 260
AMP cyclic 61
analogues of proline 288–300
analogues of lysine 288–300
anchoring fibrils 20–21
  epidermolysis bullosa 126
Anemones 82
anesthetic agents 295
animal models of hereditary disorders of collagen 135–137
Annelida 84
actinonin 303
annulus fibrosus 159
antibiotics 309
antibodies 172
$\beta_1$-anticollagenase 62, 277
antifibroblast serum 300
antigenic determinants 108–110
antigenicity of collagen 108–110
antigenicity of procollagen 108–110
antiinflammatory drugs 303–304
$\alpha_1$-antitrypsin 200, 240
aortic regurgitation 119
aqueous humor 236
arachnodactyly 119
Arthropoda 84
arthritis adjuvant 173
Ascaris lumbricoides 84
Aschelminthes 83–84
Ascidiella 86
ascorbic acid 43–44, 280–281
aspartylglycosaminuria 133
atherosclerotic plaque 190
atherosclerosis 189–192
auranofin 305
azatoprine 308
azetidine 214
L-azetidine-2-carboxylic acid 92, 289

Bacteroides gingivalis 61, 224
Bacteroides melaninogenicus 66
basement membrane
  antigenicity 109
  diabetes mellitus 102
  glomerular 246
  kidney 246
  structure 12–15
  trophoblastic 257
bile ducts 213–214
bladder urinary 247
bleomycin 144, 309
bone 149–157
bone marrow 268
bone metastases 275
Bowman's capsule 246
Bowman's membrane 238
breast cancer 273
bronchogenic carcinoma 274
Bruch's membrane 238
bupivacaine 295
butylated hydroxytoluene 98, 199

cadmium 306
Caenorhabditis elegans 84
calcitonin 154
calcium 154
calcium ethylenediaminetetraacetic acid 43, 239, 301
canaverine 72

cancer 271
carbon tetrachloride 199, 208
cardiovascular system 177–193
carpal tunnel syndrome 166
catch mechanism 85–86
cartilage 157–158
cartilagenous plate 159
catecholamines 146
cathepsin B 64, 301
cathepsin D 301
cathepsin L 64
cathepsin S 64
cathepsins 63–64
cell-mediated immunity 110
cementium 221
Cephalochordates 86
cervix 253
α-chain 5–6
chain association 42
chemonucleolysis 161
choanocytes 82
chondrocalcin 127
chondrionectin 53
chondrocyte 157
chondrodysplasias 126–128
Chordata 86
choroid 236
chronic renal failure 247
ciliary body 236
cimetidine 105
cirrhosis 214
clofenazone 303
Closteridium 66
cloxacillin 310
Cnidaria, see Coelenterata
coagulation factor V 267
coagulation factor XII 267
Coelenterata 82–83
collagen
  AB-collagen, see type V
  aging 79–80
  antigenicity 108–110
  antibodies 172
  alkaptonuria 132
  aspartylglycosaminosuria 137
  atherosclerosis 189–192
  bladder urinary 247
  bone 149–156
  bone marrow 268
  cadmium 306
  cardiovascular system 177–193
  calcitonin 154
  cancer 271
  carpal tunnel syndrome 166
  cartilage 157–158
  cathepsins 63–64
  cementium 221
  chondrodystrophies 126–128
  classification 9–10
  cornea 233–236
  cross-linking 45–49
  cutaneous diseases 227–229
  cystic fibrosis 202
  cytokines 110–111
  degradation 55–67
  dental pulp 221
  dentin 219–221
  depolymerization 57–58
  dermatitis herpetiformis 228
  dermatomyositis 144–145
  destruction lung disease 200–220
  development 77–80
  disease, see collagen disease
  distribution 7–9
  drug-induced lupus-like syndrome 145–147
  Dupuytren's contracture 165–166
  ear 241–243
  Ehlers-Danlos syndrome 113–119
  endothelial collagen, see type VIII
  emphysema 200–202
  epidermolysis bullosa 125–126
  eye 233–241
  fibrodysplasia ossificans progressiva 195
  Goodpasture syndrome 246
  growth hormone 98–99
  glucocorticoids 93–98
  glucagon 100–104
  glycosylation 103
  heart 177–186
  heart muscle 178–179
  heart values 184–186
  helical structure 42
  hemostasis 265–268
  hereditary glomerulopathy 129–131
  hereditary disorders
  hip dislocation 128
  histamine 105
  history 1–3
  homocystinuria 131–132
  hormonal regulation 87–105
  hyalinosis cutis et mucosae 134–135
  hydralazine 146–147
  hypertension 192–193
  immobilization 156
  immune response 110
  immunobiology 107–112
  inhibitors of synthesis 92
  insulin 100–104
  intima 17
  interferon 112
  interleukin 1 111

intervertebral disc 158–162
joint 166–167
keloids 229–230
kidney 245–248
Kniest dysplasia 127
lens 237
leprosy 228
lichen rubber planus 228
liver 203–217
lungs 195–202
lupus erythematosus 144–145
lymph node 269
Marfan syndrome 119–122
meniscus 167
muscular dystrophy 169
musculoskeletal system 149–175
myocardial hypertrophy 180–182
myocardial infarction 182–184
myelofibrosis 268
nail-patella syndrome 129
neoplasia 279–281
nervous system 261
nomendature 5–6
ovary 248–250
osteogenesis imperfecta 122–125
osteoporosis 155
Paget's disease 156
parathyroid hormone 154
peptidase 64–65, 145, 217, 225, 254
placenta 257–260
Peyronie's disease 248
peripheral nerve 262
pharmacology 287–310
phylogeny 81–86
placenta 257–260
porifera 82
pregnancy 250–256
prolactin 104
protcoglycan interaction 49–51
rachitis 155
regulation of metabolism 69–75
retina 238
rheumatoid arthritis 170–172
rheumatoid disorders 170–172
sarcoidosis 199, 228
scleroderma 139–144
scleroderma adultorum 228
7S collagen 216
scoliosis 161–162
scurvy 280–281
short-chain 17
skeletal muscle 167–170
skin 227–232
spongies 82
Stein-Leventhal syndrome 248
synthesis inhibitors 288–300
supramolecular structure 44
teeth 219–226
tendon 162–166
transcriptional control of synthesis 70–71
translational control of synthesis 70–71
thyrotropic hormone 87–95
thyroxine 87–95
triiodothyronine 87–95
tuberculosis 200
tumor necrosis factor 111
turnover 75–76
type I 10–12, 185, 188, 197, 204, 232, 234, 246, 251, 267, 268, 269, 304
type I-trimmer 10–12
type II 10–12, 174–175, 197, 234, 237, 242
type III 10–12, 181, 184, 188, 190, 197, 204, 229, 232, 246, 251, 257, 267, 268, 269, 304
type IV 12–15, 102, 110, 129, 130, 169, 188, 204, 216, 237, 247, 255, 258, 269, 273
type V 15–17, 188, 234, 235, 258, 263, 267, 268, 269
type VI 17–19, 188, 258, 268
type VII 20–21
type VIII 21
type IX 21–22
type X 23
type XI 24
type XII 24–25
type XIII 25
ultraviolet radiation 229
urether 247
urinary system 245–248
uterus 250–256
uvea 236–237
vessels 187–193
vitamin $D_3$ 154
vitreous body 237–238
wound healing 231
collagen disease 139–147
collagen-induced arthritis 174–175
collagenlike protein 254
collagenlike sequence proteins 25–29
collagenase
amino acid composition 59
bacterial 66–67, 268
bone 152
cervical 256
clostridial 268
corneal 239
effect of cimetidine 105
effect of gold
effect of mepyramine 105

effect of timegadine 105
epidermolysis bullosa 125–126
induced arthritis 175
inhibitors 62–63, 152, 216, 239–240, 260, 300, 301
keloids 230
leukocyte 105, 217, 278, 304
liver 209
molecular weight 58–59
ovarian 250
procollagenase 62
regulation by prostaglandins 61
saliva 225
secretion 112
serum 312
synovial 170–172
tissue enzyme 58–63
tumor-derived 276
uterine 252
colagenolysis
inductors 216
inhibitors 300
collagenolytic cathepsins 63–64, 243, 312
collagenoma 133–134
colchicine 294, 301
Cholchicum 294
Corals 82
complement 27–28
conglutinin 28
cornea 233–236, 239
scar formation 240
corneal stroma 234
cornaeal collagenase 239
corneal ulceration 239
coumaric acid 292
corpus luteum 250
cross-linking 45–49
cryptogenic fibrosing alveolitis 200
curly collagen 236
cuticle 84
cutis hyperelastica 113–119
cyclic AMP 61
cyclophosphamid 308
L-cysteine 239
cystic fibrosis 202
cytochalasin B 295
cytokines 111–112, 213
cytostatic drugs 308–309
cytotrophoblastic cells

dehydroaldehistidine 46
dehydrohistidinohydroxymerodesmosine 46, 47
3,4-dehydroproline 214, 289
dental pulp 221–222
dentin 219–221
depolymerization 57–58
dermal hypoplasia 134
dermatitis herpetiformis 228
dermatomyositis 144–145
Descement's mebrane 234
destructive lung disease 200–202
development 77–80
dexamethasone 96
diabetes mellitus 100
diabetic neuropathy 102
1,25-dihydroxycholecalciferol 154
$\alpha,\alpha'$-dipyridyl 92, 290
disc intervertebral 158–162
dislocation of the hip 128
disulfide bonding 42
dolichostenomelia 119
drug-induced lupus-like syndrome 145–147
Dupuytren's contracture 165–166
dwarfism, mice 137

ear 241–243
Echinodermata 85–86
ectopia lentis 119, 131
Ehlers-Danlos syndrome 113–119, 136
Ekman-Lobstein-Vrolik disease, see osteogenesis imperfecta
elastase 60, 201, 302
emphysema 200–202
enamel 221
endocarditis 186
endometrium 250
endotenon 163
entactin 53
epidermal growth factor 152
epidermolysis bullosa 125–126
epimysium 168
epitenon 163
17$\beta$-estradiol 256
estrogen 225
eye 233–241
exoskeleton 84

familial cutaneous collagenoma 133–134
fasciola hepatica 206–213
fetal membranes 259
fibroblast 239
fibroblast selection 141
fibril diameter 41
fibrodysplasia ossificans progressiva 135
fibrylogenesis 49–51
fibromodulin 53
fibronectin 52–53, 300
fibrosis
hepatic 206–213
pulmonary 198–200
fibrostatins 292–294

flatworms, see platyhelmintha
floppy values 186
fluke 214
fluoride 307
fragilitas ossium, see osteogenesis imperfecta
fusidic acid 310

galactosylhydroxylysyl glucotransferase 39
gastric carcinoma 277
gelatinase 64–65
genes 31–34
gingival tissue 225
glaucoma 241
Glisson's capsule 203
glomerular basement membrane 246
glomerulonephritis 247
glucagon 100–104
glucocorticoids 93–98
glycoproteins 51–53
glycosaminoglycans
  cornea 240
  interaction with collagen 49–51
  liver 240
glycosylation of collagen 103–104
glycosylation of asparagine residues 41–42
glycosylation of hydroxylysine 39–41
gold 305–306
gold thiosulfate 305
Goodpasture syndrome 246
graft-versus-host disease 198
granulation tissue 231
Graves' disease 91–92
growth hormone 98–99

heart 177–186
heart muscle 178–179
heart values 184–186
heavy metals 305 307
helical structure 6, 42
hemostatis 265–268
hepatic fibrosis 206–213
hepatocyte 203, 204
hereditary disorders 113–137
  animal models 135–137
hereditary progressive
  glomerulonephritis 129–131
hip dislocation 128
histamine 105
histidinohydroxylsinonorleucine 47
homocysteine 131
homocystinuria 131–132
hormonal regulation 87–105
hyalinosis cutis et mucosae 134
hyaluronidase 90
hydantoin 226
hydantoin-induced gingival hyperplasia 226
hydrolazine 146–147, 193
hydras 83
Hydroid 82
hydroxyallysine 46
hydroxyapaptite 150
hydroxylation of lysine 34–39
hydroxylation of proline 34–39, 95, 288
hydroxylysyl galactosyltransferase 39
hydroxylysine 34–39, 312
hydroxymerodesmosine 47
hydroxytoluene 98, 199
hydroxyproline
  fluoride 308
  myocardial infarction 184
  plants 82
  pregnancy 254
  serum 312
  thyroid diseases 88
  amniotic fluid 260
  synovial fluid 171
  unicellular organisms 82
  urinary output
    breast cancer 275
    cirrhosis 214–216
    bone 153
    dermatomyositis 144
    diabetes mellitus 100
    growth hormone 99
    glucocorticoids 93
    myocardial infarction 184
    Paget's disease 156
    pregnancy 254
    scleroderma 141
    scoliosis 161
    systemic lupus erythematosus 144
    thyroid diseases 88
    tuberculosis 200
Hydrozoa 83
Hymenomones 82
hypertension 182, 192–193
hyperthyroidism 87–95
hypoproteins 184
hypothyroidism 87–95

iatrogenic lupus, see drug-induced lupus-like
  syndrome
immobilization 156
immune response 110, 142
immunobiology 107–112
indices of collagen metabolism 311–313
indomethacin 303
infarction myocardial 182–184
inflammation 170
inhibitors of collagenolysis 216
inhibitors of collagen synthesis 288–300
insects 84

insulin 73, 100–104
insulinlike growth factor 152
interferons 112
interleukin 1 111
intervertebral disc 158–162
  age-related degeneration 161
intestitial nephritis 247
intracellular degradation of procollagen 71–74
ionizing radiation 283–285
iris 236
iron 35
isoniazid 300
isoproterenol 184

jellyfish 82
joints 166–167

keloids 229–230
keratoconus 241
kidney 245–248
kinky hair syndrome, see Menkes' hair syndrome
Kniest dysplasia 127

lamellar bone 150
laminin 53
Langer-Saldino achondrogenesis 127
lathyrism 296–298
lead 306
lectins 26–27
lens 237
leprosy 228
leukemia 278
levamisole 309
lichen ruber planus 228
limbus 236
lincomycin 310
liver 203–217
  bioecosystem 205–206
  fibrosis 206–213
  regeneration 213
localized scleroderma 143
Loligo peallii 85
Lumbrians 84
lung surfactant protein 29
lungs 195–202
  destructive disease 200–202
  fibrosis 198–200
  glucocorticoids 97
lupus erythematosis systemic 144
lymph node 269
lymphatic vessels 187
lysine analogues 288–300
lysyl hydroxylation 34–39
lysyl oxidase 48–49, 300

Macaca fascicularis 181
$\alpha_2$-macroglobulin 62, 175
malnutrition 279–280
manganese 40
mannan-binding protein 29
Marfan syndrome 119–122
Mastigophoran 82
meadow saffron 294
membrane basement 12–15
Menière disease 242
meniscus 167
Menkes' hair syndrome 136
menstrual cycle 251
meninges 261
mepyramine 105
mesoglaea 83
metals 305–307
metastases 275
Metridium dianthus 83
mollusca 95
molybdenum 306
morphea, see localized scleroderma
myometrium 251
myelofibrosis 268
myositis ossificans progressive, see fibrodysplasia ossificans progressiva
muscle skeletal 167–169
muscular dystrophies 169
musculoskeletal system 149–175
Mycobacterium butyricum 173
Mycobacterium leprae 173
Myocobacterium tuberculosis 173
myocardial hypertrophy 180–182
myocardial infarction 182–184

nail-patella syndrome 129
naproxen 304
necrobiosis herpetiformis 228
neoplasma 271–278
nephrectomy 247
Nereis virens 84
nervous system 261–263
nitroblue tetrazolium 292
nitrogen 53
nitrogen dioxide
nonenzymatic glycosylation 85
nonlamellar bone 150
nonsteroidal antiinflammatory drugs 303–304
nucleus pulposus 159
nutrition 279–281

ochronosis, see alkaptonuria
Octopus vulgaris 85
osteoarthrosis 171
osteoblast 150

osteocalcin 153
osteoclast 150, 152
osteocyte 150
osteogenesis imperfecta 122–125
  animal models 136
  type I 122–123
  type II 123–124
  type III 124–125
  type IV 125
osteolathyrism 296–298
osteonectin 153
osteo-onychodysplasia, see nail-patella syndrome
osteoporosis 155
osteopsathyrosis, see osteogenesis imperfecta
osteosarcoma 274
otitis media 243
osteosclerosis 241
overnutrition 279–280
ovary 248–250
ovulation 249
ovulatory enzyme 249
oxygen 43
2-oxyglutarate 43, 291
oxytetracycline 309
ozone 199

Paget's disease 156
parathormone, see parathyroid hormone
paraquat 199
parathyroid hormone 154
Parenti-Fraccaro achondrogenesis 127
Pelmatodydra pseudoligactis 83
peptidase collagen 64–65, 145, 217, 225, 254
peptide sygnal 34
penicillamine 298–299
perimysium 168
perineurium 262
periodontal disease 224–226
periodontal tissue 219–226
periodontium 222–223
peripheral nerve 262
perisinusoidal cell 205
Peyronie's disease 248
pharmacology of collagen 287–310
-phenentroline 290
henylbutazone 303
horbol myristate acetate 61
osphorus 154
otoreceptor cells 238
logeny of collagen 81–86
enta 257–260
ental membranes 259
ts 82
lets 265, 266
platelet-derived growth factor 152
Platylhelmintha 83
pneumonectomy 197
Porifera 82
polymyositis 169
polymorphonuclear leukocytes 240
prednisolone 98
prepro-$\alpha$-chain 5
procollagen
  antigenicity 108–110
  intracellular degradation 71–73
  genes 31–33
  C-proteinase 43, 295
  N-proteinase 43, 295
procollagenase 62
progesterone 255
prolactin 104
prolidase 217
proline hydroxylation 95
proline pool 75, 96, 210
proline analogues 288–290
prolyl 3-hydroxylase 34–39
prolyl 4-hydroxylase 34–39
propranolol 199
prostaglandins
  bone 152
  collagenase secretion regulation 61
  liver 211
  ovary 250
  synovial fluid 71
proteinase-induced emphysema 201
proteoglycans 49
pseudocapsule 272
Pseudomonas aeruginosa 66
psoriasis 228
pulmonary fibrosis 198–200
pyridoxal 300
PZ-peptidase, see collagen peptidase

rachitis 155
radiation ionizing 283–285
radiation ultraviolet 229
Rana catesbiana 90
Rana pipiens 90
renal failure 247
retinoids 301–302
respiratory system 195–202
retina 238
retro-orbital tissue 91
rheumatoid arthritis 170–172
rheumatoid disorders 170–175
  animal models 173–175
rheumatic valves 186
rickets, see rachitis
roundworms, see aschelminthes

saliva 225
samar 307
sarcoidosis 199
scar formation cornea 240
Schistosoma mansoni 212
Schlemm's canal 236, 241
Schwann cells 262
scirrhous carcinoma 274
sclera 233–236
scleroderma 139–144
animal models 143
sclerodermalike changes 144
sclerosis systemic, see scleroderma
scoliosis 161–162
scurvy 280–281
Scyphozoa 83
sea fans 83
sea star 85
sea urchins 85
serum antifibrotic 300
signal peptidase 34
signal peptide 34
silent receptors 147
silica 198
skeletal muscle 167–170
skin 140, 227–232
somatomedins 98
sponges 82
spongin B 82
Staphylococcus aureus 66
starfish sea, see sea star
Stein-Leventhal syndrome 248
Streptococus faecalis 186
Streptomyces albogriseolus 293
Streptomyces catenulae 293
stomach 274
stroma of neoplasm 272
sunscreen 229
superoxide scavanger 292
surfactant protein 29
synovial fluid 171
systemic lupus erythematosus 144–145
systemic sclerosis, see scleroderma

teeth 219–226
development 223
tendon 162–166
teratocarcinoma 274
tetracycline 309
thromboxane $A_2$ 266
thyroid gland 91–92
thyrotropic hormone 87–93
thyrotropin, see thyrotropic hormone
thyroxine 87–93
thigh skin mouse 143
timegadine 105
TIMP, see tissue inhibitor of metalloproteinases
tissue collagenase 58–63
tissue extracts 304–305
tissue inhibitor of metalloproteinases 62–63
todralazine 146
transcriptional control of collagen synthesis 70–71
translational control of collagen syndrome 70–71
Trichophyton schoenleinii 66
triiodothyronine 87–93
trophoblastic basement membrane 257
tropocollagen 42, 283
tuberculosis 200
tumor 271
tumor necrosis factor 111
tunica fibrosa 233
tunica vasculosa 236
tunicamycin 295, 302
Tunicates 86

ultraviolet radiation 229
unicellular organisms 82
urether 247
urinary bladder 247
urinary system 245–248
uterus 250–256
uvea 236–237

valves floppy 186
valves heart 184–186
vanadium 306
vessels 187–193
Vibro alginolyticus, see Achromobacter iophagus
villous fibrosis 259
vineomycin Al 293
vinyl chloride 144
vitamin $D_3$ 156

wound healing 230–232
WY-45 368

zinc 59–60

MIX
Papier aus verantwortungsvollen Quellen
Paper from responsible sources
FSC® C105338

If you have any concerns about our products,
you can contact us on
**ProductSafety@springernature.com**

In case Publisher is established outside the EU,
the EU authorized representative is:
**Springer Nature Customer Service Center GmbH**
**Europaplatz 3, 69115 Heidelberg, Germany**

Printed by Libri Plureos GmbH
in Hamburg, Germany